Inorganometallic Chemistry

MODERN INORGANIC CHEMISTRY

Series Editor: John P. Fackler, Jr.
Texas A&M University

CARBON-FUNCTIONAL ORGANOSILICON COMPOUNDS
Edited by Václav Chvalovský and Jon M. Bellama

GAS PHASE INORGANIC CHEMISTRY
Edited by David H. Russell

HOMOGENEOUS CATALYSIS
WITH METAL PHOSPHINE COMPLEXES
Edited by Louis H. Pignolet

INORGANOMETALLIC CHEMISTRY
Edited by Thomas P. Fehlner

THE JAHN-TELLER EFFECT AND
VIBRONIC INTERACTIONS IN MODERN CHEMISTRY
I. B. Bersuker

METAL INTERACTIONS WITH BORON CLUSTERS
Edited by Russell N. Grimes

MÖSSBAUER SPECTROSCOPY
APPLIED TO INORGANIC CHEMISTRY
Volumes 1 and 2 • Edited by Gary J. Long
Volume 3 • Edited by Gary J. Long and Fernande Grandjean

ORGANOMETALLIC CHEMISTRY OF THE
TRANSITION ELEMENTS
Florian P. Pruchnik
Translated from Polish by Stan A. Duraj

Inorganometallic Chemistry

Edited by
Thomas P. Fehlner
Department of Chemistry
University of Notre Dame
Notre Dame, Indiana

PLENUM PRESS • NEW YORK AND LONDON

Library of Congress Cataloging-in-Publication Data

Inorganometallic chemistry / edited by Thomas P. Fehlner.
p. cm. -- (Modern inorganic chemistry)
Includes bibliographical references and index.
ISBN 0-306-43986-7
1. Metal complexes. 2. Organometallic compounds. 3. Metal-metal bonds. I. Fehlner, Thomas P. II. Series.
QD474.I54 1992
546'.3--dc20 92-8329
CIP

ISBN 0-306-43986-7

A Division of Plenum Publishing Corporation
233 Spring Street, New York, N.Y. 10013

Printed in the United States of America

Contributors

Thomas P. Fehlner • Department of Chemistry and Biochemistry, University of Notre Dame, Notre Dame, Indiana 46556

Russell N. Grimes • Department of Chemistry, University of Virginia, Charlottesville, Virginia 22901

Royston H. Hogan • Laboratory for Electron Spectroscopy and Surface Analysis, Department of Chemistry, University of Arizona, Tucson, Arizona 85721

Catherine E. Housecroft • University Chemical Laboratory, Cambridge CB2 1EW, England

Timothy Hughbanks • Department of Chemistry, Texas A&M University, College Station, Texas 77843-3255

Dennis L. Lichtenberger • Laboratory for Electron Spectroscopy and Surface Analysis, Department of Chemistry, University of Arizona, Tucson, Arizona 85721

D. M. P. Mingos • Department of Chemistry, Imperial College of Science Technology and Medicine, London SW7 2AY, England

Robert T. Paine • Department of Chemistry, University of New Mexico, Albuquerque, New Mexico 87131

Anjana Rai-Chaudhuri • Laboratory for Electron Spectroscopy and Surface Analysis, Department of Chemistry, University of Arizona, Tucson, Arizona 85721

M. L. Steigerwald • AT&T Bell Laboratories, Murray Hill, New Jersey 07974

Preface

There is a certain fascination associated with words. The manipulation of strings of symbols according to mutually accepted rules allows a language to express history as well as to formulate challenges for the future. But language changes as old words are used in a new context and new words are created to describe changing situations. How many words has the computer revolution alone added to languages? "Inorganometallic" is a word you probably have never encountered before. It is one created from old words to express a new presence. A strange sounding word, it is also a term fraught with internal contradiction caused by the accepted meanings of its constituent parts. "Inorganic" is the name of a discipline of chemistry while "metallic" refers to a set of elements constituting a subsection of that discipline. Why then this Carrollian approach to entitling a set of serious academic papers?

Organic, the acknowledged doyenne of chemistry, is distinguished from her brother, inorganic, by the prefix "in," i.e., he gets everything not organic. Organometallic refers to compounds with carbon–metal bonds. It is simple! Inorganometallic is everything else, i.e., compounds with noncarbon–metal element bonds.

But why a new term? Is not inorganic sufficient? By virtue of training, limited time, resources, co-workers, and so on, chemists tend to work on a specific element class, on a particular compound type, or in a particular phase. Thus, one finds element-oriented chemists (e.g., boron, phosphorus), structure-oriented chemists (e.g., metallaporphyrin, cluster), or phase-oriented chemists (e.g., gas or condensed phase molecular chemists, solid state chemists). But just as there are many relationships (and differences) between carbon and other p-block element chemistries, so too there are many connections between organometallic chemistry and the analogous noncarbon element chemistry. To emphasize the essential unity of the main group (including carbon)–transition element chemistry as well as to emphasize the potential utility of element variation as a rational means of controlling chemical properties independently of compound type or phase, the term "inorganometallic" has been devised.

It explicitly accents the connection between a well-developed but still dynamic area, organometallic chemistry, and a developing area with enormous potential, inorganometallic chemistry.

In contemporary terms, inorganometallic chemistry is a flash name and the flaws in its internal structure may doom it to a transitory existence. On the other hand, inorganic, as a name for an area of chemistry, hardly does justice to the discipline and yet the name exists and will, I am sure, continue to persist. So if the term inorganometallic captures the essence of the area defined by the chemistry presented herein, it may well gain a certain acceptance despite its inadequacies.

Though explanations and apologies for the title are in order, the content of this book requires neither. The authors, experts each, have tried to capture some of the spirited developments in main group–transition element chemistry that transcend a given chemical niche. Elements as variables constitutes the underlying theme of the book; the sweep ranges from theory to structure to reactivity. Further, though emphasis is placed on molecular chemistry, the transition to the solid state is not neglected. In some ways, it is in this area that the potential for development is most exciting.

No book can be complete nor can so short a volume do justice to an area that is large even in its infancy. But it is my hope, shared I am sure by my co-authors, that what follows will serve both to highlight the essence of known inorganometallic chemistry at this time as well as to suggest directions for future exploration. If readers find that it does so, we will consider our efforts successful.

Thomas P. Fehlner
South Bend, Indiana

Contents

1. Introduction

Thomas P. Fehlner

2. Main Group Fragments as Ligands to Transition Metals

Thomas P. Fehlner

4. Bonding Connections and Interrelationships

D. M. P. Mingos

5. Experimental Comparison of the Bonding in Inorganometallic and Organometallic Complexes by Photoelectron Spectroscopy

Dennis L. Lichtenberger, Anjana Rai-Chaudhuri, and Royston H. Hogan

6. Transition Metal-Promoted Reactions of Main Group Species and Main Group-Promoted Reactions of Transition Metal Species

Russell N. Grimes

7. The Metal–Nonmetal Bond in the Solid State

Timothy Hughbanks

8. Molecular Precursors to Thin Films

M. L. Steigerwald

9. Ceramics

Robert T. Paine

1

Introduction

Thomas P. Fehlner

1. DEFINITION OF "INORGANOMETALLIC"

1.1. Literature

One definition of an organometallic compound can be found in the notice to authors of *Organometallics*. It states in part that an organometallic compound is ". . . one in which there is a bonding interaction (ionic or covalent, localized or delocalized) between one or more carbon atoms of an organic group or molecule and a main group, transition, lanthanide, or actinide metal atom (or atoms)." A very similar definition can be found in an issue of the *Journal of Organometallic Chemistry*. Hence, an inorganometallic compound can be viewed as one in which there is a bonding interaction (ionic or covalent, localized or delocalized) between one or more *p*-block elements (except carbon) of a fragment or molecule with a transition, lanthanide, or actinide metal atom (or atoms).[1] The earliest use of the term inorganometallic that I am aware of was by Richard D. Ernst in a proposal for a NSF fellowship written while still a graduate student at Northwestern University.[2] The proposed definition of an inorganometallic compound contains a certain amount of ambiguity and internal contradiction but so does the operational use of the term organometallic. That is, the statement in *Organometallics* continues: "Following longstanding tradition, organic derivatives of the metalloids (boron, silicon, germanium, arsenic, and tellurium) will be included in this definition. Papers dealing with those aspects of organophosphorus and organo-

Thomas P. Fehlner • Department of Chemistry and Biochemistry, University of Notre Dame, Notre Dame, Indiana 46556.

Inorganometallic Chemistry, edited by Thomas P. Fehlner. Plenum Press, New York, 1992.

selenium chemistry that are of interest to the organometallic chemist will also be considered." A similar set of somewhat arbitrary caveats will be found in the *Journal of Organometallic Chemistry,* and the *Dictionary of Organometallic Compounds* contains listings for all common elements except H, C, N, O, P, S, and Se. Indeed such chemical gerrymandering is required to break the intrinsic continuity of chemistry in order to provide digestible pieces containing the chemistry of closely connected compounds. This occurs in all areas of chemistry. For example, definitions of acid–base reactions are chosen to emphasize the characteristics of a given class of compounds for the useful systemization of reaction properties. Convenience is often important in choosing between the multitude of existing acid–base concepts.[3] Thus, the justification of "inorganometallic chemistry" depends more on its usefulness in the organization of existing chemistry than on whether the etymology of the word itself is correct.

Occasionally, existing definitions inhibit development by shielding practitioners of a given area from relevant developments in an area defined to be different. Organometallic chemistry grew from the fusion of metal and organic chemistries. Although its development has been traced back to the synthesis of Zeiss' salt, $K[Pt(C_2H_4)Cl_3]$, in 1827[4] it is only in the last two decades that the potential of the area was confirmed in a dramatic growth in output as well as number of practitioners. For example, the *Journal of Organometallic Chemistry* initially appeared in 1963 while *Organometallics* is only 10 years old. Other nonmetal–metal element chemistry has heretofore been carried out either under the organometallic umbrella or that of inorganic chemistry in general. But publication and presentation has been largely under the aegis of a particular main group element or under special titles such as "organometallic clusters containing main group elements" or "novel element main group ligands." In another example the word organometallic in the title is simply placed in quotation marks.[5] One aim of this book is to demonstrate that both sufficient other nonmetal–metal element chemistry exists such that there is justification for the definition of a logically coherent area called inorganometallic chemistry and that this area has a potential for development similar to that of the organometallic area a couple of decades ago.

1.2. Scope

As with organometallic chemistry the principal focus of inorganometallic chemistry is the chemistry of compounds containing nonmetal–metal bonds and, for the main part, p-block–d-block element bonds. In principle, then, this includes classical coordination compounds with two-center donor–acceptor interactions such as $[(NH_3)_6Co]^{3+}$. Because p-block elements to the right of carbon often form fragments and molecules containing available lone pairs, interactions with a metal will tend to exhibit this type of donor–acceptor coordination in competition with other types of bonding interactions. Thus, although compounds containing two-center donor–acceptor interactions are not excluded, the emphasis lies in the more complex, and often new, types of bonding interactions found in inorganometallic compounds. That is, for inorganometallic compounds containing p-block elements to the right of carbon the connection to classical coordination compounds will always be present with consequences that will be made evident in the discussion of the associated chemistry (Chapter 2).

An essential aspect of organometallic chemistry is the use of transition metals to vary the properties of the carbon fragment in a systematic fashion. For example, the reaction properties of an olefin bound to a late transition metal are significantly different from those of an olefin coordinated to an earlier transition metal.[6] Inorganometallic chemistry extends this idea and recognizes not only the variation in metal but also the variation in nonmetal element as a means for adjusting compound properties. In this way the nuclear charge of both elements formally becomes variables under the chemist's control. The systematic variation in properties possible is particularly evident in a series of isoelectronic compounds as it is in these compounds that the consequences of changes in nuclear charge are most easily appreciated.[7] Finally, although the term inorganometallic chemistry has been chosen to emphasize similarities not always recognized because of the element orientation of chemists, it is in fact the differences in compound properties that are important.

In summary, the term inorganometallic chemistry ties together much nonmetal–metal chemistry that is closely related to, yet distinctly different from, organometallic chemistry. This is expressed in the triangular relationship sketched in Figure 1 where the connection to main group element–carbon chemistry is made and the existence of mixed main group element–carbon transition metal chemistry is implied. Inorganometallic chemistry is a contemporary term and it may well never gain any general acceptance. However,

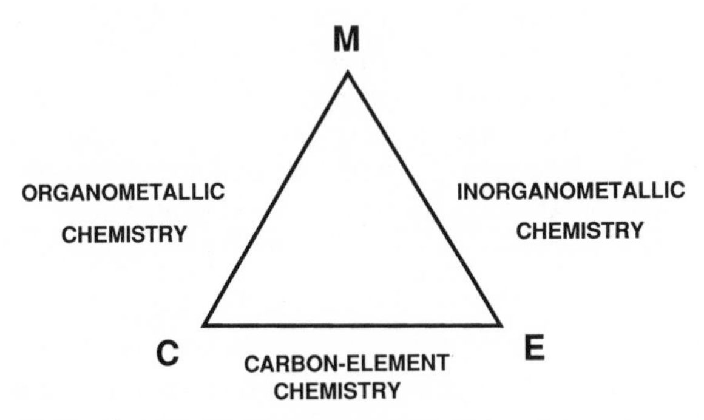

Figure 1. A representation of the formal relationships between metal, carbon, and other element chemistries which serves to define the cross-disciplinary fields.

it can be used to gather presently disparate efforts into a coherent body of information. This may well presage future development of a versatile synthetic and reaction chemistry of the nonmetal–metal bond and permit the rational choice of main group element and transition metal to tune the properties of interest in either discrete compounds or solid materials.

Before developing this theme, a brief review of some general attributes of carbon chemistry vs other main group element chemistry and the mutual effect of carbon and metal moieties on each other in an organometallic compound are presented.

2. CARBON—THE CENTRAL ELEMENT

2.1. Carbon-Element Chemistry

If chemistry is the central science, carbon is the central element of this science. Although the importance of this element in living systems cannot be denied, it is the maturity of the structural, synthetic, and reaction chemistry in the organic area relative to other element chemistries that makes it so. In addition, many of the ligands utilized by inorganic chemists are carbon based. It is for these reasons that comparison with carbon chemistry is inevitable. Unfortunately, even approaches to and models of organic systems are occasionally adopted to analyze inorganic systems without justification. The early S_N1 vs S_N2 controversy in the analysis of ligand substitution processes in classical transition metal complexes provides a good example of the difficulties thus created.[8,9] Another example is the early struggle to account for the geometries of the boranes using the two-center–two-electron bond concept.[10] Because comparison with carbon analogues is inevitable and, indeed, often enlightening, it is worthwhile reminding the reader of the essential differences between carbon and the other main group elements.

The electronegativity of carbon is nearly midway between those of fluorine and cesium, which are the elements with the highest and lowest electronegativities, respectively. As such it has little tendency to either gain or lose electrons, i.e., its bonding is highly covalent. Extraordinary, sometimes heroic, effort is required to create and stabilize carbonium ions and carbanions in quantity. Carbon has four valence electrons and four valence atomic orbitals resulting in a vast catenation chemistry and the saturated catenated structures are known for their resistance to attack by nucleophiles and electrophiles.

2.2. Other P-Block Element Chemistry

The situation is radically different no matter in which direction in the periodic table one moves from carbon. Going to the left, electronegativity falls and the number of valence orbitals energetically accessible exceeds the number of valence electrons. Going down group 14, electronegativity falls and, although the number of primary valence orbitals and valence electrons remains the same as for carbon, the availability of low-lying, empty atomic orbitals leads to more facile reactions. Going to the right, leads one to elements with higher electronegativities, with the same number of valence orbitals but more valence electrons and, in the lower rows, to elements having low-lying empty valence orbitals. Differences in effective nuclear charge for atoms of a given group superimpose on these fundamental electron, orbital considerations, changes in absolute and relative valence orbital energies thereby changing the valence orbital match with a potential binding partner. This leads to differing "oxidation state" preferences for the main group atom as well as different preferred binding stoichiometries. An interesting account of the sweep of main group chemistry reflecting these element properties has recently appeared.[11]

One consequence of these differences is a higher reactivity toward nucleophiles (elements to the left), toward electrophiles (elements to the right), and in general (elements below the first row). Further, because of complementarity in properties, there will be a tendency for elements to the right to link with elements to the left. The charge transfer nature of this interaction often leads to low reaction barriers. Consequently, kinetic control, which is a feature of carbon chemistry, often gives way to thermodynamic control and in some systems rapid equilibria make the characterization of discrete species difficult. In these cases, the concept of a unique mechanistic pathway connecting reactant and products is not terribly useful and the thermodynamics of chemical equilibria plays the predominant role in determining reaction products.[12]

2.3. *Organometallic Chemistry*

In organic chemistry, the inorganic atoms or fragments within the organic molecule often constitute the sites of functionality. In a sense, organometallic chemistry extends such functionalization to the utilization of metals although clearly the carbon–metal interaction is complex and can take many forms.[13] For example, the polarization of the metal–carbon bond gives carbanion-like character to the carbon. Coordination of alkene or alkynes induces geometric changes in the hydrocarbyl moiety reminiscent of the excited state of the free ligand. Unstable organic species are found to be "trapped" by suitable transition metal fragments. Transition metals act as sites upon which stable bonds can be broken (oxidative addition) or formed (reductive elimination). Carbon structures in the absence of functionalization or unsaturation tend to be so inert that the "activation" of CH and CC bonds is itself a vigorous area of organometallic chemistry.[14] Finally, with their larger ligand capacities, transition metals can act as gathering points for several reagents thereby facilitating reaction within the coordination sphere of the metal. These, then, are some of the features of organometallic chemistry that continue to attract chemists interested in carrying out transformations of organic moieties.

These features are just as relevant to inorganometallic chemistry. One expects to observe related behavior for nonmetals other than carbon and this is demonstrated to be true in the following chapters. But in the case of the noncarbon elements there is the possibility of an additional benefit from the point of view of the main group chemist. That is, inorganometallic complexation may change the intrinsic reactivity mentioned above that leads to chemistry determined by equilibria rather than kinetics. If so, transformations that would be impossible for the free ligand would be permitted. The idea of using metals to raise reaction barriers in order to control chemistry seems strange, but is already used in the application of organometallic reagents to organic synthesis where coordination of an organic fragment, e.g., methylene, permits convenient handling and control of the reactivity of the fragment.[15] A good example of deactivation of a main group species in this manner and its utilization in synthesis is that of $CpTiS_5$.[16] Activation of stable species toward reaction has also been effectively demonstrated and is discussed in Chapter 6.

Even though the emphasis of this book is on the interactions of nonmetal elements, excluding carbon, with metals, the comparison of inorganometallic species with similar organometallic compounds is strongly emphasized in the following chapters. This is particularly true in the case of isoelectronic pairs as the role of nuclear charge in determining structure and properties is most clearly evident in such a comparison. In many cases inorganometallic compounds might be considered the mirror images of organometallic compounds where the mirror is of the type found in an amusement park rather than in the bedroom. It is in the distortions of the chemistry found on leaving carbon that our interest lies. Further, many inorganometallic compounds will contain organometallic fragments, e.g., MCp, Cp = η^5-C_5H_5 and often the "extra" ligands on the metal and nonmetal atoms will be the same as those on the metal and carbon atoms in organometallic compounds. Thus, the relationship is made even closer.

Despite the central presence of the carbon–metal bond, the major theme of this volume is the nonmetal–metal bond. Thus, in the following, a consideration of the general characteristics of a nonmetal–metal bond raises some questions that serve as an introduction to the development of existing inorganometallic chemistry found in the succeeding chapters.

3. THE NONMETAL–METAL BOND

3.1. Metals

Although there are borderline cases, most elements are classified as either metals or nonmetals and the accepted division is shown in Figure 2.[17] The

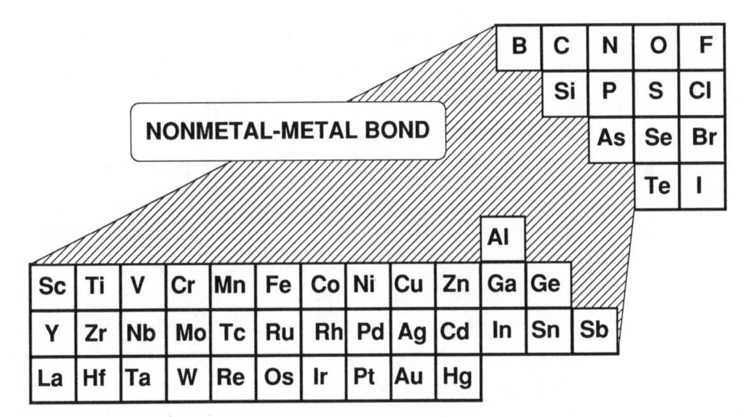

Figure 2. The formal division between the nonmetallic and metallic elements thereby defining the nonmetal–metal bond.

question of why elements exist as metals or nonmetals under ambient conditions and the conditions under which a metal–nonmetal transition can be induced have been recently discussed.[18] The elements are given the sobriquets "metal" or "nonmetal" even when the former are found in compounds that exhibit none of the properties associated with the metallic state. Hence, in one sense, the properties of the isolated nonmetal–metal bond will depend on fragment properties where the fragments are those species resulting from a simple rupture of this bond. That is, although there is a direct relationship between atomic properties and metal–nonmetal behavior in elemental forms, the effective properties of an atom center in a fragment can be altered by the other groups or ligands bound to it. Hence, there will always be a problem in separating properties generated by differences in the nonmetal and metal atoms from the modulation of those differences by the other attached groups or ligands.

3.2. *Examples of Boron and Phosphorus Element* vs. *Metal Bonding*

To illustrate some of the possibilities, the elements boron and phosphorus are considered as representative cases and the basic atomic properties contrasted with those of transition metals.

3.2.1. *Mononuclear*

As noted already, the electronegativity of boron is less than that of hydrogen as is the electronegativity of most transition metals. Because it lies to the left of carbon, boron has fewer valence electrons than valence orbitals. Hence, like a transition metal it often acts as the acid site in the formation of Lewis acid–base adducts. Even though its valence orbitals are more diffuse than those of carbon, boron exhibits a pronounced tendency for covalent bonding because of its small size. It is basically for this reason that boron does not exhibit metallic properties in its elemental forms. Although transition metals are found to form covalent bonds, they are intrinsically weaker and the strong tendency for the formation of compounds with metals in positive oxidation states is intrinsic to their chemistry. Compounds containing direct boron–metal bonds are also well known and generate significant questions.[19] Do these compounds demonstrate the conditions necessary for the direct bonding of boron and a single transition metal in a discrete molecule? Is the bonding similar to that found in metal–metal and/or boron–boron interactions? Are the properties of such a bond a simple combination of those of boron and the metal or do they exhibit hybrid properties not observed in either? Finally, how is the true nature of the boron–metal interaction separated from the effects of the other ligands or groups bound to the boron and metal atoms? That is, can one identify properties of the boron–metal bond that would be transferable to, for example, the solid state?

On the other hand, because it lies to the right of carbon, phosphorus has more valence electrons than valence orbitals. However, although it is a second-row element the electronegativity of phosphorus is greater than that of hydrogen and transition metals. It is, in fact, similar to that of carbon. Hence, phosphorus has a tendency to act as a Lewis base and in a ligand is often the atom found coordinated to a metal. But the tendency for covalent bonding by phosphorus is also well known and its interactions with metals are hardly restricted to simple acid–base interactions.[20] If one considers that under some conditions transition metals are seen to act as Lewis bases,[21] then one sees the possibility of considerable complexity in the interaction of metals and phosphorus within a single inorganometallic compound. Thus, one can pose the same questions concerning the nature of the metal–phosphorus interaction as were posed for the boron–metal bond.

3.2.2. Clusters

Leaving mononuclear metal compounds, the bonding situation within a metal cluster network must progress toward that of the metallic state as the size of the cluster approaches that of a metal crystal.[22] That is, at some point the interior bonding will dominate surface bonding and control observed properties. Catenation in main group element systems leads to elemental structures, i.e., the diamond structure might be considered the ultimate catenated carbon compound.[23] In terms of our exemplar elements, icosahedral fragments are found in elemental boron[24] and in the richness of catenated phosphorus structures one can see the structural motifs of elemental forms of phosphorus.[25] The combination of main group and metal elements in a single extended structure sets up a direct competition between the bonding tendencies of the elements. What, then, will be the resulting structures and how will the properties of the clusters differ from those with cluster frameworks composed of the pure metal or nonmetal elements? Again separation of the *exo*-cluster ligand interactions from *endo*-cluster bonding is a difficult, but necessary, task if this question is to be properly addressed. Intercomparison of isoelectronic inorganometallic (and organometallic) clusters is one unambiguous method of approaching such a separation.

3.2.3. Solids with Extended Bonding

Finally, one can envision the conversion of mono- and polynuclear metal compounds into the elemental forms of the metals. The same process can be pictured for the compounds containing main group elements. Indeed $Ni(CO)_4$ and boranes are known precursors of pure elemental nickel and boron, respectively.[26,27] A significant synthetic problem in trying to characterize the larger catenated species can be the tendency for conversion to the elemental forms or complex, extended networks. The question, then, is what occurs when inorganometallic compounds are subjected to conditions that take the analogous nonmetal or metal compounds to the elements? Do the elements segregate or form alloys? Certainly there is a great tendency for nonmetals and metals to combine, e.g., the complexities of the solid state borides and phosphides are well documented.[28] What is the role of the other ligands and can they be cleanly disposed of?

A fascinating question is whether thermodyamic products, that often can be obtained from direct combination of the elements, are produced or whether the element combinations in the compounds decomposed are predisposed to yield kinetic products of unknown structural form and properties? A mix of structural characteristics of nonmetal and metal components should permit the existence of many more structural types than are presently known. Considering that much of known molecular chemistry is kinetically controlled, i.e., positive heats of formation are common and thermodynamic instability with respect to oxide formation, for example, is the rule rather than the exception, formation of solids under kinetically controlled conditions is an attractive approach to the synthesis of new materials. As noted recently,[29] "It used to be a truism that organic chemistry was richer than inorganic chemistry because it offered the possibility of so many more different molecular structures. In fact, the single giant molecule that is the inorganic semiconductor crystal has innumerable possible arrangements, far greater than those which occur in typical organic molecules." The creation and characterization of such rearrangements constitute challenges for the synthetic, structural, and theoretical inorganic chemist.

3.3. Chapter Organization

These, then, are some of the questions raised by a general consideration of nonmetal–metal bonding in inorganometallic compounds. The answers are addressed, if not given, in the chapters that follow this one. The first section of the book, Chapters 2–6, treats discrete inorganometallic chemistry. The observed structural and chemical characteristics of mono- and dimetal inorganometallic compounds are surveyed in Chapter 2. Here the nonmetal fragment is considered as a ligand bound to a metal site. This leads naturally to the multinuclear cluster systems (Chapter 3) containing both nonmetal and metal atoms in a contiguous cluster bonding network. In Chapters 4 and 5 the complementary points of view of quantum chemical modeling and photoelectron spectroscopy are used to examine the consequences of element variation on bonding models and electronic properties. The effects of the groups and ligands bound to a given element on the valence properties of the fragment thereby generated are specifically considered in these chapters. The last chapter on molecular systems explores selected examples of the ways metals promote the interconversions of nonmetal compounds and *vice versa*.

The second section of the book explores the nonmetal–metal bond in solid state environments. As it is in the transition to the solid state that inorganometallic chemistry holds much potential, the audience addressed is primarily that composed of molecular chemists. The importance (and unimportance) of the nonmetal–metal interaction in the solid state as well as the effects of element variation are first explored via theoretical models in Chapter 7. In contrast to the discrete systems in which the boundary conditions are defined by the limits of the molecule or ion, the extended bonding in the solid state systems requires consideration of an essentially boundary-less problem. In some cases the sense of the molecular concepts carries over, in other cases the nonmetal–metal interaction is retained, and in still other cases only the atoms themselves are useful structural building blocks.

The ways in which inorganometallic molecules can be used to produce solid state materials in which the nonmetal–metal interaction is preserved and, in some cases, lost are discussed in Chapters 8 and 9. Inducing a discrete precursor to cleanly convert from a localized stable bonding system to an extended stable bonding system requires understanding of not only the stable reactant molecules and product solid state systems but also the reaction process itself. The latter is also difficult as not only must the decomposition of the molecule itself be defined but the complexities of a multiphase dynamic system must be understood as well. A recent study serves as an exemplar of a modern approach to this problem.[30] Still, in principle, the appropriate required features can be built into a precursor species. A recent exciting development is the reverse, namely, the conversion of solid state cluster moieties into discrete clusters in solution. Again a recent example illustrates the possible usefulness of this concept in the generation of new inorganometallic species.[31] Thus, these chapters return us to the molecular systems in the sense that they set goals for theoretical, structural, and synthetic chemists with implications beyond the discrete chemistry of the molecules alone.

REFERENCES

1. Fehlner, T. P. *Comments Inorg. Chem.* 1988, **7,** 307.
2. Ernst, R. D. personal communication.
3. Huheey, J. E. "Inorganic Chemistry," 3rd ed.; Harper & Row: New York, 1983; p. 286.
4. Collman, J. P.; Hegedus, L. S. "Principles and Applications of Organotransition Metal Chemistry"; University Science Books: Mill Valley, CA, 1980; p. 7.
5. Herrmann, W. A. *Angew. Chem., Int. Ed. Engl.* 1986, **25,** 56.
6. Ref. 1, p. 109.
7. Aradi, A. A.; Fehlner, T. P. *Adv. Organomet. Chem.* 1990, **30,** 189.
8. Basolo, F.; Pearson, R. G. "Mechanisms of Inorganic Reactions," 1st ed.; Wiley: New York; 1958.
9. Langford, C. H.; Gray, H. B. "Ligand Substitution Processes"; Benjamin: New York; 1965.
10. Sidgwick, N. V. "The Chemical Elements and Their Compounds"; Oxford: 1950; Vol. 1, p. 338.
11. Woollins, J. D. "Nonmetal Rings, Cages and Clusters"; Wiley: New York; 1988.
12. Van Wazer, J. R.; Moedritzer, K. *Angew. Chem., Int. Ed. Engl.* 1966, **5,** 341.
13. Coates, G. E.; Green, M. L. H.; Wade, K. "Organometallic Compounds," 3rd ed.; Methuen: London, 1968; Vol. II.
14. Graham, W. A. G. *J. Organomet. Chem.* 1986, **300,** 81.
15. Brandt, S.; Helquist, P. *J. Am. Chem. Soc.* 1979, **101,** 6473.
16. Stendel, R.; Papavassiliov, M.; Strauss, E. M.; Laitinen, R. *Angew, Chem., Int. Ed. Engl.* 1986, **25,** 99.
17. Hume-Rothery, W. "Structure of Metals and Alloys"; Institute of Metals: London; 1944.
18. Logan, D. E.; Edwards, P. P. In "The Metallic and Nonmetallic State of Matter"; Edwards, P. P.; Rao, C. N. R., Eds.; Taylor & Francis: London, 1985; p. 65.
19. Grimes, R. N. "Metal Interactions with Boron Clusters"; Plenum: New York, 1982.
20. Huttner, G.; Evertz, K. *Acc. Chem. Res.* 1986, **19,** 406.
21. John, G. R.; Johnson, B. F. G.; Lewis, J. *J. Organomet. Chem.* 1979, **181,** 143.
22. Mingos, D. M. P. *Chem. Soc. Rev.* 1986, **15,** 31.
23. DiSalvo, F. J. *Science* 1990, **247,** 649.
24. Matkovich, V. I., Ed. "Boron and Refractory Borides"; Springer-Verlag: Berlin, 1977.
25. Baudler, M. *Angew. Chem., Int. Ed. Engl.* 1987, **26,** 419.
26. Greenwood, N. N.; Earnshaw, A. "*Chemistry of the Elements*"; Pergamon: Oxford, 1984; p. 1330.
27. Bower, J. G. *Prog. Boron Chem.* 1970, **2,** 231.
28. Aronsson, B.; Lundström, T.; Rundqvist, S. "Borides, Silicides and Phosphides"; Methuen: London, 1965.
29. Yablonovitch, E. *Science* 1989, **246,** 347.
30. Bent, B. E.; Nuzzo, R. G.; Dubois, L. H. *J. Am. Chem. Soc.* 1989, **111,** 1634.
31. Rogel, R.; Corbett, J. D. *J. Am. Chem. Soc.* 1990, **112,** 8198.

2

Main Group Fragments as Ligands to Transition Metals

Thomas P. Fehlner

1. INTRODUCTION

In its broadest definition, inorganometallic chemistry appropriates a large share of both inorganic and organometallic chemistry much of which can be classified as some type of metal coordination chemistry. As there are many excellent volumes extant in which these chemistries are described,[1,2] there is no need here to repeat or summarize it. Indeed, in the following we build upon these strong foundations in order to draw new connections. But first it is necessary to distinguish the ligand-based chemistry to be discussed herein from that of classical coordination chemistry, on the one hand, and that of inorganometallic cluster chemistry, which is discussed in the chapter immediately following, on the other hand.

1.1. Scope and Definitions

The metal–ligand interactions found, for example, in a classical coordination compound such as $[Co(NH_3)_6]^{3+}$ will not be discussed in this chapter even though they contain a nonmetal–metal interaction of a type subsumed by the definition of inorganometallic chemistry. The emphasis is on compounds with metals in low oxidation states in analogy to organometallic chemistry and on unusual or complex modes of interaction of the main group

Thomas P. Fehlner • Department of Chemistry and Biochemistry, University of Notre Dame, Notre Dame, Indiana 46556.
Inorganometallic Chemistry, edited by Thomas P. Fehlner. Plenum Press, New York, 1992.

moiety, either bare or ligated element fragment or a catenated moiety, with the metal center or centers. As will be seen, the adjective "unusual" often arises from the identity of the elements interacting with the metal. In many cases formal organometallic analogs of the metal–ligand interaction exist. On the other hand, forms of metal–ligand interactions not observed in organometallic chemistry can be distinguished. In addition, the simple Lewis acid–base metal–ligand bond is not excluded from this discussion because, in the main group elements lying to the right of carbon, it competes with other bonding modes. This competition provides one gauge of the nature of the more uncommon interactions of the ligand with the metal. Further, methods of "turning off" the natural Lewis basicity of a ligand open routes to very unusual compounds.

The distinction between a metal–ligand complex and a heteroatom cluster can be a subtle one.[3] That is, in mixed main group–metal systems an ambiguity often arises in the sense that a molecule containing n metal (M) atoms and m main group (E) atoms can be equally well considered as an $n + m$ atom cluster or a metal fragment containing n metal atoms with a η^m ligand. For example, $(CO)_3FeB_4H_8$[4] can be considered as a five-atom nido (7 cluster pairs, 34 cluster valence electrons) cluster (Figure 1a) or an iron tricarbonyl fragment with a η^4-B_4H_8 4π-electron ligand isoelectronic with η^4-C_4H_4 (Figure 1b).[5] The issue is further complicated by the fact that there exist series of molecules ranging from those that are clearly metal–ligand complexes to those that are clearly clusters. A good example again comes from metalloborane chemistry where terminal $[Ir(B_5H_8)Br_2(CO)(PMe_3)_2]$,[6] bridging $[Cu(\eta^2$-$B_5H_8)(PPh_3)_2]$,[7,8] and cluster bonded $[Ir(B_5H_8)(CO)(PPh_3)_2]$,[9] metal atoms are evidenced in the structures schematically illustrated in Figure 2. The boron–metal interaction in the first is best treated as a metal–(one electron) ligand interaction while that in the last is aptly considered as an integral part of the total six-atom cluster interaction. The bridging interaction in the middle compound is intermediate in nature. As bridging hydrogens are considered

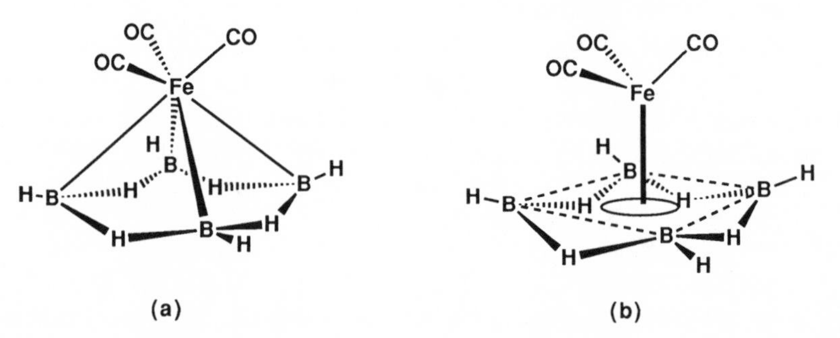

Figure 1. Two representations of the proposed structure of $(CO)_3FeB_4H_8$: (a) as a five-atom nido cluster; (b) as an 18-electron complex with a η^4-B_4H_8 4π electron ligand.

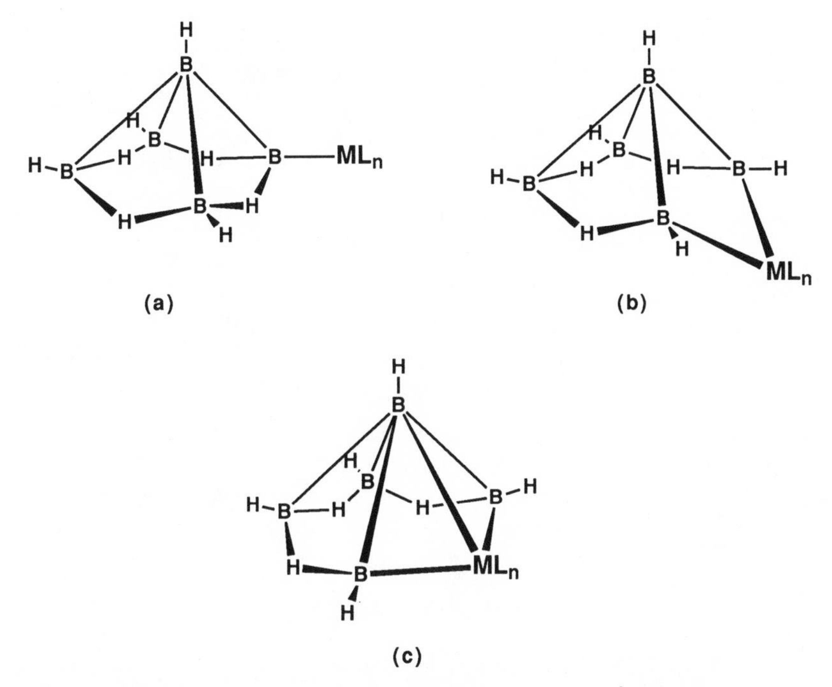

Figure 2. Structures of (a) $Ir(B_5H_8)Br_2(CO)(PMe_3)_2$, (b) $Cu(\eta^2\text{-}B_5H_8)(PPh_3)_2$, and (c) $Ir(B_5H_8)(CO)(PPh_3)_2$.

as contributing one electron each to cluster bonding,[10] the metal atom in $Cu(\eta^2\text{-}B_5H_8)(PPh_3)_2$ might also be considered as part of the cluster bonding network. On the other hand, deprotonating a borane containing B—H—B bridging hydrogens generates a site of Lewis basicity at the polyhedral edge formerly occupied by the bridging proton.[11] The resulting anion can then be considered a ligand to $[CuL_2]^+$ or $[H]^+$.

Hence, in order to meet our pedagogical needs, we arbitrarily divide the pertinent chemistry as follows. Compounds containing E_mM_n fragments with direct E–M interactions (E = main group atom, M = transition metal atom) for m = 1–6 and n = 1,2 will be considered as metal–ligand complexes. Compounds for $n > 2$ will be considered as clusters and the discussion of these species will be found in Chapter 3 immediately following this one. Further, this chapter will only treat compounds with $n = 2$ for m = 1,2. Note that the numbers n and m are used in the sense of describing the fundamental EM interaction and do not necessarily correspond to the total number of metal and main group atoms in a given compound. That is, $Fe(CO)_4(\mu\text{-}SiPh_2)_2Fe(CO)_4$[12] (Figure 3a) would be considered in the category of EM compounds while *trans*-$\{[Fe(CO)_4]_2[PCH(SiMe_3)_2]_2\}$[13] (Figure 3b) would be

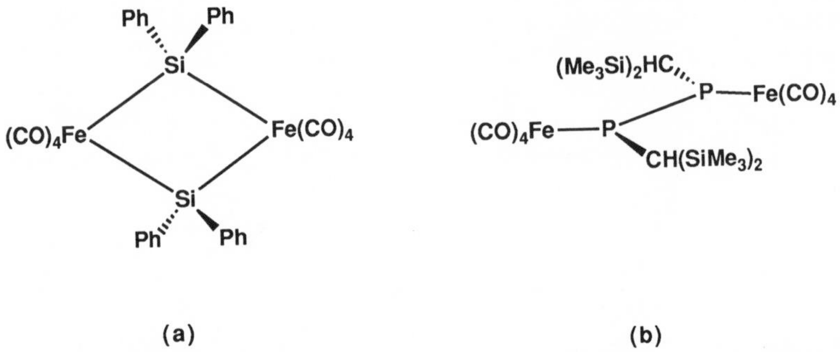

(a) (b)

Figure 3. Structures of (a) $Fe(CO)_4(\mu\text{-}SiPh_2)_2Fe(CO)_4$ and (b) *trans*-$\{[Fe(CO)_4]_2[PCH(SiMe_3)_2]_2\}$.

considered in the E_2M category below. This division has little to do with the question of what constitutes the essence of a cluster versus a metal–ligand complex. Indeed, this chapter will contain some molecules that are usefully considered as clusters, such as $(CO)_3FeB_4H_8$. Including them here merely offers more scope for logical comparison particularly with organometallic complexes.

1.2. Organization

Organization will be primarily by the element E as the primary emphasis is upon the manipulation of the main group elements by means of the transition metals. The elementary modes of E and M interactions are first illustrated with examples of compounds containing direct E—M bonds. For reasons that will be made clear below, the E—H—M bridging interaction will be considered in the category of a direct E—M bond even though much of the bonding is via the H bridge atom. Compounds with E_m–M_n interactions (m and n restricted as indicated above) will then be used to illustrate other bonding modes associated with the increased hapticity of the interactions in competition with the elementary bonding modes characteristic of the $m = 1$, $n = 1$ interactions. Examples will be preferentially selected from most highly developed systems in order to comment upon as many details of the bonding interactions as possible. Geometry is, of course, the initial source of information on bonding in complex systems, but the spatial locations of atoms only indirectly indicate the spatial distribution of electron density corresponding to the bonding network. Hence, metal–ligand interactions that appear ostensibly simple when viewed in a geometric sense alone, often reveal much more complexity when the electronic structure is directly examined. These differences are expressed in chemical reactivity albeit not always in a simple fashion. Emphasis is placed on changes with group number but also, where possible, systematic changes in properties with increasing nuclear charge for analogous compounds in a given group will be discussed.

Most of the compounds of the type considered in this chapter have been included in one or more of a number of excellent reviews already in the literature.[14] For any given set of closely related elements they provide a much more detailed and comprehensive account of the known chemistry. These reviews will be cited at the appropriate points in the chapter. This chapter only surveys a portion of the known stoichiometric and structural variety of E_mM_n metal–ligand complexes; however, the intercomparisons developed may suggest directions to presently unknown chemistry.

2. *SIMPLE COMPLEXES CONTAINING E–M_n INTERACTIONS*

In order to demonstrate the fundamental modes of nonmetal–metal binding that most closely reflect the properties of the nonmetal element, a number of relatively uncomplicated compounds containing E–M interactions are discussed in this section. The types are restricted to compounds containing EM_n units with $n = 1$ and 2. We expect these modes of nonmetal–metal binding to be found in competition with more complicated modes in the structurally more complex E_mM_n compounds discussed in succeeding sections.

2.1. *Elementary Sites of Bonding on an ER_v Fragment*

Consideration of the atomic properties of a main group element allows one to easily enumerate the direct modes of interaction expected for a nonmetal species interacting with a metal species. Ultimately they lead to the complex bonding modes as well. For simplicity, consider a ER_v main group fragment where E is a main group element and R a one-electron substituent, which in this initial discussion is taken to be H. Three properties of the element are crucial. First, elements to the left of carbon will have a low-lying empty valence orbital while those to the right will have a high-lying filled atom centered orbital. Second, the polarization of the E—H bond, which is small for C—H, will lead to hydrogens with hydridic character in going to elements to the left of carbon and hydrogens with protonic character in going to elements to the right of carbon. Third, the number of hydrogens bound to the main group atom may be said to determine the state of valence saturation. That is BH_3, CH_4, and NH_3 may be considered saturated in the sense of maximum utilization of valence electrons in 2c, 2e bonding. The acid–base adducts, $[BH_4]^-$ and $[NH_4]^+$, which are isoelectronic with CH_4, are also considered saturated in the sense of full utilization of all valence electrons and valence orbitals in forming 2c, 2e bonds. These fundamental valence considerations cause the E—H moiety to have a tendency to act as an electron acceptor or hydride donor (e.g., B—H), an electron donor or proton donor (e.g., N—H), or neither electron donor/acceptor nor hydride/proton donor (e.g., C—H).

But much more variation is possible when one leaves the first row of elements. For heavier elements, the diagonal relationship comes into play, e.g., $X(\mathrm{B}) \approx X(\mathrm{Si})$ where X is the electronegativity. The increased size permits a greater number of bonding interactions with the heavier main group atoms. In bonding models, this hypervalence is explained by the utilization of low-lying d functions on the main group atom or by multicentered bonding, e.g., 3c, 4e bonds.[15] Finally, the relative stability of "oxidation state" changes in proceeding from light to heavy elements in a given column, e.g., $PbCl_2$ is more stable than $PbCl_4$. All these factors give rise to "periodic anomalies", i.e., non-carbon-like behavior,[16] that we expect to provide interesting chemistry for inorganometallic compounds as well.

The principal modes of nonmetal–metal interactions are found in compounds possessing E–M interactions. These are: 2c, ne (two-center, n-electron) E—M bonds where $n = 2$ for normal single bonds and is higher for multiple bonds; 2c, 2e E→M donor–acceptor bonds; and 3c, ne bonds where $n = 2$ for E—H—M bridge bonds and is higher for cumulated multiple bonds, e.g., 4 for M=E=M. In the first, the E element participates in a normal, polar, single, or multiple covalent bond. For a single bond, a metal orbital with a single electron and a singly populated nonmetal element orbital which is compatible in symmetry and energy is required. In the second the E element acts as a Lewis base and the metal formally supplies an empty orbital. The

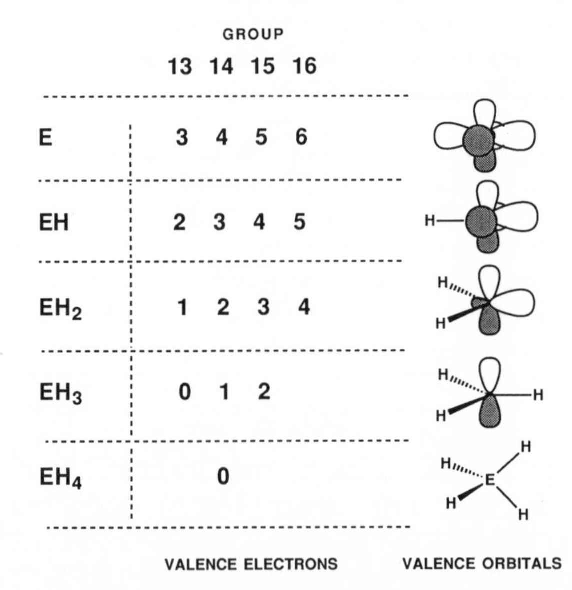

Figure 4. Schematic drawing of the frontier orbitals for EH_x fragments and electron populations for E = group 13–16 atoms.

compatibility of donor and acceptor orbitals is described by hard–soft acid–base theory[17] for normal Lewis bases. In the third various possibilities arise. For example, for an E—H—M bridge, the E—H bond may be considered to act as a Lewis base toward the metal and the metal supplies an empty orbital.[18] Coordination of a E—H bond will be ineffective unless one is dealing with an electron-rich hydride and/or an electron-poor transition metal. In the case of multiple bonding the metal supplies suitable σ and π symmetry (with respect to the E—M bond axis) orbitals and the appropriate number of valence electrons to match those of E.

Figure 4 charts the number of valence orbitals and valence electrons for EH_x fragments for groups 13–16. For our purposes the EH_x bonding fixes the hybridization of E although for a given bonding mode it will require suitable modification. For reasons that will become clear below, the bare E atom is arbitrarily sp hybridized. The number of bonding possibilities with compatible

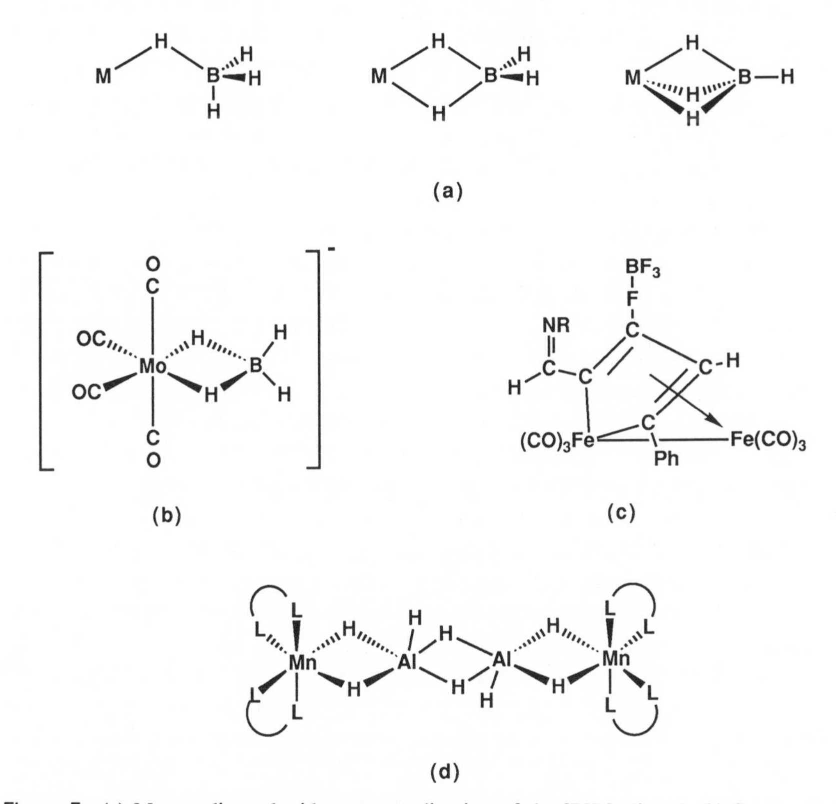

Figure 5. (a) Mono-, di-, and tridentate coordination of the $[BH_4]^-$ ligand. (b) Structure of $[Mo[(\mu\text{-}H)_2BH_2](CO)_4]^-$. (c) Structure of a $[BF_4]^-$ substituted ferracyclopentadiene complex. (d) Structure of $[(dmpe)_2Mn(\mu\text{-}H)_2Al(H)(\mu\text{-}H)]_2$.

transition metal fragments is truly enormous. Further, it is possible to combine E—M and E→M or E—M and E—H—M bonding modes in the same species. In the following, selected examples of the large number of compounds known are presented in order to illustrate the nature of the E–M interactions as a function of element group as well as period.

2.2. Metal Complexes with Saturated ER_v Ligands

2.2.1. Group 13

The saturated ER_v species will be taken to be $[ER_4]^-$. A number of reviews are available covering the compounds of boron(19,20) and the heavier elements of group 13(21) with metals. Compounds containing the BH_4 moiety exist as ionic tetrahydroborates as well as metal complexes.(22) In fact, the $[BH_4]^-$ ion is a versatile ligand binding to transition metals, lanthanides, and actinides as well as main group metals via one, two and three B–H–B interactions with the last two modes of bonding by far most commonly observed (Figure 5a). There is even one compound in which the $[BH_4]^-$ ion serves as a bidentate ligand simultaneously to two different metal centers thereby utilizing all four B—H hydrogens in bridging interactions.(23) An example of a typical complex, $[Mo[(\mu\text{-H})_2BH_2](CO)_4]^-$,(24) is shown in Figure 5b where it is seen that each BH of the ligand occupies an octahedral coordination site at the metal center. Evidence suggesting that the BH_4 moiety is reasonably considered as a ligand comes from an analysis of the fluxional behavior of transition metal borohydride complexes in the ^1H NMR. Here it is found that, with one exception,(25) bridge and terminal hydrogens attached to the boron atom scramble rapidly on the NMR time scale, i.e., M–H interactions break in preference to B–H interactions. In fact, in the case of $[Mo[(\mu\text{-H})_2BH_2](CO)_4]^-$, ^{13}C NMR studies of the metal carbonyl fragment to which the $[BH_4]^-$ ligand is bound showed that the mechanism by which the fluxional behavior takes place causes little perturbation of the metal coordination sphere. Hence, the BH_4 moiety is seen to retain its identity to a large extent in the metal complex.

An insightful analogy of the $[BH_4]^-$ ligand with an η^3-allyl ligand has been justified in terms of electron donation (four each for the anionic ligands), metrics (both have nearly the same "bite" size and cause similar reorganization of an identical metal fragment upon coordination), and frontier orbital patterns (both anions have the possibility of serving as 2-, 4- or 6-electron donors to a transition metal).(20,26) If one considers the B—H bond as a protonated "lone pair," then the function of the proton is to concentrate the electron density in space such that a single boron center can act as a multidentate ligand. Of course, in the protonation the basicity of the "lone pair" is greatly reduced and it is only for relatively electron-rich boranes that B—H metal chelation is expected. Presumably in these compounds there is little direct B—M bonding, but it is in the sense of considering B—H as a protonated lone pair that we consider the B–H–M interaction as equivalent to an interaction that might reasonably be represented as B—H→M.

The recent observation of $[BF_4]^-$ acting as a ligand to a carbon site in the organometallic molecule shown in Figure 5c via a B—F→C bridging interaction[27] is an unexpected bonding mode of the same gendre. It is reported that the $[BF_4]^-$ ion is easily displaced and, hence, may be considered as coordinating to the organometallic framework via a B—F bond. In this case, the effect of the transition metal is felt in a secondary sense as the site of $[BF_4]^-$ binding is one carbon atom removed from the metal.

Aluminum analogues exhibiting the E—H—M mode of bonding are also well known[28,29] and an example, $[(dmpe)_2Mn(\mu\text{-}H)_2Al(H)(\mu\text{-}H)]_2$,[30] is given in Figure 5d. But, in contrast to the wide range of transition metal borohydride complexes, many fewer examples of aluminum hydride complexes are known. Further, not many have been characterized crystallographically. In contrast to metal borohydrides where tetrahedral geometry around the boron atom is preserved on coordination, four-coordinate aluminum appears to be an exception, i.e., a trigonal bipyramidal geometry is found around Al in $[(dmpe)_2Mn(\mu\text{-}H)_2Al(H)(\mu\text{-}H)_2]_2$. These compound types also exhibit fluxional behavior but, in contrast to the borohydrides, terminal M—H and bridge M—H—Al readily exchange as do terminal Al—H and Al—H—Al in structures like that shown in Figure 5d. No example of gallium compounds containing E—H—M bonding have been described.

2.2.2. Group 14–16

The saturated species are EH_4, EH_3, and EH_2 for groups 14, 15, and 16, respectively. Methane is isoelectronic with $[BH_4]^-$, however the former shows little tendency for coordination to transition metals. Indeed the crux of the problem in activating C—H bonds in saturated hydrocarbons is to promote coordination which, presumably, precedes oxidative addition and rupture of the C—H bond.[31,32] A methane complex is suggested by theoretical considerations[33] and evidence for its existence stabilized at low temperature has been presented.[34] Almost everything else is a better donor than CH_4. For example, the coordination and oxidative addition of H_2 to an Ir center has been estimated to be $\approx 10^4$ faster than a C—H bond of cyclohexane.[35]

The situation is considerably different for silicon which exhibits its diagonal relationship with boron in a tendency to form Si–H–M interactions under favorable conditions.[36] A pretty example is $(\eta^6\text{-}C_6Me_6)Cr(CO)_2(H)SiHPh_2$ the structure of which is shown in Figure 6a.[37] This is one of a set of compounds which have been used to investigate the role of metal and that of substituents on metal and silicon on the nature of the Si–H–M interaction.[38] Recently, the suggestion has been made that the compound $ReH_6(PPh_3)_2(SiEt_3)$ contains a $\eta^3\text{-}H_2SiEt_3$ fragment.[39]

Complexes of EH_3, E = N, P, As, and EH_2, E = O, S, Se, are well known and the nature of the donor–acceptor bond has been thoroughly discussed in many standard inorganic texts. It is well to point out, however, that the novice chemist's notion of a donor–acceptor bond is deceptively simple. The rather complex interactions that take place are only revealed by careful study of geometries and electronic structures.[40] Even with simple acids and bases, considerable structural and electronic rearrangement takes place on formation of a donor–acceptor interaction and in the case of competition between two modes of binding a metal, one does not expect the sites to act independently. In addition this type of interaction can lead to interesting inorganometallic complexes such as that shown in Figure 6b.[41]

Interesting chemistry is observed when one of the substituents on ER_3, E = group 15, is replaced with an isolobal metal fragment.[42] This is particularly true for the heaviest elements. For example, $Co(CO)_4BiPh_2$ has been prepared and the structure of $Co(CO)_3PPh_3BiPh_2$ determined in the solid state.[43] Although the bismuth atom exhibits a pyramidal geometry, the angles are large suggesting high *s* character for the formal lone pair. Consistent with this explanation of the geometry, the compound exhibits no tendency for bridging despite all attempts to induce it. This behavior is to be contrasted with that of the earlier members of the group.

2.3. Metal Complexes with Unsaturated ER_v Ligands

The ER_v fragments other than those discussed above are unsaturated with respect to either valence electrons or valence orbitals.

2.3.1. Group 13

Many transition metal compounds containing the BR_3 fragment have been described, e.g., $Rh(BCl_3)(CO)(Cl)(PPh_3)_2$,[44,45] but no structure of a simple derivative in the solid state has been reported. Despite this, most are formulated as containing a direct B—M bond, presumably with the metal acting as a Lewis base. These conclusions are based on spectroscopic data. In the example cited, the observation of large ^{103}Rh–^{11}B coupling in the NMR spectra appears to be strong evidence for direct bonding. However, the magnitude of the coupling constant is similar to that observed for ^{31}P–^{11}B coupling and, compared to known ^{103}Rh–^{13}C coupling constants, much larger than one would expect for ^{103}Rh–^{11}B coupling. A recent measurement of ^{103}Rh–^{11}B coupling in a compound known from atom–atom distances to contain a direct Rh—B bonding interactions[46] suggests that extraordinarily large ^{103}Rh–^{11}B coupling constants are not to be expected. A strong case for a direct B—Fe bond in the adduct of BPh_3 with $[CpFe(CO)_2]^-$ has been made; however, the final isolated product is $[Ph_3BC_5H_4Fe_2(CO)_4Cp]^-$ in which the $[BPh_3]^-$ moiety has replaced a Cp hydrogen in $[CpFe(CO)_2]_2$.[47] As tricoordinate borane com-

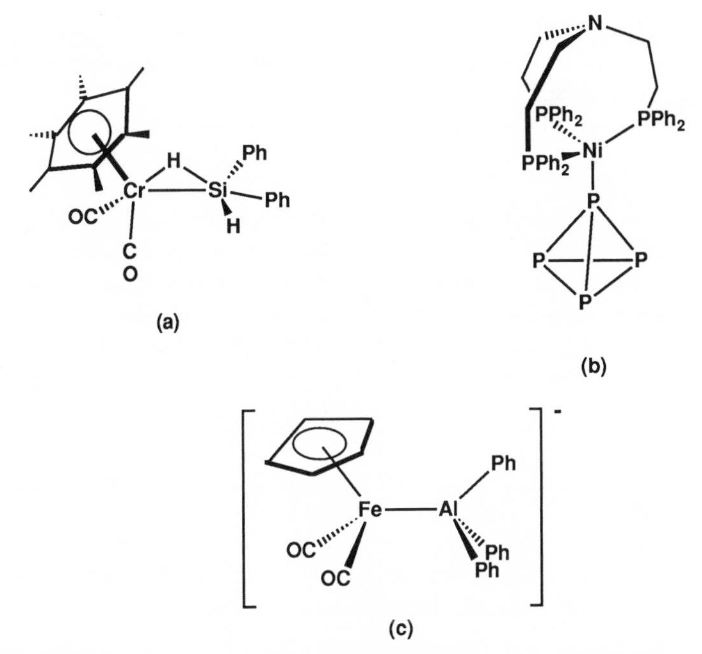

Figure 6. Structures of (a) $(\eta^6\text{-}C_6Me_6)Cr(CO)_2(H)SiHPh_2$, (b) $(np_3)Ni(\eta^1\text{-}P_4)$, and (c) $[CpFe(CO)_2AlPh_3]^-$.

pounds are strong Lewis acids and are known to coordinate to other electron-rich sites on organometallic compound,[48] the existence of a direct B—M bond in the reported compounds containing the elements of BR_3 needs to be more firmly established. However, direct B—M bonds in compounds where the boron atom is part of a polyhedral framework are well known, e.g., Figure 2c.[6]

Only few compounds containing direct, unbridged E—M bonds, E = Al, Ga, In, have been reported.[21] One, $[CpFe(CO)_2AlPh_3]^-$, has been structurally characterized in the solid state and it can be described as an $AlPh_3$ adduct of the Lewis base $[CpFe(CO)_2]^-$ (Figure 6c).[49] The Al–Fe distance is consistent with that expected for a single bond. Evidence has been presented for the Ga and In analogs as well as complexes with the metal anions, $[CpW(CO)_3]^-$, $[Co(CO)_4]^-$, and $[Mn(CO)_5]^-$. In the case of the tungsten anion clear evidence was observed for coordination at the carbonyl oxygen rather than the metal site.

A significant number of compounds containing the so-called boryl group, BR_2, have been reported.[21,50] Substituents on boron are varied but often are formal π donors. The compounds are all formulated as containing a direct, unsupported metal–boron bond, however, with one exception, none have been characterized crystallographically and known reactivity suggests metal–

boron bond cleavage is facile. The metal–boron bonding is similar to that between a metal atom bound to a borane cage as an *exo*-cage substituent replacing hydrogen.[51]

A recent example from our own laboratories is illustrative.[52] The reaction

$$Co_2(CO)_8 + 2BH_3 \cdot THF \rightarrow 2(CO)_4CoBH_2 \cdot THF + H_2$$

has been demonstrated to occur cleanly at −15 °C in THF. The very reactive $(CO)_4CoBH_2 \cdot THF$ molecule has been characterized by low-temperature ^{11}B NMR and infrared spectroscopies as well as classical chemical analysis, and the proposed structure is shown in Figure 7a. The formation of $(CO)_4CoBH_2 \cdot THF$ bears a remarkable similarity to that of $(CO)_4CoSiR_3$.[53]

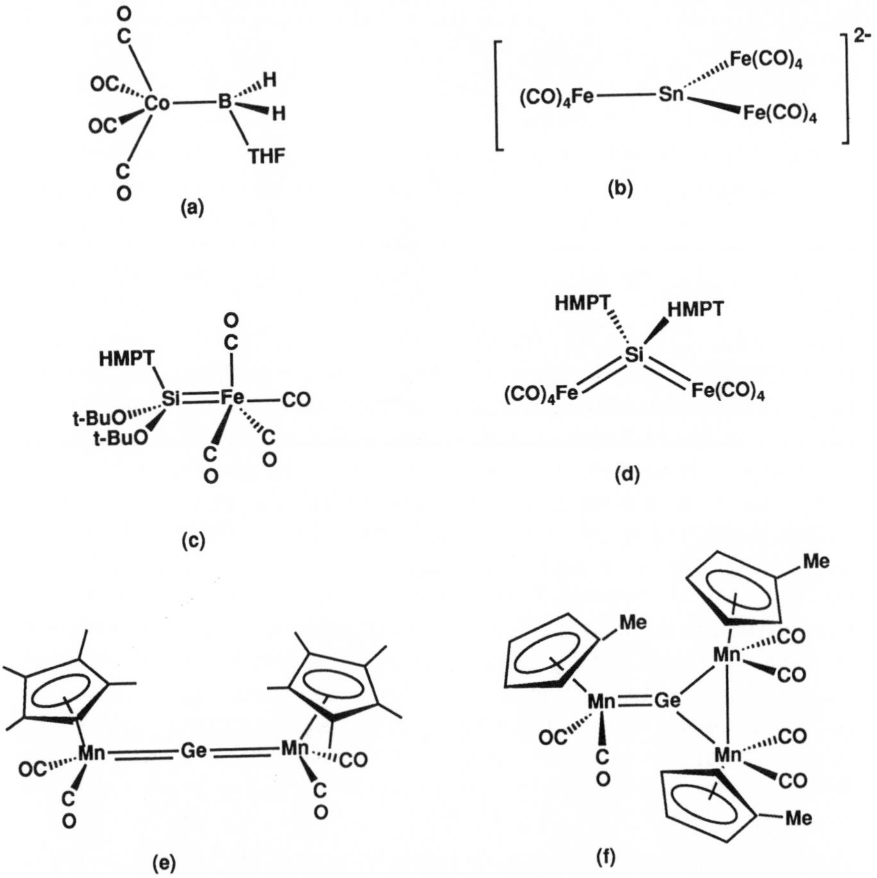

Figure 7. (a) Proposed structure of $(CO)_4CoBH_2 \cdot THF$. Structures of (b) $[\{Fe(CO)_4\}_3Sn]^{2-}$, (c) $(t\text{-}C_4H_9O)_2Si{=}Fe(CO)_4 \cdot HMPT$, (d) $[Fe(CO)_4]_2Si \cdot 2HMPT$, (e) $[Cp^*Mn(CO)_2]_2Ge$, and (f) $[(C_5H_4Me)Mn(CO)_2]_3Ge$.

Displacement of the bound THF of $(CO)_4CoBH_2 \cdot THF$ occurs with Lewis bases and the Lewis acidity of $(CO)_4CoBH_2 \cdot THF$ relative to that of $BH_3 \cdot THF$ for SMe_2 has been estimated. Displacement of $[Co(CO)_4]^-$ from $(CO)_4CoBH_2 \cdot THF$ occurs very easily, e.g., reaction with PhMgBr yields $PhBH_2$. The cobaltaborane readily accepts hydride from $[HFe_2(CO)_8]^-$ losing $[Co(CO)_4]^-$, but reduces the CO ligands of hydride free metal carbonylate anions. The compound is a very active reducing agent and above 10°C cleaves THF and condenses with hydrocarbyl and metal fragments to yield a mixture of clusters including an unusual tailed cluster $(CO)_9Co_3C(CH_2)_nOH$, $n = 4,5$. These results provide some of the first information on the effect of a direct, unsupported M—B bond on the reactivity of B—H hydrogens.

Recently the complex $[(CO)_2(\eta^1\text{-dppm})Co(\mu\text{-dppm})BH_2$ has been isolated from the reaction of $CoBr_2$ with $NaBH_4$ in the presence of dppm and CO.[54] Crystallographic characterization shows a distorted trigonal bipyramidal cobalt atom (two C, two P, and one B atoms) and a tetrahedral boron atom (one Co, two H, and one P atoms) The Co—B bond is long (2.224 Å) and is consistent with the lability found for the Co—B bond in $(CO)_4CoBH_2 \cdot THF$.

Formation of this type of species has recently also been found to occur via oxidative addition of a BH bond to a transition metal when the two R groups on boron are bulky alkyl groups. For example, the reaction of $IrH(PMe_3)_4$ with $[RBH_2]_2$, R = CMe_2CMe_2H, leads to $(PMe_3)_3Ir(H)_2BRH$ plus $Me_3P \cdot BRH_2$.[55] This reaction is basically the same as the oxidative addition of an *exo*-cage B—H bond with a late transition metal, e.g., 2-$[IrBr_2(CO)(PMe_3)_2]B_5H_8$, where the borane cage serves a similar purpose to the bulky organic ligands.[56] A closely related structurally characterized compound is $(C_6H_4O_2B)Ir(H)(Cl)(PMe_3)_3$ which contains two B—O and one B—Ir bonds. Because of the substituents on the boron atom, neither of the monoboron iridium compounds have a base coordinated to the boron atom.[57]

Only one example of a metal–ligand complex (as defined above) containing BR fragments of the type $[Mn(CO)_4L]_2BY$ has been reported.[58] This may be viewed as a dimetal substituted borane related in terms of the M—B bond to the BR_2 derivatives discussed immediately above. There is one report of the formation of a metal complex containing the BR_2 fragment formulated as a compound containing a M=B double bond.[59] However, the compound, $(CO)_4Fe{=}BNR_2$, R=Me, Et, could not be isolated in a pure form.

2.3.2. Group 14

Unsaturated fragments for this group are EH_3, EH_2, and EH. The chemistry of low valent metal complexes containing methyl, methylene, and methyne fragments and related derivatives constitutes a significant fraction of organometallic chemistry. Similar compounds can be envisioned for Si, Ge, and Sn but, in a general sense, the hydrides have an increasing tendency toward the loss of hydrogen as one moves down the group. Transition metal silyl derivatives has been the subject of recent reviews.[60,61]

For our purposes MER_3, M_2ER_2, M_3ER compounds containing no M—M bonds and 17-electron metal fragments are considered as equally representative examples of the E—M single bond and will be used as examples as circumstances demand. It should be pointed out in passing that the compounds containing several metal atoms constitute precursors to EM clusters.[62] There are large numbers of complexes containing the SiR_3 fragment directly bound to transition metals and the SiR_2 fragment bound to two metals.[63] In the case of the $SiMe_2H$ fragment bridging two W atoms, an agostic Si—H—W bridge is observed.[64] On the other hand, reactions of H_2SiR_2 (R = Ph, *t*-Bu) with $IrX(PMe_3)_4$ (X = H, Cl) yields $IrX(H)(PMe_3)_3(SiHR_2)$ with no agostic interactions.[65] Note the similarity with the formation of $(PMe_3)_3Ir(H)_2BR_2$ noted in Section 2.3.1. Generation of E—M bonds by oxidative addition has been observed to be reversible in the case of Si, e.g., the reaction of $[HRu_3(CO)_{11}]^-$ with $2Et_3SiH$ to yield $[HRu_3(CO)_{10}(SiEt_3)_2]^-$.[66] Likewise, a considerable number of related Ge and Sn compounds have been described.[67] In the case of Sn, a Cl bridged Sn—Mo bond is observed in $Cl_3SnMo(Cl)(CO)_3(dth)$.[68,69] The acidity of the Sn is important as no such interaction is observed in $Br_3GeW(Br)(CO)_3(bipy)$.[70] The relative strength of the E—M bond has been commented on.[71,72] There is a general decrease in bond energy for E—Mn in going from Si to Pb and a general increase in going from Mn to Re.

When one arrives at the heaviest elements of this series significant differences arise. The complexes $[\{Fe(CO)_4\}_3E]^{2-}$, E = Sn, Pb, with a trigonal planar arrangement of ligand around the E atoms constitute a good illustration (Figure 7b).[73] Formally, the E atom shares only six electrons and is considered to be electron deficient. In addition, the observed bond distances are interpreted as not supporting the existence of either localized or delocalized multiple bonds as a mechanism for completing the octet of the central E atom. Note was taken that an "equally probable" alternative structure containing one Fe—Fe bond and a lone pair of electrons on a pyramidal E atom is not adopted by these compounds. However, as the $Fe(CO)_4$ fragment is isolobal with CH_2, this species is analogous to the trimethylenemethane dianion which is observed stabilized as a complex of $[Fe(CO)_3]^{2+}$.[74] Hence, the presence of delocalized bonding might well be revealed by a more detailed examination of its electronic structure.

Species containing MER_3, MER_2, and MER fragments which formally represent E—M, E=M, and E≡M single, double, and triple bonds are expected to exhibit additional chemistry associated with the existence of multiple E—M bonding, i.e., carbene and carbyne chemistry differs from carbyl chemistry. In contrast to the carbenes,[75] there are very few examples of simple multiple Si—M bonds. Fully characterized neutral transition metal silylene complexes stabilized by a donor, $(t\text{-}C_4H_9O)_2Si{=}M(CO)_n \cdot HMPT$, M = Cr ($n$ = 5), Fe (n = 4), have been described and the structure of one is shown in Figure 7c.[76] Based on geometric data, the donor interaction appears to be a weak one and spectroscopic information (^{29}Si chemical shift) suggests the presence of a multiple Si—M bond. Calculations suggest the the silylene–metal bond is polar and the Si atom electron deficient. Coordination to a donor reduces the Si—M bond strength and induces a pyramidalization at the Si atoms. With fairly weak donors, substantial multiple bond character remains in the Si—M bond. Recently, evidence for the existence of base-free silylene complexes in solution, e.g., $[Cp^*(PMe_3)_2Ru{=}SiPh_2]^+$, has been reported.[77]

Considering the situation with Si, it is interesting to observe that the chemistry of the base-free heavier analogues, germylenes and stanylene, is well developed.[78] However, the element E still acts as a site of Lewis acidity and base coordination can be observed. In an example of a base free complex, $(CO)_5CrGe(Smes)_2$,[79] a Cr—Ge bond order of ≥ 1.5 has been suggested. Addition of a base to these systems induces only a slight deviation of the MER_2 fragment from planarity suggesting that E—M multiple bond character is largely retained. Little information is available on reactivity but, based on mostly negative evidence,[78] the formal ER_2 electron pair donor resists displacement from the metal by neutral bases such as phosphines.

In addition to the formal analogues of carbenes, there are characterized species containing multiple-bonded EM interactions that can be viewed as isolobal analogues of organic systems.[80,81] For example, $[Fe(CO)_4]_2Si \cdot 2HMPT$ (Figure 7d)[82] exhibits a distorted tetrahedral geometry around the silicon atom and a bonding system approaching that of allene. [The $Fe(CO)_4$ fragment is isolobal with CH_2.] Further, the tendency for hydrogen loss or oxidative addition[83] of the Ge—H bond seems to play a role in the generation of $[Cp^*Mn(CO)_2]_2Ge$ (Figure 7e) and derivatives with different substituents on the Cp ring.[84,85] In terms of electron requirements, the $Cp^*Mn(CO)_2$ fragment can be viewed as analogous to the $Fe(CO)_4$ fragment if one filled orbital of the "t_{2g}" set is used in bonding[86] but no base is necessary to stabilize the compound. This may reflect the greater stability of the divalent state for the heavier congener of E; however, see below. Again the E—M bonding has been described as analogous to the C—C bonding of allene. The lead derivative also exists.[87]

The nature of the multiple bonding in the M=E=M system has been explored with the Fenske–Hall technique for $[Cp^*Mn(CO)_2]_2Ge$.[88] This more detailed examination of the electronic structure suggests the analogy with allene is misleadingly simple. The CpCoCO fragment is isolobal with CH_2 possessing two valence electrons and two valence orbitals with one of σ and one of π symmetry with respect to a E—M bonding axis. Although one can view $CpMn(CO)_2$ as analogous to CpCoCO [or $Fe(CO)_4$] in terms of the 18-

electron rule, it has, along with the filled "t_{2g}" set, a single valence orbital with no valence electrons. That is, if the "t_{2g}" set is nonbonding, it is isolobal to $[CH_3]^+$.[89] However, the separation of frontier orbitals from the "t_{2g}" set in a transition metal fragment is only a rough approximation, albeit a very useful one, and the $CpMn(CO)_2$ fragment bonding is not restricted to one-orbital, zero-electron behavior. Thus, for example, in the "t_{2g}" set, $CpMn(CO)_2$ has two high-lying, nearly equi-energetic, filled orbitals that are mutually perpendicular and have π symmetry relative to the M—E—M axis of $[CpMn(CO)_2]_2Ge$. Hence, the bonding can be simply described in the following way. Let two sp hybrids on Ge act as donors to the σ acceptor orbitals on the two $Cp^*Mn(CO)_2$ fragments (Figure 8a). This leaves two empty $p\pi$ orbitals on Ge that can interact with both pairs of π orbitals on the metal fragments. As shown in Figure 8b, this results in a cylindrically symmetric bonding system that is consistent with the apparently low rotational barrier of $[Cp^*Mn(CO)_2]_2Ge$. The multiple bonding in this compound, then, is due to π back donation from the metal centers to Ge in which the $Cp^*Mn(CO)_2$ fragments act as three-orbital, four-electron fragments. This is probably the reason why base coordination does not take place as observed with

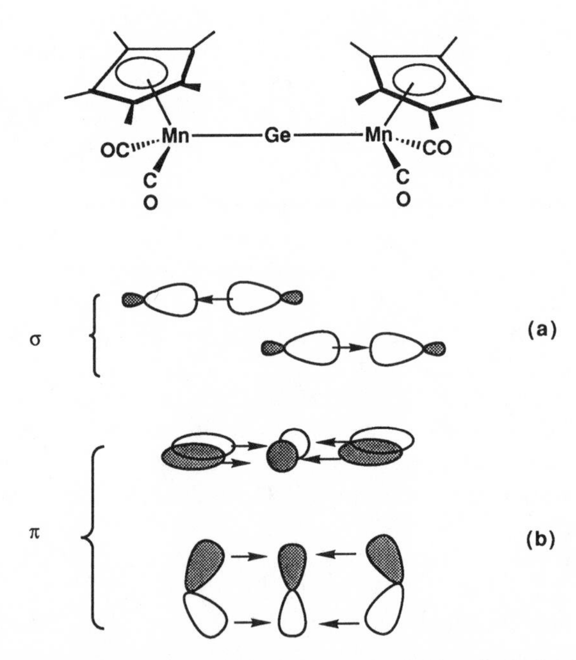

Figure 8. Schematic drawing of the principal Ge–Mn interactions in $[Cp^* Mn(CO)_2]_2Ge$. (a) Donation from filled *sp* hybrids on Ge to empty σ orbitals on the Cp^* $Mn(CO)_2$ fragments. (b) Donation from filled π orbitals on the Cp^* $Mn(CO)_2$ fragments to the empty $4p$ orbitals on the Ge atom.

$[Fe(CO)_4]_2Si \cdot 2HMPT$. An interesting counterpoint is the heteronuclear carbide complex $(Me_3CO)_3W{\equiv}C{-}Ru(CO)_2Cp$.[90]

The compounds $[(C_5H_4Me)Mn(CO)_2]_3E$, E = Ge,[84] Sn,[81] Pb[91] (Figure 7f) exhibit a trigonal coordinated E atom and, based on bond distances, one localized Mn=E double bond. Closely related compounds for Ge and Sn, but with $W(CO)_5$ metal fragments, have been characterized.[92] Note that although the geometry around E is similar to that of $[\{Fe(CO)_4\}_3E]^{2-}$ discussed above, two fewer electrons are available resulting in a M—M bond and localized double bond. Clearly, in this case the $CpMn(CO)_2$ fragments are behaving as two-orbital, two-electron fragments like $Fe(CO)_4$ and, hence, the compounds are isolobal analogues of methylene cyclopropane. This is reasonable as the $E\{(C_5H_4Me)Mn(CO)_2\}_2$ fragment presents only one empty π orbital on E to the third $(C_5H_4Me)Mn(CO)_2$ fragment.

Only a few examples of reactivity exist but one suggests that E—M multiple bonds will serve as useful reaction centers. Analogous to addition of methylene to C—M double bonds, diazomethane transfers a CH_2 fragment to the Mn=Ge double bond of $[(C_5H_5)Mn(CO)_2]_3Ge$ to yield a three-membered ring containing Mn, Ge, and C atoms.[93] On heating the product, C_2H_4 is evolved and $[(C_5H_5)Mn(CO)_2]_3Ge$ is regenerated.

2.3.3. Group 15

The unsaturated complexes are those containing ER_2, ER, and E fragments. In principle, group 15 compounds of the type MER_2, M_2ER, and

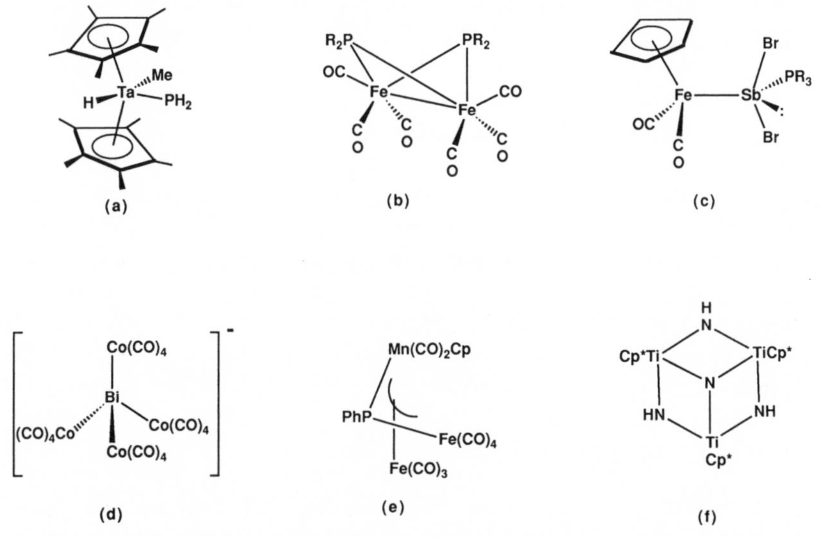

Figure 9. Structures of (a) $(Cp^*)_2Ta(PH_2)(Me)H$, (b) $[Fe(CO)_3PMe_3]_2$, (c) $Cp(CO)_2FeSbBr_2(PMe_3)$, (d) $[\{Co(CO)_4\}_4Bi]^-$, (e) $\{[Fe(CO)_4][CpMn(CO)_2]PPh\}Fe(CO)_3$, and (f) $(Cp^*Ti)_3(NH)_3N$.

M_3E containing no M—M bonds and 17-electron metal fragments are representative examples of E—M single bonds. Indeed examples of such metalated species exist, e.g., $(Cp^*)_2Ta(PH_2)(Me)H$[94] (Figure 9a). However, the existence of a lone pair on E leads to bridging modes of bonding for ER_2 fragments, e.g., $[Fe(CO)_3PMe_3]_2$ (Figure 9b) in which the lone pair on the P atom of one $Fe(CO)_3PMe_3$ fragment can be viewed as coordinated by the iron atom of the other fragment.[95] The 18-electron rule requires a Fe—Fe bond. It is interesting to note that, besides steric bulk, an important consideration in achieving low coordination numbers in metals utilizing the $[NRB(mes)_2]^-$ ligand is the reduced availability of the nitrogen lone pair for bridging interactions.[96]

Some interesting variations are exhibited by the heavier elements. Although $Cp(CO)_2FeSbMe_2$ undergoes substitution of CO at the Fe center by PMe_3 the closely related complex, $Cp(CO)_2FeSbBr_2$, undergoes addition of PMe_3 at the Sb center.[97] Substitution at Fe only occurs under more forcing conditions. Thus, replacing a Br atom of $SbBr_3$ with the 17-electron $Cp(CO)_2Fe$ fragment results in the coordination of only a single, weakly bound phosphine to Sb. The structure (Figure 9c) shows pseudo-trigonal-bipyramidal geometry around Sb indicating the presence of a stereochemically active lone pair in an equatorial position. This compound demonstrates that trivalent Sb with a σ-bonded transition metal fragment is able to function not only as a donor but also as an electron pair acceptor.

The complex $[Co(CO)_4]_3Bi$ may be considered similar to $BiCl_3$ in which the pseudo-halogen $[Co(CO)_4]^-$ replaces Cl^-.[98] In fact, this is a way to prepare the compound.[99] The same is true for $[Mn(CO)_5]_3Bi$.[100] However, $[Co(CO)_4]_3Bi$ is sufficiently acidic such that the addition of another $[Co(CO)_4]^-$ base takes place to yield $[\{Co(CO)_4\}_4Bi]^-$ whose structure is shown in Figure 9d. The net result is a paramagnetic complex which differs from normal $(ML_4)_4E$ complexes in formally having 10 electrons at the Bi center. The former lone pair of $[Co(CO)_4]_3Bi$ is unpaired by the addition of $[Co(CO)_4]^-$ and is stereochemically inactive. In the process the Co—Bi bonds increase an average of 0.14 Å.

Attaching two 16-electron metal fragments to an ER fragment leads to the generation of phosphinidene complexes, e.g., $[CpMn(CO)_2]_2PPh$, and their higher homologs.[101,102] These compounds are viewed as allyl anion analogs and the chemistry is well developed at the structural, spectroscopic, and chemical levels. Hence, they constitute a good example of the consequences of multiple E—M bonding. The ER fragment, E = P, As, Sb, Bi, has two lone pairs and an empty valence orbital. The former donate to the primary empty valence orbital of each of the two metal fragments to form the σ bonding network (see above, Section 2.3.1). The latter interacts with one of the two π symmetry filled orbitals on each of the metal fragments to form a 3c, 4e π system. The existence of such a π system is evidenced by abnormal ^{31}P NMR chemical shifts and the strong absorptions in the visible region not observed for compounds such as $CpMn(CO)_2PR_3$. It is these electronic transitions involving the π system that give rise to the intense colors of these compounds.

The low-lying, empty 3c, π orbital suggests significant Lewis acidity at the E atom and the high-lying filled 3c, π orbitals suggest allylic ligand properties. Consistent with this prediction, Lewis bases add to the central atom to form tetracoordinate, tetrahedral E atoms, and compounds like that shown in Figure 9e can be viewed as an allylic ligand, $[Fe(CO)_4][CpMn(CO)_2]PPh$, bound to an $Fe(CO)_3$ fragment.(103) Alternatively, the latter compound can be viewed as a cluster. Other reactions of this compound type have been summarized.(104) Several lead to species containing E_2 fragments and, hence, will be discussed further below. Of particular interest here is the dehalogenation of $[Cp'Mn(CO)_2]_2AsCl$ to yield $[\{Cp'Mn(CO)_2\}_2As]^+$ which is a structural analog of $[Cp^*Mn(CO)_2]_2Ge$ discussed in Section 2.3.2 (Figure 7e).(105) The phosphorus analogs have also been synthesized.(106)

The chemistry of group 15-metal complexes is not restricted to isolobal organic analogs and new, unusual compounds appear regularly even for the lighter elements. For example, the reaction of Cp^*TiMe_3 with NH_3 followed by loss of CH_4 yields $(Cp^*Ti)_3(NH)_3N$ with the symmetric, but complex, structure shown in Figure 9f.(107) The $(TiNH)_3$ ring is in a chair conformation with a pyramidal N atom bridging the three Ti atoms. This compound can be contrasted with an example of trigonal planar N,(108) i.e., $[Ir_3N(SO_4)_6(H_2O)_3]^{4-}$.

2.3.4. Group 16, 17

The unsaturated fragments here are ER and E. The chemistry of alkoxide (and amide) complexes has been reviewed(109) and interest in the former complexes is high as precursors for electronic and ceramic materials.(110) This is also true for the heavier elements(111,112) where routes to Cd—Se bonds in both solid-state as well as molecular compounds have been reported.(113) (See also Chapter 8). Complexes with bare elements are also of great interest.(80)

Compounds illustrating single E—M bonds between a variety of transition metals and group 16 elements have been characterized and exhibit a rich chemistry. For example, consider $Cp_2^*Ti(EH)_2$, E = O, S, Se,(114) and $Cp^*Re(H)(CO)_2(TeH)$.(115) The latter, shown in Figure 10a, gives rise to a number of multinuclear species with interesting structures.(116,117) A particularly fascinating set of compounds are $[\{CpMn(CO)_2\}_2EPh]^+$ complexes, E = group 16 element, which are isoelectronic with the $[CpMn(CO)_2]_2EPh$ system, E = group 15, discussed in Section 2.3.3. The structure of the $[\{CpMn(CO)_2\}_2SPh]^+$ is very similar to that of $[CpMn(CO)_2]_2PPh$.(118) An equally valid valence structure for this set of fragments is the one shown on the right side of Figure 10b, i.e., a three-membered ring. In fact this structure

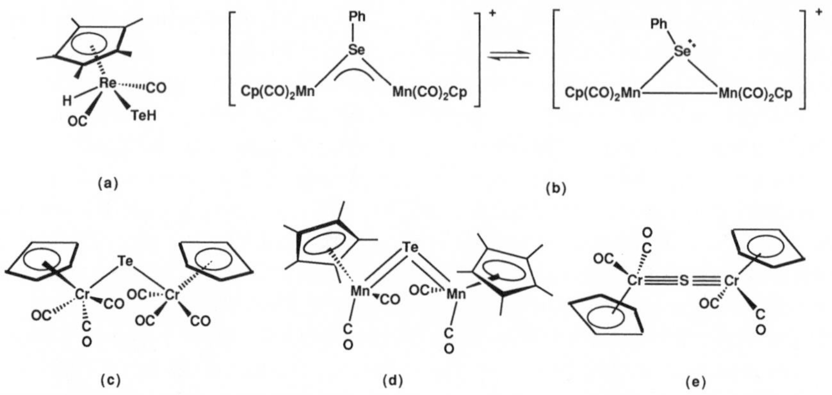

Figure 10. Structures of (a) $Cp^*Re(H)(CO)_2(TeH)$, (b) $[\{CpMn(CO)_2\}_2SePh]^+$, (c) $[CpCr(CO)_3]_2Te$, (d) $[Cp^*Mn(CO)_2]_2Te$, and (e) $[CpCr(CO)_2]_2S$.

containing a M—M bond has been observed for both E = S[119] and E = Te.[120] This suggested the existence of a valence isomerism which has now been established for the system in Figure 10b and an analysis of the equilibrium shows that the closed isomer is more stable than the open one by 2 kJ of free energy.[121] This is a beautiful example of the use of element variation to tune ΔG_0 to ≈ 0 for a structural transformation.

Of course, two single bonds to E are also possible yielding bent E bridges, e.g., $[CpCr(CO)_3]_2Te$ (Figure 10c).[122,123] With the bare elements, multiple bonding is possible with the proper choice of metal fragments. With two 16-electron fragments the bent M=E=M system is once again obtained, e.g., $[Cp^*Mn(CO)_2]_2Te$ (Figure 10d).[124] Short Mn—Te bond distances support the existence of multiple bonding. Presumably the bonding is similar to that described in Section 2.3.3 for the analogous phosphinidene complex. Indeed, the available lone pair on Te can be used to bond another 16-electron center to form $[CpMn(CO)_2]_3Te$ containing a trigonal planar Te atom.[125] Using 15-electron metal fragments the extent of multiple bonding in the M—E—M bonding system can be increased further. For example, $[CpCr(CO)_2]_2S$ (Figure 10e) has a linear M—E—M configuration and a short Cr—S distance is reported.[126] The bonding can once again be understood in terms of the diagram in Figure 8. ($[Cp^*Mn(CO)_2]_2Ge$ is isoelectronic with $[CpCr(CO)_2]_2Se$.[127] In one-electron MO schemes the electrons are added last and, hence, the same number and type of MOs will be occupied for both compounds. The polarity of the E—M bonding will be different, however, and this fact will be reflected in both the charge distributions and MO energies. Selective reduction of $[CpCr(CO)_2]_2Se$ leads to $[CpCr(CO)_2]_3Se$ with a μ_3-Se atom.[128]

The fact that group 16 atoms have a strong tendency to function as donors limits the metals to which ER and E are directly bound. The consequences of the electron-rich nature of these atoms has been discussed for oxygen.[129] In essence, the electrons on E in orbitals of π symmetry with respect to the E—M bond can interact with M in a stabilizing manner only if there are empty π symmetry orbitals on the metal to accept them. Metals with filled π symmetry orbitals give rise to destabilizing interactions with the π electrons on E, i.e., metal–ligand π antibonding interactions. This helps explain the lack of late transition metal complexes with terminal oxo ligands as well as the high polarity of the late metal alkoxide complexes. It is also responsible for the existence of analogous isolobal behavior of early transition metal alkoxide fragments with late transition metal carbonyl fragments.[130] There are fewer restrictions for the softer, heavier elements of this group and the versatility of sulfur, for example, is vividly demonstrated in metal cluster systems (Chapter 3).

A consideration of oxo complexes leads finally to the chemistry of transition metal halide complexes. The chemistry of the class of compounds with CpM—E bonds, E = group 17 elements, is rather extensive; however, space prohibits discussion. A good feeling for the behavior of the chlorides[131] and bromides[132] can be obtained from the literature. Recently, some novel chemistry involving the Cr—I bond has been reported.[133]

3. E_2M COMPLEXES CONTAINING η^2 MAIN GROUP LIGANDS

3.1. Metal Complexes with Saturated E_2R_v Ligands

Catenated main group compounds, specifically saturated E_2R_v molecules, can be expected to exhibit all the coordination modes of the EM complexes discussed in the preceding section. Hence, in discussing compounds expressing these types of coordination we can be brief simply because essentially no new EM bonding principles are evidenced. On the other hand, the added complexity of multibonding sites on such a ligand is not trivial simply because such sites do not interact with metals in an independent sense. A well-known example of how the presence of multiple bonding sites affects metal binding is the chelate effect. This is exemplified by the diphosphine ligands; however, these often have "spacer" fragments between the phosphorus atoms to give a more effective "bite" size relative to a single metal site, e.g., $R_2PCH_2CH_2PR_2$.

The utilization of multifunctional ligands of this type is pervasive in organometallic chemistry[2] and one would not initially consider analogs from group 13 to exhibit any like chemistry. The following is a good example of how a characteristic of metal–ligand bonding carries over from group to group.

Recently the ligand chemistry of bisphosphine diborane(4) neutral molecules has begun to be developed.[134,135] The B_2H_4 fragment itself can be considered a bisborane with two acceptor orbitals (Figure 11a). Coordination

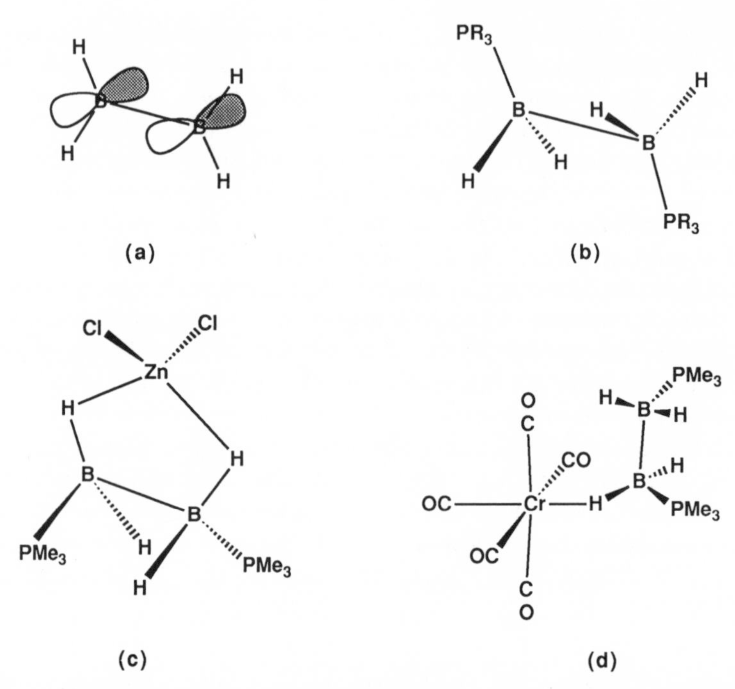

Figure 11. Schematic drawing of (a) the hypothetical B_2H_4 molecule, (b) the free $B_2H_4(PMe_3)_2$ ligand, (c) structure of the $ZnCl_2B_2H_4(PMe_3)_2$ complex, and (d) structure of the $Cr(CO)_5B_2H_4(PMe_3)_2$ complex.

of two PMe_3 ligands leads to pyramidalization at boron and an ethane-like structure (Figure 11b). A significant amount of the negative charge deposited by the base at the boron center ends up on the hydride ligands. Thus, despite being a neutral molecule, the hydrogens of $B_2H_4(PMe_3)_2$ ligand behave like those of $[BH_4]^-$ with respect to transition metals. The reaction of $B_2H_4(PMe_3)_2$ with $ZnCl_2$ yields a one-to-one complex with a structure shown in Figure 11c.[136] A similar result obtains with $Ni(CO)_4$ where $Ni(CO)_2[B_2H_4(PMe_3)_2]$ is formed, although the borane ligand is easily displaced by CO or phosphines to reform $Ni(CO)_4$ or the disubstituted $Ni(CO)_2(PR_3)_2$ complexes.[137] A closely related copper complex, $[Cu\{B_2H_4(PMe_3)_2\}_2]^+$ containing two bidentate $B_2H_4(PMe_3)_2$ ligands in a spiro structure, has been described.[138] Most recently the complexes $M(CO)_5[B_2H_4(PMe_3)_2]$ and $M(CO)_4[B_2H_4(PMe_3)_2]$, M = Cr, Mo, W, have been characterized where the $B_2H_4(PMe_3)_2$ ligand acts as a unidentate ligand in the former and a bidentate ligand in the latter.[139] In the first case coordination is via a single B—H—M bridge (Figure 11d) and similar to that observed for the mono-dentate $[BH_4]^-$ ligand (Figure 5a). Interestingly, in solution the NMR spectra indicates that the metal is rapidly moving between B—H sites on the two boron atoms.

In the same fashion that $[BH_4]^-$ mimics an allyl ligand (see Section 2.2.1 above), the $B_2H_4(PMe_3)_2$ ligands mimic a ligand like $R_2PCH_2CH_2PR_2$. That is, the $MB_2(\mu\text{-}H)_2$ five-membered ring is analogous to the MC_2P_2 five-membered ring formed by a bidentate phosphine ligand–metal complex where the μ-H atoms are playing the role of the phosphorus atoms. Again, the idea of a B—H bond serving as a protonated lone pair is a useful one. Considering the known chemistry of bisphosphine metal complexes, these observations suggest the existence of a significant undiscovered analogous chemistry for these new chelating ligands.

Such chemistry is not restricted to analogues of bidentate ligands or even to small borane fragments. For example, in the metalloborane $Mn(CO)_3B_8H_{13}$ the metal fragment is found outside the borane cage coordinated via three of the *exo*-cluster B—H bonds such that the geometry around the metal atom is approximately octahedral.[140] In effect the B_8H_{13} fragment is acting as a tridentate, chelating ligand not unlike $[B_3H_8]^-$ which is discussed in another context in Section 5.1 below. The same mode of coordination is found in the so-called *exo-nido*-rhodacarboranes which is a metallocarborane cluster compound that plays a role in reactions catalyzed by these species.[141]

3.2. Metal Complexes with Unsaturated E_2R_v Ligands

The characterization of a side-on bound complex of an olefin with a transition metal center, e.g., $Fe(CO)_4(\eta^2\text{-}C_2H_4)$, gave rise to the Dewar–Chatt–Duncanson model for π-bonding.[142,143] The development of this class of compounds was one driving force for the rapid expansion of organometallic chemistry in the last twenty years. Unsaturation with respect to the E—E bond is a feature of E_2R_v ligands that provides an analogous electron-rich site for binding electrophilic metal fragments. Indeed examples of metal–element bonding similar to the π-bonding of organic ligands exist for the elements on both sides of and below carbon. These compounds constitute a strong argument for a parallel development of the associated inorganometallic chemistry in the future. However, the η^2-binding sites of the inorganic ligands will be in competition with the other bonding modes described in Section 2. Additionally, the ligands used to provide isolable multiple-bonded compounds also reduce the tendency toward coordination. This requires very different synthetic strategies. In the following, some of the basic interactions of unsaturated E_2R_v ligands are described along with compounds demonstrating evidence of competition between electron-rich sites on main group ligands.

3.2.1. Group 13

The interactions of polyboranes and polyborane fragments with transition metals has been reviewed several times recently.[144–150] Diborane is isoelectronic with C_2H_4 but its low Lewis basicity is reflected in its high ionization potential (12.6 eV) relative to that of C_2H_4 (10.5 eV). However, the model in which B_2H_6 is considered as an ethylenic $[B_2H_4]^{2-}$ with protonated double bonds[151] formally permits a $[B_2H_5]^-$ anion to function as a η^2-ligand to a metal (Figure 12a). This idea has been fully developed for the set of compounds $K[M(CO)_4(\eta^2\text{-}B_2H_5)]$, M = Fe, Ru, Os and $M'(\eta^5\text{-}C_5H_5)(CO)_2(\eta^2\text{-}B_2H_5)$, M' = Fe, Ru.[152] The molecular structure of $Fe(\eta^5\text{-}C_5H_5(CO)_2(\eta^2\text{-}B_2H_5)$ has been determined in the solid state and is shown in Figure 12b. It can be seen that the $CpFe(CO)_2$ fragment occupies the position of a bridging proton in diborane(6) such that in a geometric sense the metal complex is a true analog of $Fe(CO)_4(\eta^2\text{-}C_2H_4)$. The characterization of $Cp_2(H)Mo(\eta^2\text{-}B_2H_5)$ provides an example with a group 6 metal while $Cp^*(PMe_3)Ru(\eta^2\text{-}B_2H_7)$ exhibits a similar MB_2 triangle but with both M—B and B—B edges bridged with hydrogens.[153]

The nature of the metal–boron interaction has been explored more fully utilizing UV-photoelectron spectroscopy to directly probe the valence-level energies and Fenske–Hall calculations to compare the nature of the molecular orbitals involved in metal–boron and metal–carbon bonding.[154] Consistent with the lower nuclear charge of boron relative to carbon, the filled donor orbital of the $[B_2H_5]^-$ fragment (the analog of the π orbital C_2H_4) is a better match for the metal acceptor orbitals while the unfilled acceptor orbital of

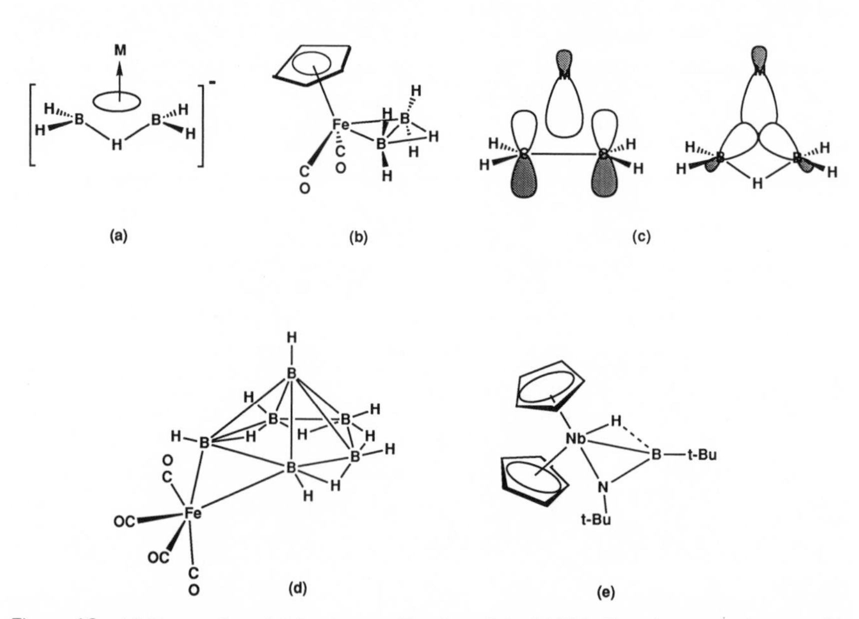

Figure 12. (a) Proposed model for the coordination of the $[B_2H_5]^-$ ligand to a metal center. (b) The structure of $Fe(\eta^5\text{-}C_5H_5)(CO)_2(\eta^2\text{-}B_2H_5)$. (c) Comparison of the coordination of C_2H_4 and $[B_2H_5]^-$ to the $Fe(CO)_4$ fragment. (d) Proposed structure of $Fe(CO)_4(\eta^2\text{-}B_6H_{10})$. (e) Structure of $Cp_2Nb(H)[\eta^2\text{-}B(t\text{-}Bu)N(t\text{-}Bu)]$.

the $[B_2H_5]^-$ fragment (the analog of the π^* orbital C_2H_4) is a much poorer match for the metal donor orbitals. Consequently, metal-to-ligand back donation is considered to be much less important in the case of the metalloborane. However, the presence of the bridging proton in the $[B_2H_5]^-$ fragment plays an important role in enhancing the metal–boron bonding interaction. This can be seen by comparing the σ interactions of C_2H_4 and $[B_2H_5]^-$ with a metal fragment orbital. In the former case, the σ-bonding electron density lies mainly outside of the triangle defined by the carbon and metal atoms. As shown in Figure 12c, the bridging hydrogen forces effective sp^3 hybridization on the boron atoms thereby "focusing" the donor orbital pair on the σ acceptor orbital of the metal atom. This enhances the net overlap and the bonding density lies mainly within the triangle defined by the boron and metal atoms. In a sense, this mechanism replaces the back-bonding interaction that is largely lost in going from C_2H_4 to $[B_2H_5]^-$.

The bonding exhibited in $Fe(\eta^5\text{-}C_5H_5)(CO)_2(\eta^2\text{-}B_2H_5)$ is an example of a more general $\eta^2\text{-}E_2\text{–}M$ interaction found in metalloborane chemistry. The skeletal B—B edges formed by removal of bridging B—H—B hydrogens as protons exhibit the capacity for η^2-bonding interactions with Lewis acids including transition metals. Even neutral polyhedral boranes with unbridged edges having partial multiple bond character, such as B_6H_{10},[155] form olefin analogues, e.g., $Fe(CO)_4(\eta^2\text{-}B_6H_{10}$[156] (Figure 12d). As with the B_2M system, we expect the η^2-borane ligand to act mainly as a σ donor to the metal atom. Again, the cage structure and the tendency toward sp^3 hybridization at boron (tetrahedral coordination) gives a directional character to the bonding that strengthens the σ bonding and makes up for the lack of significant π back-bonding with the metal atom.

The B—N fragment is isoelectronic with C—C and a considerable body of chemistry on systems containing this "polar acetylene" has been developed.[157] An interesting mixed group 13–15 η^2-EE′ ligand is found in $Cp_2Nb(H)[\eta^2\text{-}B(t\text{-}Bu)N(t\text{-}Bu)]$. The structure is shown in Figure 12e.[158] Although the Nb–N distance is shorter than the Nb–B distance, a comparison of the geometric parameters with those of an authentic N metalated aminoborane shows that $Cp_2Nb[\eta^2\text{-}B(t\text{-}Bu)N(t\text{-}Bu)]$ can be considered an authentic π-bound acetylene analog. Both the small increase in B–N distance on coordination as well as the E–E′–R angles suggest that back-bonding is less important here than in similar acetylene complexes. Based on distances, the Nb—H hydrogen exhibits a weak interaction with the boron atom.

No aluminum compounds of the type discussed in this section have been reported.

3.2.2. Group 14

It is only recently that three examples of $\eta^2\text{-}Si_2$ bonding to metals analogous to that exhibited by C_2H_4 have been reported. In the first, evidence has been obtained for the coordination of $mes_2Si{=}Simes_2$ to $Hg(OCOCF_3)_2$; however, the compound is unstable above 0 °C and difficult to characterize.[159] This compound was obtained directly from the bulky ligand stabilized disilene and the presence of these ligands may well inhibit the metal–silicon interaction. More recently, two relatively unhindered disilene metal complexes were prepared using a strategy in which formation of a X—M—Si—Si—X open intermediate precedes elimination of 2X and ring closure. This leads to stable compounds. In one approach with X = H, a Pt bound η^2-$Si_2(i\text{-}Pr)_4$ was observed and the proposed structure shown in Figure 13a.[160] In the second, X = Cl yielding $Cp_2W(\eta^2$-$Si_2Me_4)$[161] and, for Cl—M—C—Si—Cl, $Cp_2W(\eta^2$-$CH_2SiMe_2)$.[162] A crystal structure of $Cp_2W(\eta^2$-$Si_2Me_4)$ (Figure 13b) gives significant geometric information on the nature of the main group–transition metal interaction. In analogy with the changes observed in olefin structure on coordination to a metal center, the Si–Si distance in $Cp_2W(\eta^2$-$Si_2Me_4)$ [2.260(2) Å] lies between the values expected for a single Si—Si bond (2.35 Å) and double Si—Si bond (2.14 Å).[163,164] Further, the pyramidalization at silicon suggests hybridization somewhere between sp^2 and sp^3. Thus, $Cp_2W(\eta^2$-$Si_2Me_4)$ can be considered either a π complex in the sense of the Dewar–

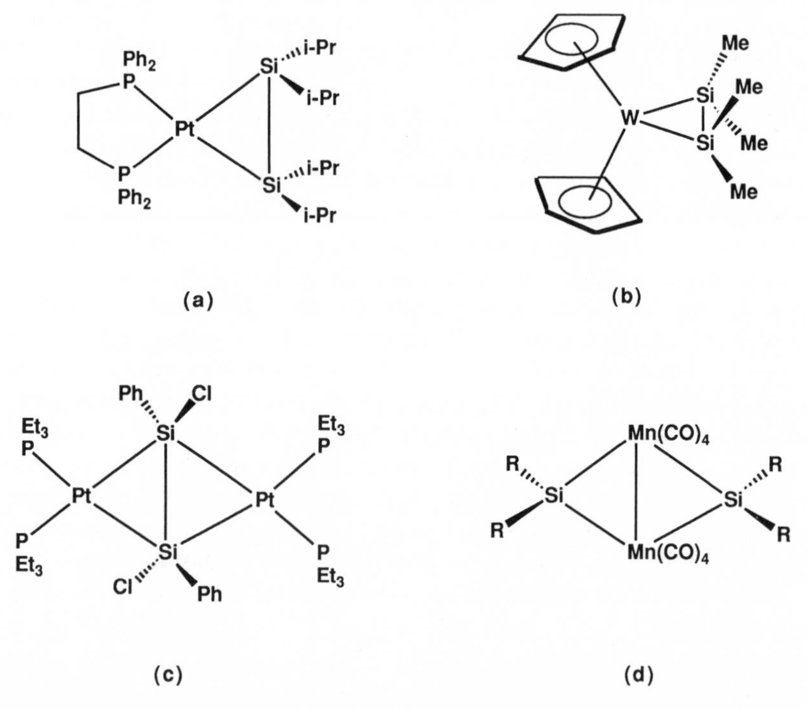

Figure 13. (a) Proposed structure Pt (dppe) $\{\eta^2$-$Si_2(i\text{-}Pr)_4\}$. (b) Structure of $Cp_2W(\eta^2$-$Si_2Me_4)$. (c) Structure of $[(Et_3P)_2Pt(SiPhCl)]_2$. (d) Structure of $Mn_2(CO)_8(SiPh_2)_2$.

Chatt–Duncanson model or a MSi_2 metallacycle. As observed with olefin complexes,[165] the variation between these two extremes should depend on the metal fragment and the substituents on the disilene.

Some aspects of the reactivity of the Pt-bound disilene shown in Figure 13a have been reported.[166] Hydrogenolysis leads to the cleavage of the Si—Si bond rather than the Pt—Si bond which is known to be labile to hydrogenolysis under mild conditions.[167] Reaction with O_2 leads to insertion into the Si—Si bond yielding a PtSiOSi metallacycle. The treatment of the disilene complex with phosphine donor resulted in decomposition with no evidence of the release of the free disilene. This suggests that the η^2-disilene fragment is more strongly bound to the metal center than is the equivalent η^2-alkene. Although substituted amines failed to reaction or caused decomposition, the reaction with NH_3 led to insertion of an NH fragment into the Si—Si bond to give a nearly quantitative yield of the isoelectronic analogue of the product of the reaction with O_2.

An interesting variation on this theme is found in the compounds $[(Et_3P)_2Pt(SiPhX)]_2$ where X = H, Cl. These, and the mixed H, Cl dimer, have the basic structure shown in Figure 13c.[168] If it were not for the very short Si–Si distance [2.575(15)–2.602(4) Å] which lies on the long side of the range of known Si—Si single bonds, these compounds would be considered in Section 2.3.2. The rather long Pt–Si distances and the acute Si—Pt—Si angle suggest that Si—Si bonding is an integral part of the bonding scheme. Hence, it has been suggested that these compounds are examples of the disilene, Ph(X)Si=Si(X)Ph, coordinated to two $Pt(PEt_3)_2$ fragments. This is substantiated by the characterization of the *cis* isomer which has a bent "butterfly" Pt_2Si_2 skeleton.[169]

A MO model of the bonding in this and a related system has been described.[170] For our purposes the bonding can be described in terms of what might be called a three-center Dewar–Chatt–Duncanson model. Consider the molecules as constructed from two PtL_2 fragments interacting with a disilene. The frontier orbitals of a PtL_2 fragment consist of a filled *dp* hybrid in the plane of the PtL_2 fragment with π symmetry with respect to a Pt–M axis and an unfilled *ds* hybrid with σ symmetry with respect to the same axis (Figure 14a)[171] while those of the disilene are the filled π and unfilled π^* orbitals. A σ interaction between the fragments involves the filled Si=Si bonding orbital and the negative combination of the empty σ symmetry PtL_2 orbitals. This bonding 3c,2e interaction (Figure 14b) utilizes two of six available valence electrons. A π interaction involves the unfilled Si=Si antibonding orbital and the negative combination of the filled π symmetry PtL_2 orbitals and results in a bonding 3c,2e interaction. The remaining two electrons occupy a nonbonding MO. (The net 3c,4e π interaction is shown in Figure 14b.) Not shown in the simplified model in Figure 14b is the fact that the MO calculations[170] indicate that the low-lying σ Si—Si antibonding fragment orbital

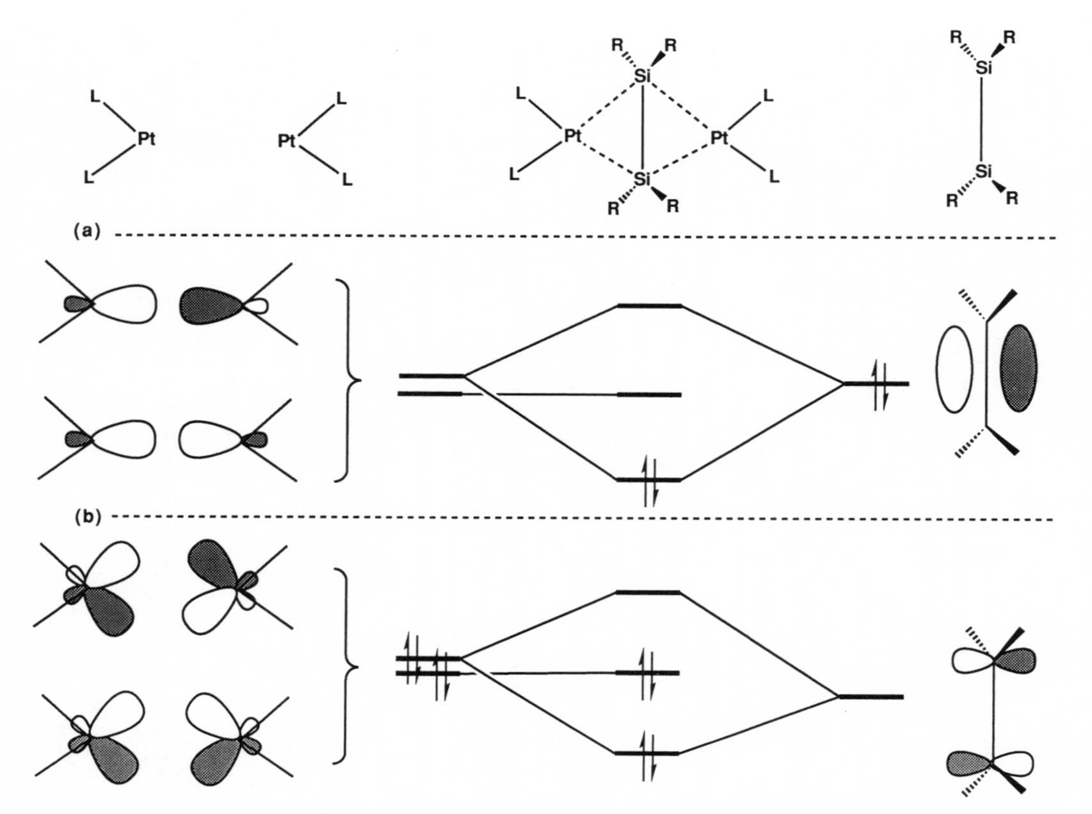

Figure 14. Representation of the bonding between Pt and Si in $[(Et_3P)_2Pt(SiPhCl)]_2$. (a) Interaction of the unfilled σ orbitals of the PtL_2 fragments with the filled Si—Si π bonding orbital. (b) Interaction of the filled π orbitals of the PtL_2 fragments with the unfilled Si—Si π^* orbital. Note that the two diagrams are separate and the energy scales are arbitrary.

mixes with the filled nonbonding π MO to produce significant net bonding character. Formally, this allows one to consider the disilene as a 4e donor and this behavior constitutes one of the fascinating differences between the organometallic and inorganometallic analogs. In this sense, $[(Et_3P)_2Pt(SiPhX)]_2$ can be considered analogous to $[Cp^*Mn(CO)_2]_2As_2$ (Figure 17b) discussed in Section 3.2.3 below. However, consistent with our discussion in Section 2.2.2 of the interactions of saturated carbon and silicon species with metals, one might expect to observe coordination of an olefin to two transition metal centers if the metal is rather electropositive. Indeed, the ethylene fragment in $Zr_2X_6(PEt_3)_4(CH_2{=}CH_2)$, X = Br, Cl, exhibits η^2-coordination to two metal centers.(172)

In the discussion of the bonding of $[(Et_3P)_2Pt(SiPhX)]_2$,(170) attention is drawn to the curious, but interesting, parallels in the bonding of $Mn_2(CO)_8(SiPh_2)_2$ (Figure 13d).(173) Although the detailed bonding interactions are somewhat more complex, $Mn_2(CO)_8(SiPh_2)_2$ can be viewed as a triply bonded $Mn_2(CO)_8$ moiety [the $Mn(CO)_4$ fragment is isolobal with CH] coordinated to two silene fragments. As a bridging silene is equivalent to a carbonyl in terms of counting electrons, the net Mn—Mn bond order is the same as that in $Mn_2(CO)_{10}$, i.e., one. In essence, $[(Et_3P)_2Pt(SiPhX)]_2$ can be viewed as an electron-rich doubly bonded main group E_2 fragment coordinated by two metal fragments acting as σ acids and π bases while $Mn_2(CO)_8(SiPh_2)_2$ can be viewed as an electron-poor triply bonded transition metal M_2 fragment coordinated by two main group fragments acting as σ bases and π acids. That is, they are inside-out versions of each other.(174,175) This curious parallelism will be encountered again in group 15 complexes.

The chemistry of compounds containing the Ge=Ge double bond has been reviewed(176) but no examples of π coordination to transition metals have been reported.

3.2.3. Group 15

At the present time, the coordinated η^2-E_2 ligands of group 15, with the exception of carbon, have the most fully developed structure and chemistry. A number of contemporary reviews(177–184) have appeared that cover not only the η^2-E_2 ligands of this section but also the μ-η^2-E_2 ligands and the η^p-rings of the following two sections. Just for arsenic alone 100 complexes are known and about half of these have been crystallographically characterized.(183)

An early example of a coordinated diazene π complex is $(t\text{-BuNC})_2Ni(\eta^2\text{-}N_2Ph_2)$.(185) The geometric changes in the N_2Ph_2 ligand on coordination suggest strong back donation with approximately tetrahedral geometry (including the lone pair) around the nitrogen atoms. The incorporation of sterically demanding substituents allows ligands containing P=P and As=As double bonds to be isolated and structurally characterized.(186,187) Coordination of

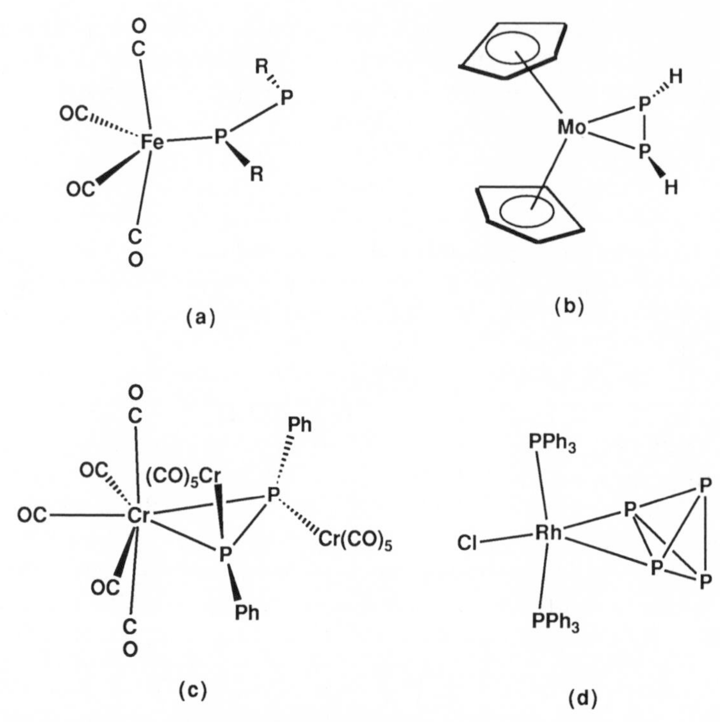

Figure 15. Structures of (a) $Fe(CO)_4\{\eta^1\text{-}P_2(2,4,6\text{-}(Bu)_3C_6H_2\}$, (b) $Cp_2Mo(\eta^2\text{-}P_2H_2)$, (c) $Cr(CO)_5(\eta^2\text{-}PhP{=}PPh)[Cr(CO)_5]_2$, and (d) $Rh(PPH_3)_2(Cl)(\eta^2\text{-}P_4)$.

the free doubly bonded ligands with transition metal fragments often leads to η^1-E bonding through the lone pairs on the E atom (Figure 15a) rather than η^2-bonding.[188] Despite this tendency, the longest established mode of diphosphene–metal bonding is of the η^2-type as exemplified by $Cp_2Mo(\eta^2\text{-}P_2H_2)$ in Figure 15b.[189] However, as with the diselene complexes discussed above, the synthetic route is an indirect one. All combinations of η^1- and η^2-bonding are known. For example, both types of E—M bonding are demonstrated by compounds such as $Cr(CO)_5(\eta^2\text{-}PhP{=}PPh)[Cr(CO)_5]_2$ shown in Figure 15c.[190] Heating this compound results first in the loss of the η^2-bonding $Cr(CO)_5$ fragment. This suggests that of the two types of sites on the ligand the multiple bond coordinates less strongly to a $Cr(CO)_5$ fragment. Indeed, the η^2-coordinated P=P bond is reactive. For example, CH_3COOH and CH_3OH add across a coordinated P=P double bond.[191]

Similar structures have been observed for the arsenic analogues[192] and, for both elements, competition between η^1- and η^2-coordination depends on a balance between ligand bulk, E=E bond lengths and the steric demands of the ML_n Lewis acid.

More complex phosphorus ligands exhibit η^2-bonding as well. For example, the P_4 molecule coordinates to transition metals in this fashion as demonstrated by the structure of $Rh(PPh_3)_2(Cl)(\eta^2\text{-}P_4)$ (Figure 15d).[193] The increased length of the P–P edge coordinated to the metal as well as quantum chemical calculations support the analogy between the η^2-bonded P_4 molecule and a similarly bound olefin. The "back-bonding" interaction is attributed to a three orbital–four electron interaction between a filled t_2 and unfilled t_1 on the P_4 fragment (both of which have b_2 symmetry in the complex) and a filled d_{yz} metal orbital of b_2 symmetry in the RhP_2 plane of the η^2-interaction. Both the principle donor and the "back-bonding" interactions reduce the P—P bond order of the edge attached to the metal atom.

This bonding complexity extends to the η^1-coordination mode as well. For example, it would be incorrect to conclude from the geometric description of $Fe(CO)_4(P_2R_2)$ above (Figure 15a) that the E–M interaction consists simply in a σ donor–acceptor bond with the main group species acting as the base. The compound, $Cr(CO)_5(P_2H_2)$, that is a model for a known complex,[194] has been investigated theoretically.[195] As the point established by this study is a general one, the pertinent interaction diagram is sketched in Figure 16. Here it is seen that, although the primary interaction does result from the metal acceptor orbital and the symmetric "lone pair" combination MO of the E=E fragment, two other interactions are important. In one, a filled metal fragment orbital interacts in a stabilizing fashion with the empty an-

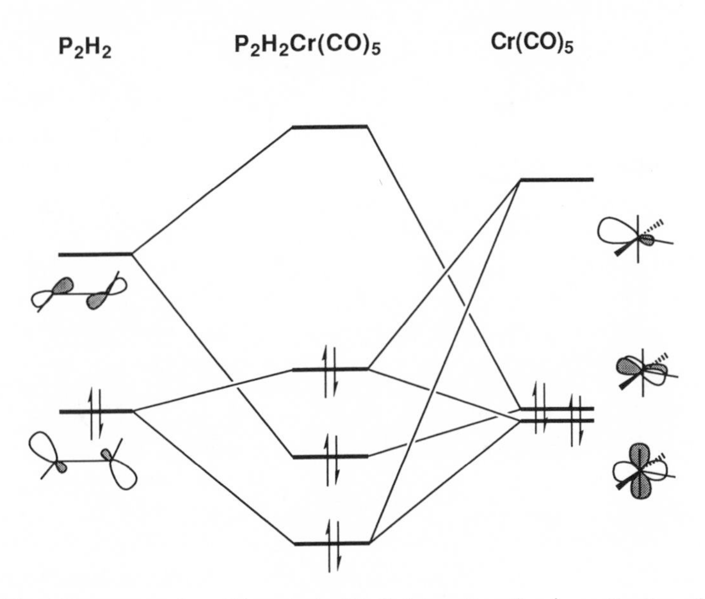

Figure 16. MO representation of the principal P–Cr interactions for η^1-coordination of P_2H_2 to the $Cr(CO)_5$ fragment.

tibonding π MO of the E=E fragment while in the other a filled metal fragment orbital interacts in a destabilizing fashion with a filled ligand fragment orbital. The net result is that the HOMO of the complex lies at higher energy than the HOMO of the free base and considerable charge is transferred from the metal fragment into the empty orbital of the P_2H_2 ligand. Thus, the formation of an η^1-complex from a E=E ligand is not simple in an electronic sense.

For group 15, a large number of alkyne–metal complex analogues have been characterized for the substituent free ligand E≡E, primarily for E = P, As. Perhaps the most varied behavior is exhibited by arsenic. The compound $[CpMn(CO)_2]_4As_2$ (Figure 17a) might be considered to be an As_2 fragment exhibiting η^1-coordination to four separate metal fragments.[196] As the $CpMn(CO)_2$ fragment is a σ acceptor and π donor (see section 2.3.2), the compound actually is better viewed as two coupled arsinidene fragments with 3c,4e bonding. In this view the As–As interaction corresponds to a single bond.

This point is emphasized by a compound which is related in terms of composition, $[Cp^*Mn(CO)_2]_2As_2$, but which exhibits a structure inconsistent with the As_2 fragment acting as a simple σ donor (Figure 17b).[197] Here the

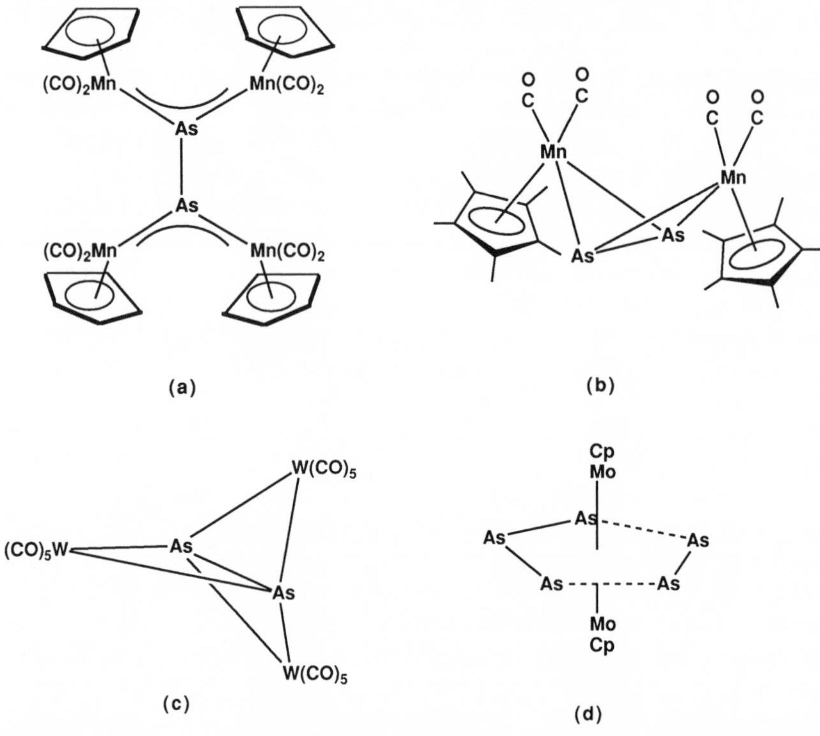

Figure 17. Structures of (a) $[CpMn(CO)_2]_4As_2$, (b) $[Cp^*Mn(CO)_2]_2As_2$, (c) $[W(CO)_5]_3As_2$, and (d) $(CpMo)_2As_5$.

metal fragments are η^2-bound and the As≡As species acts as a formal 4π electron donor. As the $Cp^*Mn(CO)_2$ fragment can be isolobal with the $(Et_3P)_2Pt$ fragment, this complex is the triple bond analogue of the disilene derivatives $[(Et_3P)_2Pt(SiPhX)]_2$ which were discussed in Section 3.2.2 above. On this basis, we expect coordination to reduce the As≡As bond order by one. Indeed, the As–As distance is in the accepted double-bond range. Further, this would suggest the possibility of coordination of yet another metal fragment and this is observed in $[W(CO)_5]_3As_2$ shown in Figure 17c.(198) The As–As distance increases further but only slightly (0.06 Å). A bonding model has been proposed(198) which again suggests the π acceptor characteristics of the As≡As molecule are important in allowing it to act as an η^2-ligand to three separate metal fragments. Interestingly, the antimony(199) and bismuth(200) analogues of this compound have been characterized as well.

There are sufficient examples of group 15 compounds to compare η^2-coordination as one goes from N to Sb. One expects to find the differences to lie principally in the nature of the donor and acceptor orbitals of the E≡E molecule. For N_2 the highest-energy donor orbital is a σ orbital with rather high ionization potential (15.6 eV) and the π orbital lies 1.1 eV even lower in energy. Hence, N_2 is expected to function as a σ-donor, albeit a relatively poor one, and $\{(Cp^*)_2ZrN_2\}_2(\mu\text{-}N_2)$(201) plus $\{Cp_2Ti(PMe_3)\}_2(\mu\text{-}N_2)$(202) are examples of molecules with N_2 ligands acting as σ donors. There are two examples, however, of N_2 acting effectively as an η^2-ligand in the manner of the structure in Figure 17b(203,204) and the bonding has been investigated.(205) Although the HOMO–LUMO gap in N_2 is large, the energy of the LUMO (which is the degenerate set of π^* orbitals) is relatively low lying with respect to potential donor orbitals. In essence, the N_2 molecule acts primarily as a strong π acceptor to the electron-rich metal fragments. When one goes to P_2 and heavier analogs, three significant changes take place. All the filled orbitals rise in energy (the first ionization potential of P_2 is 10.8 eV), the gap between the σ and π orbitals decreases, and the HOMO–LUMO gap becomes smaller. The net result is that the E_2 molecule becomes a better π donor while retaining π-acceptor capabilities. This, then, leads to the rich chemistry that has been reported.

One more compound deserves to be mentioned in this section. Just as $[(Et_3P)_2Pt(SiPhX)]_2$ has an inside-out analog in $Mn_2(CO)_8(SiPh_2)_2$, so too does $[W(CO)_5]_3As_2$ have an inside-out analog in $(CpMo)_2As_5$ shown in Figure 17d.(206) This compound might be considered as an example of a triple decker sandwich compound (see Section 5.3 below) but the significant distortions in the As_5 ring and an Mo–Mo distance intermediate between that accepted for single and double bonds suggest partitioning of the As_5 unit into $\mu\text{-}(\eta^2\text{-}As_3)$ and $\mu\text{-}(\eta^2\text{-}As_2)$ ligands. (As such the compound belongs in Section 4.3 below where similar examples are discussed.) The compound has been viewed as isolobal with bicyclo[1.1.1]pentane(80) and the characterization of a phosphorus analog with the three $W(CO)_5$ fragments replaced with two $SiMe_2$ and one $Pt(PPh_3)_2$ fragments is an amusing mixed system of the same type.(207) Curiously, if in the latter compound the $Pt(PPh_3)_2$ fragment is replaced with a $W(CO)_5$ fragment, coordination of the tungsten atom is η^1 to one phosphorus. Indeed, a second $W(CO)_5$ fragment can be added in like fashion to the other phosphorus atom thereby showing the adjustable isolobal behavior exhibited by transition metal fragments. A theoretical study reports the electronic structure of these and related systems in more detail.(208)

3.2.4. Group 16

The bare E_2 molecules of group 16 have a degenerate HOMO which is π antibonding between the E atoms and a formal E=E double bond. The first ionization potential is considerably lower than that of the corresponding E_2 molecules of group 15 (15.6 eV for N_2 vs 12.1 eV for O_2). These differences are much smaller in the heavier congeners and those of group 16 exhibit modes of coordination ostensibly similar to those of RE=ER for group 15 where R is a one-electron substituent. That is $[W(CO)_5]_3Te_2$ (Figure 18a)(209) is analogous to $[W(CO)_5]_3R_2Sb_2$(209) in that formally four electrons from the π system are used to form donor–acceptor bonds with two of the $W(CO)_5$ fragments (see Figure 15c also). Note the difference between the Te_2 compound and the As_2 compound in Figure 17c. The two additional electrons in the group 16 ligand has caused two η^2-interactions to become η^1 interac-

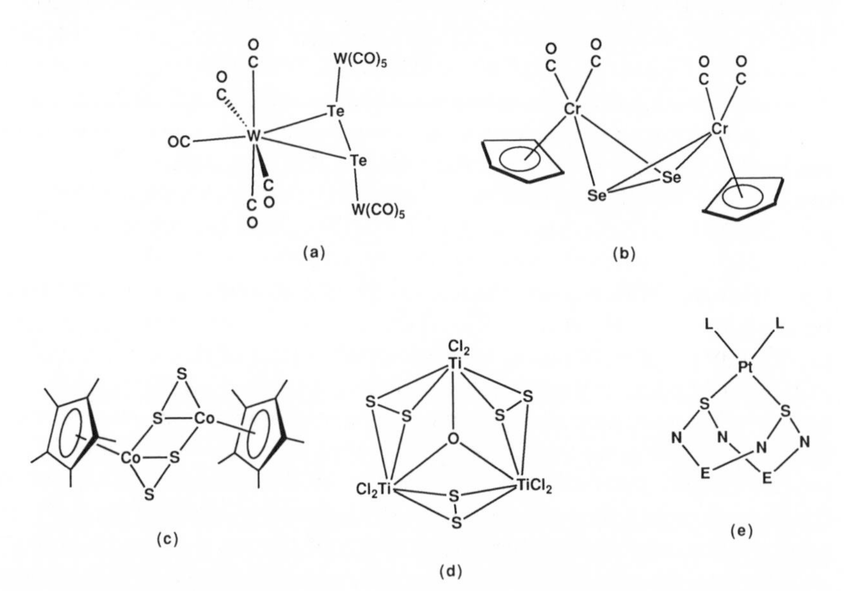

Figure 18. Structures of (a) $[W(CO)_5]_3Te_2$, (b) $[CpCr(CO)_2]_2Se_2$, (c) $[Cp^*CoS_2]_2$, (d) $[TiCl_2S_2]_3O$, and (e) $Pt(PPh_3)_2(1,5\text{-}E_2N_4S_2)$, E = Me_2NC.

tions. Further, $[CpCr(CO)_2]_2Se_2$ (Figure 18b)[80] is similar to $[CpMn(CO)_2]_2As_2$ (Figure 17b) except that now the metal fragment brings one less electron to the match.

A variety of η^2-bonded S_2 molecules are known with and without added η^1-metal coordination.[22] The cobalt and titanium complexes shown in Figures 18c[210] and 18d[211] are illustrative of this point. It has been pointed out that the HOMO and LUMO of 1,5-$E_2N_4S_2$, E = Ph_2P, Me_2NC, are associated with the sulfur atoms but differ in character from the σ and σ^* orbitals of a disulfide linkage.[212] The suggestion that they are isolobal with the π and π^* orbitals of a simple, electron-deficient olefin is backed up with observed chemistry. Consistent with these views, reaction of the ligand with a source of metal fragments produces an interesting variant of η^2-S_2 bonding to a metal. The complex, $Pt(PPh_3)_2$(1,5-$E_2N_4S_2$), E = Me_2NC, has the structure shown in Figure 18e.[213] Indeed, the metal complexes of sulfur–nitrogen ligands constitutes a fascinating area of inorganometallic chemistry in itself.[214]

4. E_2M_2 COMPLEXES CONTAINING μ-η^2 MAIN GROUP LIGANDS

Although not required by the designation μ-η^2, we limit discussion here to compounds containing a M—M bond as without this bond we have considered the complexes as containing two η^2-E_2M interactions. Examples of these complexes were discussed in Section 3.

4.1. Group 13

The chelating ability of the B—H bond is demonstrated once more in the five known examples of η^2-B_2H_6 ligands bridging a M_2 fragment. The oldest structurally characterized example is $HMn_3(CO)_{10}B_2H_6$ in which all six BH bonds are used in Mn–H–B interactions.[215] More recently $Fe_2(CO)_6B_2H_6$,[216] $Cp^*_2Ta_2(\mu\text{-}Br)_2(B_2H_6)$ and $Cp^*_2Ta_2(B_2H_6)_2$[217] and $Cp^*Nb_2(B_2H_6)_2$[218] have been characterized. The B_2H_6 ligand in the first can be viewed as using three B—H in M–H–B interactions while the third and fourth use four B—H bonds to chelate the dimetal fragment. The second exists as a mixture of two tautomers utilizing three and four B—H bonds, respectively, to interact with the dimetal fragment. The structure of one tautomer of $Cp^*_2Ta_2(\mu\text{-}Br)_2(B_2H_6)$ in the solid state is shown in Figure 19a. Formally, one can consider the bridging ligand as the dianion $[B_2H_6]^{2-}$, which is isoelectronic with C_2H_6. This is clearly appropriate in the examples with four M–H–B interactions as the ligand has an ethane-like structure (Figure 11b). It is also appropriate in the cases where a B–H–B interaction is present in the static structure. That is, an analysis of the fluxional behavior of $Fe_2(CO)_6B_2H_6$ allowed the protons to be grouped into sets of 2H and 4H with the latter being

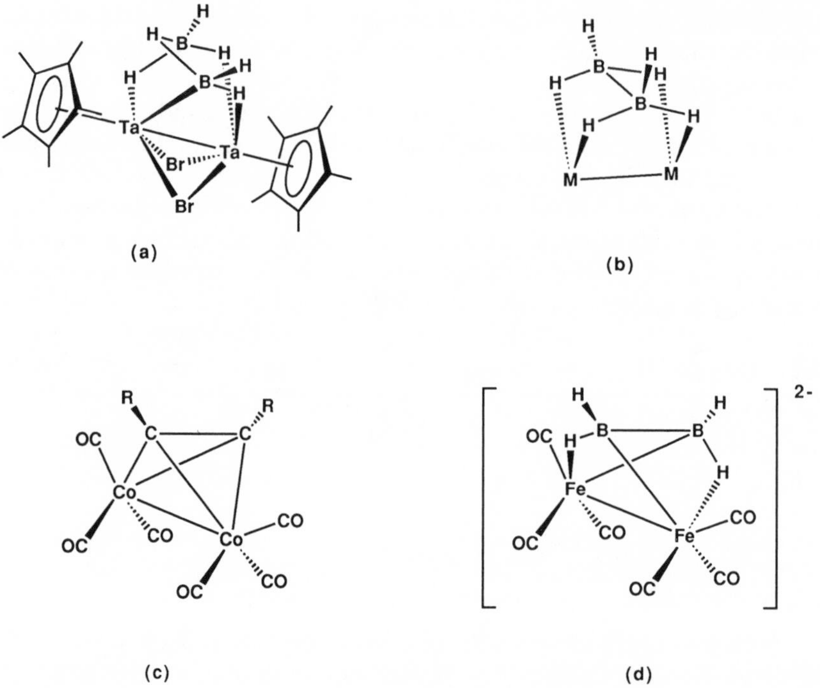

(a) (b)

(c) (d)

Figure 19. (a) Structure of $Cp^*_2Ta_2(\eta\text{-}Br)_2(B_2H_6)$. (b) Tautomeric structure of the B_2H_6 ligand. (c) Structure of $Co_2(CO)_6C_2(t\text{-}Bu)_2$. (d) Proposed structure of $[Fe_2(CO)_6B_2H_4]^{2-}$.

associated with the ligand–metal interaction.[216] Hence, the difference is more apparent than real.

4.2. Group 14

The chemistry of dimetal fragments bridged with alkynes has been extensively studied[2] and a well-known example of this complex type, $Co_2(CO)_6C_2(t\text{-}Bu)_2$, is shown in Figure 19c.[219] The electronic structure associated with these "perpendicular" acetylene complexes as well as the associated "parallel" complexes (metallacycles) has been discussed.[220] There are no similar complexes with bridging olefins nor with bridging silene or silynes. These observations, as well as comparison with the group 13 compounds (Section 4.1), raises some interesting questions. The bridged diiron complex, $Fe_2(CO)_6B_2H_6$, has been doubly deprotonated and spectroscopic information suggest the structure in Figure 19d.[221] As the $[B_2H_4]^{2-}$ ligand is isoelectronic with C_2H_4 it suggests that $Fe_2(CO)_6C_2H_4$ should have a bridged structure with two C—H—Fe bridging hydrogens. Although C–H–M interactions are not uncommon,[222] this compound has not been reported.

Considering the more favorable situation with respect to Si–H–M interactions (see Section 2.2.2), complexes like $Fe_2(CO)_6Si_2H_4$ constitute attractive synthetic objectives. Likewise, one possible precursor to the E_2M_2 class is $Rh_2(SiEtH)_2(CO)_2(dppm)_2$ with Si–Si and Rh–Rh distances of 2.85 and 2.814 Å, respectively.[223] Loss of H_2 could yield a complex isolobal with tetrahedrane. On the other hand, the Si–Si distance is much shorter than in typical bis-(μ-SiR_2) structures (and sum of van der Waals radii) even though larger than the accepted bonding range (see Section 3.2.2). If an Si–Si interaction exists, it would be an expression of a difference between carbon and the heavier group 14 elements.

4.3. Group 15

With group 15, we find a generous selection of μ-η^2-complexes including compounds containing elements beyond the first-row element. Beginning with E_2R_4 ligands, an example of a "parallel" complex of As_2R_4 is shown in Figure 20a in which the ligand acts as a simple bidentate σ donor ligand to the individual metal sites.[224] Moving to E_2R_2 ligands, the structure of $Fe_2(CO)_6P_2(t\text{-Bu})_2$ in Figure 20b represents an example of a diphosphene acting as a formal 6-electron donor to the dimetal fragment.[225]

Examples of compounds containing a μ-η^2-bound E_2 ligand, albeit with additional structural complexity, range over E = N,[226] P,[227] As,[228] Bi.[229] The structure of the phosphorus derivative is illustrated in Figure 20c. These compounds are analogues of $Co_2(CO)_6C_2R_2$ and require that the E_2 ligands act as formal 4-electron donors. This implies the existence of nonbonding pairs on the E atoms and coordination to acidic transition metals has been observed in the cases of E = P, As (Figure 20d).[230] The entire E_2M_2 complex can act as a bidentate ligand and this results in some spectacular molecular architecture, e.g., Figure 20e.[231] The donor characteristics depend on the metal fragment as well as the E atom. For example, the $Co_2(CO)_6P_2$ cluster can be synthesized with two $Cr(CO)_5$ fragments bound to the phosphorus atoms[232] while the $Co_2(CO)_6As_2$ analog shows weak mono-coordination.[233] Replacing a CO ligand on each cobalt atom with phosphine increases the basicity of the arsenic atoms as expected.

Metal–metal bonded dimetal fragments with two μ-η^2-P_2 ligands have been characterized. These are related to $(CpMo)_2As_2As_3$ (Figure 17d), which was discussed briefly in Section 3.2.3. In the first, *cis*-$[(Cp^*)(CO)MoP_2\{Cr(CO)_5\}]_2$ (Figure 21a), the P_4 fragment is a planar unit forming a trapezoid with two P–P distances appropriate for μ-η^2-P_2 bonding (2.07 Å), one long P–P distance (2.85 Å) and one very long P–P distance (3.96 Å).[234] In a closely related compound, $[(\eta^5\text{-}C_5Me_4Et)RhP_2]_2$ (Figure 21b), the P_4 unit is rectangular with P–P distances of 2.05 ηA and 2.84 ηA.[235] This raises the question of whether the P_4 fragment should be viewed as two P_2 ligands or a four-membered ring

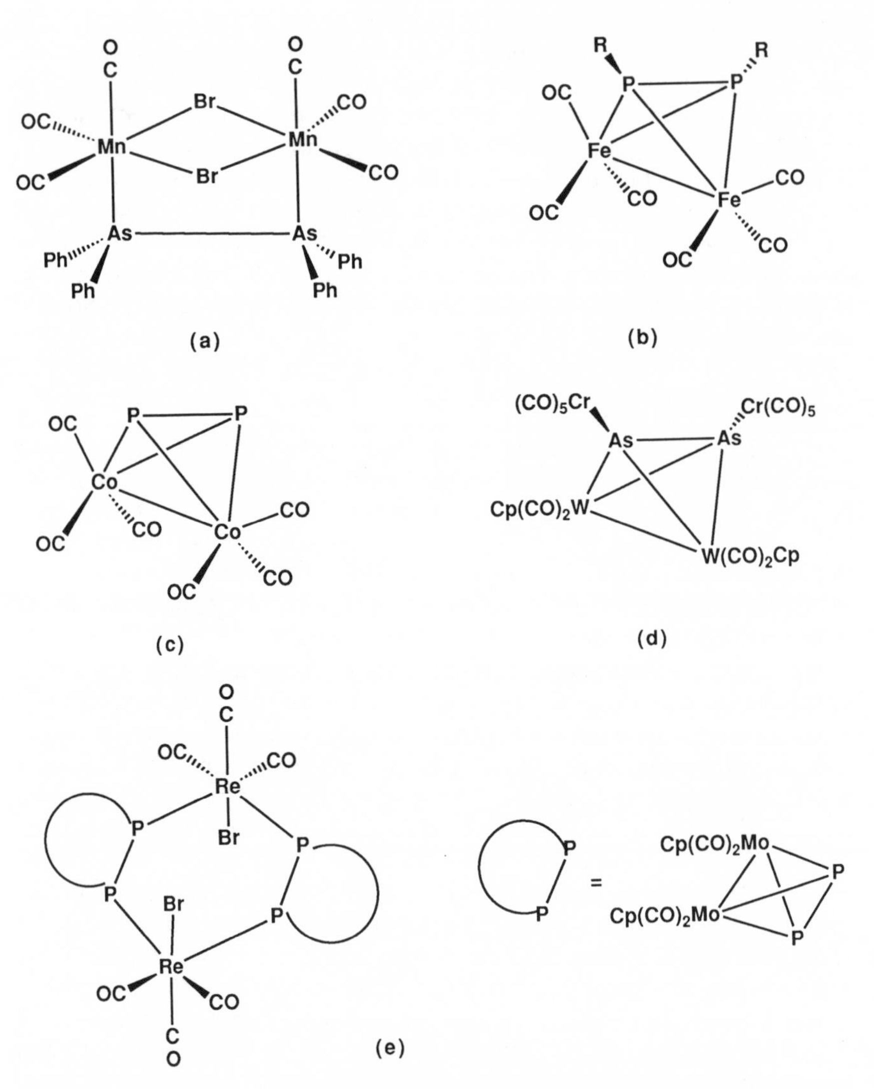

Figure 20. Structures of (a) $[Mn(CO)_3Br(AsPh_2)]_2$, (b) $Fe_2(CO)_6P_2(t\text{-}Bu)_2$, (c) $Co_2(CO)_6P_2$, (d) $\{Cp(CO)_2W\}_2\{AsCr(CO)_5\}_2$, and (e) $\{Re(CO)_3Br\}_2\{[Cp(CO)_2Mo]_2P_2\}_2$.

sandwiched between two metals (see also Section 5.1 below). The similarity of the long P–P distance to that of the short side of the P_4 trapezoid in the Mo compound as well as the long Rh–Rh distance relative to that observed in authentic tripledeckers (Section 5.1) lead the authors to suggest that this compound should be regarded as having two 4-electron μ-η^2-P_2 ligands. Even so, clearly the P_4 fragment in this compound is approaching a ring sandwiched between two metal fragments. The factors favoring intact vs cleaved rings

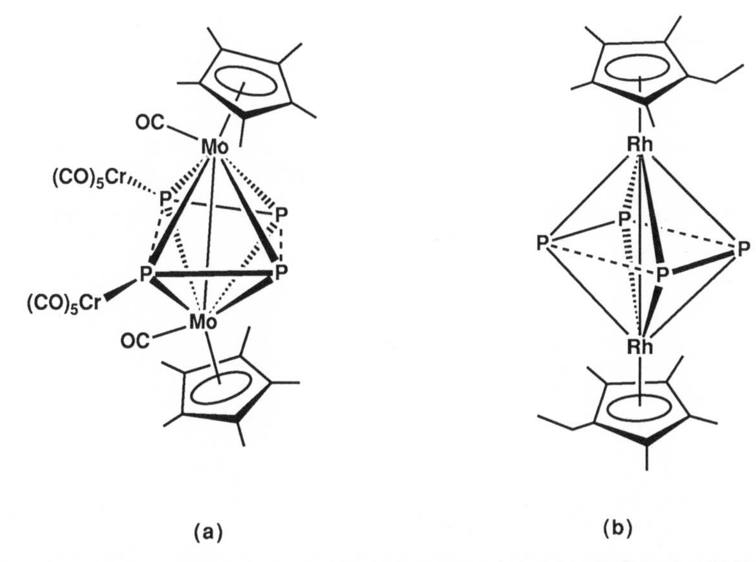

Figure 21. Structures of (a) *cis*-$[(Cp^*)(CO)MoP_2\{Cr(CO)_5\}]_2$ and (b) $[(\eta^5\text{-}C_5Me_4Et)RhP_2]_2$.

have been discussed.[220] however, only $CpCrP_4CrCp$ with an intact P_4 ring was considered.

Some closely related hybrids present some very interesting structures as well as chemistry. For example, the phospha- and arsaalkynes might be viewed as partly organometallic and partly inorganometallic, e.g., $Co_2(CO)_6(PCMe)$, and have analogous structures.[236] The group 15 element retains its capability of acting as a Lewis base to another transition metal fragment. A comparison of phospha- and arsaalkyne complexes of $Co_2(CO)_6$ demonstrated that the donor ability of the phosphorus center is greater than that of the arsenic.[236] This chemistry is discussed in the general reviews already mentioned[180] as well as specific reviews.[237,238] A cationic complex containing an *N*-aryldiaza ligand (PhN_2) bound in a μ-η^2-mode to a dicobalt fragment has been synthesized and characterized.[239] Although not crystallographically characterized, an iminoborane ligand [(*t*-Bu)BN(*t*-Bu)] μ-η^2-bound to a $Co_2(CO)_6$ fragment has been reported.[240] As already mentioned, iminoboranes are isoelectronic to alkynes and the chemistry of these species suggests considerable scope for inorganometallic compounds.[157]

4.4. Group 16

The sulfur analogue of $Fe_2(CO)_6P_2R_2$ (Figure 20b) and $Co_2(CO)_6C_2R_2$ (Figure 19c) is $Fe_2(CO)_6S_2$[241] and it has a similar structure (Figure 22a).[242] However, the electronic structure of this compound, which has been the subject of several studies,[243–245] of which the latest[246] appears most definitive, exhibits

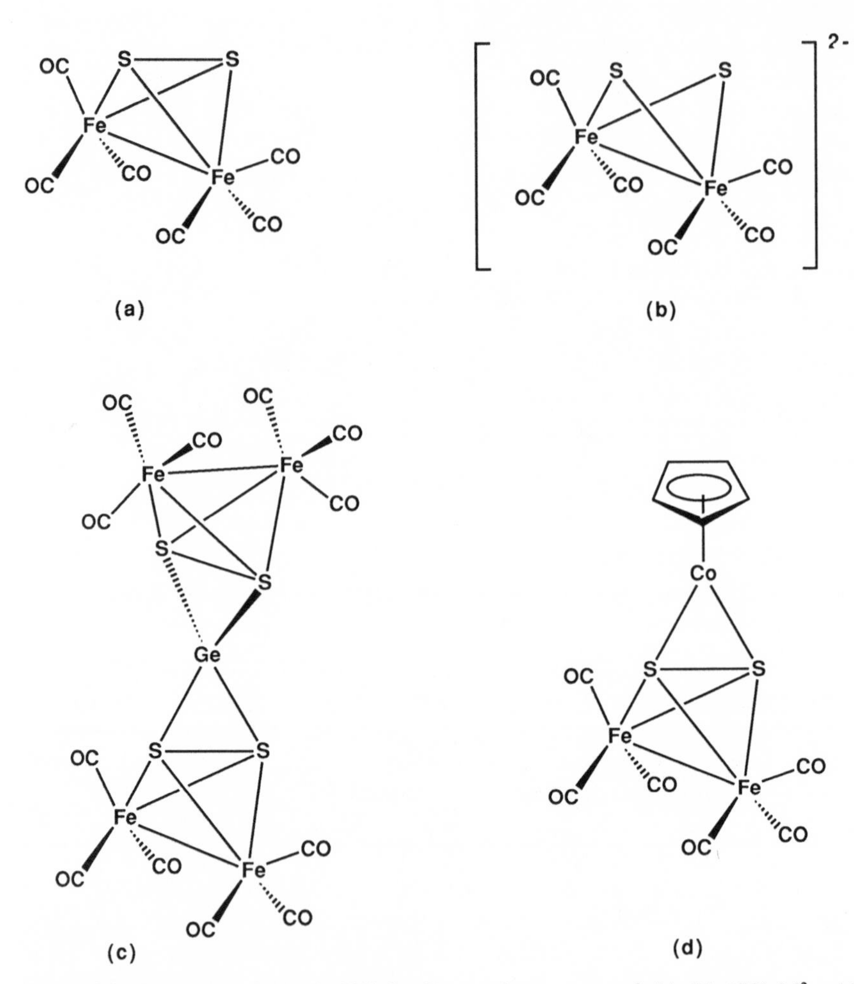

Figure 22. Structures of (a) $Fe_2(CO)_6S_2$. Proposed structures of (b) $[Fe_2(CO)_6S_2]^{2-}$, (c) $[Fe_2(CO)_6S_2]_2Ge$, and (d) $CpCo[Fe_2(CO)_6S_2]$.

a significant difference. In all these compounds the HOMO corresponds to an orbital associated with the M—M bond. As one goes from $Co_2(CO)_6C_2R_2$ to $Fe_2(CO)_6S_2$ the LUMO changes from M—M antibonding to S—S antibonding. Recall that with η^2-ligands, E = N, P, As (Section 3.2.3), increasing the electronegativity of the E atoms causes both the filled and unfilled ligand orbitals to fall in energy. Thus, in going from $Co_2(CO)_6C_2R_2$ to $Fe_2(CO)_6S_2$, the metal orbitals rise in energy while the E_2 orbitals fall in energy and in the process the LUMO changes from primarily metal to mostly main group character. This has significant chemical consequences in that, like a disulfide, reduction of $Fe_2(CO)_6S_2$ leads to the dianion and cleavage of the S—S rather than Fe—Fe bond (Figure 22b).[247] Indeed this dianion has a rich chemistry and can be used as a ligand for both main group and transition metals as shown in Figures 22c and 22d. The selenium[248] and tellurium[249] analogues of $Fe_2(CO)_6S_2$ are also known. Consistent with the behavior of the sulfur analogue, $Fe_2(CO)_6Se_2$ exhibits reactivity typical of diselenides and forms similar compounds.[250] The tellurium compound is thermally labile which is attributed to strain resulting from the disparity in the requirements for μ-η^2-bonding and the size of the nonmetal atoms.[249]

The S_2 moiety is an extremely versatile ligand. Figure 23a shows an inorganometallic complex in which S_2 ligands exhibit μ-η^1- and μ-η^2-bonding to a single dimetal fragment[251,252] while Figure 23b shows S_2 ligands with perpendicular and parallel μ-η^2-bonding to a single dimetal fragment.[253] The theoretical significance of this versatility has been discussed and connections to solid-state analogs established.[254] The versatility of the E_2 ligand is hardly restricted to sulfur as demonstrated by the μ-η^2-Te_2 ligand in $Fe_2(CO)_6Te_3]^{2-}$ (Figure 23c).[255] Mixed compounds, such as $Fe_2(CO)_6HNS$ (Figure 23d), are also well characterized.[256]

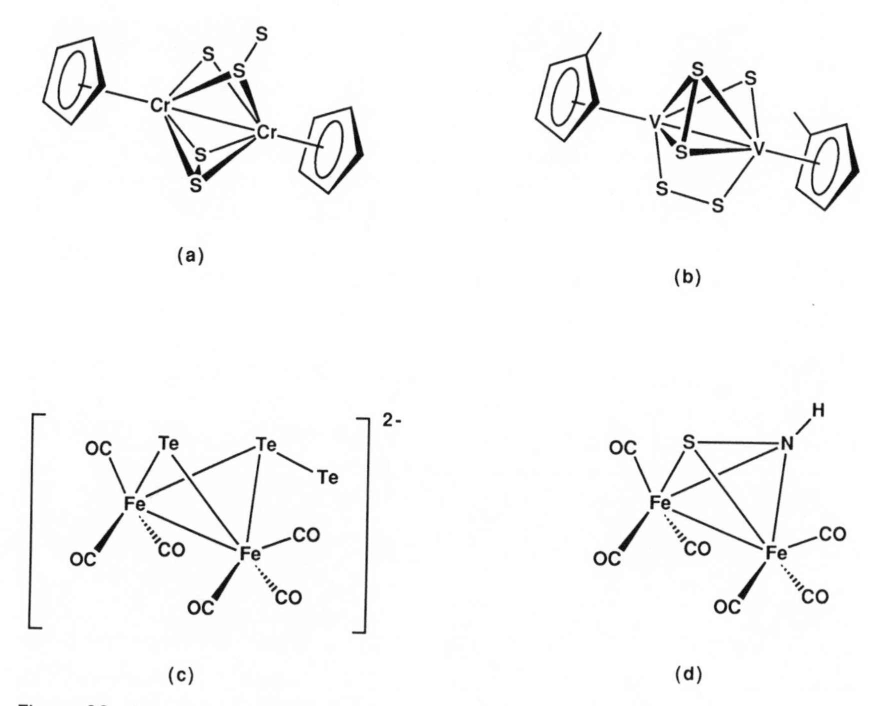

Figure 23. Structures of (a) $[CpCrS_2]_2S$, (b) $[Cp'VS_2]_2S$, (c) $[Fe_2(CO)_6(Te)(Te_2)]^{2-}$, and (d) $Fe_2(CO)_6HNS$.

5. E_mM COMPLEXES CONTAINING η^m MAIN GROUP LIGANDS

The synthesis and structural characterization of ferrocene was another landmark in the explosive growth of organometallic chemistry in the fifties and sixties.[257] These sandwich complexes have analogs in inorganometallic chemistry, albeit less numerous and, presently, with little known reaction chemistry. Further, the judicious replacement of carbon in rings by various main group atoms leads to structural chemistry which, while clearly related to that of the organometallic complexes, has another dimension.

In this section we compare the structures and some properties of main group ring systems that form π sandwich complexes with transition metals. The mode of presentation is by ring size with examples containing homonuclear rings of the various elements as well as heteronuclear rings containing carbon and other elements. Examples of stoichiometric organometallic analogs with nonsandwich structures will be pointed out as well.

5.1. E_3 Rings with η^3-Bonding

The π system of an E_3 ring system consists of a single bonding orbital (a_2'' in D_{3h} symmetry) and an antibonding pair (e'' symmetry). Hence, an important interaction in the stabilization of η^3-E_3 ligands with partially filled e'' orbitals is donation from the ring into compatible energy orbitals on the metal center.[258]

We begin with the organometallic prototype. As shown in Figure 24a, the structure of $Co(CO)_3C_3Ph_3$ can be considered as a C_3 ring π bound to a $Co(CO)_3$ fragment.[259] As such it constitutes a 3π electron donor to the 15-electron metal fragment. Note that this compound is also a member of the set of four-atom, cluster compounds $(CR)_{4-n}[Co(CO)_3]_n$ related by the isolobal analogy. The CpNi analog of $Co(CO)_3C_3Ph_3$ has also been structurally characterized.[260]

There is no known example of an Si_3 ring with η^3 bonding.

Although no example of a simple B_3H_3 ring bound to a transition metal is known, $Mn(CO)_3B_3H_8$[261] has been fully characterized and its structure is shown in Figure 24b. As there are three B—H—Mn bridging hydrogens one might consider the B_3H_8 fragment of $Mn(CO)_3B_3H_8$ as a tridentate chelating ligand in the spirit of the discussions of Sections 2.2.1 and 3.1. This view is reinforced by the observation that this compound reversibly adds CO to yield $Mn(CO)_4B_3H_8$ having an open "butterfly" structure.[262] However, if we think of removing the B—H—M bridging hydrogens as protons (B—H—M hydrogens have considerable Brönsted acidity[263]) then the resulting η^3-$[B_3H_5]^{3-}$ ligand would be a 5π electron donor to the 13-electron $Mn(CO)_3$ fragment. The bridging hydrogens on the $[B_3H_5]^{3-}$ fragment can

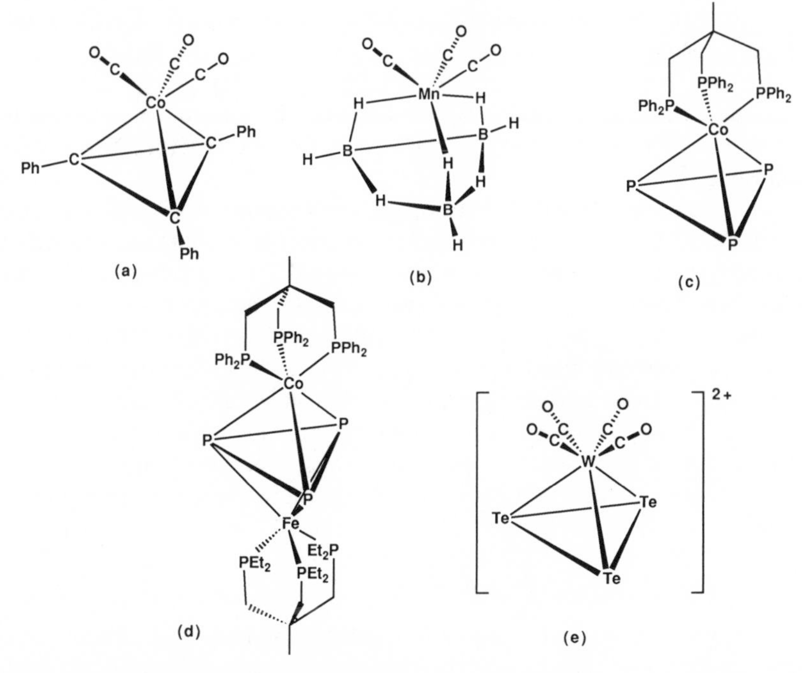

Figure 24. Structures of (a) $Co(CO)_3C_3Ph_3$, (b) $Mn(CO)_3B_3H_8$, (c) $Co(triphos)P_3$, (d) $Co(triphos)P_3Fe(etriphos)$, and (e) $[W(CO)_4Te_3]^{2+}$.

be thought to play a role similar to that of the bridging hydrogen on the $[B_2H_5]^-$ ligand (Section 3.2.1). That is, even in larger, planar π ligands, the presence of B—H—B hydrogens tends to "focus" the boron atom hybrids on the side of the ring opposite the B—H—B sites toward the metal atom.[154] The compound $Mn(CO)_3B_3H_8$ is isolobal with $Fe_2(CO)_6B_2H_6$ (Section 4.1) as well as with the cluster $Fe_3(CO)_9BH_5$ (Chapter 3).

The chemistry of η^3-E_3 ligands, E = group 15, has been summarized.[184] Reaction of P_4 with appropriate metal fragment sources produces η^3-P_3 complexes such as $Co(triphos)P_3$ shown in Figure 24c.[264] The arsenic analog, $Co(CO)_3As_3$,[265] has been known for an even longer time. Formally the ring is a 3π electron donor to the 15-electron metal fragments. This requires that each main group atom have a formal lone pair occupying the position of the external substituents in the carbon and boron examples above. An MO model of the bonding[184] is analogous to that for organometallic η^3-C_3R_3 bonding. Consistent with this analysis, as well as the behavior of P_2 ligands (Section 3.2.3), the cyclotriphosphorus ligand can coordinate from one to three other metal fragments in a η^1 fashion in addition to η^3-coordination to one metal atom. Finally, the P_3 ligand is found η^3-coordinated to two separate metal fragments in a multidecker sandwich fashion in, for example, Co(triphos)P_3Fe(etriphos) (Figure 24d).[266] This compound has a total of 30 valence electrons and various mixed metal triple deckers having valence electron counts ranging from 30 to 34 are known. The rationale supporting the existence of a range of stable electron counts for this structural form is well established.[267]

One example of η^3-E_3, E = group 16, coordination is known. This compound, $[W(CO)_4Te_3]^{2+}$, was prepared in an attempt to coordinate $[Te_4]^{2+}$ to a tungsten metal center as a six-electron donor.[268] The structure is shown in Figure 24e. The $[Te_3]^{2+}$ ligand is viewed as a 4π electron donor to the 14-electron $W(CO)_4$ fragment. As with the smaller group 16 ligands discussed in previous sections, the filled π bonding levels lie at low energy relative to the metal valence levels and donation from a half-filled antibonding π level of the $[Te_3]^{2+}$ ligand to the metal is thought to provide significant stabilization of the complex. The electronic structure of the hypothetical metal complex of an η^3-O_3 ligand has been explored[258] and there are many similarities to the model used to describe the bonding of the η^3-P_3 ligand to a metal.

5.2. E_4 Rings with η^4-Bonding

The π system of an E_4 ring system consists of a bonding orbital (a_{2u} in D_{4h} symmetry), a nonbonding pair (e_g symmetry), and a high-lying antibonding orbital (a_u symmetry). As with the three-membered ring, an important interaction in the stabilization of η^4-E_4 ligands is that of the e_g set with compatible empty orbitals on the metal center.[269]

The structure of CpCo(η^4-C_4H_4) is shown in Figure 25a.[270] Formally the C_4H_4 ligand acts as a 4π electron donor toward the 14-electron CpCo fragment. The electronic structure of the η^4 C_4 bonding has been extensively studied and justifies the qualitative conclusion noted immediately above.[271] The development of the chemistry of metal bound cyclobutadiene also constitutes one important component in the rapid growth and development of organometallic chemistry.[272] There are no known heavier analogs, however, the *closo*-clusters $[Cr(CO)_3Sn_9]^{4-}$ and $[Cr(CO)_3Pb_9]^{4-}$ might be considered as examples of η^4-E_4 bonding of the heavier group 14 elements.[273,274]

The compounds $Fe(CO)_3$(η^4-B_4H_8)[275] and CpCo(η^4-B_4H_8)[276] constitute two characterized examples of a B_4H_8 ring η^4-bound to a transition metal fragment that have been characterized and the structure of CpCo(η^4-B_4H_8), determined in the solid state,[277] is shown in Figure 25b. In contrast to the B_3 system above, these compounds have unbridged E–M interactions and are more easily identified with the cyclobutadienyl analogs. Again the B—H—B bridges are thought to play a role analogous to the one that the B—H—B bridge plays in the $[B_2H_5]^-$ ligand (Section 3.2.1).[154] Indeed, the electronic structure of $Fe(CO)_3$(η^4-B_4H_8) has been directly compared with that of

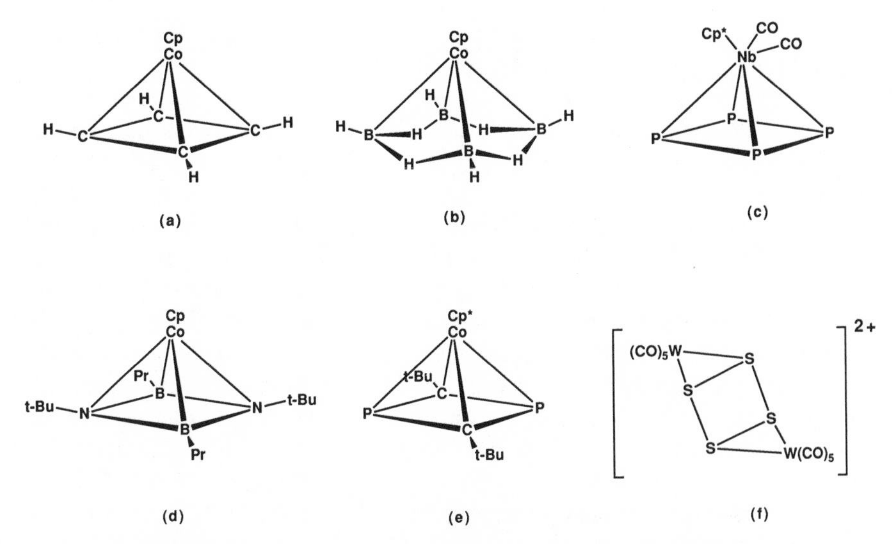

Figure 25. Structures of (a) $CpCo(\eta^4\text{-}C_4H_4)$, (b) $CpCo(\eta^4\text{-}B_4H_8)$, (c) $Cp^*(CO)_2Nb(\eta^4\text{-}P_4)$, (d) $CpCo[\eta^4\text{-}\{B(Pr)N(t\text{-}Bu)\}_2]$, (e) $Cp^*Co[\eta^4\text{-}\{PC(t\text{-}Bu)\}_2]$, and (f) $[\{W(CO)_5\}_2S_4]^{2+}$.

$Fe(CO)_3(\eta^4\text{-}C_4H_4)$ utilizing the technique of UV-photoelectron spectroscopy.[278] In addition, there have been some similarities noted between the photochemical induced reactions of $Fe(CO)_3(\eta^4\text{-}B_4H_8)$ and $Fe(CO)_3(\eta^4\text{-}C_4H_4)$ with alkynes.[279]

There is one structurally characterized example of a $\eta^4\text{-}E_4$ group 15 ligand, i.e., $Cp^*(CO)_2Nb(\eta^4\text{-}P_4)$, with a structure shown in Figure 25c.[280] Like C_4H_4, the P_4 ligand formally acts as a 4π electron donor to a 14-electron metal fragment. Consistent with this, the P_4 ring is planar and only slightly distorted from a square shape; hence it is considered a cyclobutadiene analog. The early transition metal is probably required for the same reasons discussed in connection with η^2-bonding of the more electronegative E_2 ligands (Section 3.2.3). Evidence for a possible triple-decker compound, $Cp^*Co(\mu,\eta^4\text{-}P_4)CoCp^*$, has been presented[178] but, in the absence of a crystal structure, formulation as a $Cp^*Co(\mu,\eta^2\text{-}P_2)_2CoCp^*$ complex cannot be eliminated (Section 4.3).

The $\eta^4\text{-}B_2N_2$ diazadiboretidine transition metal complexes are also interesting inorganometallic analogs. One example, $CpCo[\eta^4\text{-}\{B(Pr)N(t\text{-}Bu)\}_2]$, is shown in Figure 25d.[281] Crystal structure determinations of the $Fe(CO)_3$ and $W(CO)_4$ derivatives show nonplanar four-membered rings with M–N distances about 0.1 Å shorter than M–B distances. This is similar to the differences in covalent radii and does not rule out η^4-bonding. However, the mechanism for achieving η^4-coordination is not a simple one. It is suggested that the nitrogen centers in the ring are the principle donors to the metal but, as indicated by the ^{11}B NMR chemical shifts, there is extensive back donation from the metal into the $p\pi$ orbitals of the boron atoms.

Although isoelectronic with nitriles, the chemistry of phosphaalkynes resembles that of alkynes. This is understandable in that the electronegativity of phosphorus is much closer to that of carbon than that of nitrogen. Consistent with these facts, another η^4-$E_2E'_2$ system has been prepared from the free E≡E′ phosphaalkyne ligand. The complexes Cp*M[η^4-{PC(*t*-Bu)}$_2$], M = Co, Rh, Ir,[282] and Cp*Co[η^4-{PC(*t*-Bu)}$_2$] were reported at about the same time.[283] The Cp*Co derivative has been crystallographically characterized (Figure 25e) thereby allowing η^4-$E_2E'_2$ coordination to be distinguished from other possibilities, e.g., bis η^2-$E_2R'_2$. The ring is planar and all P—C bond distances are equal and considerably longer than those in the free P≡C(*t*-Bu) ligand. Hence, the geometric consequences of the metal to ring bonding in this compound are more closely related to those in the organometallic analog than are those of the coordinated η^4-{B(Pr)N(*t*-Bu)}$_2$ ring discussed immediately above. An interesting compound containing the same C_2P_2 ring is [Rh(η^4-{PC(*t*-Bu)}$_2$(η^5-$C_2B_9H_{11}$)]$^-$[284] in which the [η^5-$C_2B_9H_{11}$]$^{2-}$ ligand can be considered as a [Cp]$^-$ ligand analog.[285] In this case a phosphorus atom in the ring was shown to act as a base to the [$AuPPh_3$]$^+$ and [Co(CO)$_2$(η^4-C_4R_4)]$^+$ fragments.

No group 16 compound of this type has been reported. Perhaps the closest example is [{W(CO)$_5$}$_2S_4$]$^{2+}$ which has the structure shown in Figure 25f.[286] The electron-rich sulfur ring of this compound appears to create problems in achieving η^4-coordination.

5.3. E_5 Rings with η^5-Bonding

The π system of an E_5 ring system consists of a bonding orbital (a''_2 in D_{5h} symmetry), a less bonding pair (e''_1 in D_{5h} symmetry), and a high-lying antibonding pair of orbitals (e''_2 in D_{5h} symmetry). Again, an important interaction in the stabilization of η^5- E_5 ligands is an interaction of the e''_1 set with compatible empty orbitals on the metal center.

The structure of ferrocene, represented with a convenient, if asymmetric, notation, is shown in Figure 26a. The chemistry of ferrocene is vast[2] and, at minimum, suggests the existence of related chemistries for the inorganometallic analogs.

A metalloborane analog of ferrocene, CpFe(η^5-B_5H_{10}), with the proposed structure shown in Figure 26b has been reported.[287] The η^5-B_5H_{10} ligand acts as a formal 5π electron donor to the CpFe fragment. As with the B_4H_8 ligand, it is the B–B edges that are bridged with hydrogens. The exploration of the chemistry of this metalloborane awaits the development of a more practical synthetic route.

A particularly interesting example of an inorganometallic compound is "inorganic ferrocene," Fe[(NMe)$_2$(BPh)$_2$N]$_2$[288]; however, it was not crystal-

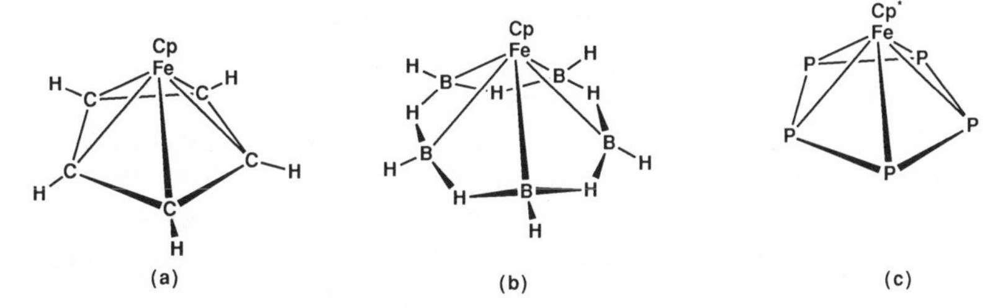

Figure 26. (a) Structure of $CpFe(\eta^5\text{-}C_5H_5)$. (b) Proposed structures of $CpFe(\eta^5\text{-}B_5H_{10})$ and (c) $Cp^*Fe(\eta^5\text{-}P_5)$.

lographically characterized. As with the smaller ring systems, a number of $C_nE_{5\text{-}n}$ ring systems have been studied. One of these is the 1,2-azaborolinyl ligand in which a C_2 fragment of the cyclopentadienyl ligand is formally replaced with BN.[289] The examination of the effect of this nuclear charge change on the electronic structure of these sandwich compounds provides additional information on the ring–metal interaction. Other such hybrids are the C_4BR_5 ring which is a formal 4π electron donor to a metal fragment,[290] the $C_3B_2H_5$ ring which is a formal 3π electron donor,[291] and the $C_2B_3H_5$ ligand which is a formal 2π electron donor.[292] These ligands, in contrast to C_5H_5 itself, are particularly effective μ,η^5-ligands bridging between two metal centers in multidecker sandwich complexes. The 1,2,5-thiadiborolene ligand ($C_2R_2B_2R'_2S$) is also a 4π electron ligand that has been used effectively to form multidecker sandwich complexes.[293]

In the case of group 15, $Cp^*Fe(\eta^5\text{-}P_5)$ (Figure 26c) forms from the reaction of P_4 with $[Cp^*Fe(CO)_2]_2$[294] or directly from the reaction of the free $[P_5]^-$ ligand.[295] As with the borane, the $\eta^5\text{-}P_5$ ligand acts as a 5π electron donor to the Cp*Fe fragment. The free ligand, itself, exhibits properties consistent with aromatic character.[295] Hence, $Cp^*Fe(\eta^5\text{-}P_5)$ can be considered a true analog of ferrocene. The geometric features of the P_5 ring remain to be determined; however, these parameters are known for the closely related triple-decker, $Cp^*Cr(\mu,\eta^5\text{-}P_5)CrCp^*$.[296] The crystallographic data on the latter compound, although not as accurate as might be desired, show a regular pentagon for the P_5 ring. A stoichiometric analog of this triple-decker, $(CpMo)_2(\mu,\eta^2\text{-}As_2)(\mu,\eta^2\text{-}As_3)$ (Figure 17d), was discussed in Section 3.2.3.

5.4. E_6 Rings with η^6-Bonding

The π system of a E_6 ring system consists of a bonding orbital (a_{2u} in D_{6h} symmetry), a less bonding pair (e_{1g} symmetry), a higher-lying antibonding pair of orbitals (e_{2u} symmetry), and a high-lying antibonding orbital (b_{2g} symmetry). As with all the rings, an important interaction in the stabilization of

η^6-E_6 ligands is an interaction of the e_{1g} set with compatible empty orbitals on the metal center.[297]

There are numerous examples of η^6-coordination of C_6 rings but inorganometallic examples of η^6-E_6 coordination of homonuclear rings are limited to group 15. These are triple-decker complexes containing η^6-P_6 rings as the central deck. One is the 28 valence electron sandwich Cp*Mo(μ,η^6-P_6)MoCp* (Figure 27a)[298] while the other is the 26 valence electron sandwich Cp*Nb(μ,η^6-P_6)NbCp′, Cp′ = $C_5Me_4Et_2$.[280] The latter P_6 ring contains two long and four short P–P distances and is considered to be Cp′Nb(μ,η^3,η^3-P_6)NbCp′, i.e., the $[P_6]^{2-}$ ligand is formally represented as two allyl-like $[P_3]^-$ fragments joined together. A similar structural distortion observed in the 26 valence electron complex Cp′V(η^6-C_6H_6)VCp′ has been interpreted in like fashion.[299] In going from Cp*Mo(μ,η^6-P_6)MoCp* to the 24-electron Cp*Ti(P_6)TiCp* the planarity of the P_6 ring is lost and the structure has been described as a distorted dimetallaphosphacubane.[300] However, as shown in Figure 27b, the P_6 ligand can also be viewed as serving as a tridentate ligand to each metal atom utilizing alternate phosphorus atoms. The difference in phosphorus environments for these P_6 ligands results in striking differences in ^{31}P NMR shifts.

Borazine, which is isoelectronic with benzene, coordinates to transition

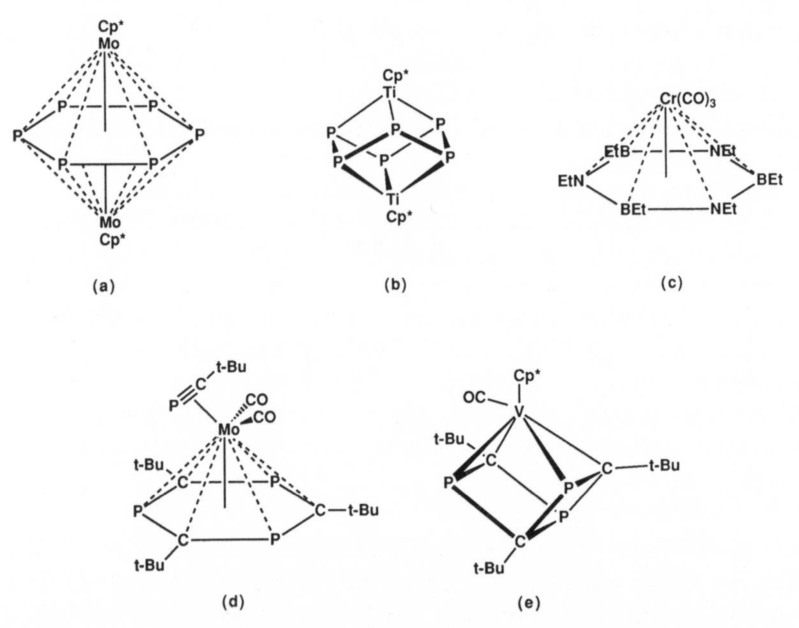

Figure 27. Structures of (a) Cp*Mo(μ,η^6-P_6)MoCp*, (b) Cp*Ti(P_6)TiCp*, (c) $Cr(CO)_3$(η^6-$B_3N_3Et_6$), (d) η^2-PC(*t*-Bu)$(CO)_2$Mo[η^6-{PC(*t*-Bu)}$_3$] (proposed), and (e) Cp*V(CO)[PC(*t*-Bu)]$_3$.

metal fragments, e.g., $Cr(CO)_3(\eta^6\text{-}B_3N_3Et_6)$ (Figure 27c),[301] but, similarly to the smaller BN ligands, one can question whether coordination is via nitrogen lone pairs or whether it is effectively η^6 as with benzene itself. Here the situation is relatively clear. Although there is a C_{3v} out-of-plane distortion caused by Cr–B and Cr–N distances that differ proportionately to the difference in B and N covalent radii, the N_2BC and B_2NC fragments are approximately planar. This suggests effective sp^2 hybridization at B and N and, thus, η^6-coordination rather than localized N—M σ donation.[302] However, the intraring distances of the borazine ring do not change significantly on coordination, suggesting a fairly weak interaction with the metal center. The mechanism for achieving η^6-coordination may well be the same as that suggested for the B_2N_2 ring (Section 5.1.2), i.e., donation by the nitrogen centers combined with strong back-donation from the metal to the boron centers.

As with the smaller rings, the replacement of one or more carbon atoms of the benzene ring with E atoms leads to considerable variety in the types of η^6-complexes that can be constructed, e.g., C_5BH_6, borabenzene is an η^6, 5π electron donor, and the $(\eta^6\text{-}C_5BH_6)Fe$ $(\eta^6\text{-}C_5BH_6)$ complex is analogous to ferrocene.[303] The C_5H_5As ligand, on the other hand, is a 6π electron donor and forms the analogue of bisbenzene chromium, $(\eta^6\text{-}C_5H_5As)\text{-}Cr(\eta^6\text{-}C_5H_5As)$.[304]

Similarly with the four-membered ring, the P≡C(*t*-Bu) ligand associates on a transition metal center to form an apparent η^6-coordinated ring, i.e., η^2-$PC(t\text{-}Bu)(CO)_2Mo[\eta^6\text{-}\{PC(t\text{-}Bu)\}_3]$.[305] The proposed structure, which is based principally on an interpretation of the ^{31}P NMR, is shown in Figure 27d. A related approach with a different metal fragment source leads to two complexes shown by X-ray crystallographic studies to contain coordinated valence isomers of 2,4,6-tritertbutyl-1,3,5-triphosphabenzene, $Cp^*V(CO)[PC(t\text{-}Bu)]_3$.[306] The structure of the one containing a Dewar benzene-like structure is shown in Figure 27e. A similar reaction with PhC≡CPh yields coordinated $\eta^6\text{-}C_6Ph_6$; hence, in this instance the phosphaalkyne demonstrates a distinct difference from the behavior of an alkyne. Metal–ligand coordination very similar to that found in $Cp^*V(CO)[PC(t\text{-}Bu)]_3$ is observed in $[Cr(CO)_3(As_7)]^{3-}$ in which a norbornadiene-like $[As_7]^-$ ligand acts as a 4-electron donor to the 14-electron $[Cr(CO)_3]^{2-}$ fragment.[307]

This tendency for the coordinated group 15 ligand to adopt a structural form which is a high-energy valence tautomer of the equivalent organic ligand is nicely illustrated by a four-atom system.[308] Formal two-electron reduction of P_4 results in the rupture of a P—P bond and the structure shown in Figure 28a. An alternate valence structure is that shown in Figure 28b, i.e., an analog of 1,3-butadiene. The coordination of the first valence tautomer to a metal is similar to that discussed for $Rh(PPh_3)_2Cl(P_4)$ shown in Figure 15d and two additional examples are now known, i.e., $(Cp'')_2M(P_4)$, M = Zr, Hf.[308] Coordination of the second valence tautomer has also been observed in the

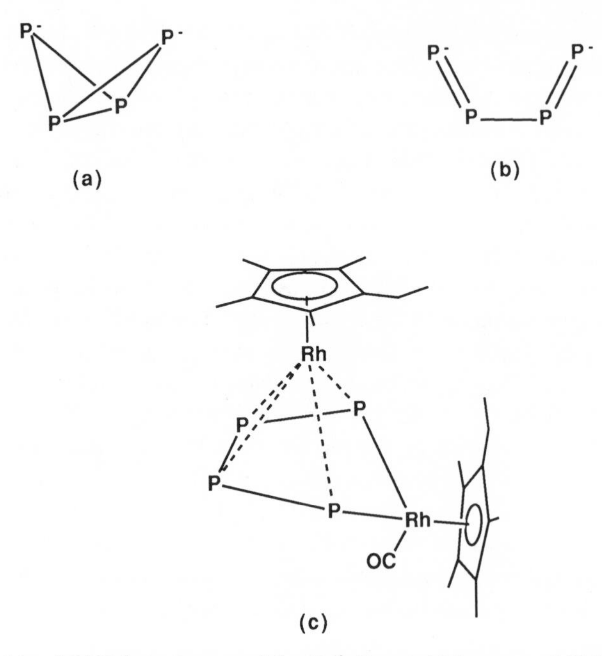

Figure 28. (a) and (b) Valence isomers of the $[P_4]^{2-}$ ligand. (c) Structure of $(Cp^{*\prime}Rh)_2(CO)(P_4)$.

dimetal complex $(Cp^*Rh)_2(CO)(P_4)$ with an observed structure shown in Figure 28c. The compound mimics 1,3-butadiene in its coordination to the $Cp^{*\prime}Rh$ fragment. One wonders if the CO ligand could be dissociated thereby producing an analog of $CpCo[\eta^5\text{-}C_4H_4CpCo]$ containing a coordinated metalole. To complete the circle, note that the borane ligand in Ir $(\eta^4\text{-}B_4H_9)(CO)\{P(CH_3)_2C_6H_5\}_2$ is an analog of 1,3-butadiene.[309]

6. CONCLUSIONS

In the midst of writing this chapter, it became more and more clear that the area being surveyed could easily fill a chapter several times larger than the one originally contemplated. Hence, a number of exciting areas have been left out or only briefly mentioned, for example, metallacycles containing main group elements other than carbon,[310] or the use of early transition metal alkoxide fragments as mimics of metal carbonyl fragments,[311] or the trimetaphosphate ligand as an inorganic analogue of the cyclopentadienyl ligand.[312] In addition, the mixed carbon-element and mixed element compounds were only discussed when appropriate examples of the single element compounds were few or missing. The sheer number of compounds touched upon in this chapter prevented any consistent description of chemistry; however, in most cases the chemistry characteristic of these inorganometallic compounds is only now beginning to emerge. Despite these caveats, the array of compounds presented does indeed support the thesis that inorganometallic chemistry has a bright future and will, eventually, give the synthetic chemist another avenue of control over structure and reactivity, namely, via element substitution.

ACKNOWLEDGMENTS

The suggestions and comments of Prof. Claire A. Tessier, Mr. Vance Kennedy, and Dr. B. H. S. Thimmappa were very helpful. Our work has been generously supported by the National Science Foundation.

ABBREVIATIONS:

bipy = 2,2′-bipyridine
Cp = η^5-C_5H_5
Cp′ = η^5-C_5H_4Me
Cp″ = 1,3-$C_5H_3(t\text{-Bu})_2$
Cp* = η^5-C_5Me_5
Cp*′ = η^5-C_5Me_4Et
dppm = $Ph_2PCH_2PPh_2$
dppe = $Ph_2CH_2CH_2PPh_2$
dmpe = $Me_2CH_2CH_2PMe_2$
dth = dithiahexane
etriphos = $MeC(CH_2PEt_2)_3$
HMPT = $P(NMe_2)_3$
mes = 2,4,6-$Me_3C_6H_3$
$(np)_3$ = $N(CH_2CH_2PPh_2)_3$
triphos = $MeC(CH_2PPh_2)_3$
t-Bu = *tert*-butyl

REFERENCES

1. "Gmelin Handbook of Inorganic Chemistry"; Springer-Verlag: New York 1924–1991.
2. Wilkinson, G.; Stone, F. G. A.; Abel, E. W., Eds. "Comprehensive Organometallic Chemistry"; Pergamon: Oxford, 1982.
3. Housecroft, C. E.; Fehlner, T. P. *Inorg. Chem.* 1982, **21,** 1739.
4. Greenwood, N. N.; Savory, C. G.; Grimes, R. N.; Sneddon, L. G.; Davison, A.; Wreford, S. S. *Chem. Commun.* 1974, 718.
5. Wade, K. *Adv. Inorg. Chem. Rad. Chem.* 1976, **18,** 1.
6. Churchill, M. R.; Hackbarth, J. J.; Davison, A.; Traficante, D. D.; Wreford, S. S. *J. Am. Chem. Soc.* 1974, **96,** 4041.
7. Brice, V. T.; Shore, S. G. *Chem. Commun.* 1970, 1312.
8. Greenwood, N. N.; Howard, J. A.; McDonald, W. S. *J. Chem. Soc., Dalton Trans.* 1976, 37.
9. Greenwood, N. N.; Kennedy, J. D.; McDonald, W. S.; Reed, D.; Staves, J. *J. Chem. Soc., Dalton Trans.* 1979, 117.
10. Mingos, D. M. P.; Johnston, R. L. *Struct. Bonding* 1987, **68,** 29.
11. Shore, S. G. In "Boron Hydride Chemistry"; Muetterties, E. L., Ed.; Academic Press: New York, 1975; p. 79.
12. Carré, F. H.; Moreau, J. J. E. *Inorg. Chem.* 1982, **21,** 3099.
13. Flynn, K. M.; Olmstead, M. M.; Power, P. P. *J. Am. Chem. Soc.* 1983, **105,** 2085.
14. Roesky, H. W., Ed. "Rings, Clusters and Polymers of Main Group and Transition Elements"; Elsevier: New York, 1989.

15. Shriver, D. F.; Atkins, P. W.; Langford, C. H. "Inorganic Chemistry"; Freeman: New York, 1990; p. 81.
16. Huheey, J. E. "Inorganic Chemistry," 3rd. ed.; Harper & Row: New York, 1983; p. 841.
17. Pearson, R. G. *Inorg. Chem.* 1988, **27,** 734.
18. Parry, R. W.; Edwards, L. J. *J. Am. Chem. Soc.* 1959, **81,** 3554.
19. Gilbert, K. B.; Boocock, S. K.; Shore, S. G. In "Comprehensive Organometallic Chemistry": Wilkinson, G.; Stone, F. G. A.; Abel, E. W., Eds.; Pergamon: Oxford, 1982; Vol. 6, p. 879.
20. Marks, T. J.; Kolb, J. R. *Chem. Rev.* 1977, **77,** 263.
21. Boocock, S. K.; Shore, S. G. In "Comparative Organometallic Chemistry"; Wilkinson, G.; Stone, F. G. A.; Abel, E. W., Eds.; Pergamon: Oxford, 1982; Vol. 6, p. 947.
22. Greenwood, N. N.; Earnshaw, A. "Chemistry of the Elements"; Pergamon: Oxford, 1984.
23. Rhodes, L. F.; Venanzi, L. M.; Sorato, C.; Albinati, A. *Inorg. Chem.* 1986, **25,** 3335.
24. Kirtley, S. W.; Andrews, M. A.; Bau, R.; Grynkewich, G. W.; Marks, T. J.; Tipton, D. L.; Whittlesey, B. R. *J. Am. Chem. Soc.* 1977, **99,** 7154.
25. Frost, P. W.; Howard, J. A. K.; Spencer, J. L. *Chem. Commun.* 1984, 1362.
26. Marks, T. J.; Kennelly, W. J.; Kolb, J. R.; Shimp, L. A. *Inorg. Chem.* 1972, **11,** 2540.
27. Mirkin, C. A.; Lu, K.-L.; Geoffroy, G. L.; Rheingold, A. L. *J. Am. Chem. Soc.* 1990, **112,** 461.
28. Barron, A. R.; Wilkinson, G. *Polyhedron* 1986, **5,** 1897.
29. Bulychev, M. *Polyhedron* 1990, **9,** 387.
30. Howard, C. G.; Girolami, G. S.; Wilkinson, G.; Thornton-Pett, M.; Hursthouse, M. B. *J. Chem. Soc., Dalton Trans.* 1985, 921.
31. Janowicz, A. H.; Bergman, R. G. *J. Am. Chem. Soc.* 1982, **104,** 352.
32. Hoyano, J. K.; Graham, W. A. G. *J. Am. Chem. Soc.* 1982, **104,** 3723.
33. Saillard, J.-Y.; Hoffmann, R. *J. Am. Chem. Soc.* 1984, **106,** 2006.
34. Rest, A. J.; Whitwell, I.; Graham, W. A. G.; Hoyano, J. K.; McMaster, A. D. *J. Chem. Soc., Dalton Trans.* 1987, 1181.
35. Graham, W. A. G. *J. Organomet. Chem.* 1986, **300,** 81.
36. Schubert, U. *Adv. Organomet. Chem.* 1990, **30,** 151.
37. Schubert, U.; Müller, J.; Alt, H. G. *Organometallics* 1987, **6,** 469.
38. Schubert, U.; Scholz, G.; Müller, J.; Ackermann, K.; Wörle, B.; Stansfield, R. F. D. *J. Organomet. Chem.* 1986, **306,** 303.
39. Luo, X.-L.; Baudry, D.; Boydell, P.; Charpin, P.; Nierlich, M.; Ephritikhine, M.; Crabtree, R. H. *Inorg. Chem.* 1990, **29,** 1511.
40. Gutmann, V. "The Donor–Acceptor Approach to Molecular Interactions"; Plenum: New York, 1978.
41. Dapporto, P.; Midollini, S.; Sacconi, L. *Angew. Chem., Int. Ed. Engl.* 1979, **18,** 469.
42. Caminade, A.-M.; Veith, M.; Huch, V.; Malisch, W. *Organometallics* 1990, **9,** 1798.
43. Calderazzo, F.; Poli, R.; Pellizzi, G. *J. Chem. Soc., Dalton Trans.* 1984, 2535.
44. Powell, P.; Nöth, H. *Chem. Commun.* 1966, 637.
45. Lehmann, D. D.; Shriver, D. F. *Inorg. Chem.* 1974, **13,** 2203.
46. Khattar, R.; Puga, J.; Fehlner, T. P. *J. Am. Chem. Soc.* 1989, **111,** 1877.
47. Burlitch, J. M.; Burk, J. H.; Leonowicz, M. E.; Hughes, R. E. *Inorg. Chem.* 1979, **18,** 1702.
48. Horwitz, C. P.; Shriver, D. F. *Adv. Organomet. Chem.* 1984, **23,** 219.
49. Burlitch, J. M.; Leonowicz, M. E.; Petersen, R. B.; Hughes, R. E. *Inorg. Chem.* 1979, **18,** 1097.
50. Schmid, G. *Angew. Chem., Int. Ed. Engl.* 1970, **9,** 819.
51. Gaines, D. F.; Iorns, T. V. *Inorg. Chem.* 1968, **5,** 1041.
52. Basil, J. D.; Aradi, A. A.; Bhattacharyya, N. K.; Rath, N. P.; Eigenbrot, C.; Fehlner, T. P. *Inorg. Chem.* 1990, **29,** 1260.
53. Chalk, A. J.; Harrod, J. F. *J. Am. Chem. Soc.* 1967, **89,** 1640.

54. Elliot, D. J.; Levy, C. J.; Puddephatt, R. J.; Holah, D. G.; Hughes, A. N. Magnuson, V. R.; Moser, I. M. *Inorg. Chem.* 1990, **29,** 5015.
55. Baker, R. T.; Ovenall, D. W.; Calabrese, J. C.; Westcott, S. A.; Taylor, N. J.; Marder, T. B. *J. Am. Chem. Soc.* 1990, **112,** 9399.
56. Churchill, M. R.; Hackbarth, J. J.; Davison, A.; Traficante, D. D.; Wreford, S. S. *J. Am. Chem. Soc.* 1974, **96,** 4041.
57. Merola, J. S.; Knorr, J. R. "Abstracts 199th Am. Chem. Soc. Meeting," 1990; INORG 392.
58. Nöth, H.; Schmid, G. *Angew. Chem.* 1963, **75,** 861.
59. Schmid, G.; Petz, W.; Nöth, H. *Inorg. Chim. Acta* 1970, **4,** 423.
60. Aylett, B. J. *Adv. Inorg. Chem. Radiochem.* 1982, **25,** 327.
61. Tilly, T. D. In "The Chemistry of Organic Silicon Compounds"; Patai, S.; Rappoport, Z., Eds.; Wiley: New York, 1989; Part 2, p. 1415.
62. Whitmire, K. H. *J. Coord. Chem.* 1988, **17,** 95.
63. Buckingham, J., Ed. "Dictionary of Organometallic Compounds"; Chapman & Hall: London, 1984.
64. Brooks, E. H.; Cross, R. J. *Organomet. Chem. Rev. (A)* 1970, **6,** 227.
65. Kennedy, V. O.; Zarate, E. A.; Tessier, C. A.; Youngs, W. J. "Abstracts XXIII Organosilicon Symposium," 1990; P39.
66. Süss-Fink, G. *Angew. Chem., Int. Ed. Engl.* 1982, **21,** 73.
67. MacKay, K. M.; Nicholson, B. K. In "Comprehensive Organometallic Chemistry"; Wilkinson, G.; Stone, F. G. A.; Abel, E. W., Eds.; Pergamon: Oxford, 1982; Vol. 6, p. 1043.
68. Pomeroy, R. K.; Sams, J. R.; Tsin, T. B.; *J. Chem. Soc., Dalton Trans.* 1975, 1216.
69. Anderson, R. A.; Einstein, F. W. B. *Acta Cryst. (B)* 1976, **32,** 966.
70. Cradwick, E. M.; Hall, D. *J. Organomet. Chem.* 1970, **25,** 91.
71. Burnham, R. A.; Stobart, S. R. *J. Chem. Soc., Dalton Trans.* 1977, 1489.
72. Clark, H. C.; Rake, A. T. *J. Organomet. Chem.* 1974, **82,** 159.
73. Cassidy, J. M.; Whitmire, K. H. *Inorg. Chem.* 1989, **28,** 2494.
74. Magyar, E. S.; Lillya, C. P. *J. Organomet. Chem.* 1976, **116,** 99.
75. Dötz, K. H. "Transition Metal Carbene Complexes"; Verlag Chemie: Weinheim, 1983.
76. Zybill, C.; Müller, G. *Organometallics* 1988, **7,** 1368.
77. Straus, D. A.; Zhang, C.; Quimbita, G. E.; Grumbine, S. D.; Heyn, R. H.; Tilley, T. D.; Rheingold, A. L.; Geib, S. J. *J. Am. Chem. Soc.* 1990, **112,** 2673.
78. Petz, W. *Chem. Rev.* 1986, **86,** 1019.
79. Jutzi, P.; Steiner, W.; Koenig, E.; Huttner, G.; Frank, A.; Schubert, U. *Chem. Ber.* 1978, **111,** 606.
80. Herrmann, W. *Angew. Chem., Int. Ed. Engl.* 1986, **25,** 56.
81. Herrmann, W. A. In "Rings, Clusters and Polymers of Main Group and Transition Elements"; Roesky, H. W., Ed.; Elsevier: New York, 1989; p. 345.
82. Zybill, C.; Wilkinson, D. L.; Müller, G. *Angew. Chem., Int. Ed. Engl.* 1988, **27,** 583.
83. Rieviére, P.; Rivére-Baudet, M.; Satgé, J. In "Comprehensive Organometallic Chemistry"; Wilkinson, G.; Stone, F. G. A.; Abel, E. W., Eds.; Pergamon: Oxford, 1982; Vol. 2, p. 399.
84. Gäde, W.; Weiss, E. *J. Organomet. Chem.* 1981, **213,** 451.
85. Korp, J. D.; Bernal, I.; Hörlein, R.; Serrano, R.; Herrmann, W. A. *Chem. Ber.* 1985, **118,** 340.
86. Hoffmann, R. *Angew. Chem., Int. Ed. Engl.* 1982, **21,** 711.
87. Herrmann, W. A.; Kneuper, H.-J.; Herdtweck, E. *Angew. Chem., Int. Ed. Engl.* 1985, **24,** 1062.
88. Kostic, N. M.; Fenske, R. F. *J. Organomet. Chem.* 1982, **233,** 337.
89. Albright, T. A.; Burdett, J. K.; Whangbo, M. H. "Orbital Interactions in Chemistry"; Wiley: New York, 1985; p. 404.
90. Latesky, S. L.; Selegue, J. P. *J. Am. Chem. Soc.* 1987, **109,** 4731.

91. Kneuper, H.-J.; Herdtweck, E.; Herrmann, W. A. *J. Am. Chem. Soc.* 1987, **109,** 2508.
92. Huttner, G.; Weber, U.; Sigwarth, B.; Scheidsteger, O.; Lang, H.; Zsolnai, L. *J. Organomet. Chem.* 1985, **282,** 331.
93. Herrmann, W. A.; Weichmann, J.; Küsthardt, U.; Schäfer, A.; Hörlein, R.; Hecht, C.; Voss, E.; Serrano, R. *Angew. Chem., Int. Ed. Engl.* 1983, **22,** 979.
94. Parkin, G.; Bercaw, J. E. *Polyhedron* 1988, **7,** 2053.
95. Bor, G. *J. Organomet. Chem.* 1975, **94,** 181.
96. Bartlett, R. A.; Chen, H.; Power, P. *Angew. Chem., Int. Ed. Engl.* 1989, **28,** 316.
97. Kaul, H.-A.; Gressinger, D.; Malisch, W.; Klein, H.-P.; Thewalt, U. *Angew. Chem., Int. Ed. Engl.* 1983, **22,** 60.
98. Norman, N. C. *Chem. Soc. Rev.* 1988, **17,** 269.
99. Etzrodt, G.; Boese, R.; Schmid, G. *Chem. Ber.* 1979, **112,** 2574.
100. Wallis, J. M.; Müller, G.; Schmidbaur, H. *Inorg. Chem.* 1987, **26,** 458.
101. Huttner, G.; Evertz, K. *Acc. Chem. Res.* 1986, **19,** 406.
102. Huttner, G. *Pure Appl. Chem.* 1986, **58,** 585.
103. Huttner, G.; Mohr, G.; Frank, A. *Angew. Chem., Int. Ed. Engl.* 1976, **15,** 682.
104. Huttner, G.; Lang, H. In "Rings, Clusters and Polymers of Main Group and Transition Elements"; Roesky, H. W., Ed.; Elsevier: New York, 1989; p. 409.
105. Strube, A.; Huttner, G.; Zsolnai, L. *Angew. Chem., Int. Ed. Engl.* 1988, **27,** 1529.
106. Strube, A.; Heuser, J.; Huttner, G.; Lang, H. *J. Organomet. Chem.* 1988, **356,** C9.
107. Roesky, H. W.; Bai, Y.; Noltemeyer, M. *Angew. Chem., Int. Ed. Engl.* 1989, **28,** 754.
108. Ciechanowicz, M.; Griffith, W. P.; Pawson, D.; Skapski, A. C. *Chem. Commun.* 1971, 876.
109. Bryndza, H. E.; Tam, W. *Chem. Rev.* 1988, **88,** 1163.
110. Bradley, D. C. *Chem. Rev.* 1989, **89,** 1317.
111. Murray, S. G.; Hartley, F. R. *Chem. Rev.* 1981, **81,** 365.
112. Atherton, M. J.; Holloway, J. H. *Adv. Inorg. Chem. Radiochem.* 1979, **22,** 171.
113. Struczynski, S. M.; Brennan, J. G.; Steigerwald, M. L. *Inorg. Chem.* 1989, **28,** 4431.
114. Bottomley, F.; Drummond, D. F.; Egharevba, G. O.; White, P. S. *Organometallics* 1986, **5,** 1620.
115. Herrmann, W. A.; Hecht, C.; Herdwerk, E.; Kneuper, H.-J. *Angew. Chem., Int. Ed. Engl.* 1987, **26,** 132.
116. Draganjac, M.; Rauchfuss, T. B. *Angew. Chem., Int. Ed. Engl.* 1985, **24,** 742.
117. Wachter, J. *J. Coord. Chem.* 1987, **15,** 219.
118. Braunwarth, H.; Huttner, G.; Zsolnai, L. *Angew. Chem., Int. Ed. Engl.* 1988, **27,** 698.
119. Burckett-St. Laurent, J. C. T. R.; Caira, M. R.; English, R. B.; Haines, R. J.; Nassimbeni, L. R. *J. Chem. Soc., Dalton Trans.* 1977, 1077.
120. Huttner, G.; Schuler, S.; Zsolnai, L.; Gottlieb, M.; Braunwarth, H.; Minelli, M. *J. Organomet. Chem.* 1986, **299,** C4.
121. Braunwarth, H.; Ettel, F.; Huttner, G. *J. Organomet. Chem.* 1988, **355,** 281.
122. Herrmann, W. A.; Rohrmann, J.; Ziegler, M. L.; Zahn, T. *J. Organomet. Chem.* 1984, **273,** 221.
123. Tremel, W.; Hoffmann, R.; Jemmis, E. D. *Inorg. Chem.* 1989, **28,** 1213.
124. Herrmann, W. A.; Hecht, C.; Ziegler, M. L.; Balbach, B. *Chem. Commun.* 1984, 868.
125. Heberhold, M.; Reiner, D.; Neugebauer, D. *Angew. Chem., Int. Ed. Engl.* 1983, **22,** 59.
126. Greenhough, T. J.; Kolthammer, B. W. S.; Legzdins, P.; Trotter, J. *Inorg. Chem.* 1979, **18,** 3543.
127. Herrmann, W. A.; Rohrmann, J.; Nöth, H.; Narula, C. K.; Bernal, I.; Draux, M. *J. Organomet. Chem.* 1985, **284,** 1989; 1985, **289,** C26.
128. Herrmann, W. A.; Rohrmann, J.; Herdtweck, E.; Bock, H.; Veltmann, A. *J. Am. Chem. Soc.* 1986, **108,** 3134.
129. Mayer, J. M. *Comments Inorg. Chem.* 1988, **8,** 125.

130. Chisholm, M. H.; Clark, D. L. *Comments Inorg. Chem.* 1987, **6,** 23.
131. Messerle, L. *Chem. Rev.* 1988, **88,** 1229.
132. Morse, D. B.; Hendrickson, D. N.; Rauchfuss, T. B.; Wilson, S. R. *Organometallics* 1988, **7,** 496.
133. Morse, D. B.; Rauchfuss, T. B.; Wilson, S. R. *J. Am. Chem. Soc.* 1990, **112,** 1860.
134. Kameda, M.; DePoy, R. E.; Kodama, G. In "Boron Chemistry"; Hermanek, S.; Ed.; World Scientific: Singapore, 1987; p 104.
135. Kodama, G. In "Advances in Boron and the Boranes"; Liebman, J. F.; Greenberg, A.; Williams, R. E., Eds.; VCH: New York, 1980; p. 105.
136. Snow, S. A.; Shimoi, M.; Ostler, C. D.; Thompson, B. K.; Kodama, G.; Parry, R. W. *Inorg. Chem.* 1984, **23,** 511.
137. Snow, S. A.; Kodama, G. *Inorg. Chem.* 1985, **24,** 795.
138. Shimoi, M.; Katoh, K.; Tobita, H.; Ogino, H. *Inorg. Chem.* 1990, **29,** 814.
139. Shimoi, M.; Katoh, K.; Ogino, H. *Chem. Commun.* 1990, 811.
140. Calabrese, J. C.; Fischer, M. B.; Gaines, D. F.; Lott, J. W. *J. Am. Chem. Soc.* 1974, **96,** 6318.
141. Knobler, C. B.; Marder, T. B.; Mizusawa, E. A.; Teller, R. G.; Long, J. A.; Behnken, P. E.; Hawthorne, M. F. *J. Am. Chem. Soc.* 1984, **106,** 2990.
142. Dewar, M. J. S. *Bull. Soc. Chim. Fr.* 1951, **18,** C71.
143. Chatt, J.; Duncanson, L. A. *J. Chem. Soc.* 1953, 2939.
144. Grimes, R. N., Ed. "Metal Interactions with Boron Clusters"; Plenum: New York; 1982.
145. Kennedy, J. *Prog. Inorg. Chem.* 1986, **34,** 211; 1984, **32,** 519.
146. Housecroft, C. E.; Fehlner, T. P. *Adv. Organomet. Chem.* 1982, **21,** 57.
147. Schmid, G. *Angew. Chem., Int. Ed. Engl.* 1978, **17,** 392.
148. Housecroft, C. E. *Polyhedron* 1987, **6,** 1935.
149. Gilbert, K. B.; Boocock, S. K.; Shore, S. G. *Comp. Organometal. Chem.* 1982, **6,** 879.
150. Greenwood, N. N. *Chem. Soc. Rev.* 1984, **13,** 353.
151. Pitzer, K. S. *J. Am. Chem. Soc.* 1945, **67,** 1126.
152. Coffey, T. J.; Medford, G.; Plotkin, J.; Long, G. J.; Huffman, J. C.; Shore, S. G. *Organometallics* 1989, **8,** 2404.
153. Brebenik, P. D.; Green, M. L. H.; Kelland, M. A.; Leach, J. B.; Mountford, P.; Stringer, G.; Walker, N. M.; Wong, L. L. *Chem. Commun.* 1988, 799.
154. DeKock, R. L.; Deshmukh, P.; Fehlner, T. P.; Housecroft, C. E.; Plotkin, J. S.; Shore, S. G. *J. Am. Chem. Soc.* 1983, **105,** 815.
155. Lipscomb, W. N. In "Boron Hydride Chemistry"; Muetterties, E. L., Ed.; Academic Press: New York, 1963; p. 39.
156. Davison, A.; Traficante, D. D.; Wreford, S. S. *J. Am. Chem. Soc.* 1974, **96,** 2802.
157. Paetzold, P. *Adv. Inorg. Chem.* 1987, **31,** 123.
158. Bulak, E.; Herberich, G. E.; Manners, I.; Mayer, H.; Paetzold, P. *Angew. Chem., Int. Ed. Engl.* 1988, **27,** 958.
159. West, R. *Angew. Chem., Int. Ed. Engl.* 1987, **26,** 1201.
160. Pham, E. K.; West, R. *J. Am. Chem. Soc.* 1989, **111,** 7667.
161. Berry, D. H.; Chey, J. H.; Zipin, H. S.; Carroll, P. J. *J. Am. Chem. Soc.* 1990, **112,** 452.
162. Koloski, T. S.; Carroll, P. J.; Berry, D. H. *J. Am. Chem. Soc.* 1990, **112,** 6405.
163. Fink, M. J.; Michalczyk, J. J.; Haller, K. J.; West, R.; Michl, *J. Organometallics* 1984, **3,** 793.
164. Masamune, S.; Murakami, S.; Snow, J. T.; Tobita, H.; Williams, D. J. *Organometallics* 1984, **3,** 333.
165. Collman, J. P.; Hegedus, L. S. "Principles and Applications of Organotransition Metal Chemistry"; University Science Books: Mill Valley, CA, 1980.
166. Pham, E. K.; West, R. *Organometallics* 1990, **9,** 1517.

167. Chatt, J.; Eaborn, C.; Ibekwe, S. D.; Kapoor, P. N. *Chem. Commun.* 1970, 1343.
168. Zarate, E. A.; Tessier-Youngs, C. A.; Youngs, W. J. *J. Am. Chem. Soc.* 1988, **110,** 4068.
169. Zarate, E. A.; Tessier-Youngs, C. A.; Youngs, W. J. *Chem. Commun.* 1989, 577.
170. Anderson, A. B.; Shiller, P.; Zarate, E. A.; Tessier-Youngs, C. A.; Youngs, W. J. *Organometallics* 1989, **8,** 2320.
171. Evans, D. G.; Mingos, D. M. P. *J. Organomet. Chem.* 1982, **240,** 321.
172. Cotton, G. A.; Kibala, P. A. *Polyhedron* 1987, **6,** 645; *Inorg. Chem.* 1990, **29,** 3192.
173. Simon, G. L.; Dahl, L. F. *J. Am. Chem. Soc.* 1973, **95,** 783.
174. Cowley, A. H.; Giolando, D. M.; Nunn, C. M.; Pakulski, M.; Westmoreland, D.; Norman, N. C. *J. Chem. Soc., Dalton Trans.* 1988, 2127.
175. Fehlner, T. P. *Comments Inorg. Chem.* 1988, **7,** 307.
176. Barrau, J.; Escudié, J.; Sateé, J. *Chem. Rev.* 1990, **90,** 283.
177. Cowley, A. H. *Polyhedron* 1984, **3,** 389.
178. Scherer, O. J. *Angew. Chem., Int. Ed. Engl.* 1985, **24,** 924.
179. Scherer, O. J. *Angew. Chem., Int. Ed. Engl.* 1990, **29,** 1104.
180. Cowley, A. H.; Norman, N. C. *Prog. Inorg. Chem.* 1986, **34,** 1.
181. Huttner, G. *Pure Appl. Chem.* 1986, **58,** 585.
182. Scherer, O. J. *Comments Inorg. Chem.* 1987, **6,** 1.
183. Dimaio, A.-J.; Rheingold, A. L. *Chem. Rev.* 1990, **90,** 169.
184. DiVaira, M.; Sacconi, L. *Angew. Chem., Int. Ed. Engl.* 1982, **21,** 330.
185. Dickson, R. S.; Ibers, J. A. *J. Am. Chem. Soc.* 1972, **94,** 2988.
186. Yoshifuji, M.; Shima, I.; Inamoto, N.; Hirotsu, K.; Higuchi, T. *J. Am. Chem. Soc.* 1981, **103,** 4587.
187. Cowley, A. H.; Lasch, J. G.; Norman, N. C.; Pakulski, M. *J. Am. Chem. Soc.* 1983, **103,** 5506.
188. Cowley, A. H.; Kilduff, J. E.; Lasch, J. G.; Norman, N. C.; Pakulski, M.; Ando, F.; Wright, T. C. *J. Am. Chem. Soc.* 1983, **105,** 7751.
189. Green, J. C.; Green, M. L. H.; Morris, G. E. *Chem. Commun.* 1974, 212.
190. Borm, J.; Zsolnai, L.; Huttner, G. *Angew. Chem., Int. Ed. Engl.* 1983, **22,** 977.
191. Huttner, G.; Borm, J.; Zsolnai, L. *J. Organomet. Chem.* 1986, **304,** 309.
192. Weber, L.; Bungardt, D.; Müller, A.; Bögge, H. *Organometallics* 1989, **8,** 2800.
193. Ginsberg, A. P.; Lindsell, W. E.; McCullough, K. J.; Sprinkle, C. R.; Welch, A. J. *J. Am. Chem. Soc.* 1986, **108,** 403.
194. Flyn, K. M.; Hope, H.; Murray, B. D.; Olmstead, M. M.; Power, P. P. *J. Am. Chem. Soc.* 1983, **105,** 7750.
195. Schugard, K. A.; Fenske, R. F. *J. Am. Chem. Soc.* 1985, **107,** 3384.
196. Huttner, G.; Sigwarth, B.; Scheidsteger, O.; Zsolnai, L.; Orama, O. *Organometallics* 1985, **4,** 326.
197. Herrmann, W. A.; Koumboris, B.; Zahn, T.; Ziegler, M. L. *Angew. Chem., Int. Ed. Engl.* 1984, **23,** 812.
198. Sigwarth, B.; Zsolnai, L.; Berke, H.; Huttner, G. *J. Organomet. Chem.* 1982, **226,** C5.
199. Huttner, G.; Weber, U.; Sigwarth, B.; Scheidsteger, O. *Angew. Chem., Int. Ed. Engl.* 1982, **21,** 215.
200. Huttner, G.; Weber, U.; Zsolnai, L. *Z. Naturforsch. B* 1982, **37,** 707.
201. Manriquez, J. M.; Sanner, R. D.; Marsh, R. E.; Bercaw, J. E. *J. Am. Chem. Soc.* 1976, **98,** 3042.
202. Berry, D. H.; Procopio, L. J.; Carroll, P. J. *Organometallics* 1988, **7,** 570.
203. Jonas, K. *Angew. Chem., Int. Ed. Engl.* 1973, **12,** 997.
204. Jonas, K.; Brauer, D. J.; Krüger, C.; Roberts, P. J.; Tsay, Y.-H. *J. Am. Chem. Soc.* 1976, **98,** 74.
205. Goldberg, K. I.; Hoffman, D. M.; Hoffman, R. *Inorg. Chem.* 1982, **21,** 3863.

206. Rheingold, A. L.; Foley, J. J.; Sullivan, P. J. *J. Am. Chem. Soc.* 1982, **104,** 4727.
207. Driess, M.; Fanta, A. D.; Powell, D.; West, R. *Angew. Chem., Int. Ed. Engl.* 1989, **28,** 1038.
208. Tremel, W.; Hoffmann, R.; Kertesz, M. *J. Am. Chem. Soc.* 1989, **111,** 2030.
209. Scheidsteger, O.; Huttner, G.; Dehnicke, K.; Pebler, J. *Angew. Chem., Int. Ed. Engl.* 1985, **24,** 428.
210. Brunner, H.; Janietz, N.; Meier, W.; Sergeson, G.; Wachter, J.; Zahn, T.; Ziegler, M. L. *Angew. Chem., Int. Ed. Engl.* 1985, **24,** 1060.
211. Müller, U.; Krug, V. *Angew. Chem., Int. Ed. Engl.* 1988, **27,** 293.
212. Boeré, R. T.; Cordes, A. W.; Craig, S. L.; Oakley, R. T.; Reed, R. W. *J. Am. Chem. Soc.* 1987, **109,** 868.
213. Chivers, T.; Dhathathreyan, K. S.; Ziegler, T. *Chem. Commun.* 1989, 86.
214. Chivers, T.; Edelmann, F. *Polyhedron* 1986, **5,** 1661.
215. Kaesz, H. D.; Fellman, W.; Wilkes, G. R.; Dahl, L. F. *J. Am. Chem. Soc.* 1965, **87,** 2756.
216. Jacobsen, G. B.; Andersen, E.; Housecroft, C. E.; Hong, F.-E.; Buhl, M. L.; Long, G. J.; Fehlner, T. P. *Inorg. Chem.* 1987, **26,** 4040.
217. Ting, C.; Messerle, L. *J. Am. Chem. Soc.* 1989, **111,** 3449.
218. Bell, R. A.; Cohen, S. A.; Doherty, N. M.; Threldel, R. S.; Bercaw, J. E. *Organometallics* 1986, **5,** 972.
219. Cotton, F. A.; Jamerson, J. D.; Stults, B. R. *J. Am. Chem. Soc.* 1976, **98,** 1774.
220. Hoffman, D.; Hoffmann, R.; Fisel, C. R. *J. Am. Chem. Soc.* 1982, **104,** 1982.
221. Meng, X., Ph.D. Thesis, Univ. Notre Dame, Notre Dame, IN, 1990.
222. Brookhart, M.; Green, M. L. H.; Wong, L.-L. *Prog. Inorg. Chem.* 1988, **36,** 1.
223. Eisenberg, R.; Wang, W.-D. *J. Am. Chem. Soc.* 1990, **112,** 1833.
224. Calderazzo, F.; Poli, R.; Vitale, D.; Korp, J. D.; Bernal, I.; Pelizzi, G.; Atwood, J. L.; Hunter, W. E. *Gazz. Chim. Ital.* 1983, **113,** 761.
225. Vahrenkamp, H.; Wolters, D. *Angew. Chem., Int. Ed. Engl.* 1983, *22,* 154.
226. Tasi, M.; Powell, A. K.; Vahrenkamp, H. *Angew. Chem., Int. Ed. Engl.* 1989, **28,** 318.
227. Campana, C. F.; vizi-Orosz, A.; Palyi, G.; Markó, L.; Dahl, L. F. *Inorg. Chem.* 1979, **18,** 3054.
228. Foust, A. S.; Campana, C. F.; Sinclair, J. D.; Dahl, L. F. *Inorg. Chem.* 1979, **18,** 3047.
229. Arif, A. M.; Cowley, A. H.; Norman, N. C.; Pakulski, M. *J. Am. Chem. Soc.* 1985, **107,** 1062.
230. Huttner, G.; Sigwarth, B.; Scheidsteger, O.; Zsolnai, L.; Orama, O. *Organometallics* 1985, **4,** 326.
231. Scherer, O. J.; Sitzmann, H.; Wolmershäuser, G. *Angew. Chem., Int. Ed. Engl.* 1984, **23,** 968.
232. Lang, H.; Zsolnai, L.; Huttner, G. *Angew. Chem., Int. Ed. Engl.* 1983, **22,** 976.
233. Müller, M.; Vahrenkamp, H. *J. Organomet. Chem.* 1983, **253,** 95.
234. Scherer, O. J.; Sitzmann, H.; Wolmershäuser, G. *J. Organomet. Chem.* 1986, **309,** 77.
235. Scherer, O. J.; Swarowsky, M. *Angew. Chem., Int. Ed. Engl.* 1988, **27,** 405.
236. Seyferth, D.; Merola, J. S.; Henderson, R. S. *Organometallics,* 1982, **1,** 859.
237. Regitz, M.; Binger, P. *Angew. Chem., Int. Ed. Engl.* 1988, **27,** 1484.
238. Nixon, J. F. *Chem. Rev.* 1988, **88,** 1327.
239. DeBloix, R. E.; Rheingold, A. L.; Samkoff, D. E. *Inorg. Chem.* 1988, **27,** 3506.
240. Paetzold, P.; Delpy, K. *Chem. Ber.* 1985, **118,** 2552.
241. Hieber, W.; Gruber, J. *Z. Anorg. Allg. Chem.* 1958, **296,** 91.
242. Wei, C. H.; Dahl, L. F. *Inorg. Chem.* 1964, **4,** 1.
243. Teo, B. K.; Hall, M. B.; Fenske, R. F.; Dahl, L. F. *Inorg. Chem.* 1979, **101,** 6550.
244. Andersen, E. L.; Fehlner, T. P.; Foti, A. E.; Salahub, D. R. *J. Am. Chem. Soc.* 1980, **102,** 7422.

245. Van Dam, H.; Stufkens, D. J.; Oskam, A.; Doran, M.; Hillier, I. H. *J. Electron Spectrosc. Relat. Phenom.* 1980, **21,** 47.
246. DeKock, R. L.; Baerends, E. J.; Hengelmolen, R. *Organometallics* 1984, **3,** 289.
247. Seyferth, D.; Henderson, R. S.; Song, L.-C. *Organometallics* 1982, **1,** 125.
248. Campana, C. F.; Lo, F. Y.-K.; Dahl, L. F. *Inorg. Chem.* 1979, **18,** 3060.
249. Lesch, D. A.; Rauchfuss, T. B. *Inorg. Chem.* 1981, **20,** 3583.
250. Seyferth, D.; Henderson, R. S. *J. Organomet. Chem.* 1981, **204,** 333.
251. Brunner, H.; Wachter, J.; Guggolz, E.; Ziegler, M. L. *J. Am. Chem. Soc.* 1982, **104,** 1765.
252. Brunner, H.; Kauermann, H.; Meier, W.; Wachter, J. *J. Organomet. Chem.* 1984, **263,** 183.
253. Bolinger, M. C.; Rauchfuss, T. B.; Rheingold, A. L. *Organometallics* 1982, **1,** 1551.
254. Tremel, W.; Hoffmann, R.; Jemmis, E. D. *Inorg. Chem.* 1989, **28,** 1213.
255. Eichhorn, B. W.; Haushalter, R. C.; Merola, J. S. *Inorg. Chem.* 1990, **29,** 728.
256. Herberhold, M.; Bühlmeyer, W. *Angew. Chem., Int. Ed. Engl.* 1984, **23,** 80.
257. Wilkinson, G. *J. Organomet. Chem.* 1975, **100,** 273.
258. Shen-shu, S.; Hoffmann, R. *J. Mol. Sci.* 1983, **1,** 1.
259. Chiang, T.; Kerber, R. C.; Kimball, S. D.; Lauher, J. W. *Inorg. Chem.* 1979, **18,** 1687.
260. Tuggle, R. M.; Weaver, D. L. *Inorg. Chem.* 1971, **10,** 1504.
261. Hildebrandt, S. J.; Gaines, D. F.; Calabrese, J. *Inorg. Chem.* 1978, **17,** 790.
262. Gaines, D. F.; Hildebrandt, S. J. *Inorg. Chem.* 1978, **17,** 794.
263. Fehlner, T. P. *New J. Chem.* 1988, **12,** 307.
264. Ghilardi, C. A.; Midollini, S.; Orlandini, A.; Sacconi, L. *Inorg. Chem.* 1980, **19,** 301.
265. Foust, A. S.; Foster, M. S.; Dahl, L. F. *J. Am. Chem. Soc.* 1969, **91,** 5631.
266. Bianchini, C.; DiVaira, M.; Meli, A.; Sacconi, L. *Inorg. Chem.* 1981, **20,** 1169.
267. Lauher, J. W.; Elian, M.; Summerville, R. H.; Hoffmann, R. *J. Am. Chem. Soc.* 1976, **98,** 3219.
268. Gaggiani, R.; Gillespie, R. J.; Campana, C. F. Kolis, J. W. *Chem. Commun.* 1987, 485.
269. Cotton, F. A. "Chemical Applications of Group Theory"; Wiley: New York, 1971.
270. Amiet, R. G.; Pettit, R. *J. Am. Chem. Soc.* 1968, **90,** 1059.
271. Hall, M. B.; Hillier, I. H.; Connor, J. A.; Guest, M. F.; Lloyd, D. R. *Mol. Phys.* 1975, **30,** 839.
272. Pettit, R. *J. Organomet. Chem.* 1975, **100,** 205.
273. Eichhorn, B. W.; Haushalter, R. C.; Pennington, W. R. *J. Am. Chem. Soc.* 1988, **110,** 8704.
274. Eichhorn, B. W.; Haushalter, R. C. *Chem. Commun.* 1990, 937.
275. Greenwood, N. N.; Savory, C. G.; Grimes, R. N.; Sneddon, L. G.; Davison, A.; Wreford, S. S. *Chem. Commun.* 1974, 718.
276. Miller, V. R.; Weiss, R.; Grimes, R. N. *J. Am. Chem. Soc.* 1977, **99,** 5646.
277. Venable, T. L.; Sinn, E.; Grimes, R. N. *J. Chem. Soc., Dalton Trans.* 1984, 2275.
278. Ulman, J. A.; Andersen, E. L.; Fehlner, T. P. *J. Am. Chem. Soc.* 1978, **100,** 456.
279. Fehlner, T. P. *J. Am. Chem. Soc.* 1980, **102,** 2644.
280. Scherer, O. J.; Vondung, J.; Wolmershäuser, G. *Angew. Chem., Int. Ed. Engl.* 1989, **28,** 1355.
281. Paetzold, P.; Delpy, K.; Boese, R. *Z. Naturforsch.* 1988, **43b,** 839.
282. Hitchcock, P. B.; Maak, J. J.; Nixon, J. F. *Chem. Commun.* 1986, 737.
283. Binger, P.; Milczarek, R.; Mynott, R.; Regitz, M.; Rösch, W. *Angew. Chem., Int. Ed. Engl.* 1986, **25,** 644.
284. Care, H. F.; Howard, J. A. K.; Pilotti, M. U.; Stone, F. G. A.; Szameitat, J. *Chem. Commun.* 1989, 1409.
285. Hawthorne, M. F. *J. Organomet. Chem.* 1975, **100,** 97.
286. Collins, M. J.; Gillespie, R. J.; Kolis, J. W.; Sawyer, J. F. *Inorg. Chem.* 1986, **25,** 2057.
287. Weiss, R.; Grimes, R. N. *Inorg. Chem.* 1979, **18,** 3291.
288. Nöth, H.; Regent, W. *Z. Anorg. Allg. Chem.* 1967, **352,** 1.

289. Schmid, G. *Comments Inorg. Chem.* 1985, **4,** 17.
290. Herberich, G. E. In "Comprehensive Organometallic Chemistry"; Wilkinson, G.; Stone, F. G. A.; Abel, E. W., Eds.; Pergamon: Oxford, 1982; Vol. 1, p. 381.
291. Edwin, J.; Bochmann, M.; Böhm, M. C.; Brennan, D. E.; Geiger, W. E.; Drüger, C.; Pebler, J.; Pritzkow, H.; Siebert, W.; Swiridoff, W.; Wadepohl, H.; Weiss, J.; Zenneck, U. *J. Am. Chem. Soc.* 1983, **105,** 2582.
292. Grimes, R. N. *Coord. Chem. Rev.* 1979, **28,** 47.
293. Siebert, W. *Adv. Organomet. Chem.* 1980, **18,** 301.
294. Scherer, O. J.; Brück, T. *Angew. Chem., Int. Ed. Engl.* 1987, **26,** 59.
295. Baudler, M.; Akpapoglou, S.; Ouzounis, D.; Wasgestian, F.; Meingke, B.; Budzikiewicz, H. Münster, H. *Angew. Chem., Int. Ed. Engl.* 1988, **27,** 280.
296. Scherer, O. J.; Schwalb, J.; Wolmershäuser, G.; Kaim, W.; Gross, R. *Angew. Chem., Int. Ed. Engl.* 1986, **25,** 363.
297. Elian, M.; Chen, M. M. L.; Mingos, D. M. P.; Hoffmann, R. *Inorg. Chem.* 1976, **15,** 1148.
298. Scherer, O. J.; Sitzman, H.; Wolmershäuser, G. W. *Angew. Chem., Int. Ed. Engl.* 1985, **24,** 351.
299. Angermund, K.; Claus, K. H.; Goddard, R.; Krüger, C. *Angew. Chem., Int. Ed. Engl.* 1985, **24,** 237.
300. Scherer, O. J.; Swarowsky, H.; Wolmershäuser, G.; Kaim, W.; Kohlmann, S. *Angew. Chem., Int. Ed. Engl.* 1987, **26,** 1153.
301. Huttner, G.; Krieg, B. *Chem. Ber.* 1972, **105,** 3437.
302. Lagowski, J. J. *Coord. Chem. Rev.* 1977, **22,** 185.
303. Ashe, A. J., III; Butler, W.; Sandford, H. F. *J. Am. Chem. Soc.* 1979, **101,** 7066.
304. Elschenbroich, C.; Kroher, J.; Marsa, W.; Wünsch, M.; Ashe, A. J., III *Angew. Chem., Int. Ed. Engl.* 1986, **25,** 571.
305. Barron, A. R.; Cowley, A. H. *Angew. Chem., Int. Ed. Engl.* 1987, **99,** 986.
306. Milczarek, R.; Rüsseler, W.; Binger, P.; Jonas, K.; Angermund, K.; Krüger, C.; Regitz, M. *Angew. Chem., Int. Ed. Engl.* 1987, **99,** 987.
307. Eichhorn, B. W.; Haushalter, R. C.; Huffman, J. C. *Angew. Chem., Int. Ed. Engl.* 1989, **28,** 1032.
308. Scherer, O. J.; Swarowsky, M.; Swarowsky, H.; Wolmershäuser, G. *Angew. Chem., Int. Ed. Engl.* 1988, **27,** 694.
309. Boocock, S. K.; Toft, M. A.; Inkrott, K. E.; Hsu, L.-Y.; Huffman, J. C.; Folting, K.; Shore, S. G. *Inorg. Chem.* 1984, **23,** 3084.
310. Roesky, H. W. In "Rings, Clusters and Polymers of Main Group and Transition Elements"; Roesky, H. W., Ed.; Elsevier: New York, 1989; p. 369.
311. Chisholm, M. H.; Clark, D. L.; Hampden-Smith, M. J.; Hoffman, D. H. *Angew. Chem., Int. Ed. Engl.* 1989, **28,** 444.
312. Klemperer, W. G.; Lockledge, S. P.; Rosenberg, F. S.; Wang, R. C. "Abstracts 200th Am. Chem. Soc. Meeting," 1990; INORG 333.

3

Transition Metal–Main Group Cluster Compounds

Catherine E. Housecroft

1. INTRODUCTION

In the previous chapter, compounds in which main group fragments functioned as ligands toward low-oxidation-state transition metal atoms were discussed. In contrast to, but also to compliment the former, the present chapter focuses on those compounds which classify as "clusters." In keeping with Chapter 2, the word "cluster" is used here to refer to a compound with a core formulation of M_nE_m (M = low-oxidation-state transition metal, E = main group element) in which $n \geq 3$ and $m \geq 1$. In the light of this definition, the majority of clusters described within this chapter may be classified as being *metal-rich.*

Over the past ten to fifteen years, the chemical literature has witnessed a phenomenal growth in the area of transition metal cluster synthesis and certainly one reason for this has been the search for catalytically active species.(1) However, whether molecular clusters are catalytically active or not (and most appear not to be), even the most superficial glance through relevant journals is sufficient to convince the reader that transition metal clusters are structurally intriguing in their own right; they are far more than simple aggregates of metal atoms clothed in π-acceptor ligands such as CO, Cp, or phosphines. The structural variation exemplified by M_n-clusters is impressive; n metal atom fragments [e.g., $M(CO)_3$ units] will combine together in a

Catherine E. Housecroft • University Chemical Laboratory, Cambridge CB2 1EW, England.
Inorganometallic Chemistry, edited by Thomas P. Fehlner. Plenum Press, New York, 1992.

geometry determined by the number of valence electrons available and, thus, a change in overall charge (viz. oxidation or reduction) or the addition of extra ligands (e.g., H = 1 electron donor, CO = 2 electron donor) results in rearrangement of the metal skeleton.[2] Add to the metal cluster a main group fragment, and the scope for structural variation and reactivity widens. It is convenient to divide main group fragments into two groups; carbon-based and non-carbon-based. The presence of a metal–carbon bond implies the formation of an organometallic compound while any other metal–main group element bond generates an *inorganometallic* compound. Two particular transition metal units, triangular-M_3 and butterfly-M_4 (Figure 1), occur over and over again in organometallic cluster chemistry, and, as we shall see, are also recurrent features of inorganometallic clusters. The study of metal-mediated reactions of small organic and inorganic molecules is of particular interest if the metal framework coincides with a site available on a transition metal surface.[3] This is the case for both the M_3- and M_4-units as illustrated in Figure 1. At the molecular level, *exo*-ligands usually take the place of adjacent metal atoms present in the bulk metal, although for some high nuclearity clusters the M_3- or M_4-unit may be surrounded by other M atoms as well as by *exo*-ligands. Examples of two such compounds[4,5] are shown in Figure 2.

In an organometallic cluster, transformation of the organic moiety may be brought about by a change in the number of valence electrons supplied by the metal framework. For example, the loss of a carbonyl ligand reduces the cluster electron count by two. The cluster has two options: it may regain its original electron count either by interacting with a new two-electron donor or by the activation of, for example, a C—H bond (see Figure 3); or the metal framework may rearrange so as to adopt a structure that is satisfied with two fewer valence electrons. It is the option of facilitating C—H bond activation or the metal-mediated coupling of organic fragments that attracts the interest of many cluster chemists.[6] Figure 3 illustrates the specific example

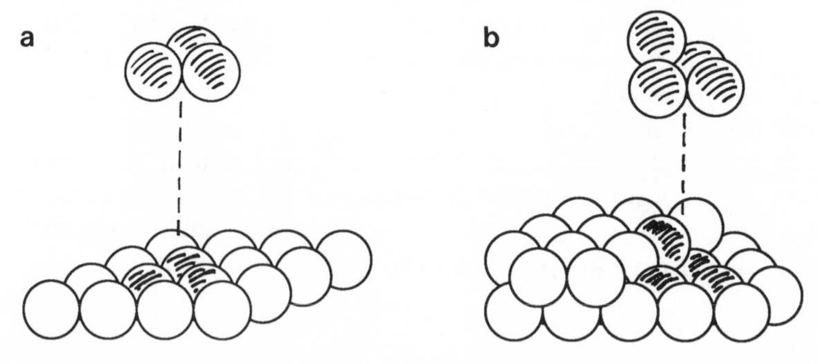

Figure 1. Triangular M_3- and butterfly M_4-skeletons and analogous sites on (a) flat and (b) stepped metal surfaces.

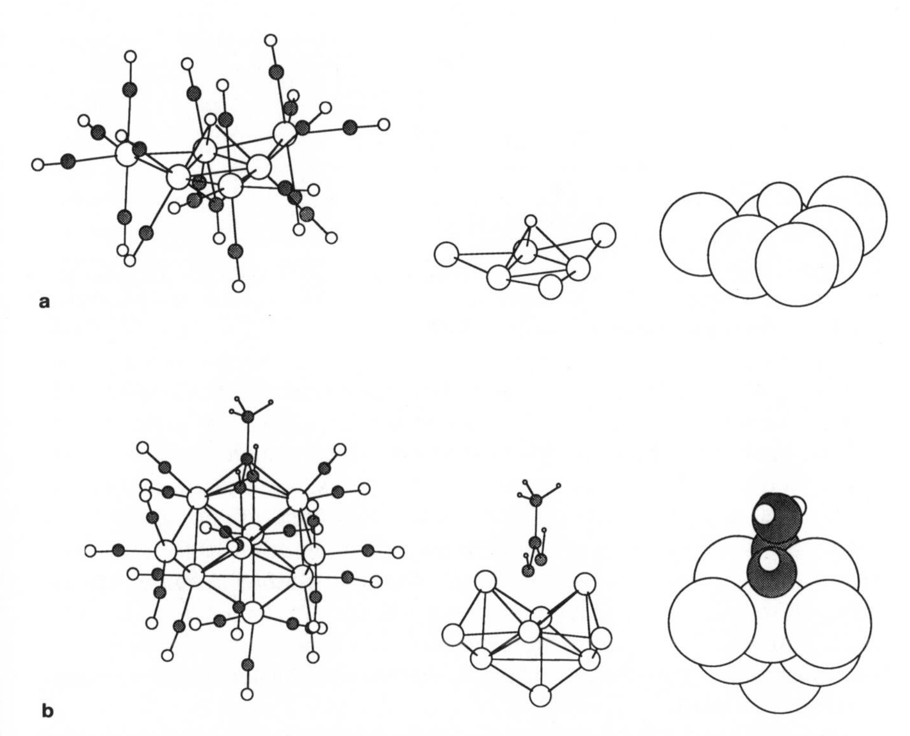

Figure 2. Structures of (a) $Os_6(CO)_{18}(\mu_3\text{-}CO)(\mu_3\text{-}O)$ and (b) $[Os_9(CO)_{21}\{CHC(Me)CH\}]^-$, with and without carbonyl ligands to emphasize the analogy with metal surface species.

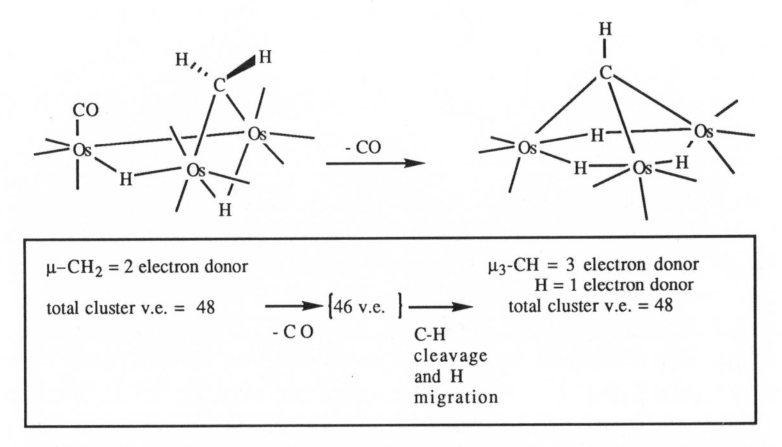

Figure 3. C—H activation concomitant with CO loss from $H_2Os_3(CO)_{10}(\mu\text{-}CH_2)$.

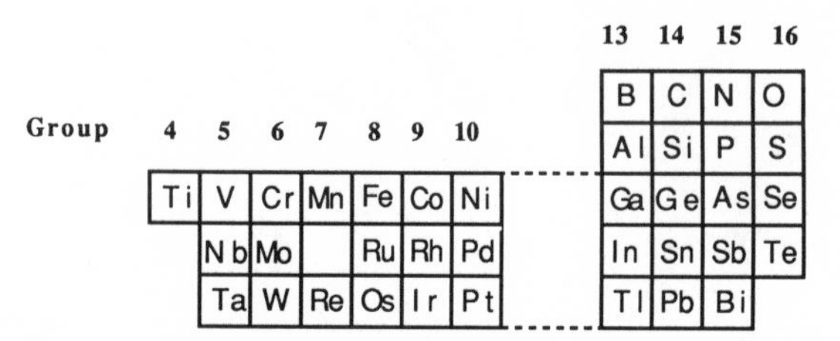

Figure 4. Transition metal–main group cluster compounds: this reduced Periodic Table indicates those transition metals and main group elements to which the following discussion is restricted, the restriction being a function of areas so far explored.

of the transformation of the bridging methylene group in $H_2Os_3(CO)_{10}(\mu$-$CH_2)$[7] into a capping methylidyne fragment in $H_3Os_3(CO)_9(\mu_3$-$CH)$.

Numerous examples of activation of inorganic fragments induced by the interaction of the latter with transition metal atoms have now been reported. In the discussion that follows, which considers the literature up to mid-1990, the scope of such chemistry will be appraised in addition to the structural aspects of the growing classes of inorganometallic clusters. Compounds are arranged in order of Periodic group and by main group element in descending order within the group and, where appropriate, organometallic analogs of inorganometallic clusters are presented. The range of cluster components surveyed is illustrated in Figure 4.

2. GROUP 13

2.1. Clusters Containing Metal–Boron Bonds

The synthesis and characterization of metalloborane clusters has been an area of active interest for many years, and when viewed as a derivative area of borane chemistry it is not at all surprising that the vast majority of metalloboranes known so far are boron-rich.[8] The chemistry of metal-rich metalloboranes is now recognized as an emerging and productive field of research.[9,10] Being adjacent to carbon in the Periodic Table, boron has the potential to mimic carbon in its interactions with transition metal atoms. On the other hand, since a group 13 element possesses only three valence electrons and is expected to function as a Lewis acid, a borane fragment may exhibit significantly different structural properties and reaction chemistry from an isoelectronic organic counterpart.[10,11] Two basic reactions are commonly exploited in organometallic chemistry: C—H bond activation and C—C coupling. The analogous pathways in metalloborane chemistry, viz. B—H bond activation and B—B bond formation, are both known. Activation of B—H bonds occurs in, for example, the formation of the ruthenaboride cluster $HRu_6(CO)_{17}B$ by the pyrolysis of $Ru_3(CO)_{12}$ with $BH_3 \cdot THF$[12] or

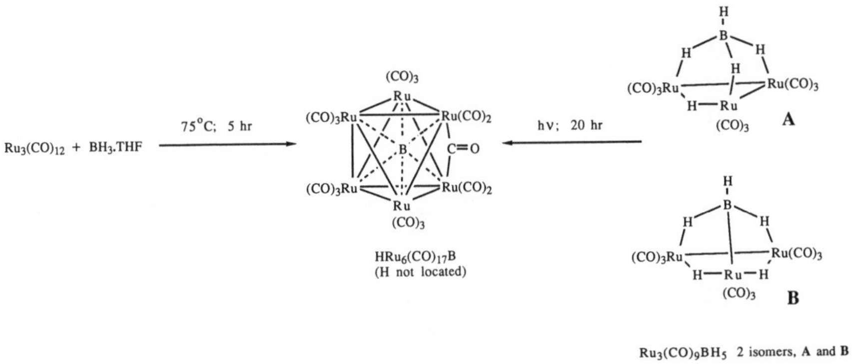

$Ru_3(CO)_{12}$ + BH_3.THF —75°C; 5 hr→ $HRu_6(CO)_{17}B$ (H not located) ←hν; 20 hr— $Ru_3(CO)_9BH_5$ 2 isomers, A and B

Figure 5. Activation of B—H bonds in the formation of $HRu_6(CO)_{17}B$ by two independent routes.

the photolysis of $Ru_3(CO)_9BH_5$[13] as illustrated in Figure 5. Reactions which illustrate boron–boron bond formation occurring at multimetal centers are the syntheses of $(CpTa)_2(B_2H_6)_2$ from $(CpTa)_2(\mu\text{-}X)_4$ (X = halide) and $[BH_4]^-$,[14] and of $HRu_3(CO)_9B_2H_5$ from $Ru_3(CO)_{12}$ and $BH_3 \cdot THF/Li[Et_3BH]$.[15] Cluster formation by the aggregation of originally mononuclear metal as well as boron precursors may involve concomitant B—B and M—M bond formation as exemplified in Figure 6[16,17]; related transformations occurring at mononuclear centers were provided in the preceding chapter.

Metalloborane and boride clusters of type M_nE_m ($n \geq 3$) fall into seven main structural categories (**1** to **7** in Figure 7) and the predominance of the

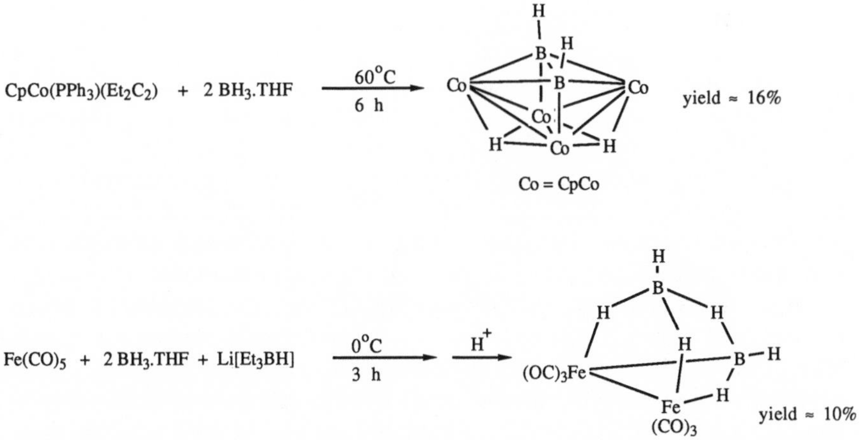

$CpCo(PPh_3)(Et_2C_2)$ + 2 BH_3.THF —60°C, 6 h→ yield ≈ 16%; Co = CpCo

$Fe(CO)_5$ + 2 BH_3.THF + Li[Et_3BH] —0°C, 3 h→ —H^+→ yield ≈ 10%

Figure 6. Homologation of $BH_3 \cdot THF$ accompanying metal–metal bond formation during cluster aggregation.

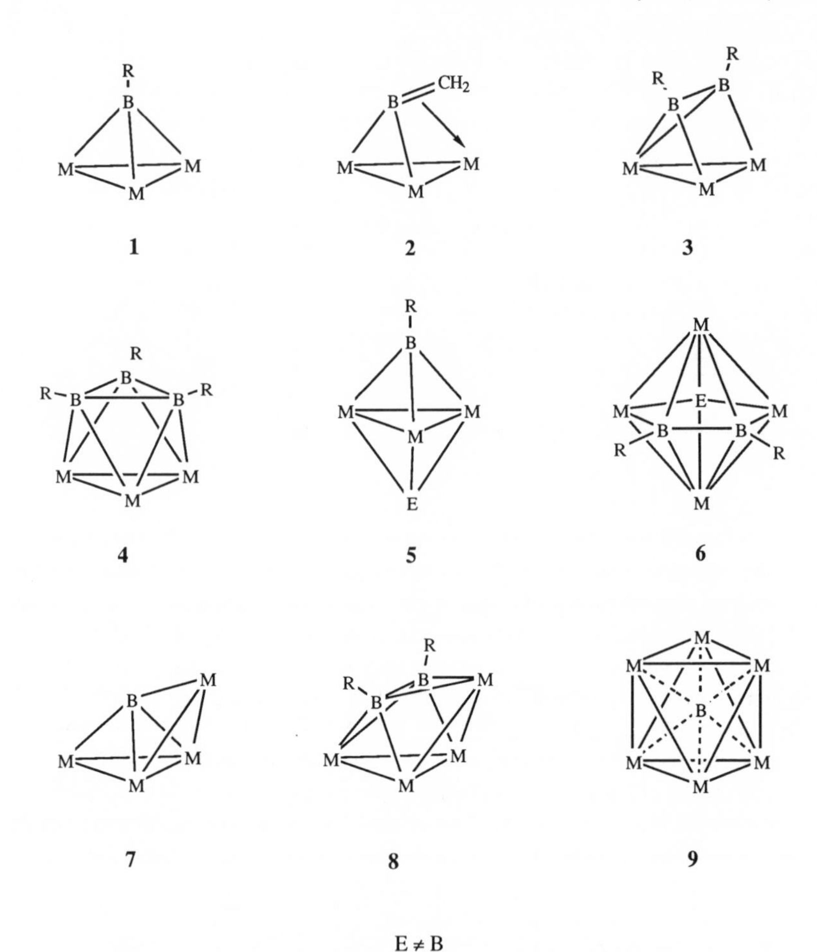

Figure 7. Structural types for clusters containing M–B interactions.

M_3-triangle and the M_4-butterfly frameworks is clearly apparent. Apart from the introduction of gold in the form of *endo*-gold(I) phosphine fragments, clusters containing M–B interactions are restricted to metals from the iron and cobalt triads. Selectivity in synthetic routes is difficult to achieve; for example, the reaction of $Fe(CO)_5$ with $BH_3 \cdot THF$ and $Li[Et_3BH]$ followed by acidification leads to $Fe_2(CO)_6B_2H_6$ in addition to the type **1** cluster $HFe_3(CO)_9BH_3R$ (R = Me, Et) and hydrocarbyl clusters $H_3Fe_3(CO)_9CR$ (R = H, Me,[20–22] Et), with ratios of products being critically dependent upon exact reaction conditions.[17,18] Addition of $Fe_2(CO)_9$ to the reaction mixture encourages the formation of the type **7** cluster, $HFe_4(CO)_{12}BH_2$.[19] However, the unsubstituted cluster $HFe_3(CO)_9BH_4$ eludes this particular reaction and the conjugate base is, instead, best prepared by treating $[(CO)_4FeC(O)Me]^-$ with $BH_3 \cdot THF$.[18] The aforementioned triiron products form isoelectronic pairs (Figure 8) and the striking feature that differentiates the ferraboranes from their organometallic analogs is the location of the *endo*-hydrogen atoms. The observed trend, brought about in part by the contraction of the tangential $2p$ atomic orbitals of the capping atom E, is that Fe–H–B interactions are favored over Fe–H–Fe, but Fe–H–Fe is preferable to Fe–H–C.[23] Allowing for the presence of one extra *endo*-H in the ferraborane, the anions $[Fe_3(CO)_9EH_3]^-$ (E = C or BH) possess comparable structures in solution (Figure 9). Upon protonation, the kinetic product in *each* case exhibits an additional Fe–H–E interaction. For E = BH, the kinetic product is also the thermodynamic one (although at room temperature the four *endo*-hydrogen atoms are fluxional on the NMR timescale),[18] but for E = C, proton migration from Fe–H–E to Fe–H–Fe sites occurs to give $H_3Fe_3(CO)_9CH$ (Figure 9).[24]

In contrast to the methods described for the triiron systems, syntheses of triruthenium and triosmium boron-capped clusters use precursors in which the metal triangle is already present. $Ru_3(CO)_9BH_5$ is a major product in the reaction of $Ru_3(CO)_{12}$ with $BH_3 \cdot THF$ and $Li[Et_3BH]$.[25] In solution, $Ru_3(CO)_9BH_5$ exists as a mixture of two isomers, **A** and **B**, which differ from one another only in the location of one *endo*-hydrogen atom (right-hand side of Figure 5). At room temperature and in hexane solution **A** and **B** are present in approximately equal amounts.[25] The ^1H NMR signature of **A** is consistent with this isomer being isostructural with $HFe_3(CO)_9BH_4$, the structure of which has been crystallographically confirmed.[18,26] The process that interconverts **A** and **B** at $T > 373$ K is presumably similar to that which exchanges *endo*-H atoms in $HFe_3(CO)_9BH_4$ at $T \geq 353$ K.[18,26] Interestingly, the osmaborane $Os_3(CO)_9BH_5$ has not, as yet, been reported; the first example of a type **1** cluster containing direct Os—B bonds was $H_3Os_3(CO)_9BCO$, formed

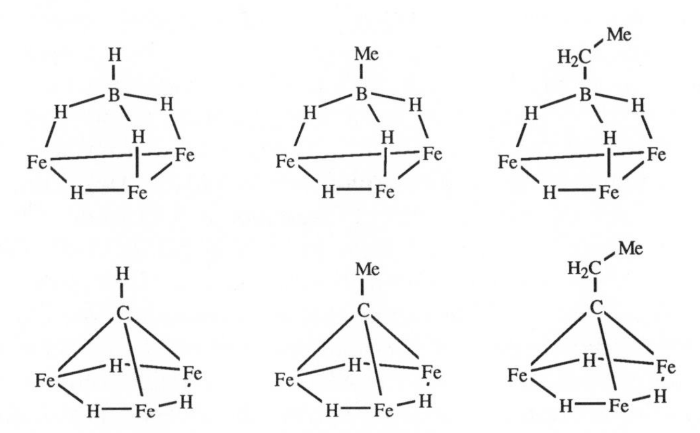

Figure 8. Structures of isoelectronic pairs of Fe_3E clusters where E = B or C.

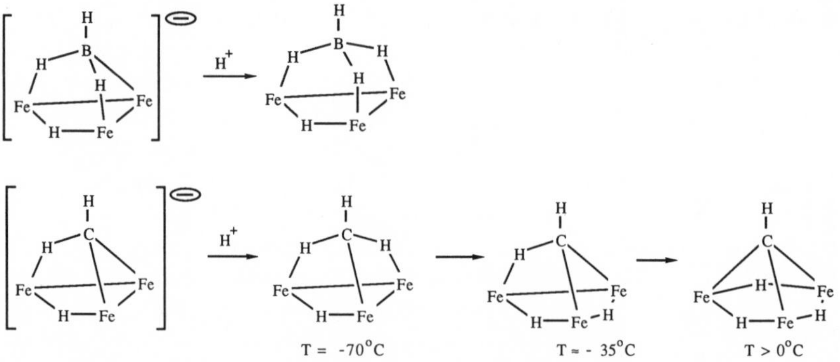

Figure 9. Protonation of $[Fe_3(CO)_9EH_3]^-$ (E = C or BH): kinetic and thermodynamic products.

when the unsaturated triosmium hydride $H_2Os_3(CO)_{10}$ is treated with B_2H_6 in the presence of $BH_3 \cdot NEt_3$.[27] The aminoborane adduct plays a critical role in the formation of the boron-capped cluster. As Figure 10 illustrates, replacement of amine by THF swings the reaction pathway in favor of a carbon-capped system, but even more intriguing is the fact that three of the $\{H_3Os_3(CO)_9CO\}$ units are coordinated to a central boroxine ring.[27,28] This system provides a useful starting point for the synthesis of several triosmium methylidyne clusters (Figure 10).[28]

Structural details and the reactivity of $H_3Os_3(CO)_9BCO$ are of interest particularly when compared with those of $[H_xM_3(CO)_9CCO]^{2-x}$ (M = Fe, x = 2; M = Ru, x = 0, 1, 2; M = Os; x = 0, 2).[29–39] The BCO vector in $H_3Os_3(CO)_9BCO$ lies approximately perpendicular to the Os_3-plane[27] and mimics the orientation of the capping CCO (ketenylidene)[40] group in $H_2Os_3(CO)_9CCO$.[30] A comparative study of the bonding in these two clusters indicates that the BCO fragment interacts more strongly with the triosmium framework than does the CCO group.[29] Crystallographic characterization of the dianions $[M_3(CO)_9CCO]^{2-}$ (M = Fe,[31] Ru,[32] Os[33]) reveals a structural variation (Figure 11) which significantly affects the reactivity of these clusters.[32–35,38,39,41] Analysis of the bonding in these anions shows that a change in the geometry of the carbonyl ligands in going from M = Ru to M = Fe or Os redirects the frontier orbitals of the $\{M_3(CO)_9\}$-framework in such a way as to encourage the tilting of the capping CCO group. This in turn exposes the α-carbon atom and leads to each of $[Fe_3(CO)_9CCO]^{2-}$ and $[Os_3(CO)_9CCO]^{2-}$ exhibiting a quite different pattern of reactivity from that of $[Ru_3(CO)_9CCO]^{2-}$.[33] The capping ketenylidene group provides a wealth of chemistry and derivatives of the family of clusters $[H_xM_3(CO)_9CCO]^{2-x}$ include methylidyne, alkylidyne, vinylidene, and acetylide compounds.[33,40] Related to this are transformations involving the tricobalt supported ketenylidene, $[Co_3(CO)_9CCO]^+$,[42,43] and although the structurally similar cluster

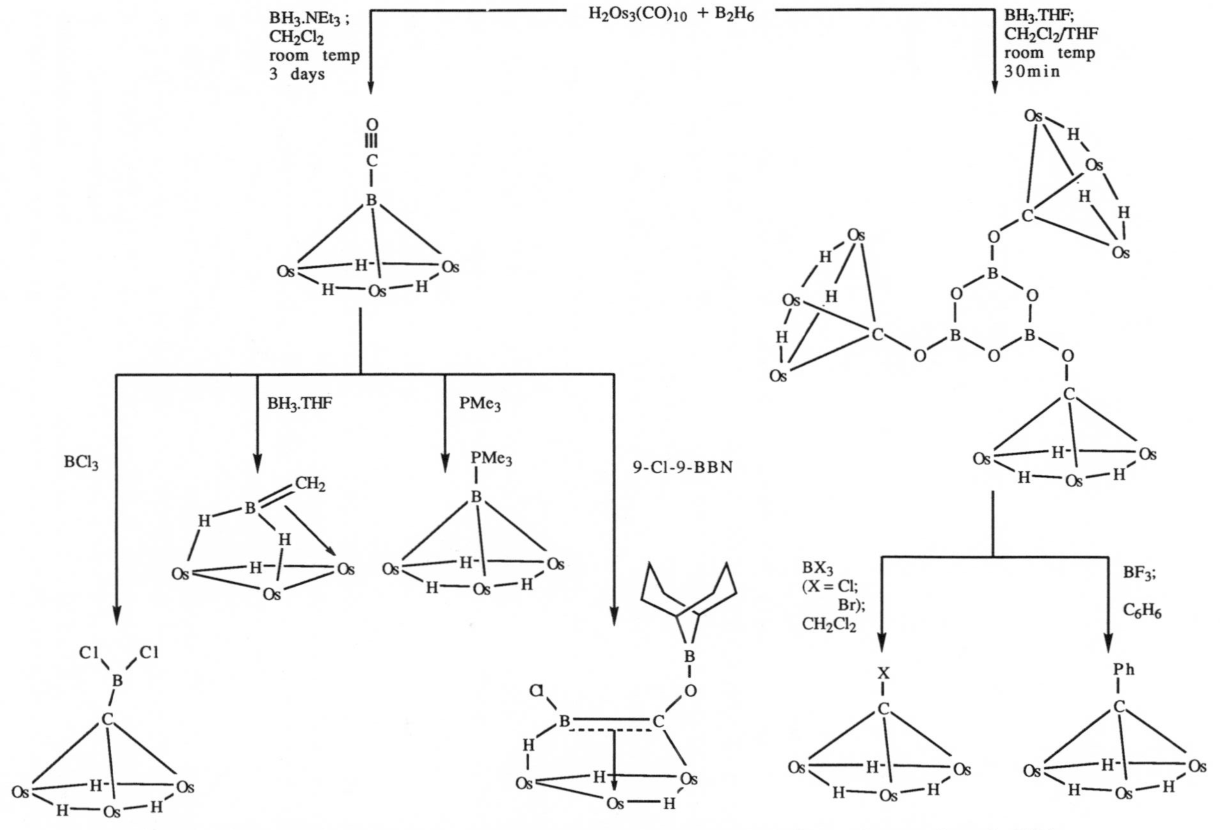

Figure 10. Syntheses and reactions of triosmium borylidyne and methylidyne clusters; Os = $Os(CO)_3$.

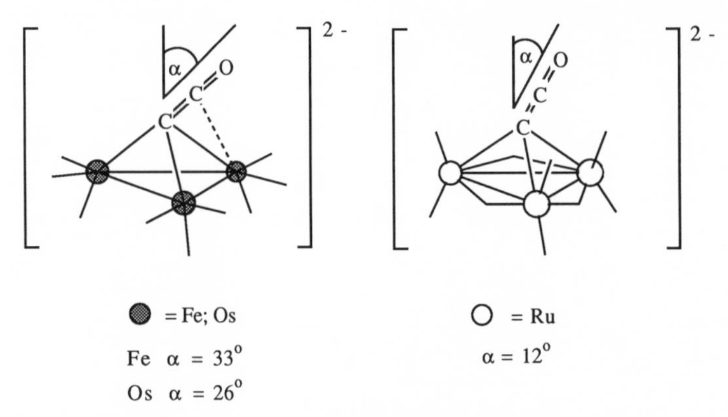

Figure 11. Structures of the series of ketenylidene dianions $[M_3(CO)_9CCO]^{2-}$ (M = Fe, Ru, Os).

$Co_3(CO)_9BNEt_3$ has been reported, no detailed information about its properties is available.[44] The M_3CCO-unit is also a useful building block for cluster expansion reactions, in particular the synthesis of mixed metal carbido-clusters.[38,39] It is reasonable to suppose that $H_3Os_3(CO)_9BCO$ should exhibit a rich chemistry, and indeed this is unfolding as detailed below.[27,45,47–50] Other related boron-containing systems of the type $[H_xM_3(CO)_9BCO]^{3-x}$ (M = Os, x = 1–3; M = Fe, Ru, x = 0–3) may be postulated but are, to date, unexplored.

The relative reactivity of boron- and osmium-bound CO in $H_3Os_3(CO)_9BCO$ has been tested in the reaction with trimethylphosphine; quantitative displacement of CO from boron is observed[27,45] and the product, $H_3Os_3(CO)_9BPMe_3$, has been crystallographically characterized.[45] By way of comparison, the reaction of $[HFe_3(CO)_9BH_3]^-$ with PMe_2Ph illustrates that the competition of boron vs metal atom as a site of Lewis base attack is strictly dependent upon the concentration of phosphine. At low $[PMe_2Ph]$, substitution of an iron attached carbonyl group occurs while at high $[PMe_2Ph]$ the cluster is degraded with formation of $PhMe_2P \cdot BH_3$.[46]

Three further transformations of the μ_3-BCO group in $H_3Os_3(CO)_9BCO$ are shown in Figure 10. Reaction with BCl_3 induces an exchange of the B and C atoms of the original capping BCO and generates a μ_3-$CBCl_2$ cap. Although tilted through 15° from vertical, the C–B vector does not lean far enough to enable the boron atom to interact with the osmium triangle.[47] Thus, $H_3Os_3(CO)_9CBCl_2$ is still related to the type **1** cluster. On the other hand, in $H_3Os_3(CO)_9BCH_2$ (now boron rather than carbon capped), the tilt of the B–C vector is 60° and this is sufficient to produce a new bonding mode for the capping group[48] thereby leading to the type **2** cluster shown in Figure 7. $H_3Os_3(CO)_9BCH_2$ has been termed a "borylidene" cluster[48] and has an

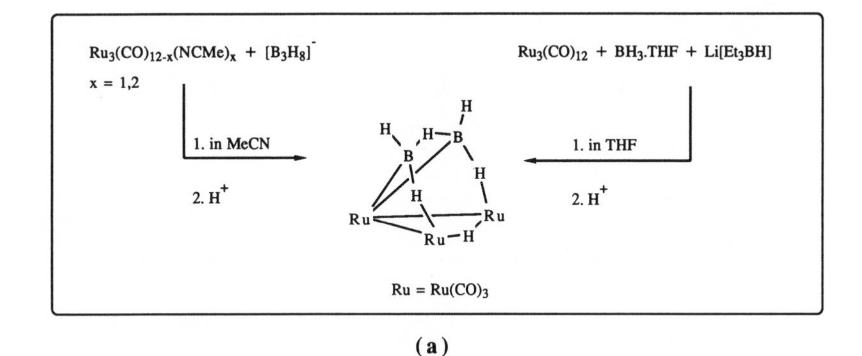

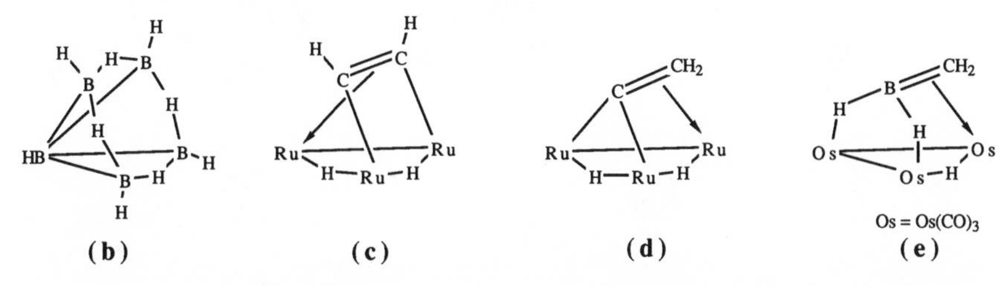

Figure 12. (a) Synthetic routes to and proposed structure of $HRu_3(CO)_9B_2H_5$; (b) structure of B_5H_9; (c) structure of $H_2Ru_3(CO)_9C_2H_2$ illustrating the "parallel" mode for the unsaturated hydrocarbon ligand; (d) structure of $H_2Ru_3(CO)_9C{=}CH_2$ illustrating the "perpendicular" mode (viz. vinylidene) for the hydrocarbon ligand; (e) structure of the "borylidene" cluster $H_3Os_3(CO)_9BCH_2$.

isoelectronic counterpart in $H_2Os_3(CO)_9CCH_2$,[51,52] a cluster in which the triosmium platform supports a vinylidene fragment, in the so-called "perpendicular mode" (Figure 12d). In terms of reactivity studies, the same is true for a comparison of $B{=}CH_2$ vs $C{=}CH_2$ as we saw for BCO vs CCO, namely, that while metal coordinated vinylidene ligands exhibit a rich chemistry,[53] that of the $B{=}CH_2$ group remains to be detailed. The Os_3BC-core is also present in $H_3Os_3(CO)_9C(OBC_8H_{14})BCl$, but in this case the RCBR′ unit bonds in a "parallel" fashion to give a species derived from a type **3** cluster (Figures 10 and 12c).[50] The parallel and perpendicular modes are discussed further below.

The cluster $HRu_3(CO)_9B_2H_5$[15,54] exemplifies the type **3** system. This metalloborane is prepared via either the homologation of $BH_3 \cdot THF$ or the degradation of the $[B_3H_8]^-$ anion (Figure 12a) and may be viewed either as a metal-rich derivative of B_5H_9 (Figure 12b)[15] or as an inorganic analog of a metal supported unsaturated hydrocarbon (Figure 12c). As the former, $HRu_3(CO)_9B_2H_5$ follows as a member of a series of group 8 metalloboranes[55–58] which are formally derived from pentaborane(9) in which BH fragments are

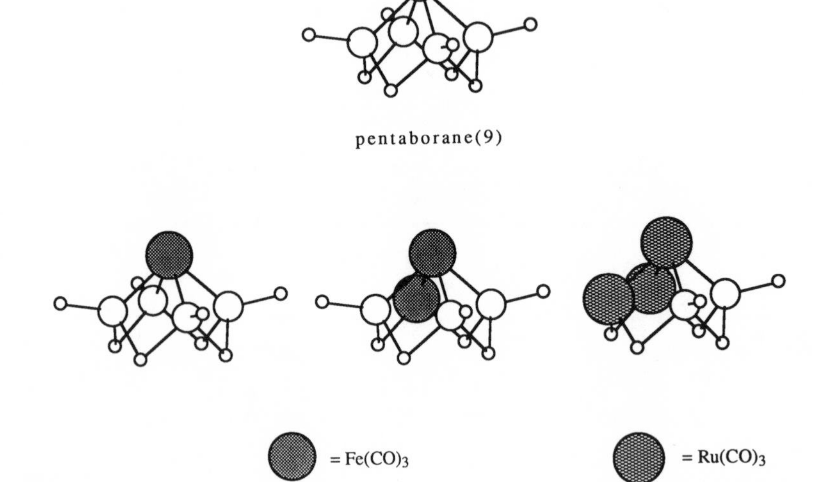

Figure 13. Isolobal relationship between *nido*-B_5H_9 and the metalloborane clusters $Fe(CO)_3B_4H_8$, $Fe_2(CO)_6B_3H_7$, and $HRu_3(CO)_9B_2H_5$.

successively replaced by isolobal $M(CO)_3$ (M = Fe or Ru) units as illustrated in Figure 13.[59,60] The structural relationship that exists between $HRu_3(CO)_9B_2H_5$ and $H_2Ru_3(CO)_9C_2H_2$ underlines the different bonding requirements of boron and carbon atoms. The organometallic cluster exists in two isomeric forms (the so-called "parallel" and "perpendicular" modes of bonding for the organic fragment as shown in Figures 12c and 12d),[52,61–63] only one of which is mimicked by the triruthenaborane.[15] This result illustrates the fact that each boron atom tends to populate a site in which it can use its three valence electrons to greatest effect, viz. two cluster vertices; the formation of a "$B{=}BH_2$" unit is clearly not favorable. Significantly, the borylidene cluster $H_3Os_3(CO)_9BCH_2$, being intermediate in nature between $HRu_3(CO)_9B_2H_5$ and $H_2Ru_3(CO)_9C_2H_2$, chooses to adopt a structure analogous to the second isomer of the organometallic cluster (Figures 12d vs 12e) and with a μ_2,η^2-$B{=}CH_2$ in preference to a μ_2,η^2-$C{=}BH_2$ fragment. The presence of a terminal BX_2 group requires π-stabilization, and this is available in $H_3Os_3(CO)_9CBCl_2$ (Figure 10) and in the type 7 cluster $HFe_4(CO)_{12}CBH_2$ (Figure 14).[64,65] In the former, the halide substituents provide π-electron density while in the latter, the cluster itself is the source; in the hypothetical "$H_3M_3(CO)_9B{=}BH_2$" no π-electrons could readily be donated to the terminal BH_2 unit. Like the B_2H_5-unit in $HRu_3(CO)_9B_2H_5$, the RCBCl-fragment in $H_3Os_3(CO)_9C(OBC_8H_{14})BCl$ chooses to adopt the "parallel" mode.[50]

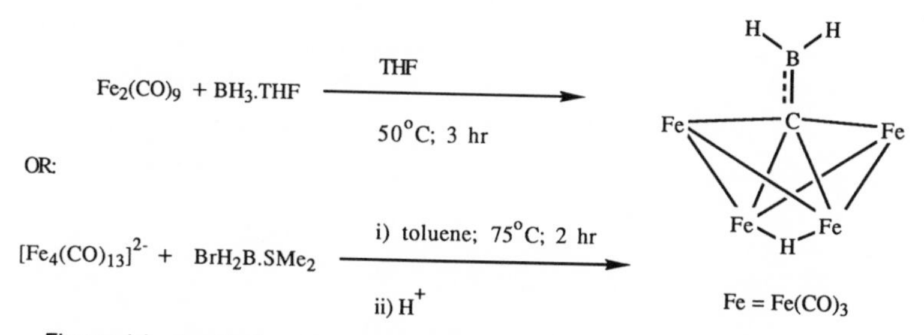

Figure 14. Synthetic routes to and structure of the cluster $HFe_4(CO)_{12}CBH_2$.

Deprotonation of $HRu_3(CO)_9B_2H_5$ occurs via loss of the Ru—H—Ru *endo*-hydrogen atom, thereby leaving the B_2H_5-fragment intact.[54]

Trimetallic frameworks support B_3-cluster fragments in addition to the B_n-units ($n = 1,2$) described above. Cluster type **4** is exemplified by 1,2,3-$(CpCo)_3B_3H_5$,[66–68] 1,2,3-$(CpCo)_3(\mu_3\text{-}CO)B_3H_3$ (Figure 15a)[69,70] and 1,2,3-$(CpCo)_2Fe(CO)_4B_3H_3$ (Figure 15b).[71] The first two of these compounds have been characterized crystallographically and the Co_3 and B_3-triangles are mutually staggered in each one. The metal–main group interaction may be considered in terms of a transition metal π-complex involving a cycloborenyl ligand and, as such, would be analogous to that in mononuclear cyclopropenyl complexes of the type $L_xM(\eta^3\text{-}C_3R_3)$.[72] Alternatively, the metalloboranes may be described as cluster compounds. The distinction between a complex and a cluster is not always clear-cut,[11] but in this case a cluster classification is more reasonable. A complex constructed upon an M_3-triangle would require 48 valence electrons. The $\{(CpCo)_3(H)_2\}$, $\{(CpCo)_3(\mu_3\text{-}CO)\}$, and $\{(CpCo)_2Fe(CO)_2(\mu\text{-}CO)_2\}$ fragments each contributes 44 electrons, leaving the $\eta^3\text{-}B_3H_3$ ligand (with assumed localized 2-center 2-electron B—B σ-bonds) to function as a 4π-electron donor, which, for this neutral fragment, is difficult to rationalize. On the other hand, PSEPT[73] requires that a closed octahedral cluster possesses seven pairs of bonding electrons. Each of the metalloboranes 1,2,3-$(CpCo)_3B_3H_5$, 1,2,3-$(CpCo)_3(\mu_3\text{-}CO)B_3H_3$, and 1,2,3-$(CpCo)_2Fe(CO)_4B_3H_3$ fulfills this requirement and each is thus more readily classified as a cluster rather than a complex. If classed as an inorganometallic *complex*, the related compound 1,2,3-$Fe_3(CO)_9B(H)C(H)C(Me)$, shown in Figure 15c, presents a similar electron-counting problem.[65] This chiral cluster is synthesized in $\approx 5\%$ yield from the reaction of $[Fe_4(CO)_{13}]^{2-}$ with $BrH_2B \cdot SMe_2$, the major product being $HFe_4(CO)_{12}CBH_2$ (Figure 14).

With the exception of a report of $Co_3(CO)_9BNEt_3$ and $Co_6(CO)_{16}B$,[44] the exploration of metal-rich cobaltaboranes containing direct Co–B interactions has been a recent development[10] and has provided two new metalloborane cluster structures,[74,75] namely, classes **5** and **6**. In each characterized

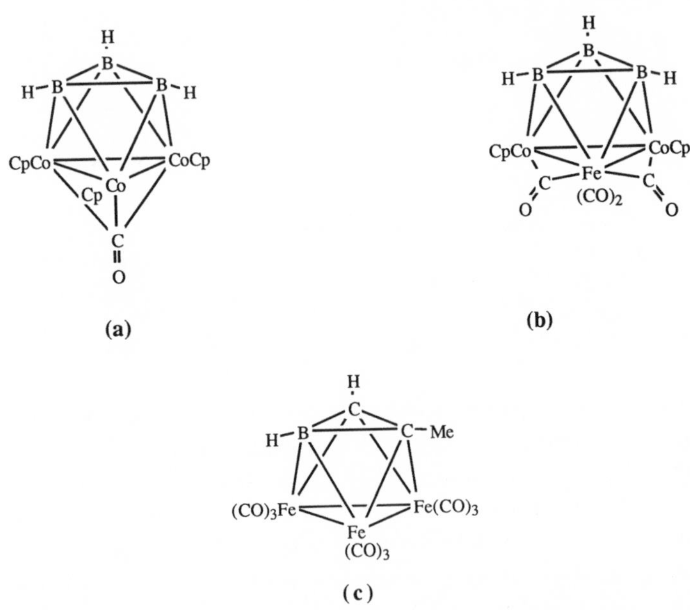

Figure 15. (a) Structure of 1,2,3-$(CpCo)_3(\mu_3\text{-}CO)B_3H_3$; (b) proposed structure of 1,2,3-$(CpCo)_2Fe(CO)_4B_3H_3$; (c) structure of 1,2,3-$Fe_3(CO)_9B(H)C(H)C(Me)$.

example, the second main group element labelled E in Figure 7 is a PPh group. Products formed from the combination of $CpCoL_2$ (L = two-electron ligand such as PPh_3 or C_2Et_2) and $BH_3 \cdot THF$ depend critically upon the conditions under which the reaction is carried out.[10,16,74–76] By reacting $CpCo(PPh_3)_2$ with $BH_3 \cdot THF$ in a 1:2 molar ratio at ambient temperature, $(CpCo)_3(\mu_3\text{-}BPh)(\mu_3\text{-}PPh)$, structure type **5**, is formed in $\approx$7% yield.[74] Note that the borane precursor has a dual role: (i) phosphine abstractor and (ii) source of a cluster fragment.[10] With a reactant ratio of 1:2.5, aggregation of the cluster fragments proceeds further to give the type **6** cluster $(CpCo)_4(PPh)B_2H_2$.[75] These cobaltaboranes along with examples of closely related organometallic clusters[77,78] are shown in Figure 16.

Type **7** metalloborane clusters are presently restricted to tetrairon and tetraruthenium clusters. Although the structural characterization of $HFe_4(CO)_{12}BH_2$ (Figure 17a) gave unambiguous evidence for the involvement of an M_4-butterfly framework,[79,80] such a structure had previously been proposed from spectroscopic data for another member of the series, viz. $HRu_4(CO)_{12}BH_2$ (Figure 17b).[54] The presence of the Ru_4-butterfly framework has more recently been verified indirectly from the structural elucidation of $HRu_4(CO)_{12}Au_2(PPh_3)_2B$ (Figure 17c).[81] At this point it is instructive to digress from the main theme of this section in order to examine the concept and application of another aspect of the isolobal analogy.

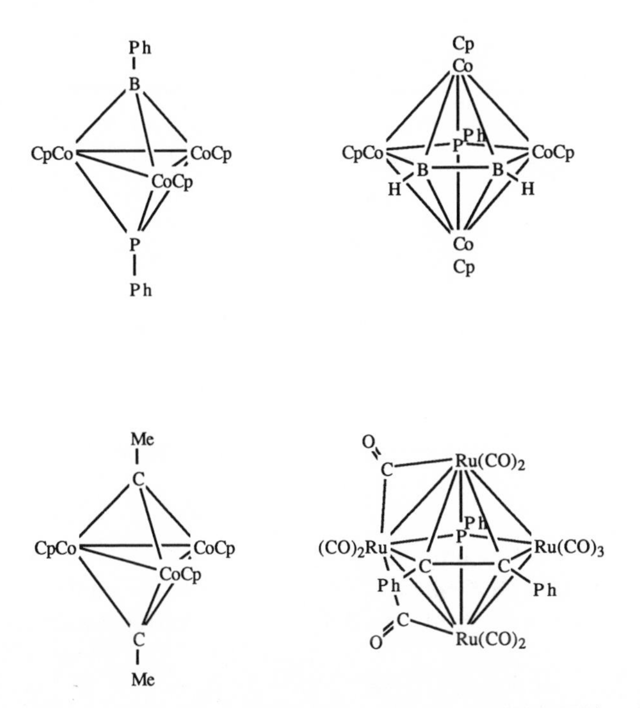

Figure 16. Structures of the *closo*-cobaltaboranes $(CpCo)_3(\mu_3\text{-}BPh)(\mu_3\text{-}PPh)$ and $(CpCo)_4(PPh)B_2H_2$ and isoelectronic or pseudo-isoelectronic organometallic counterparts.

A gold(I) phosphine fragment is formally isolobal with a proton, having a single frontier orbital of σ-symmetry which has similar bonding requirements to the 1s AO of the proton.[82–85] Replacement in clusters of *endo*-hydrogen atoms by $AuPR_3$ fragments often gives a derivative, the structure of which replicates that of the parent cluster. However, the tendency for the gold atoms to aggregate together[83] may override a strict isostructural analogy[86–88]; this point is neatly made in a study of clusters in the series $Fe_4(CO)_{12}BAu_x(PR_3)_xH_{3-x}$ (x = 0–3) and discussed below.[86,87] Significantly though, the cluster *core* appears to be quite resistant to changes in the actual locations of the *endo*-H or $AuPR_3$ groups.[89] Hence, crystallographic characterization of a gold(I) phosphino-derivative of a transition metal cluster may provide convincing evidence for the geometry of the cluster core in the parent hydrido-species. The characterization of $HRu_4(CO)_{12}Au_2(PPh_3)_2B$, supported by a comparison of IR and multinuclear NMR spectroscopic data for members of the series $HRu_4(CO)_{12}Au_x(PPh_3)_xBH_{2-x}$ (x = 0–2), allows extrapolation to the parent $HRu_4(CO)_{12}BH_2$ and confirms a type 7 structure for these compounds.[81]

The chemistry of $HFe_4(CO)_{12}BH_2$ is now quite well developed and se-

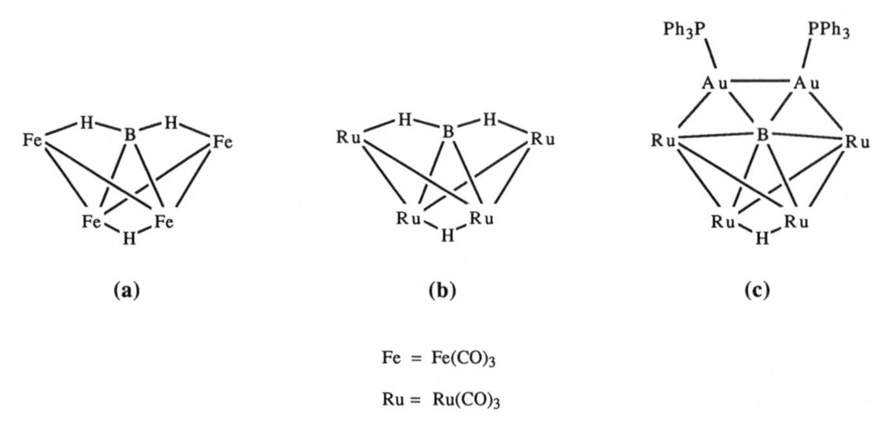

Figure 17. Structures of (a) $HFe_4(CO)_{12}BH_2$; (b) $HRu_4(CO)_{12}BH_2$; and (c) $HRu_4(CO)_{12}Au_2(PPh_3)_2B$.

lected reactions are summarized in Figures 18 and 19. Synthesis of $HFe_4(CO)_{12}BH_2$ is achieved nonselectively by the reaction of $Fe(CO)_5$, $Fe_2(CO)_9$, $BH_3 \cdot THF$, and $Li[Et_3BH]$[90] or by treatment of $Fe_2(CO)_6B_2H_6$ with $Fe_2(CO)_9$ in pentane.[79] Its conjugate base is formed quantitatively in the designed expansion of $[HFe_3(CO)_9BH_3]^-$ using $Fe_2(CO)_9$ as a source of the additional $Fe(CO)_3$ fragment; this reaction is driven by the expulsion of H_2 gas (Figure 20).[80,91] Although the molecular formula $HFe_4(CO)_{12}BH_2$ tends to suggest that this compound is a metallo*borane,* the environment of the boron atom is unusual since it carries no *exo*-hydrogen atoms. Analysis of the bonding in $HFe_4(CO)_{12}BH_2$, carried out at the Fenske–Hall level,[80] indicates that each hydrogen atom interacts directly with the metal butterfly core; i.e., they are truly *endo* and thus the compound may be classed as a metallo*boride.*[79,80] Not surprisingly then, the bonding in $HFe_4(CO)_2BH_2$ may be compared with that in related butterfly carbide clusters.[92,93] Complete exposure of the boron atom is achieved through deprotonation to the trianion $[Fe_4(CO)_{12}B]^{3-}$ (Figure 18),[94] a cluster[95] which has a structurally characterized isoelectronic carbido analog, $[Fe_4(CO)_{12}C]^{2-}$.[96,97] Phosphine substitution of up to two carbonyl ligands in the monoanion $[HFe_4(CO)_{12}BH]^-$ occurs readily and causes concomitant rearrangement of the *endo*-hydrogen atoms.[90,98] Expansion of the metal cage leads to the conversion of an exposed boron atom (type **7** cluster) to an encapsulated one and a true metalloboride cluster (type **9**) as illustrated by the reaction of $[HFe_4(CO)_{12}BH]^-$ with a source of "$\{Rh(CO)_2\}^+$" fragments to give $[Fe_4Rh_2(CO)_{16}B]^-$ (Figure 18).[99] Closely related organometallic compounds are $[Fe_6(CO)_{16}C]^{2-}$ [100] and $[Fe_3Rh_3(CO)_{15}C]^-$.[101] Structure type **9** is also exemplified by the boride $HRu_6(CO)_{17}B$ (Figure 5)[12] and, presumably, by its conjugate base $[Ru_6(CO)_{17}B]^-$.[102] The former has been crystallographically character-

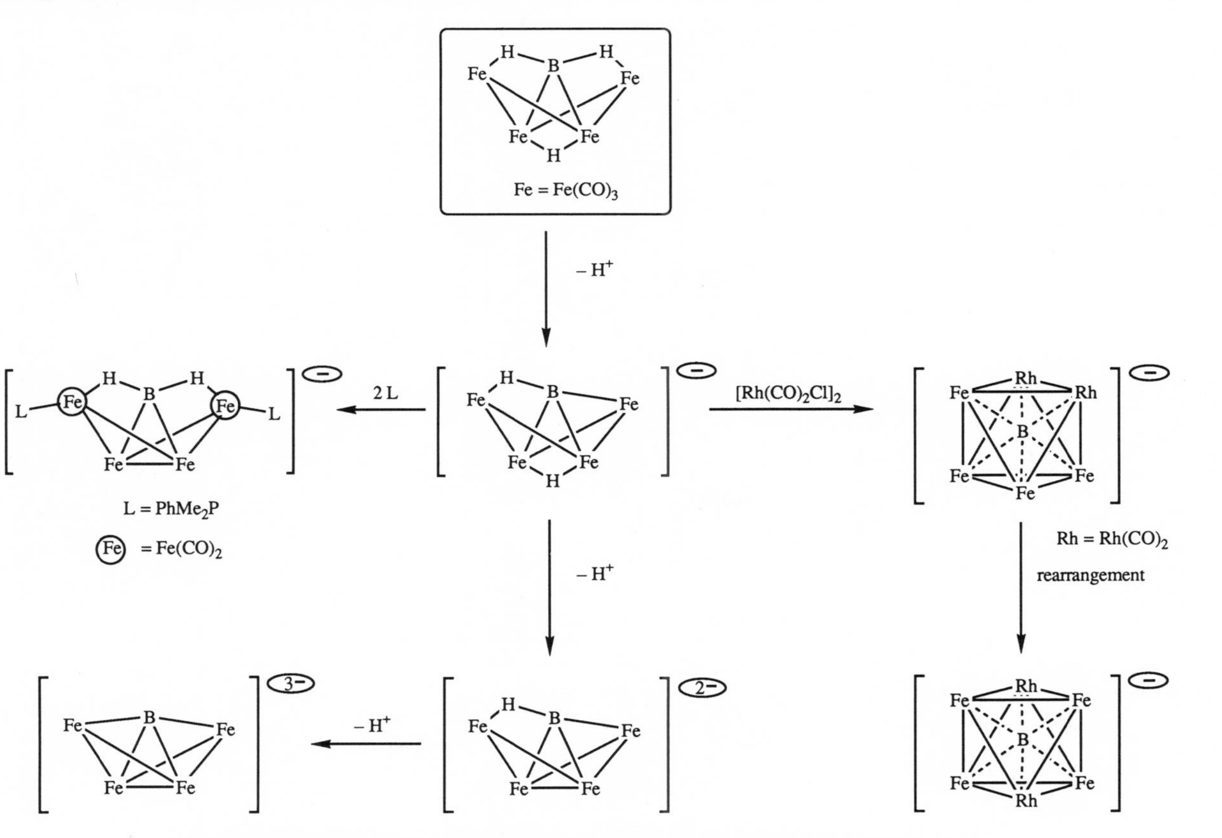

Figure 18. Deprotonation of $HFe_4(CO)_{12}BH_2$ and selected reactions of the monoanion.

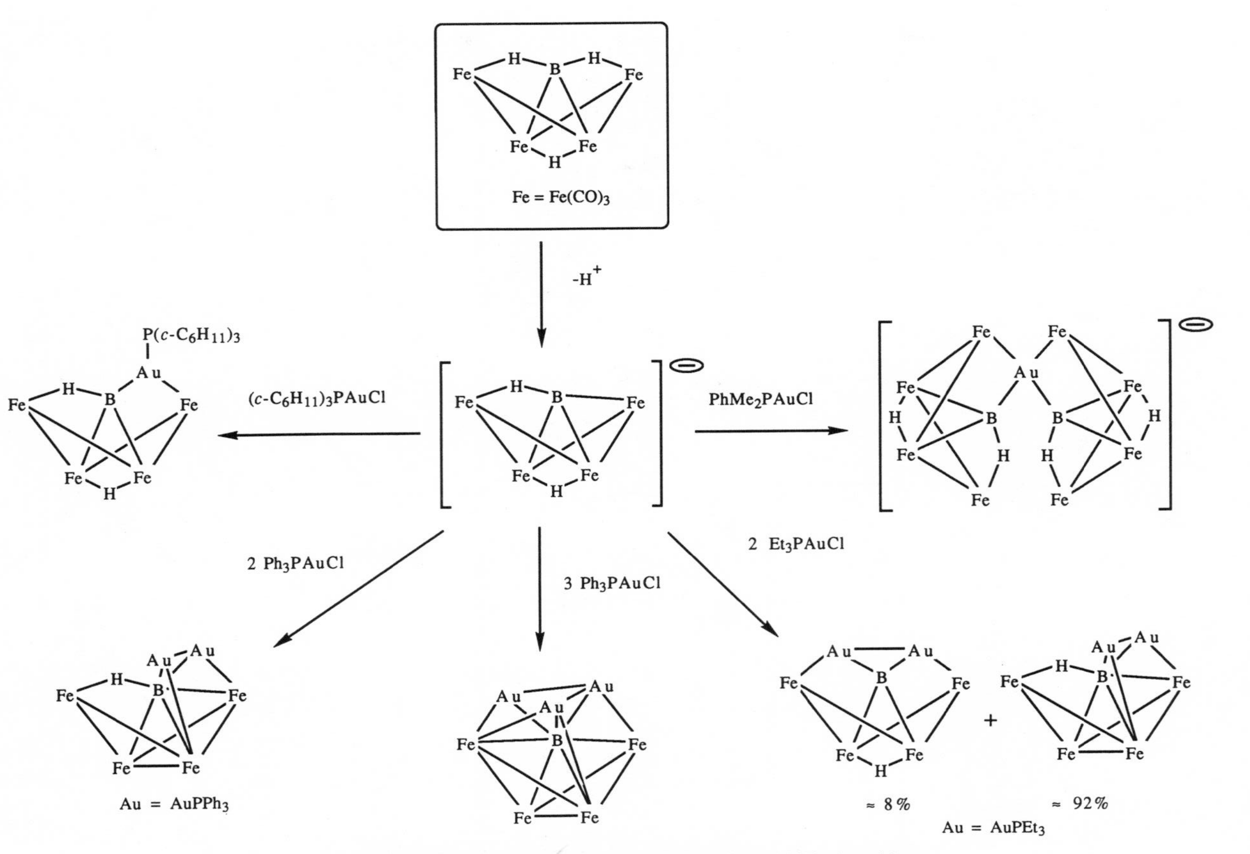

Figure 19. Reactions of $[HFe_4(CO)_{12}BH]^-$ with gold(I) phosphines.

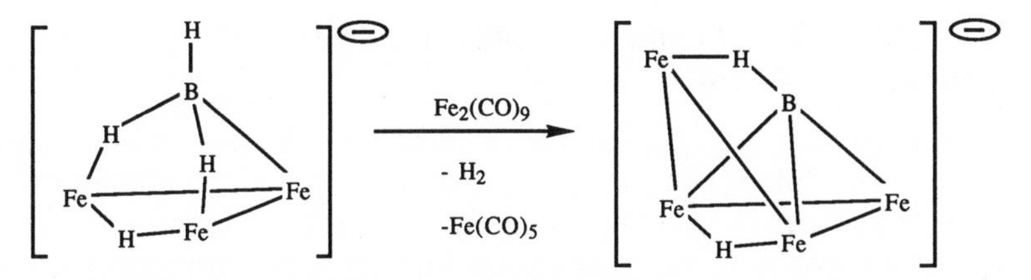

Figure 20. Expansion of $[HFe_3(CO)_9BH_3]^-$ to $[HFe_4(CO)_{12}BH]^-$.

ized and exhibits a structure which is directly analogous to that of the isoelectronic $Ru_6(CO)_{17}C$.(103)

The reactions of $[HFe_4(CO)_{12}BH]^-$ with gold(I) phosphine chlorides provide novel auraferraboranes,(86,87,89,104–107) each of which retains the original Fe_4B-cluster core (Figure 19): a fascinating structural variation is achieved by varying the phosphine substituent and/or the molar ratio of cluster anion: metal halide. The two isomers illustrated in Figure 19 (bottom right) for $Fe_4(CO)_{12}Au_2(PEt_3)_2BH$ interconvert in solution; the isomer distribution is controlled by the steric requirements of the PEt_3 group and may be altered by substituting a different phosphine. A small phosphine favors the more symmetrical structure (left-hand isomer in Figure 19).(89,104) So far, reactivity studies of the borides $[Fe_4(CO)_{12}BH_x]^{x-3}$ have given some exciting results as Figures 18 and 19 testify and, if isoelectronic relationships can act as a guide, then reactions(108–113) of the butterfly carbides $[Fe_4(CO)_{12}CH_x]^{x-2}$ indicate that a wealth of chemistry awaits discovery. Differences between boron and carbon containing butterflies are to be expected as the reaction of $HRu_4(CO)_{12}BH_2$ with PhC≡CPh illustrates.(114) This reaction, shown in Figure 47, will be discussed further in Section 4.1.

The interaction of an acetylene with a transition metal butterfly framework gives rise to homo- and heterometallic clusters such as $Co_4(CO)_{10}$-C_2H_2,(115) $Ru_4(CO)_{12}C_2Ph_2$,(116) $Ru_3Fe(CO)_{12}C_2Ph_2$,(117) and $[CpWRu_3$-$(CO)_{10}C_2Me_2]^-$.(118) Replacing C by B gives rise to the type **8** cluster and this class is represented by $H_2(CpCo)_4B_2H_2$ (Figure 6).(16) Although both are supported by an M_4-butterfly skeleton, a boron atom in a type **8** cluster is quite distinct from one in cluster type **7** since the former bears a terminal hydrogen atom (metalloborane) and the latter, as we have seen, is boridic in nature.

Before leaving the metal–boron bond, it is worth mentioning the role that ^{11}B NMR spectroscopy has played in the characterization of metalloborane and boride clusters. As the degree of metal–boron bonding increases, the ^{11}B NMR resonance shifts to progressively lower field. The chemical shift for an ^{11}B nucleus is extremely sensitive to its environment, and a parameterized model for calculating values of δ ^{11}B in ferraborane compounds has been developed.(119,120)

2.2. Clusters Containing Metal–Aluminum, Gallium, or Indium Bonds

The chemistry of M_nE_m clusters in which E = Al, Ga, or In is indeed sparce.[121] The failure of such compounds to appear in the literature does not seem to reside in the fact that aluminum, gallium, and indium containing groups are unable to function as cluster fragments. All three elements are known to be incorporated into borane or carbaborane clusters, for example, $B_4C_2H_6EMe$ (E = Ga, In)[122] and $B_9C_2H_{11}E(Et)$ (E = Al,[123–125] Ga[123]). In each compound, the ER group interacts with the open face of the *nido*-carbaborane cage. An interesting indication of the ability of the heavier group 13 elements to form clusters is the aggregation of InCp* units to form the closed, octahedral cage $Cp^*_6In_6$. Indium(I) chloride reacts with LiCp* in diethyl ether to give Cp*In; this crystallizes as the hexameric clusters, but a volatile nature implies only a limited stability.[126] A comparison of $Cp^*_6In_6$ with *closo*-$[B_6H_6]^{2-}$ has been made; the latter obeys PSEPT but the former, like the halogenated boranes B_6X_6, is satisfied with only six (instead of seven) pairs of bonding electrons.

Type **1** clusters in which the boron atom has been replaced by a heavier group 13 congener have been reported; tetrahedral "$Co_3(CO)_{12}Al$" and "$Co_3(CO)_{12}Ga$" have been proposed[127,128] as products of the reactions of $AlCl_3$ or $GaCl_3$ with $[Co(CO)_4]^-$. However, closed cage structures are not consistent with a consideration of PSEPT. It has been suggested that open structures of the type $E\{Co(CO)_4\}_3$ (E = Al, Ga) are more likely.[121] The capping mode of bonding has, however, been confirmed crystallographically in the tetrarhenium cluster $Re_4(CO)_{12}\{InRe(CO)_5\}_4$ shown in Figure 21a[129]; the analogous compound $Re_4(CO)_{12}\{GaRe(CO)_5\}_4$ has been pro-

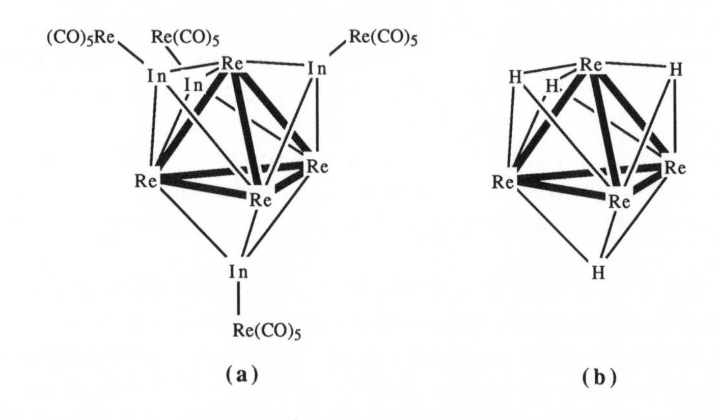

Figure 21. A comparison of the structures of (a) $Re_4(CO)_{12}\{InRe(CO)_5\}_4$ and (b) $H_4Re_4(CO)_{12}$.

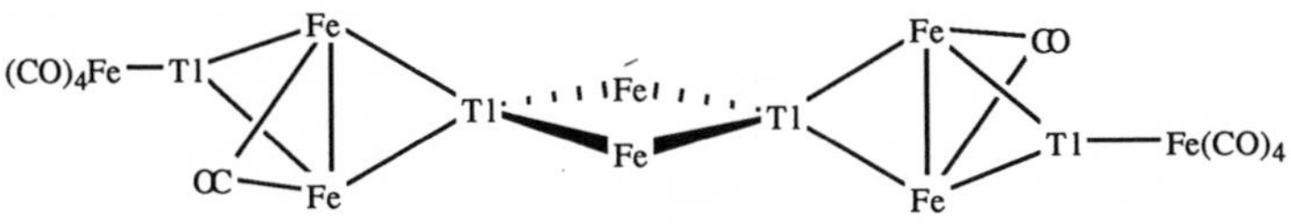

$[Tl_4Fe_8(CO)_{30}]^{4-}$

$[Tl_6Fe_{10}(CO)_{36}]^{6-}$

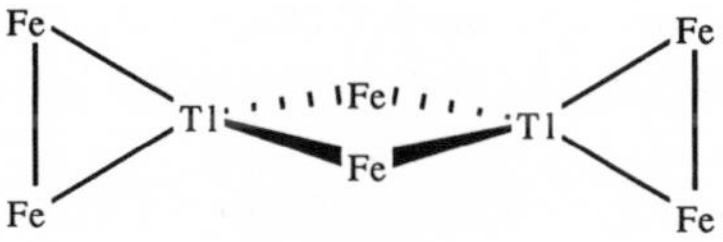

$[Tl_2Fe_6(CO)_{24}]^{2-}$

Figure 22. Some examples of the core structures of open iron–thallium clusters.

posed.[130] A structural relationship appears to exist between $Re_4(CO)_{12}$-$\{ERe(CO)_5\}_4$ (E = In, Ga) and $H_4Re_4(CO)_{12}$ (Figure 21b)[131] if the $ERe(CO)_5$-fragments are considered as replacing the *endo*-hydrogen atoms.

Open clusters (Figure 22) containing direct iron–thallium bonds have been synthesized over recent years,[121,132] but extention to closed clusters has yet to be reported.

3. GROUP 14: CLUSTERS CONTAINING METAL–SILICON, GERMANIUM, TIN, OR LEAD BONDS

Despite the extremely wide range of characterized organometallic cluster compounds, there are comparatively few clusters incorporating the heavy group 14 elements. Although the open clusters[133–135] shown in Figure 23 are not strictly within the cluster definitions laid out in Section 1, their structures are included for two purposes. First, the spirocyclic units are comparable with those observed in the iron thallium clusters illustrated in Figure 22. Second,

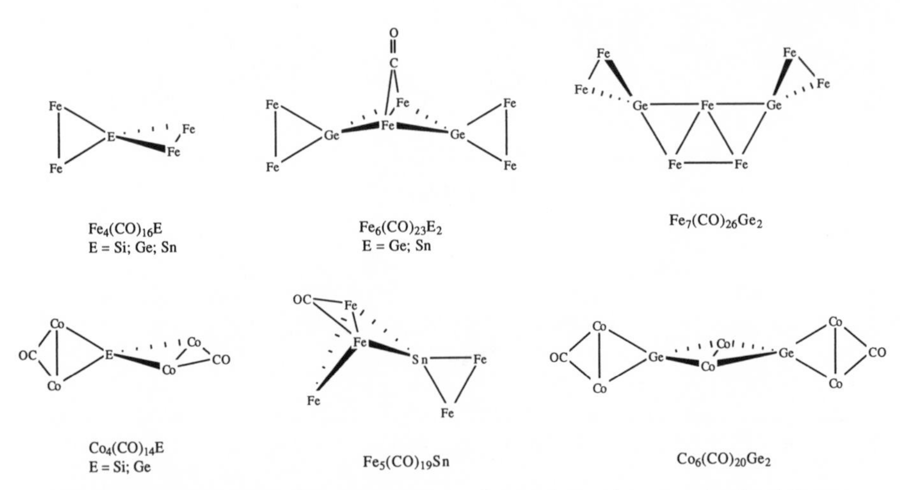

Figure 23. Some examples of the core structures of open clusters incorporating bonds between transition metals and group 14 elements.

condensation of some of these open frameworks to give closed species has been observed (see below), and this suggests that further transformations of this type are possible.

Clusters incorporating elements in group 14 from silicon downwards fall into the classes **10** to **17** illustrated in Figure 24. With the exception of **10** which is analogous to **1**, the structural types are different from those summarized for boron in Figure 7. Significantly, and presumably as a consequence of steric factors, these group 14 elements have not been observed as residing within the interstices of butterfly-M_4, square pyramidal-M_5, or octahedral-M_6 cages as, quite commonly, is the smaller carbon atom.[108,109]

Clusters **10** to **13** constitute a series, each member of which exhibits one or two μ_3-E fragments. In type **10** clusters, the capping group 14 atom interacts further through a localized bond to an alkyl or aryl group, R [e.g., $Co_3(CO)_9GeR$ (R = Me, ^{t}Bu, Ph),[136] $H_3CpMoCoFe(CO)_8Ge^tBu$[137]]. Thermolysis of the closed tetrahedral cluster $Co_3(CO)_9GeR$ under a pressure of carbon monoxide leads to $Co_2(CO)_7GeR\{Co(CO)_4\}$, an open molecule with a pendant $Co(CO)_4$ group attached to the now μ_2-Ge atom (Figure 25).[138] A similar framework is present in $Co_2(CO)_6\{\mu\text{-Ge(Me)Co}(CO)_4\}_2$.[139] Addition of phosphine to $Co_3(CO)_9GeR$ leads to either an addition or substitution product as shown in Figure 25.[138] In types **11** and **13** compounds, in addition to capping the M_3-triangle, the group 14 atom is bonded to a metal carbonyl or cyclopentadienyl fragment [e.g., $Co_3(CO)_9SiCo(CO)_4$,[140] $[Fe_3(CO)_{10}GeFe(CO)_4]^{2-}$,[141] $Co_3(CO)_9GeCo(CO)_4$,[142] $Co_3(CO)_9GeMn(CO)_5$,[143] $Co_3(CO)_9GeFe(CO)_2Cp$,[136] $Fe_3(CO)_9\{SnFe(CO)_2Cp\}_2$[144]]. Note that, due to the number of valence electrons, the *exo*-E—M bond contrasts with the dative E → M interactions

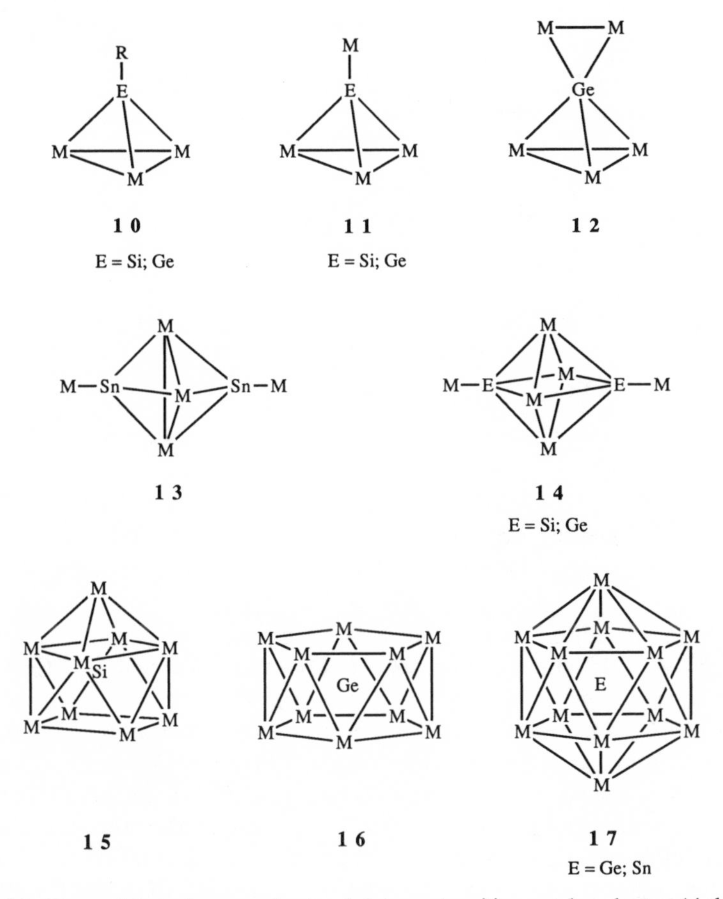

Figure 24. Types of cluster incorporating bonds between transition metals and group 14 elements from Si to Sn.

observed for capping group 15 or 16 atoms (see Sections 4 and 5). Consequently, the μ_3-Ge or Si atom can stabilize a 17-electron transition metal fragment while a μ_3-E (E = group 15 or 16) stabilizes a 16-electron unit. Synthetic approaches usually utilize the reactions of group 14 halides or hydrides with metal carbonyl anions or neutral carbonyl complexes as the selected reactions in Figure 26 demonstrate. Substitution of metal fragments in the basal metal triangle can occur without disruption of the transition metal–main group interaction, for example, reaction of $Co_3(CO)_9GeR$ reacts with $\{CpW(CO)_3\}_2$ to give $CpWCo_2(CO)_8GeR$.[137] Such reactions confirm the "clamping effect" that a main group fragment has with respect to its supporting transition metal framework.[145] In organometallic cluster chemistry, similar

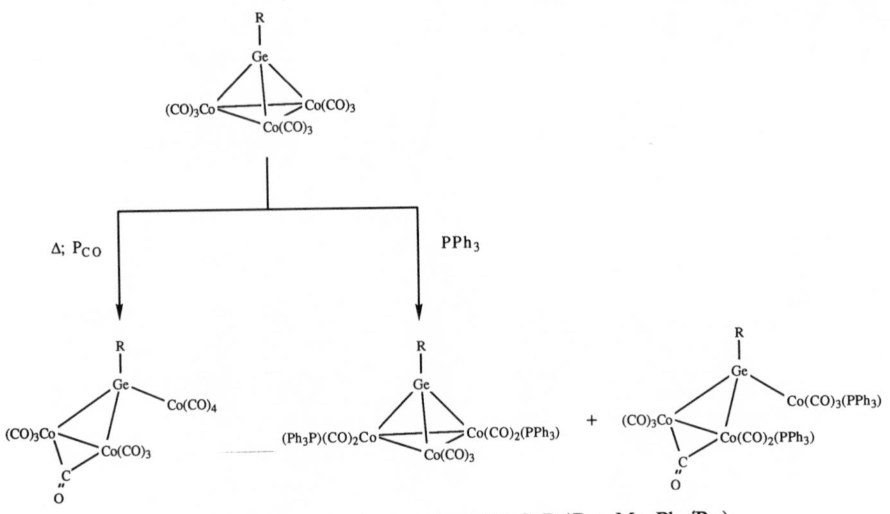

Figure 25. Some reactions of $Co_3(CO)_9GeR$ (R = Me, Ph, ^{t}Bu).

substitution reactions have been observed, one example being the reaction of $[Fe_3(CO)_9CCO]^{2-}$ (Figure 11) with $Co_2(CO)_8$ to give $[CoFe_2(CO)_9CCO]^-$.[146]

The reaction of Ge_2H_6 with $Co_2(CO)_8$ gives the open clusters $Co_4(CO)_{14}Ge$ and $Co_6(CO)_{20}Ge_2$. Pyrolysis of the latter at 60 °C results in decarbonylation and the formation of $Co_4(CO)_{11}\{GeCo(CO)_4\}_2$, a closed type **14** cluster, and (in trace amounts only) $Co_3(CO)_9GeCo(CO)_4$ (Figure 27).[147] Clearly, such condensation reactions have great potential in the synthesis of closed M_nE_m (E = Si, Ge, Sn) clusters. For silicon, a type **14** cluster is formed by the reaction of Si_2H_6 with $Co_2(CO)_8$; $Co_4(CO)_{11}\{SiCo(CO)_4\}_2$ is isostructural with the germanium cluster illustrated in Figure 27.[148]

The type **12** cluster, like a member of class **14**, contains a 5-coordinate germanium atom and has just one representative, namely, the anion $[Co_3(CO)_9GeCo_2(CO)_6]^-$.[149] In a sense, the main group atom is interstitial although the cage surrounding it is not continuous.

Clusters in which either tin or lead atoms or fragments interact with three or more transition metal atoms are quite rare, but are a source of current interest. Even the simple M_3E-tetrahedral core, common for most elements E, is represented in only one set of related compounds for E = Sn[150] and not at all for Pb. The tin containing compounds are shown in Figure 28 and are related to the type **10** cluster. Unlike most of the examples in this chapter, platinum is the constituent metal. The cluster cation $[(\mu\text{-}Ph_2PCH_2PPh_2)_3Pt_3(SnF_3)]^+$ is unusual both for its M_3Sn-core and for the presence of a metal–SnF_3 interaction.[150] The omission of any clusters exhibiting a closed M_3Pb-core is worth examination. An open molecule, $Pb\{Fe(CO)_4\}_3$ (Figure 29a),

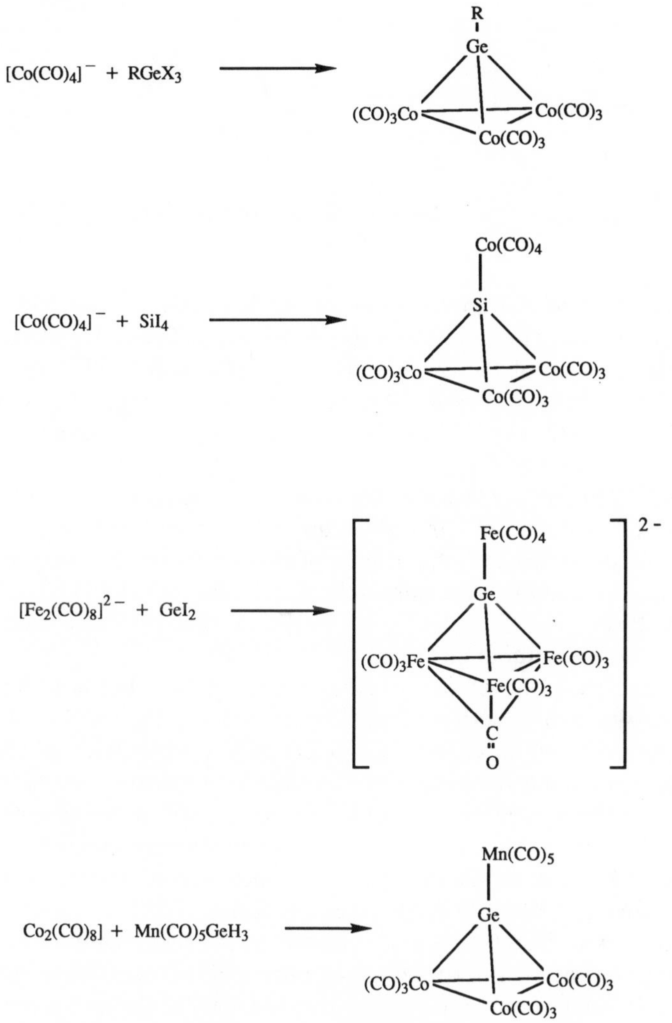

$[Co(CO)_4]^- + RGeX_3 \longrightarrow$ R–Ge / $(CO)_3Co$, $Co(CO)_3$, $Co(CO)_3$

$[Co(CO)_4]^- + SiI_4 \longrightarrow$ $Co(CO)_4$–Si / $(CO)_3Co$, $Co(CO)_3$, $Co(CO)_3$

$[Fe_2(CO)_8]^{2-} + GeI_2 \longrightarrow$ $[Fe(CO)_4$–Ge / $(CO)_3Fe$, $Fe(CO)_3$, $Fe(CO)_3$ / C=O$]^{2-}$

$Co_2(CO)_8] + Mn(CO)_5GeH_3 \longrightarrow$ $Mn(CO)_5$–Ge / $(CO)_3Co$, $Co(CO)_3$, $Co(CO)_3$

Figure 26. Selected syntheses of clusters of types **10** and **11**.

has been structurally characterized as has its tin analog and the environment around each main group atom is trigonal planar.[151] In considering if a closed cluster with an M_3Pb-core is a valid synthetic target, one should ask the question: is the lead atom too large to span a triangle of metal atoms, i.e., are the mutual requirements of the Pb—M and M—M bond lengths compatible with an M_3Pb tetrahedral core? In the light of related transition metal–bismuth

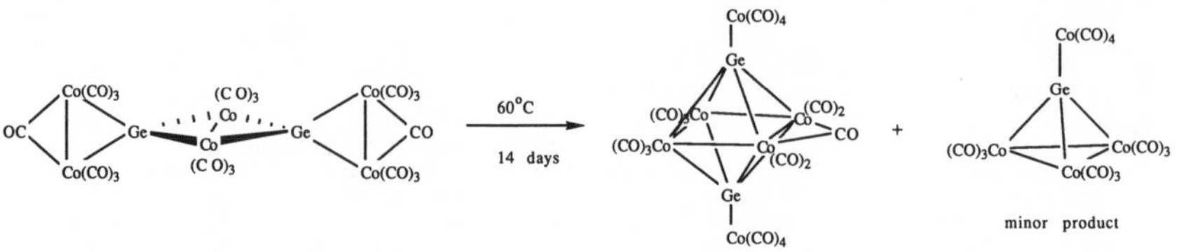

Figure 27. Condensation products formed from $Co_6(CO)_{20}Ge_2$.

chemistry, there would appear to be no steric problem, even with an M_3-platform comprising first-row transition metal atoms. Bismuth and lead atoms are similar in size; in the *closed* cluster $[Fe_3(CO)_{10}Bi]^-$ (Figure 29b and discussed more fully in Section 4.4), the Fe—Bi bond lengths lie in the range 2.648(2) and 2.652(1) Å[152] and these values compare with an average value of 2.625(5) Å in $Pb\{Fe(CO)_4\}_3$.[151] Thermolysis of the related $Bi\{Co(CO)_4\}_3$[153] (Figure 29c) leads to the closed $Co_3(CO)_9Bi$[154,155] (see Section 4.4). Thus, precedence for tetrahedral clusters incorporating large main group elements certainly exists. Some degree of cluster closure is observed during the reversible oxidation of the spirocyclic clusters $[Fe_2(CO)_8E\{Fe(CO)_4)\}_2]^{2-}$ (E = Sn, Pb) to $Fe_4(CO)_{16}E$ (Figure 30); note that the metal–metal bond formation occurs without CO loss.[141,156]

The encapsulation of silicon, germanium, and tin, but not lead, atoms within transition metal cages is feasible provided that the framework is large enough. In the dianionic cluster, $[Co_9(CO)_{21}Si]^{2-}$, which is a type **15** cluster, the silicon atom resides within a capped square antiprismatic cage but interacts essentially with only the eight cobalt atoms that constitute the antiprism (Figure 31).[157] Interestingly, the cluster is paramagnetic possessing 129 valence electrons; the characteristic count for a capped square antiprism is 130 as demonstrated by $[Rh_9(CO)_{21}P]^{2-}$ (see Section 4.2),[158] a cluster which is isostructural with $[Co_9(CO)_{21}Si]^{2-}$. Recently, the first examples of clusters containing completely interstitial germanium and tin atoms have been characterized. In each example, the cavity that holds these atoms is a pentagonal antiprismatic one. The reaction of $[Ni_6(CO)_{12}]^{2-}$ with $GeCl_4$ in THF leads to $[Ni_{12}(CO)_{22}Ge]^{2-}$ (type **17** cluster) with $[Ni_{10}(CO)_{20}Ge]^{2-}$ (type **16** cluster) as a secondary product.[159] On exposure to carbon monoxide, the ico-

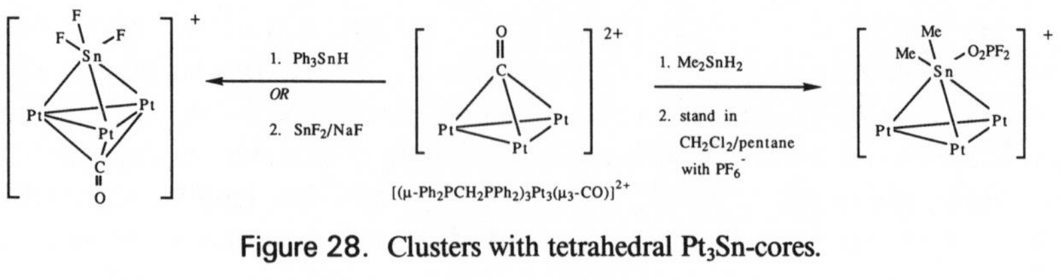

Figure 28. Clusters with tetrahedral Pt_3Sn-cores.

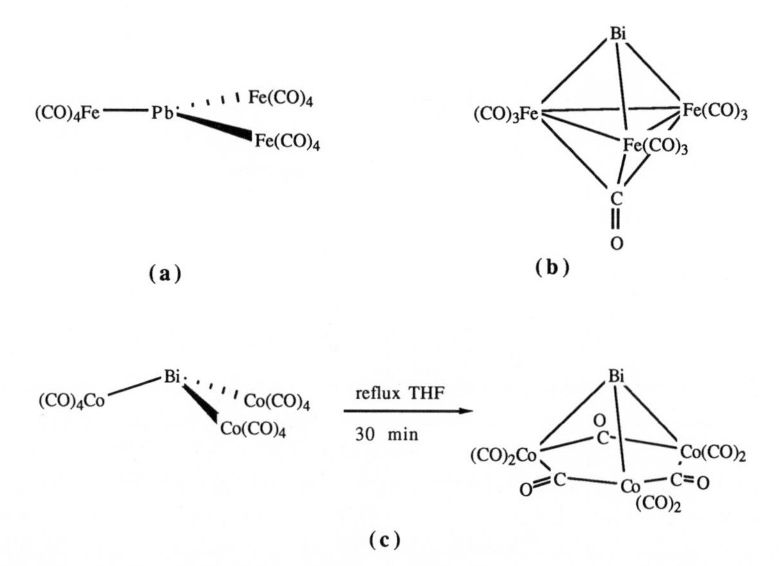

Figure 29. Relationships between ME_3-type clusters with first-row transition metal and large main group atoms: (a) synthesis and structure of $Pb\{Fe(CO)_4\{_3$; (b) structure of $[Fe_3(CO)_{10}Bi]^-$; (c) conversion of $Bi\{Co(CO)_4\}_3$ to $Co_3(CO)_9Bi$.

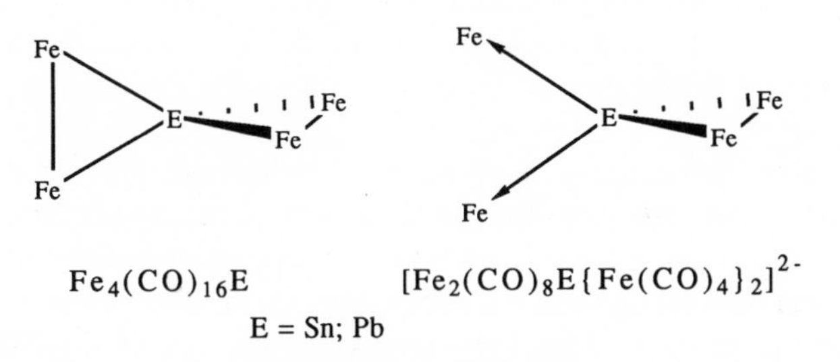

Figure 30. Reversible oxidation of $[Fe_2(CO)_8E\{Fe(CO)_4)\}_2]^{2-}$ (E = Sn, Pb).

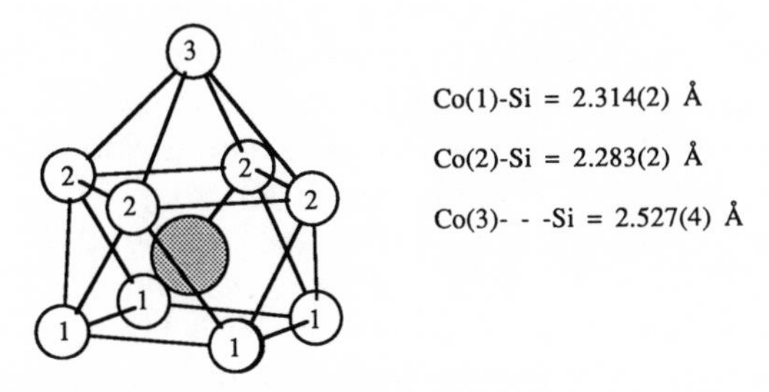

Figure 31. Structure of the cluster core of $[Co_9(CO)_{21}Si]^{2-}$.

sahedral cluster degrades to give $[Ni_{10}(CO)_{20}Ge]^{2-}$. Treatment of $[Ni_6(CO)_{12}]^{2-}$ with hydrated $SnCl_2$ gives $[Ni_{12}(CO)_{22}Sn]^{2-}$, but exposure to CO in this case fails to provide the type **16** species and causes degradation to lower-nuclearity clusters instead.[159]

4. GROUP 15

4.1. Clusters Containing Metal–Nitrogen Bonds

The chemistry of cluster coordinated nitrogen atoms and fragments containing the nitrogen atom has developed significantly in recent years.[121,160] Molecular clusters in which a metallic framework supports NH_x fragments represent models for surface species involved in, for example, the Haber process; cluster reactions facilitating N—H bond activation or N—N bond formation or cleavage are described in the following section. Figure 32 illustrates the cage types exhibited by transition metal cluster imides and nitrides and a comparison with the structures shown in Figure 7 illustrates some similarities between M_nN_m and M_nB_m clusters and, of course, between both of these and organometallic (M_nC_m) species. In the syntheses of metalloimides, the coordinated nitrosyl ligand is often a key starting material. Hence, in addition to studies of the product clusters, the mechanisms of the transition metal cluster mediated transformations of the NO group has relevance to the treatment of atmospheric pollutants to which nitric oxide is a major contributor.[161]

Like metalloboranes of type **1**, metalloimides of type **18** are less well documented than their organometallic analogs, viz. the alkylidyne clusters. The M_3N-core generally exists with the nitrogen atom bonded in a localized manner to a terminal group, e.g., hydrogen atom, alkyl or aryl group. The first such clusters to be characterized were $Fe_3(CO)_{10}NH$[162] and $Ru_3(CO)_{10}(NOH)$.[163] However, the availability of five valence electrons (and therefore the localization of one lone pair) does permit the formation of a partially open and planar M_3N-core as illustrated in $(^iPrO)_{12}Mo_4(N)_2$[164] and $Cp_3Mo_3(N)(O)(CO)_4$[165] (Figure 33). The latter molecule may be prepared along with related tungsten derivatives by the reaction of $CpM(CO)_2(NO)$ and $Cp_2M_2(CO)_6$ (M = Mo and W).[166] One important synthetic strategy that leads to type **18** clusters is the reduction of nitrosyl clusters. Figure 34 illustrates one example; both amido and imido clusters result from the reduction of the bridging NO group, initially incorporated into the cluster via the reaction of $[HRu_3(CO)_{10}(\mu\text{-}CO)]^-$ with NO^+. Although the reaction may be made selective in the sense that the μ_3-NH group can be converted into the μ_2-NH_2, the process is not reversible in favor of the capping imide.[167,168] The mechanism of this reduction and related reactions has been thoroughly investigated.[169] The nitronium ion is also the

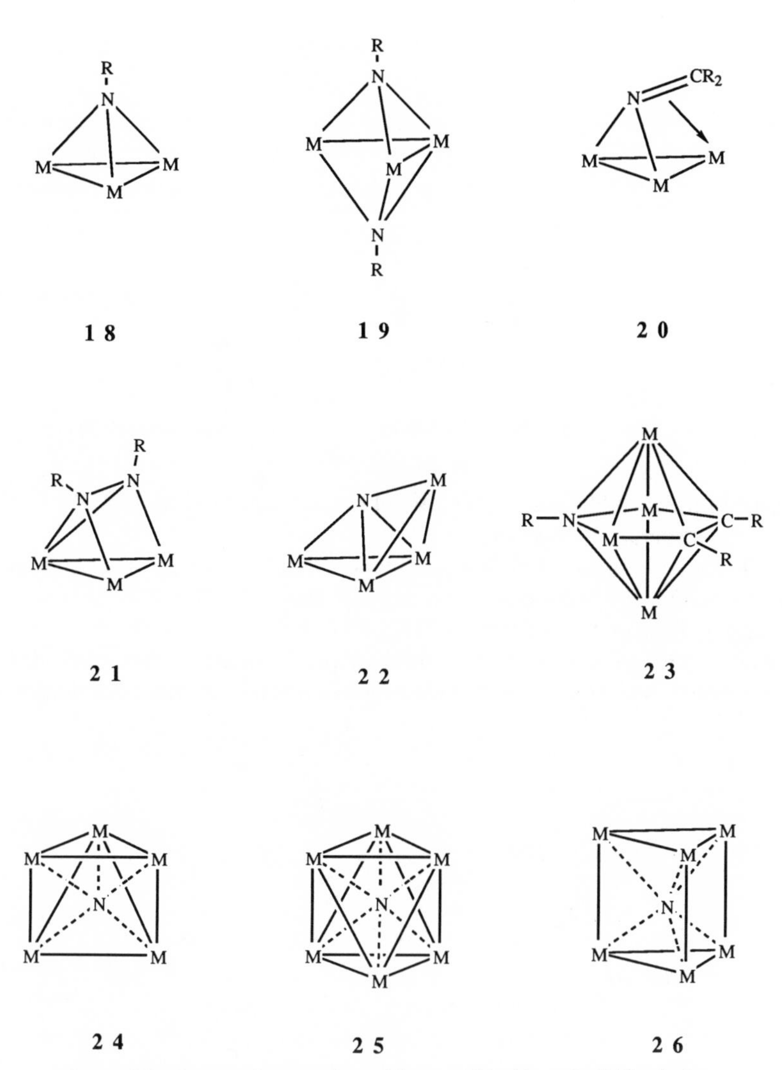

Figure 32. Structural types of transition metal imido and nitrido clusters.

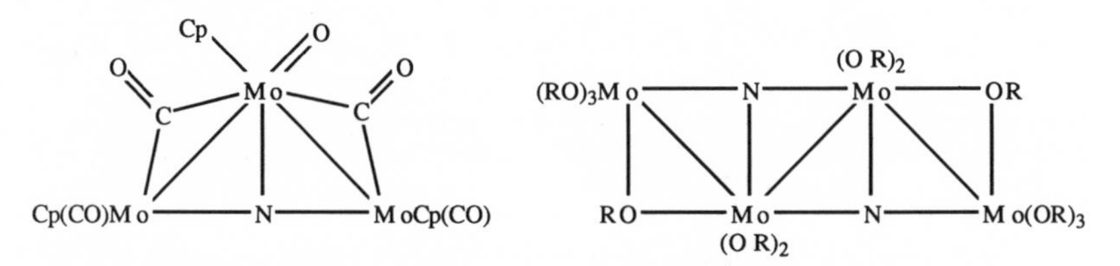

Figure 33. Structures of $Cp_3Mo_3(N)(O)(CO)_4$ and $(^iPrO)_{12}Mo_4(N)_2$.

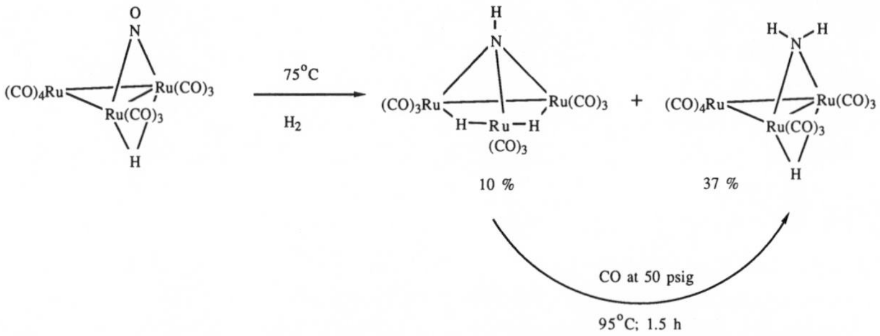

Figure 34. Synthesis of $H_2Ru_3(CO)_9NH$ by reduction of a coordinated nitrosyl ligand.

source of the μ_3-NH group in the mixed metal cluster $FeCo_2(CO)_9NH$ which has been structurally characterized.[169]

Terminally substituted clusters of type **18** are now quite well documented. One in which an NO bond is retained is $Ru_3(CO)_{10}NOH$. The cluster is formed by protonation of $[Ru_3(CO)_{10}NO]^-$ but the choice of acid is critical as Figure 35 indicates; note that the two products are in fact tautomers of one another. $Ru_3(CO)_{10}NOH$ (the kinetic product of the protonation) is

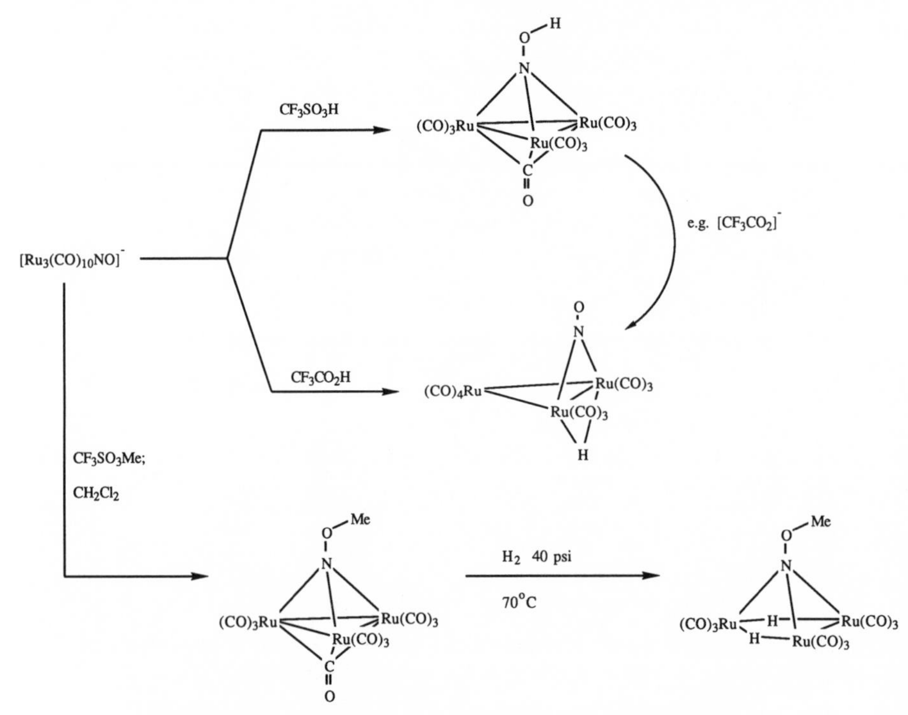

Figure 35. Protonation and methylation of $[Ru_3(CO)_{10}NO]^-$.

stable in inert solvents but will convert to $HRu_3(CO)_{10}(\mu\text{-NO})$ in the presence of anions such as the $[CF_3CO_2]^-$ ion.[163] A related system $Ru_3(CO)_{10}NOMe$ (Figure 35) has been crystallographically characterized.[163] The capping μ_3-NOH group (formed by the reversible protonation of a μ_3-NO ligand) is also observed supported upon a trimanganese framework in $[(\eta^5\text{-}C_5H_4Me)_3Mn_3(\mu\text{-}NO)_3(\mu_3\text{-}NOH)]^+$.[170,171] Transformations involving this system are summarized in Figure 36 and the system may be considered as a cluster model for the conversion of NO to NH_3 on a metal surface. A parallel therefore exists between this and the model for CO to CH_4 conversion illustrated in Fe_4C-butterfly clusters by Shriver *et al.*[172,173]

Of the type **18** clusters containing an alkyl and aryl substituted imido group, probably the most commonly studied are those with a μ_3-NPh cap. Some synthetic routes[174–176] to $Ru_3(CO)_{10}NAr$ (Ar = Ph or substituted phenyl) are illustrated in Figure 37 and competitive reactions are also indicated; the reaction of a nitrosoarene with $Ru_3(CO)_{12}$ provides the most efficient route to $Ru_3(CO)_{10}NPh$.[174] The balance between the production of types **18** and **19** is a delicate one and is critically dependent upon solvent and temperature. At $T \leq 60$ °C in THF, the reaction of $Ru_3(CO)_{12}$ with PhNO proceeds in favor of $Ru_3(CO)_{10}NPh$ while in hexane, or in refluxing THF, $Ru_3(CO)_9(NPh)_2$ predominates.[174] Analogous osmium compounds have also been synthesized.[174] Reaction of $Fe_3(CO)_{12}$ or $Ru_3(CO)_{12}$ with azobenzenes generates the type **19** cluster $M_3(CO)_9(NAr)_2$.[177] The cluster skeleton **19** has been structurally confirmed, for example, in $Fe_3(CO)_9(NMe)_2$[178] and $Ru_3(CO)_9(NPh)_2$.[179] Selective methods of converting clusters **18** to **19**

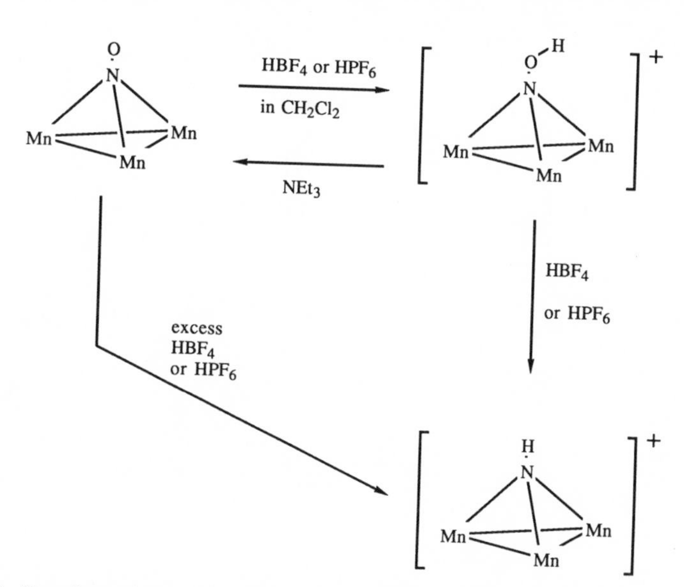

Figure 36. Transformations involving the μ_3-NO ligand in $(\eta^5\text{-}C_5H_4Me)_3Mn_3(\mu\text{-}NO)_3(\mu_3\text{-}NO)$.

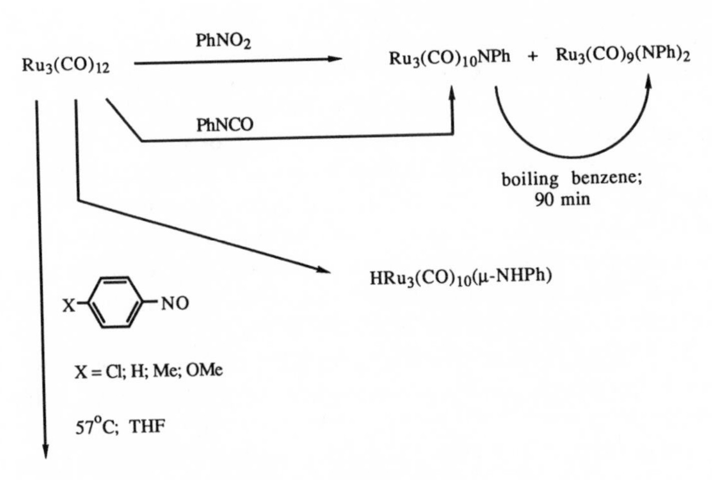

Figure 37. Preparation of $Ru_3(CO)_{10}NAr$ (Ar = Ph or substituted phenyl) and competitive formation of $Ru_3(CO)_9(NAr)_2$.

are illustrated in Figure 38[174,180]; *p*-substituted arenes are used to create unsymmetrical clusters.

Returning to clusters of type **18**, a variety of N-substituted clusters have been structurally characterized, for example, $Fe_3(CO)_9NSiMe_3$,[181]

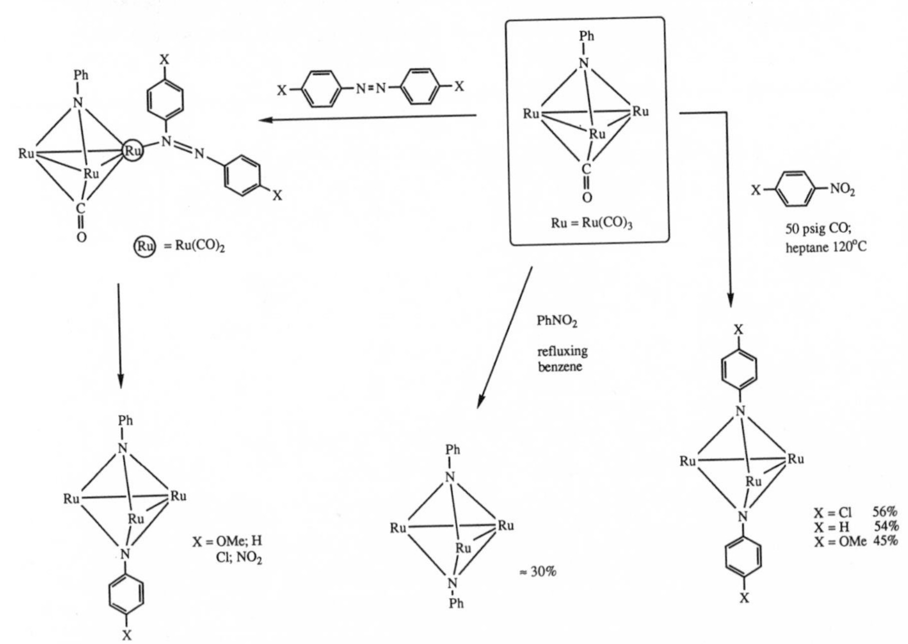

Figure 38. Methods of conversion from $Ru_3(CO)_{10}NAr$ to $Ru_3(CO)_9(NAr)_2$.

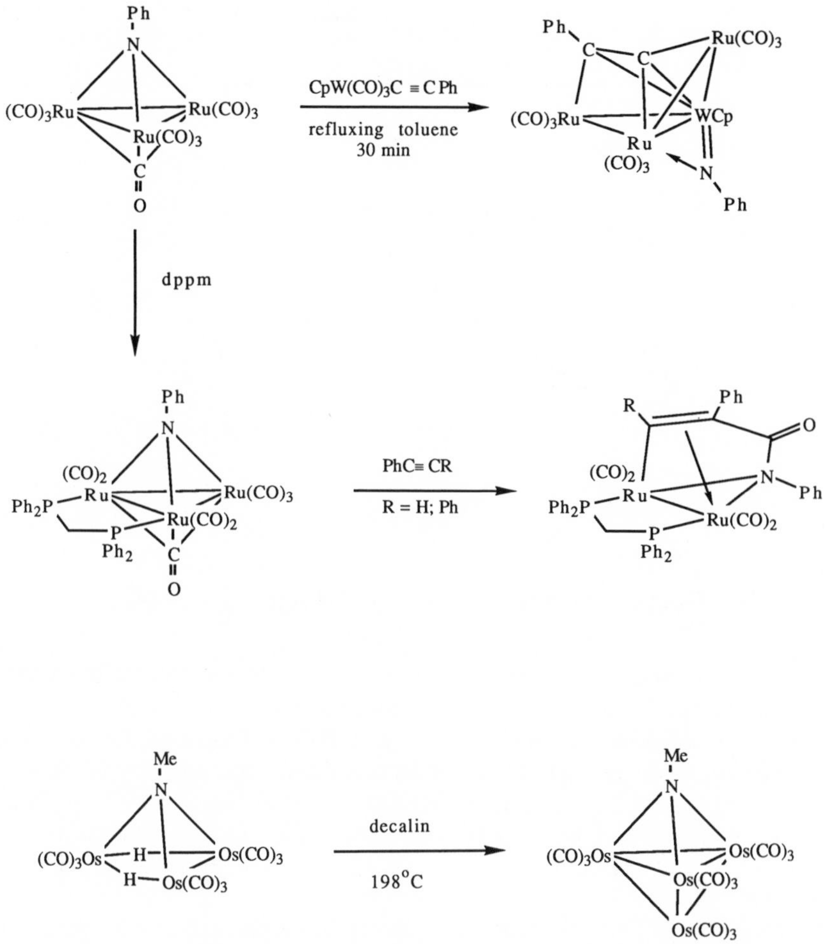

Figure 39. Selected reactions of $Ru_3(CO)_{10}NPh$ and related compounds.

$Ru_3(CO)_7(C_6H_6)NPh$,[182] $H_2Os_3(CO)_9NPh$.[183] In Figure 39, some selected reactions of the type **18** cluster are presented. A recent and novel reaction of $Ru_3(CO)_{10}NPh$ is that with $CpW(CO)_3C{\equiv}CPh$. Cluster expansion accompanies the conversion of the μ_3-NPh cluster fragment into a bridging imido ligand having a $W{=}N(Ph) \rightarrow Ru$ bonding mode.[184] Cluster expansion, but this time with the retention of the μ_3-NR fragment, is achieved after the thermolysis of $H_2Os_3(CO)_9NMe$ (Figure 39).[185] Two carbonyl ligands in $Ru_3(CO)_{10}NPh$ may be substituted by a dppm ligand[186] and the resulting cluster undergoes an interesting degradation reaction in which the metalloimido cluster is converted in a metallopyrrolidone complex (Figure 39); the product has been crystallographically characterized.[187]

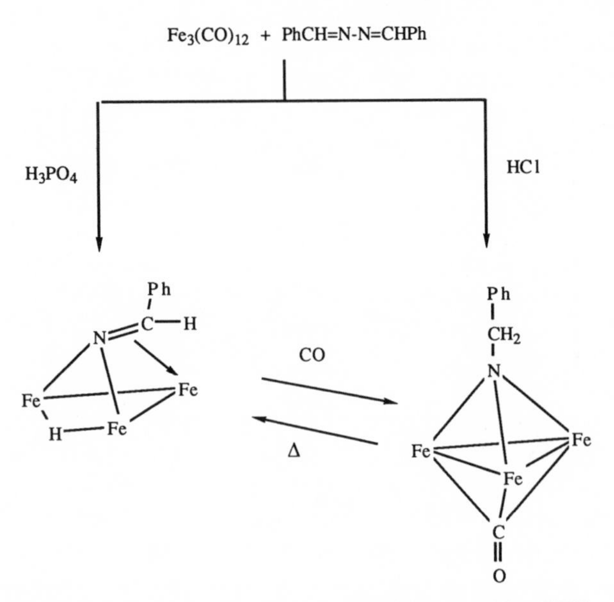

Figure 40. Reaction of $Fe_3(CO)_{12}$ with PhCH=N—N=CHPh.

The type **20** cluster is clearly analogous to **2** and to organometallic clusters such as $[HFe_3(CO)_9C{=}CH_2]^-$[188,189] and $H_2Os_3(CO)_9C{=}CH_2$.[51,52] Reaction of $Fe_3(CO)_{12}$ with PhCH=N-N=CHPh results in the formation of the type **20** cluster $HFe_3(CO)_9N{=}CHPh$, although this pathway is sensitive to the acid used in the protonation step as shown in Figure 40.[190] A direct comparison may be made between the interconversions of $HFe_3(CO)_9N{=}CHPh$ and $Fe_3(CO)_{10}NCH_2Ph$ and $[HFe_3(CO)_9C{=}CH_2]^-$ and $[Fe_3(CO)_{10}CCH_3]^-$.[189,191]

The cluster $Fe_3(CO)_9N_2R_2$ is a type **21** molecule and its core structure is analogous to that in $HRu_3(CO)_9B_2H_5$[15,54] and the "parallel" isomer of

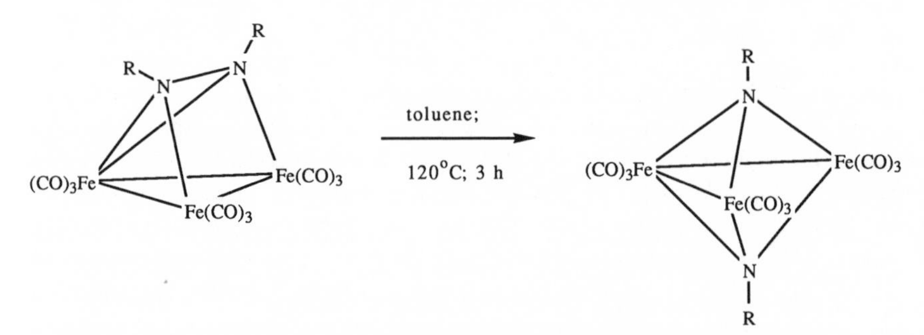

Figure 41. N—N bond cleavage in a type **21** cluster.

$H_2Ru_3(CO)_9C_2H_2$ (see Section 2.1). $Fe_3(CO)_9N_2R_2$ (R = Et; ^nPr) has been prepared by the reaction of $Fe_3(CO)_{12}$ with *cis*-azoalkanes or of $Fe(CO)_3(cyclo\text{-}C_8H_{14})_2$ with *trans*-azoalkanes, and crystallographic characterization of $Fe_3(CO)_9N_2Et_2$ has confirmed the "parallel" mode of bonding (Figure 41).[192] The metal supported RN=NR fragment is cleaved upon thermolysis and a type **19** cluster results. The temperature is critical; decomposition occurs if the thermolysis is carried out at 80 °C while at 120 °C, $Fe_3(CO)_9(NR)_2$ is obtained in 68% yield.[192]

The type **22** cluster is now well established but, in the main, only for the group 8 transition metals. For M = Fe, Ru, or Os, and assuming the incorporation of $M(CO)_3$ fragments, the 62 electrons required for a neutral butterfly cluster are provided with the addition of only one *endo*-hydrogen atom. For example, the series of isoelectronic compounds $HFe_4(CO)_{12}BH_2$,[79,80] $HFe_4(CO)_{12}CH$,[193,194] and $HFe_4(CO)_{12}N$[195] show progressive loss of hydrogen atoms from around the main group atom and, therefore, a progressively more exposed interstitial atom. The need for *endo*-hydrogen atoms can be completely eliminated by altering one group 8 to a group 9 metal atom as, for example, in the cluster $Ru_3Co(CO)_{12}N$,[196] which is isoelectronic (valence electrons only) with $HRu_4(CO)_{12}N$.[197,198] Synthesis of type **22** metallonitrides commonly utilizes the nitronium ion or else coordinated nitrosyl or isocyanate ligands as the source of the interstitial nitrogen atom, and some synthetic routes are summarized in Figure 42. The first butterfly nitride to be prepared and characterized was $HFe_4(CO)_{12}N$.[195,199,200] In using $[Fe_2(CO)_8]^{2-}$ and NO^+ as the precursors to $[Fe_4(CO)_{12}N]^-$, the temperature of the reaction is critical; below 130 °C, a nitrosyl complex is obtained and above 130 °C, the higher nuclearity cluster anion $[Fe_5(CO)_{14}N]^-$ is produced.[195] In the reaction of $[Fe(CO)_3(NO)]^-$ with $Fe_3(CO)_{12}$, the yield of the butterfly anion is optimized at 50% when the reagents are combined in a ratio of 2:3.[200] Protonation of $[Fe_4(CO)_{12}N]^-$ results not only in the formation of the desired $HFe_4(CO)_{12}N$ (43%) but also in the formation of the metalloimide $Fe_3(CO)_{10}NH$.[200] A detailed study of the protonation of $[Ru_4(CO)_{12}N]^-$ and of the mixed metal cluster anion $[FeRu_3(CO)_{12}N]^-$ (see below) has illustrated that the protonation first occurs at the nitrido atom to give $M_4(CO)_{12}NH$ (Figure 42).[201] This cluster then rearranges to give the stable isomer $HM_4(CO)_{12}N$ but, in the presence of CO, there is competitive formation of the imido cluster $M_3(CO)_{10}NH$.[198] The N-protonated species appear as intensely colored intermediates with lifetimes of hours, $t_{1/2}$ depending upon the nature of the metal atoms and *exo*-ligands.[198]

As Figure 42 shows, one route to $HM_4(CO)_{12}N$ (M = Ru; Os) involves the reaction of the hydrido cluster anion $[H_3M_4(CO)_{12}]^-$ with $NOBF_4$.[197,202] It has been proposed that for both ruthenium and osmium, the reaction proceeds via the nitrosyl cluster $H_3M_4(CO)_{12}(\mu\text{-}NO)$.[202] This product has been isolated and structurally characterized[197] for M = Os. Elimination of a mole

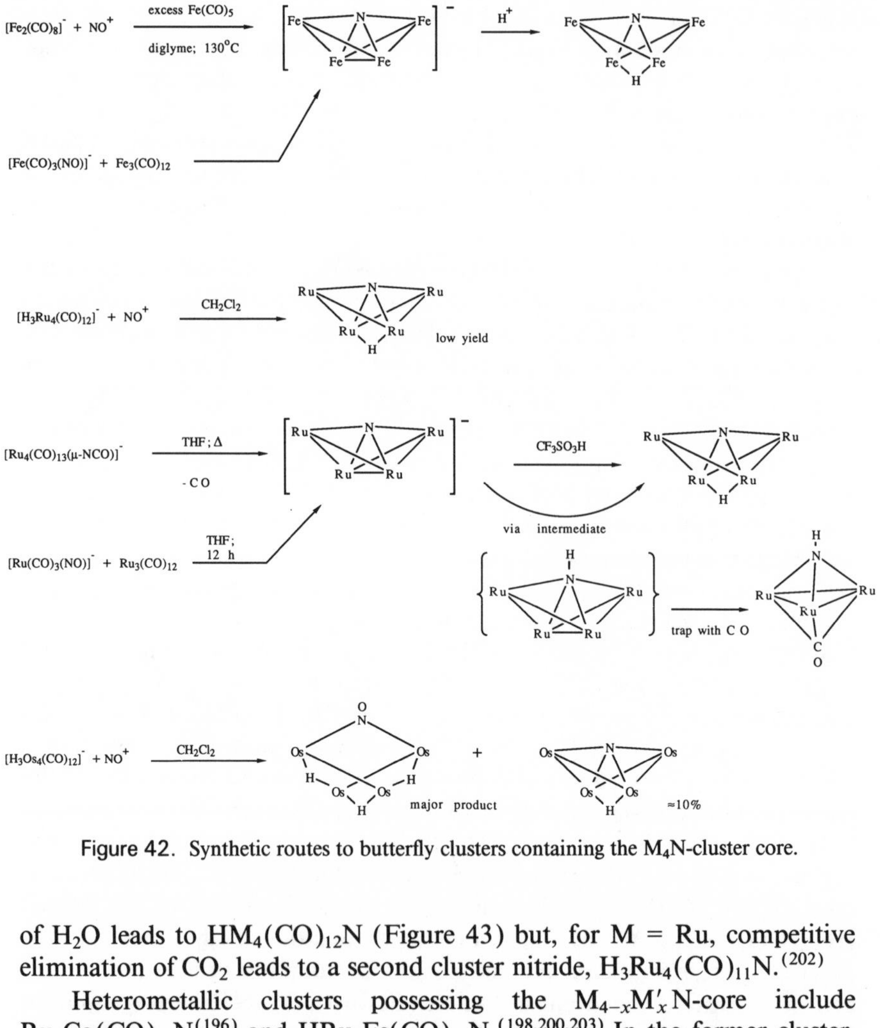

$[Fe_2(CO)_8]^- + NO^+$ —(excess $Fe(CO)_5$; diglyme; 130°C)→ $[Fe_4N]^-$ —(H^+)→ HFe_4N

$[Fe(CO)_3(NO)]^- + Fe_3(CO)_{12}$ → $[Fe_4N]^-$

$[H_3Ru_4(CO)_{12}]^- + NO^+$ —(CH_2Cl_2)→ HRu_4N low yield

$[Ru_4(CO)_{13}(\mu\text{-}NCO)]^-$ —(THF; Δ; - CO)→ $[Ru_4N]^-$ —(CF_3SO_3H)→ HRu_4N

$[Ru(CO)_3(NO)]^- + Ru_3(CO)_{12}$ —(THF; 12 h)→ $[Ru_4N]^-$

via intermediate $\{Ru_4NH\}$ —(trap with C O)→ $Ru_4(NH)(CO)$

$[H_3Os_4(CO)_{12}]^- + NO^+$ —(CH_2Cl_2)→ $H_3Os_4(NO)$ major product + HOs_4N ≈10%

Figure 42. Synthetic routes to butterfly clusters containing the M_4N-cluster core.

of H_2O leads to $HM_4(CO)_{12}N$ (Figure 43) but, for M = Ru, competitive elimination of CO_2 leads to a second cluster nitride, $H_3Ru_4(CO)_{11}N$.[202]

Heterometallic clusters possessing the $M_{4-x}M'_xN$-core include $Ru_3Co(CO)_{12}N$[196] and $HRu_3Fe(CO)_{12}N$.[198,200,203] In the former cluster, no X-ray diffraction data are available and the exact location of the cobalt atom in the butterfly framework (viz. wing-tip or hinge site) is not known. However, the conjugate base of $HRu_3Fe(CO)_{12}N$ and a substituted derivative thereof, $[Ru_3Fe(CO)_{10}\{P(OMe_3)_3\}_2N]^-$, have been the subjects of structural determinations.[198,203] Both structures are disordered. In $[Ru_3Fe(CO)_{10}$-$\{P(OMe_3)_3\}_2N]^-$, iron and ruthenium atoms are disordered over the two hinge sites,[198] but in $[Ru_3Fe(CO)_{12}N]^-$ the disorder involves iron/ruthenium occupancies in both the hinge and wing-tip sites.[203] The cluster exists in two isomeric forms (Figure 44a) which are in equilibrium ($\Delta H = 14.6 \pm 4.2$ kJ mol^{-1}; $\Delta S = -13 \pm 2$ eu).[203] Such examples of metal framework isomer-

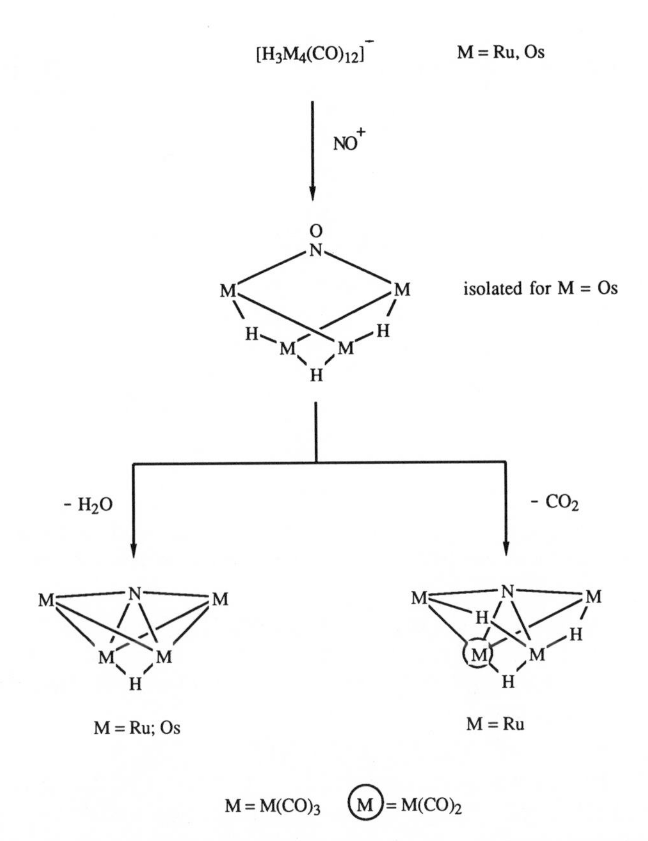

Figure 43. Proposed pathway to the formation of type **22** clusters from $[H_3M_4(CO)_{12}]^-$ (M = Ru, Os).

izations are relatively rare, and indeed the present case represents the first documented example of a wing-tip/hinge atom interconversion in a metal butterfly cluster. Once protonated, the cluster exhibits a preference for the iron atom to be accommodated in a wing-tip site.[198] This latter observation for neutral $HRu_3Fe(CO)_{12}N$ parallels the pattern of iron/ruthenium site distribution in the isoelectronic heterometalloborane, $HRu_3Fe(CO)_{12}BH_2$ (Figure 44b). This cluster has recently been prepared by the photolysis of $Ru_3(CO)_9BH_5$ with $Fe(CO)_5$ and appears to show no skeletal isomerism.[204] In the related compound $Ru_3Fe(CO)_{12}(AuPPh_3)N$ (see below), the iron atom also exhibits a site preference, being disordered over the two wing-tip sites.[205]

Two molecules which are closely related to the type **22** cluster are $Ru_4(CO)_{12}N(\mu\text{-}NO)$ and $Ru_4(CO)_{12}N(\mu\text{-}NCO)$.[206] The syntheses and structures of these compounds are summarized in Figure 45. Since NO and NCO are both three-electron donors, each cluster possesses 64 valence elec-

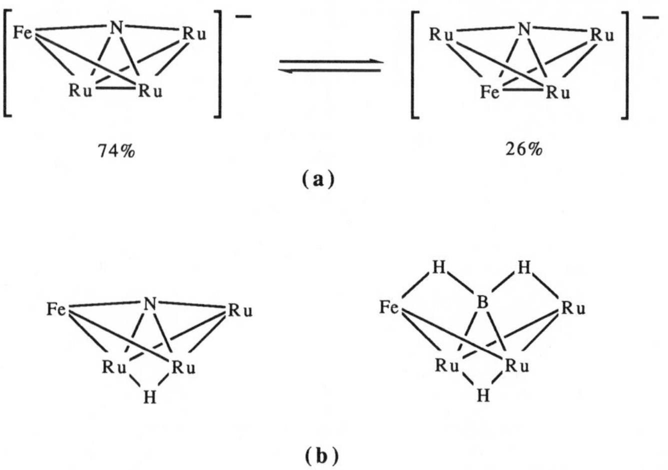

Figure 44. (a) Skeletal isomerization in $[Ru_3Fe(CO)_{12}N]^-$ in contrast to (b) the preferred arrangement of Fe and Ru atoms in $Ru_3Fe(CO)_{12}N$ and the isoelectronic $HRu_3Fe(CO)_{12}BH_2$.

trons, two too many for a butterfly framework. The observation that, in each case, the hinge Ru–Ru edge is quite open [average of 3.244(1) Å for $Ru_4(CO)_{12}N(\mu\text{-}NO)$ and $Ru_4(CO)_{12}N(\mu\text{-}NCO)$] is thus predicted on the basis of the electron count.

The reactivity of $[M_{4-x}M'_x(CO)_{12}N]^-$ (M = Ru, M' = Fe; x = 0,1,4) toward Ph_3PAuCl has been explored.[205] The fact that the parent neutral compound $HM_{4-x}M'_x(CO)_{12}N$ possesses only a single *endo*-hydrogen atom restricts the range of possible products; thus, in contrast to the reactions illustrated in Figure 19, digold- and trigold-derivatives of the cluster nitride are not expected and indeed are not observed (Figure 46). Significantly, however, crystallographic characterization of $Ru_4(CO)_{12}(AuPPh_3)N$ reveals a different siting for the gold(I) phosphine fragment from that observed in either of the isoelectronic compounds $HRu_4(CO)_{12}(AuPPh_3)C$[207]† or $HRu_4(CO)_{12}$-$(AuPPh_3)BH$.§ In related auraferraboranes, it has been shown[86,87] that there is a tendency for the $AuPR_3^+$ electrophiles to interact with the boron atom thereby generating up to three M—Au(PR_3)—B interactions per Fe_4B-cluster and no M—Au(PR_3)—M bridges.[105] Stable compounds based upon the type **22** cluster core all exhibit a bare nitrido atom and the gold(I) phosphine derivatives are no exception. Bearing the isolobal principle in mind

† The iron analog $HFe_4(CO)_{12}(AuPPh_3)C$ shown in Figure 46 has also been structurally characterized; see Reference 208.

§ The structure of $HRu_4(CO)_{12}(AuPPh_3)BH$ has been proposed from spectroscopic data (Reference 81) and by a comparison with $HFe_4(CO)_{12}\{AuP(o\text{-}C_6H_4Me)_3\}BH$, the structure of which has been determined[209] Spectroscopic data for $HFe_4(CO)_{12}(AuPPh_3)BH$ support the structure shown in Figure 46b; see Reference 210.

$$Ru_3(CO)_{10}(\mu\text{-}NO)_2 \xrightarrow[110^{\circ}C;\ 9\ h]{CO\ \ 1\ atm} \underset{5\%}{Ru_3(CO)_{12}} + \underset{30\%}{Ru_4(CO)_{12}(\mu\text{-}NO)N} + \underset{5\%}{Ru_4(CO)_{12}(\mu\text{-}NCO)N} + \underset{5\%}{Ru_3(CO)_{10}(\mu\text{-}NO)(\mu\text{-}NCO)}$$

(a)

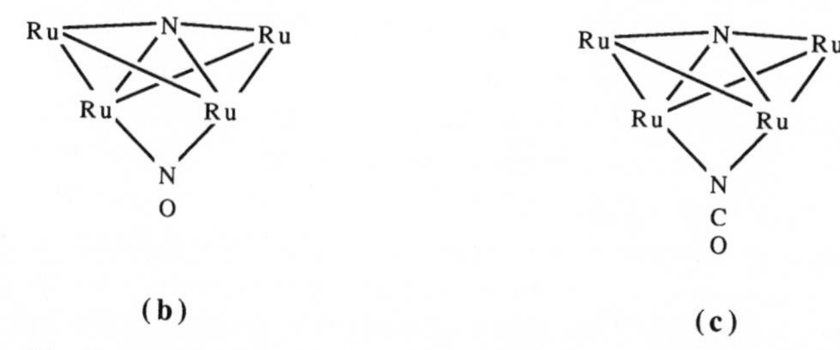

Figure 45. Syntheses and structures of $Ru_4(CO)_{12}N(\mu\text{-}NO)$ and $Ru_4(CO)_{12}N(\mu\text{-}NCO)$.

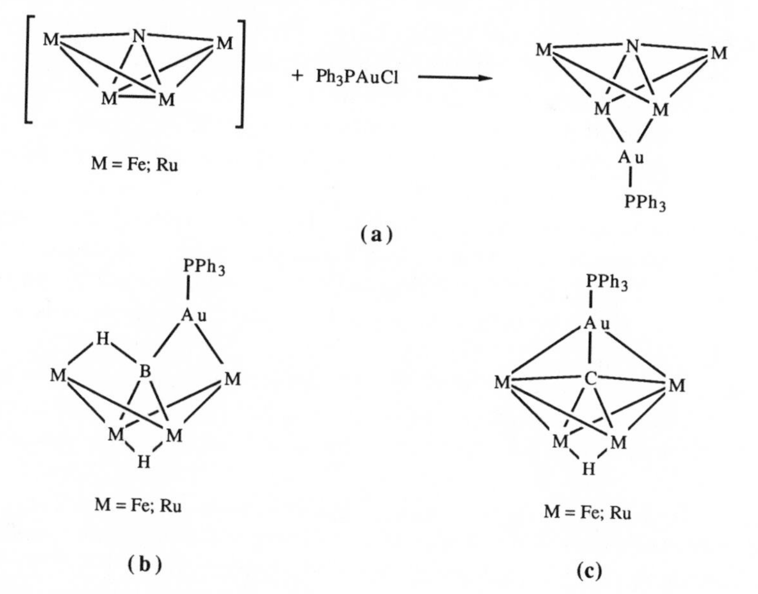

Figure 46. Synthesis of $M_4(CO)_{12}(AuPPh_3)N$ (M = Fe, Ru) and structural comparisons with isoelectronic analogs.

(see Section 2.1), this is not surprising since no examples of 3-center 2-electron M—H—N bridges have been documented and so M—Au(PR_3)—N interactions are, perhaps, not expected; bridging sites tend to be preferred over terminal ones for the gold(I) phosphines and thus an N-bonded terminal $AuPR_3$ group will be less favored than a bridging M—Au(PR_3)—M mode.

The reaction of $[Ru_4(CO)_{12}N]^-$ with PhC≡CPh in the presence of acid leads to insertion of the alkyne into the Ru_{hinge}—Ru_{hinge} bond. The product, illustrated in Figure 47c, is a type **23** cluster and incorporates an imido group; viz. during the reaction, the nitrido atom is drawn out of the interstice of the Ru_4-butterfly and becomes a cluster vertex carrying a terminal hydrogen atom.[(211)] In Figure 47 the reactions of diphenylacetylene with three isoelectronic M_4-butterfly clusters are compared. Each proceeds in a significantly different manner. One Ru_{hinge}—Ru_{wing} edge in $HRu_4(CO)_{12}BH_2$ is opened as the acetylene inserts, and there is concomitant B—H bond activation and transfer of one cluster hydrogen atom to the organic molecule to give $HRu_4(CO)_{12}B(H)C(Ph)CHPh$ in which a B—C bond has been formed.[(114)] With $H_2Ru_4(CO)_{12}C$, carbon–carbon coupling occurs in addition to hydrogen atom transfer to generate $HRu_4(CO)_{12}CC(Ph)CHPh$ in which the organic fragment is supported *above* the Ru_4-cluster; unlike the metalloborane and nitride, there is no significant perturbation of the Ru_4-framework in the case of the carbide cluster.[(212)] This series of reactions illustrates beautifully the significant effect that a main group atom can have in determining the course of reaction. Of course, the role of the *endo*-hydrogen atoms should not be forgotten; as stressed earlier in this chapter, changing the main group atom runs hand in hand with altering the number of *endo*-hydrogen atoms if a given cluster geometry is to be preserved.

On passing from clusters of type **22** to **24**, **25**, and **26** the nitrogen atom is progressively encapsulated within a metal skeleton. In **24**, the nitrogen atom lies within bonding contact of all the metal atoms in the square pyramidal M_5-framework; for example, in $[Ru_5(CO)_{14}N]^-$, the nitrogen atom is situated 0.21(2) Å below the square plane, with Ru_{apex}–N = 2.14(2) Å and an average Ru_{base}–N = 2.03(2) Å.[(201)] In $HFe_5(CO)_{14}N$, the N lies closer to the plane, viz. 0.093 Å,[(195)] in keeping with the fact that the Fe_5-framework is smaller than the Ru_5-cage (i.e., Fe–N < Ru–N distance). The anion $[Fe_5(CO)_{14}N]^-$ has also been fully characterized.[(213)] The formation of such high nuclearity clusters often accompanies that of the butterfly clusters **22**, and may predominate if conditions are made harsher. For example, the reaction of $[Fe_2(CO)_8]^{2-}$ with NO^+ gives $[Fe_4(CO)_{12}N]^-$ at 130 °C (Figure 42) but $[Fe_5(CO)_{14}N]^-$ if $T > 130$ °C.[(195)] Reduction of $[Fe_5(CO)_{14}N]^-$ using $Li[Et_3BH]$ leads to $[HFe_5(CO)_{13}N]^{2-}$.[(195)] The group 8 transition metal M_5N-nitrides are structurally analogous to the isoelectronic carbides $Fe_5(CO)_{15}C$[(214)] and $Ru_5(CO)_{15}C$.[(215)] Although $Os_5(CO)_{15}C$ is also fully characterized,[(216)] its nitrido analog has yet to be reported. The heterometallic cluster

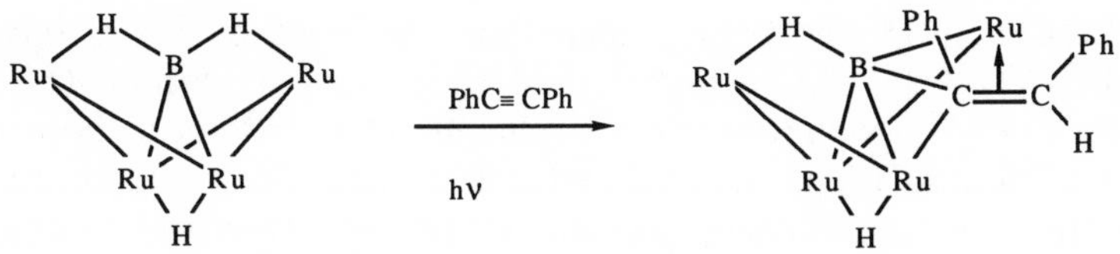

Ru H B H Ru PhC≡CPh hν Ru Ru H

(a)

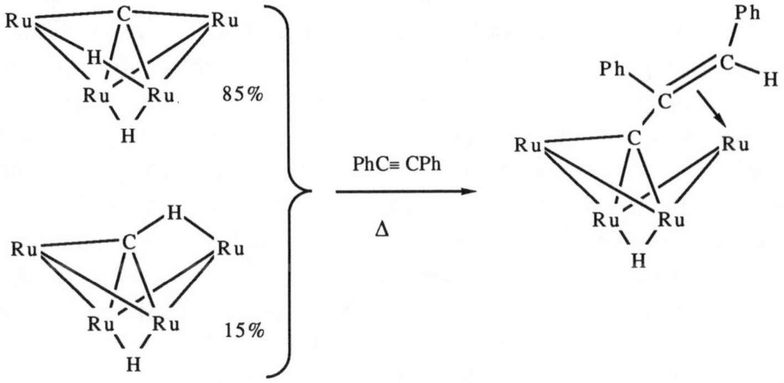

Ru C Ru H Ru Ru H 85% PhC≡CPh Δ Ph C H Ph C C Ru Ru Ru Ru H

Ru C H Ru Ru Ru H 15%

(b)

Ru N Ru Ru Ru PhC≡CPh H^+ H N Ru Ru Ru Ru C C Ph Ph

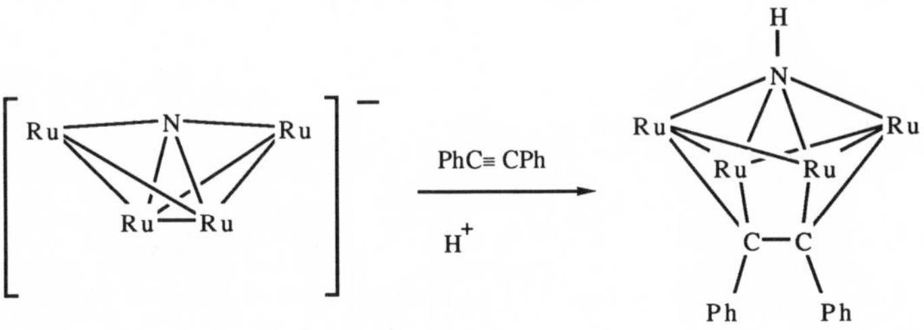

(c)

Figure 47. Contrasting reactions of PhC≡CPh with isoelectronic clusters (a) $HRu_4(CO)_{12}BH_2$, (b) $H_2Ru_4(CO)_{12}C$, (c) $[Ru_4(CO)_{12}N]^-$.

$[Ru_4Fe(CO)_{14}N]^-$ has been prepared in low yield by the pyrolysis of $[Ru_3Fe(CO)_{12}N]^-$ and in higher yield by prolonged reaction (50 h) of $[Ru_3Fe(CO)_{12}N]^-$ with $Ru_3(CO)_{12}$ in refluxing THF.(203) Due to the increased degree of encapsulation compared to that in clusters of group **22**, the nitrogen atom in type **24** clusters appears to be less reactive. Reaction of $[Ru_5(CO)_{14}N]^-$ with CO does not occur except at high pressures.(201) With $Ru_3(CO)_{12}$, cluster expansion occurs quantitatively within 24 h to yield the type **25** anion, $[Ru_6(CO)_{16}N]^-$.(201)

Hexanuclear cluster nitrides fall into two types, **25** and **26**. The octahedral array of metal atoms in $[Ru_6(CO)_{16}N]^-$ was inferred initially from spectroscopic data and from the fact that such a geometry is consistent with the 86 valence electrons possessed by the anion.(201,217) The structure has recently been confirmed by X-ray crystallography, (Figure 48).(218) The synthetic route to $[Ru_6(CO)_{16}N]^-$ is unusual since it utilizes the azide ion as a source of an interstitial nitrogen atom (Figure 48).(201,217) For a first-row main group atom, both the octahedral and triginal prismatic interstices are sterically acceptable; this is not the case for larger atoms as will be illustrated in, for example, Section 4.2. The anions $[Co_6(CO)_{15}N]^-$ and $[Rh_6(CO)_{15}N]^-$ both possess 90 valence electrons and this is consistent with a trigonal prismatic cage (**26** in Figure 32). Both clusters may be formed by the reactions (illustrated for cobalt) shown in Figure 49,(219,220) but attempts to prepare the iridium analog via the reaction of $Ir_4(CO)_{12}$ with PPN$[NO_2]$ lead instead to $[Ir_4(CO)_{11}(NCO)]^-$.(220) The elimination of two moles of carbon monoxide from $[Co_6(CO)_{15}N]^-$ reduces the valence electron count by four and permits the conversion of the type **26** into a type **25** cluster,(221) viz. $[Co_6(CO)_{13}N]^-$ which is isoelectronic and isostructural with the carbido species $[Co_6(CO)_{13}C]^{2-}$,(222) (Figure 49). A rather beautiful transformation involves the aggregation of three Co_6N-cages to give $[Co_{14}(CO)_{26}(N)_3]^{3-}$ as shown in

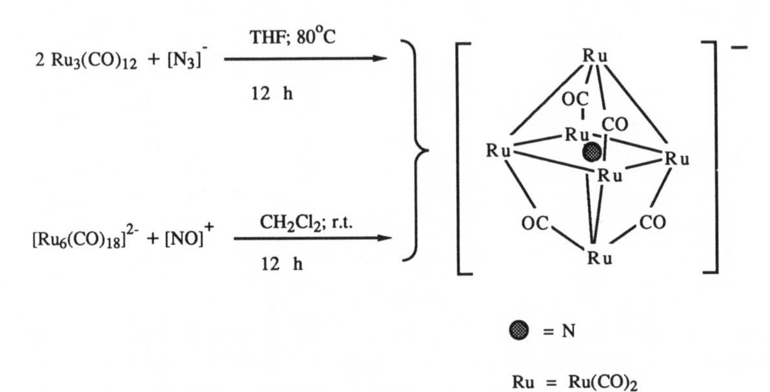

Figure 48. Formation and structure of $[Ru_6(CO)_{16}N]^-$.

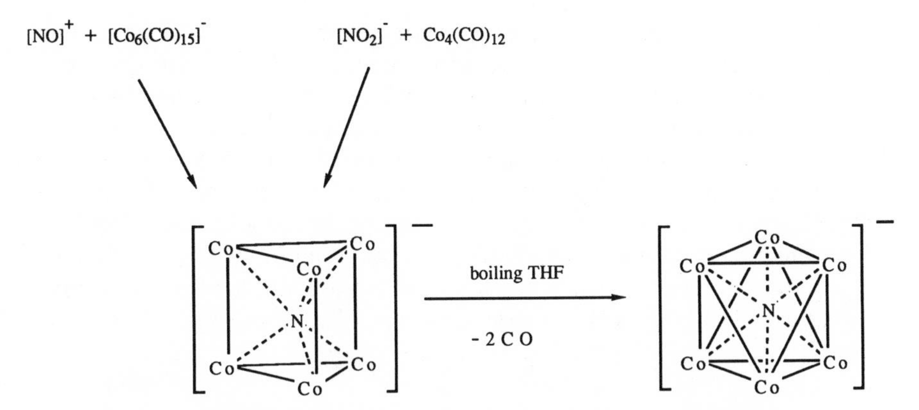

Figure 49. Formation of $[Co_6(CO)_{15}N]^-$ and its conversion to $[Co_6(CO)_{13}N]^-$.

Figure 50. The cluster is somewhat distorted with the two centered hexagonal layers being folded rather than planar; the three nitrogen atoms occupy alternate trigonal prismatic interstitial sites.[223]

In Section 2.1, the importance of ^{11}B NMR spectroscopy to the development of metalloborane and boride cluster chemistry was noted. Just as significant has been the role of the ^{15}N nucleus.[160,224] A relationship exists between the ^{13}C and ^{15}N NMR chemical shifts for carbon and nitrogen atoms which reside in structurally similar environments and this clearly aids the characterization of new nitrido clusters.

4.2. Clusters Containing Metal–Phosphorus Bonds

Moving from nitrogen to phosphorus obviously introduces two factors which will influence cluster structure: the increased size of a phosphorus atom in relation to nitrogen, and the possibility of increased coordination number. Examples in the previous section illustrated that the nitrogen atom, despite

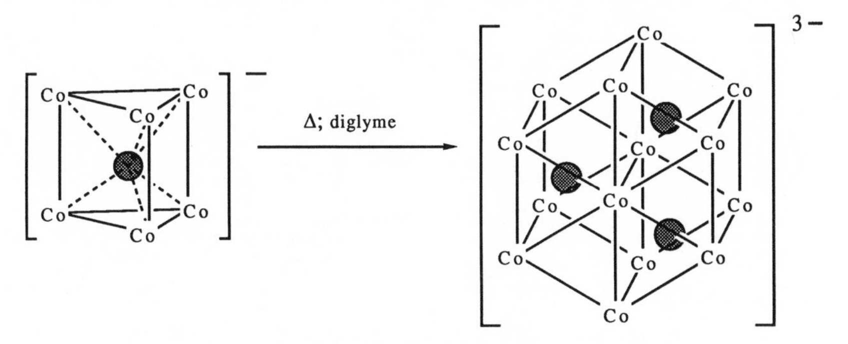

Figure 50. Aggregation of three type **26** clusters, $[Co_6(CO)_{13}N]^-$.

having available only 2*s* and 2*p* valence atomic orbitals, can interact with up to six metal atoms by virtue of delocalized bonding.[218,225] One would anticipate an even greater structural variation for cluster bound phosphorus atoms and phosphorus containing fragments. This is indeed the case as Figure 51 illustrates. The principal modes of interaction may be classified as being either μ_3- or μ_4-capping, or μ_6- or μ_8-interstitial. Note that the vast range of transition metal phosphido clusters which incorporate μ-PR_2 bridging ligands are omitted from the discussion with the exception of that found in the type **29** cluster.

The occurrence of M_nP_m clusters in which a μ_3-phosphorus atom may be either naked (type **27**) or carrying an *exo*-substituent (types **28** and **30**)

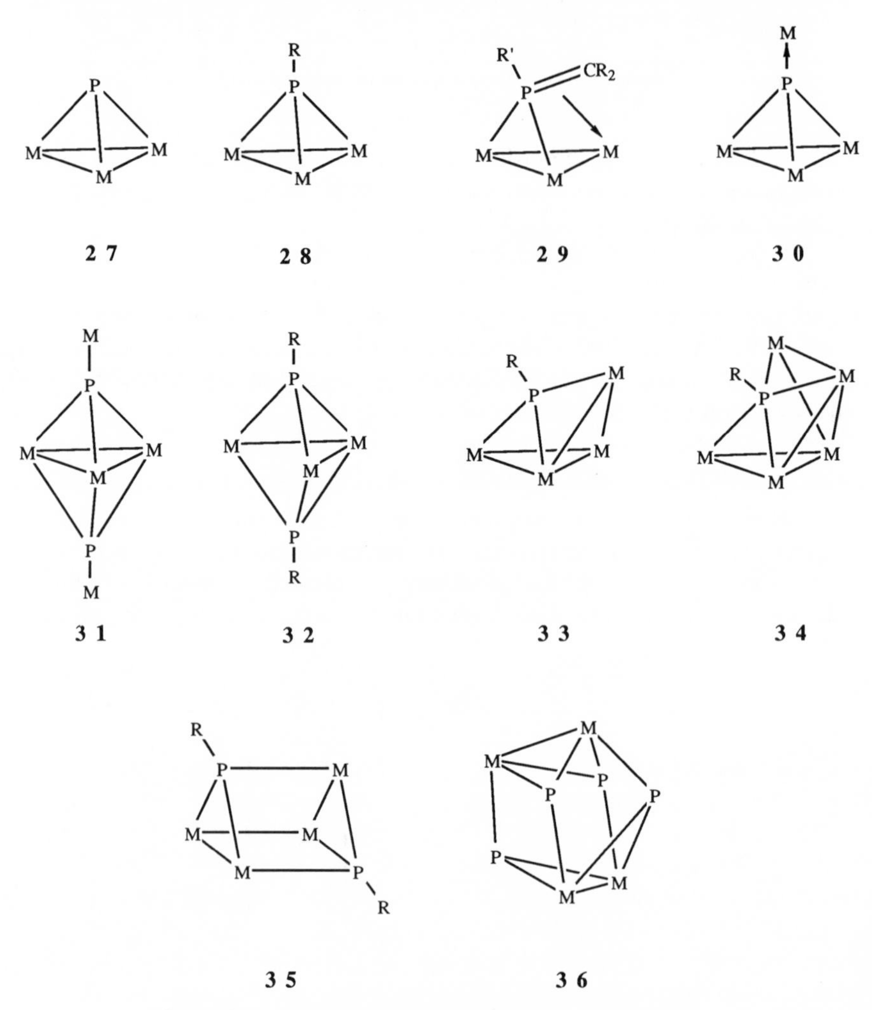

Figure 51. Structural types for phosphorus containing clusters.

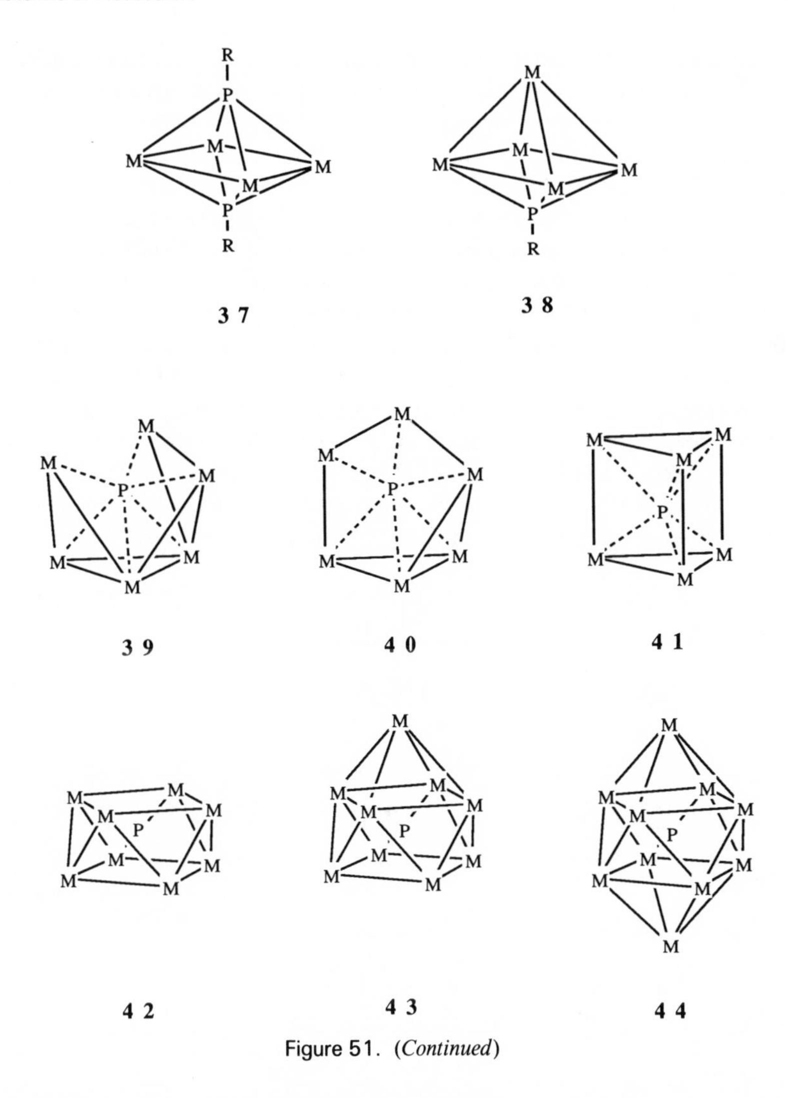

Figure 51. (*Continued*)

contrasts with the preferences encountered so far for group 13 and 14 elements, for nitrogen, (see Figures 7, 24, and 32) and also with structures for clusters of the later group 15 elements and those in group 16 (see Figures 69 and 85). Earlier elements tend not to be naked unless they are interstitial, while later elements sport an *exo*-lone pair which may or may not be used to form a coordinate bond to a second metal fragment. The type **27** cluster is exemplified by $Co_3(CO)_9P$,[226] synthesized by the reaction of $Co_2(CO)_8$ with white phosphorus or by reacting phosphorus(III) halides with $[Co(CO)_4]^-$ (Figure 52). Trimerization of $Co_3(CO)_9P$ with concomitant loss of CO is spontaneous in hexane solution.[226] The tetrahedral core of $Co_3(CO)_9P$ has been implied on the basis of comparison with arsenic analog $Co_3(CO)_9As$, which also undergoes trimerization.[227] Recently, structural details of $\{Co_3(CO)_8As\}_3$ (see Section 4.3) have been published,[228] thereby providing evidence for the proposed structure of $\{Co_3(CO)_8P\}_3$ and, by implication, for $Co_3(CO)_9P$. The Lewis basicity of the capping phosphorus atom in $Co_3(CO)_9P$ is evident in reactions of the cluster with 16-electron metal fragments such as $\{Fe(CO)_4\}$ to form a type **30** cluster.[229] Such products are formed indirectly by reaction of $Co_2(CO)_8$ with L_nMPX_3 (L_nM = $CpMn(CO)_2$, $Cr(CO)_5$, or $W(CO)_5$; X = Br, Cl).[230] Thus, the *exo*-lone pair in **27** is an active Lewis base; this feature will be addressed again in Sections 4.3 and 4.4 since, as the Lewis basicity of the μ_3-E atom decreases on descending group 15, the stability of clusters analogous to type **27** increases; for example, $Co_3(CO)_9Bi$ does not trimerize.[155] A cluster closely related to $Co_3(CO)_9P$ and which shows a similar chemistry is $Co_2(CO)_6C(R)P$, which contains a tetrahedral Co_2CP-core. On standing in THF solution, trimerization to $\{Co_2(CO)_5C(R)P\}_3$ is observed.[231]

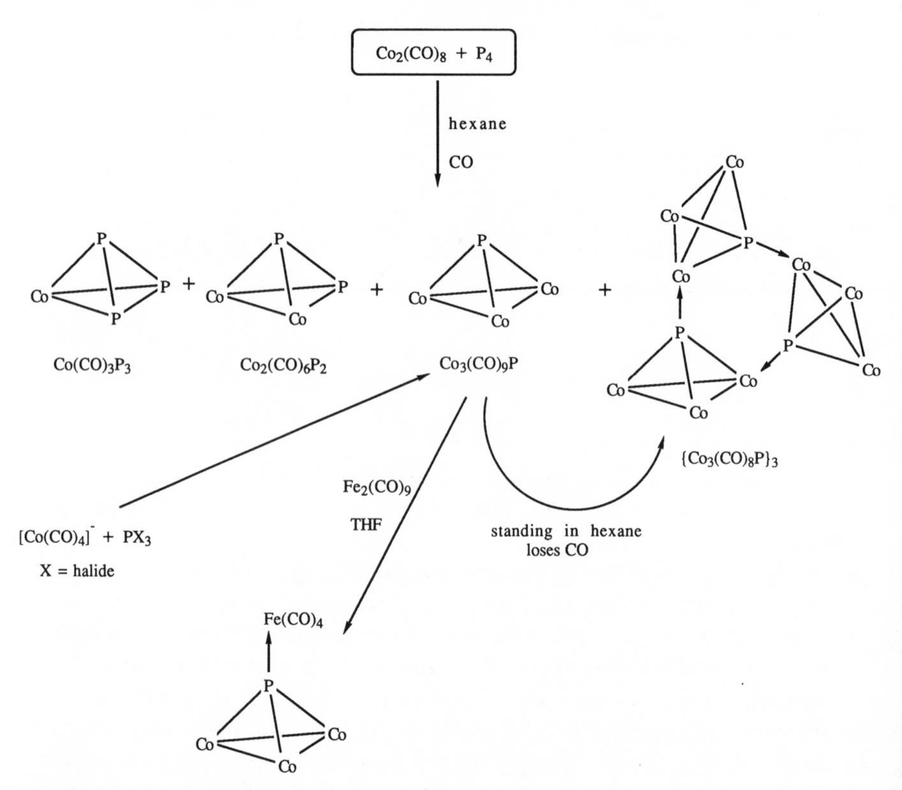

Figure 52. Cobalt phosphorus containing clusters arising from the reaction of $Co_2(CO)_8$ with elemental phosphorus.

In Figure 52, the reaction of $Co_2(CO)_8$ with P_4 is seen to give two products which do not strictly fall within the definitions of the contents of this chapter. However, as members of the series of clusters with tetrahedral cores $Co_{4-n}P_n$ (n = 0–4), they are worthy of inclusion. Again, the Lewis basicity of the phosphorus atoms is evident in that $Co_2(CO)_6P_2$ forms the derivative $Co_2(CO)_6\{PCr(CO)_5\}_2$ in which the tetrahedral Co_2P_2-core is retained.[232,233]

In terms of core structure, the type **28** cluster appears to be analogous to **1**, **10**, and **18**, although as far as recognizing directly related compounds, only cluster types **18** and **28** will be structurally similar (capping group is NR or PR, a 4-electron fragment); types **10** (capping group is the 3-electron fragment ER where E = group 14 element) and **1** (capping group is BR, a 2-electron unit) require additional numbers of *endo-* or *exo-*ligands. Thus, for example, $H_2Ru_3(CO)_9PPh$,[234,235] $H_2Ru_3(CO)_9NPh$,[175] $H_3Ru_3(CO)_9$-CMe,[236] and $H_4Ru_3(CO)_9BH$[25] form a series of clusters each of which comprises an M_3E-tetrahedron and is *E*-substituted but the number of *endo*-hydrogen atoms differs as a function of the number of valence electrons provided by the group ER. This clearly influences reactivity patterns (compare the sequence illustrated in Figure 47). Synthetic routes to the type **28** cluster are varied, but one common strategy is to make use of readily activated P—H bonds. Thus, for example, reaction of $Ru_3(CO)_{12}$ with a primary phosphine would systematically lead first to a terminally substituted phosphine derivative, then to a bridging phosphido (μ-PPhH) group, and finally to a μ_3-PPh fragment. Analogous osmium systems have been reported.[237] In practice, specificity of product is difficult as shown by the range of products illustrated in Figure 53.[238] Changing to the secondary phosphine PPh_2H gives a variety of Ru_3-based products plus those created by *in situ* cluster condensations.[239] However, here the ratio of reagents is critical; $H_2Ru_3(CO)_9PPh$ is not formed in the 1:1 reaction but is when the ratio $Ru_3(CO)_{12}$:PPh_2H is 1:2. For a 1:3 reaction, products become dependent upon the time of reaction.[239] Related to this strategy is an investigation of the reactivity of $H_2Os_3(CO)_9PPh(C_6H_4)$ (Figure 54)[240]; this cluster exhibits a cyclometalated phenyl substituent[241] which eliminates benzene upon reaction with dihydrogen. The yield of the type **28** cluster is comparable with that obtained either from direct reaction of $Os_3(CO)_{12}$ with PPh_2H,[235] or of $Os_3(CO)_{11}(NCMe)$ with $PPhH_2$ as shown in Figure 54.[242]

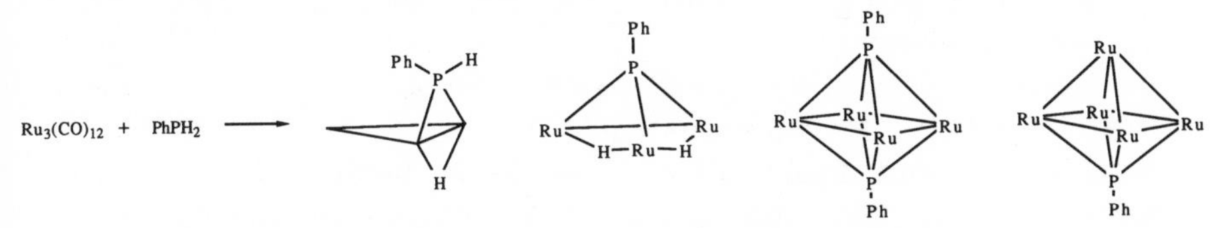

Figure 53. Products of the reaction of $Ru_3(CO)_{12}$ with $PhPH_2$.

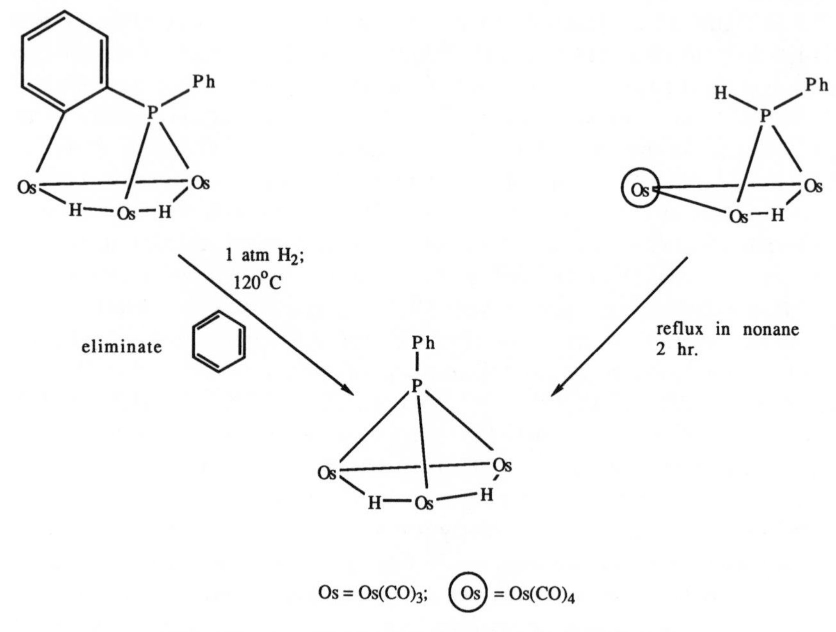

Figure 54. Routes to the type **28** cluster, $H_2Os_3(CO)_9PPh$.

A novel route to cluster class **28** involves cleavage of a bidentate bis-(phosphino) ligand; bis(diphenylphosphino)methane is progressively activated until P—C bond fission with concomitant C—H bond formation occurs to yield a terminal Ph_2PMe ligand and a capping PPh group (Figure 55).[243]

In the introduction, the idea of surface analogies was introduced. Few raft clusters exist, but one that does and, moreover, incorporates a type **28** cluster unit, is $H_2Os_6(CO)_{18}(\mu_3\text{-PPh})$. Interestingly the PPh fragment is ligated over a corner Os_3-site rather than the central one adopted by the μ_3-O atom represented in Figure 2. This presumably is a consequence of the successive closure of a terminal → bridging → capping phosphorus atom; significant perturbation to the Os_6-platform occurs with a "fold" developing in the raft-surface; the internal dihedral angle measured at the point of "fold" is 72.4° (Figure 56).[244] Another variation in cluster **28** is observed in $Co_4(PPh_3)_4(\mu_3\text{-PPh})_4$, a cluster with a tetrahedral Co_4-core, each triangular face being capped by a PPh fragment.[245]

The conversion of the type **28** to **29** cluster is achieved in the case of $Fe_3(CO)_{10}PR$ (R = aryl) going to $Fe_3(CO)_{10}P(R){=}CH_2$ (Figure 57a).[246] Note that, compared to cluster types **2** and **20**, the phosphorus atom bears a terminal substituent, another example of the increased connectivity which is typical of the heavier *p*-block elements. A related species in which the phosphorus atom interacts with the M_3-metal framework to a lesser extent than

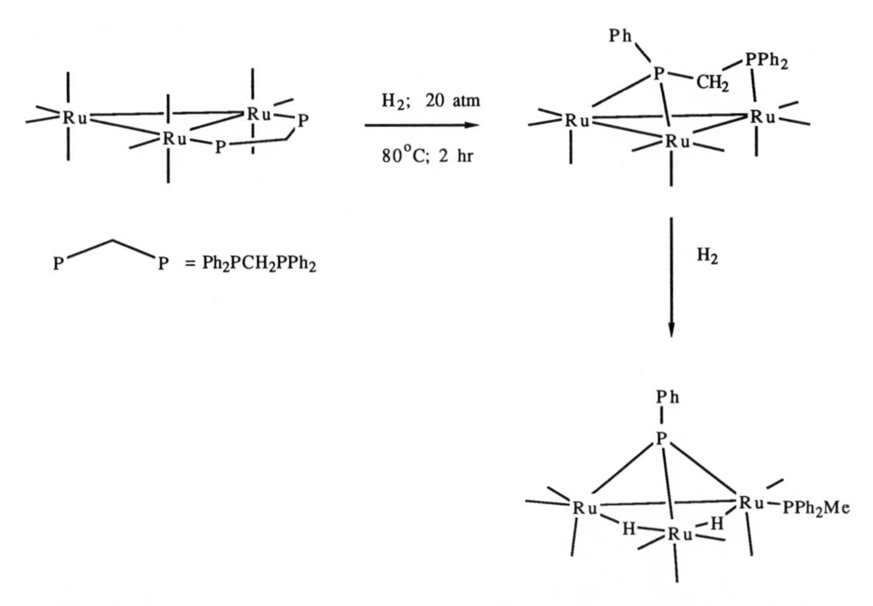

Figure 55. Formation of a type **28** cluster via cleavage of a bidentate phosphino ligand.

in cluster type **29** is $[Fe_3(CO)_9PC(Me){=}CPhH]^-$ shown in Figure 57b.[247] Note that this synthetic reaction apparently involves P—C cleavage since $\{PC(Me){=}CPhH\}$ is formed from $Ph_2PC{\equiv}CMe$.

Like other capping main group fragments discussed in the previous sections, μ_3-PR units tend to "clamp" the metal framework together during reactions thereby allowing metal–metal bond opening and closing to occur but preventing competitive degradation pathways.[248] One example though in which the M—P bonds are directly affected is in the reaction of the $M_3(\mu_3$-PR) cluster with alkynes.[249] This provides a particularly good example of product dependence upon relative P—M, P—C, and M—C bond energies, or rather, since the energetic preferences in the case of M = Fe appear slight,

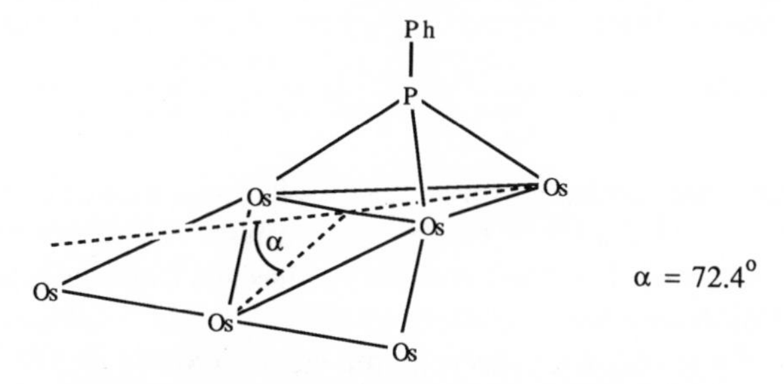

Figure 56. The type **28** cluster supported on an Os_6-raft.

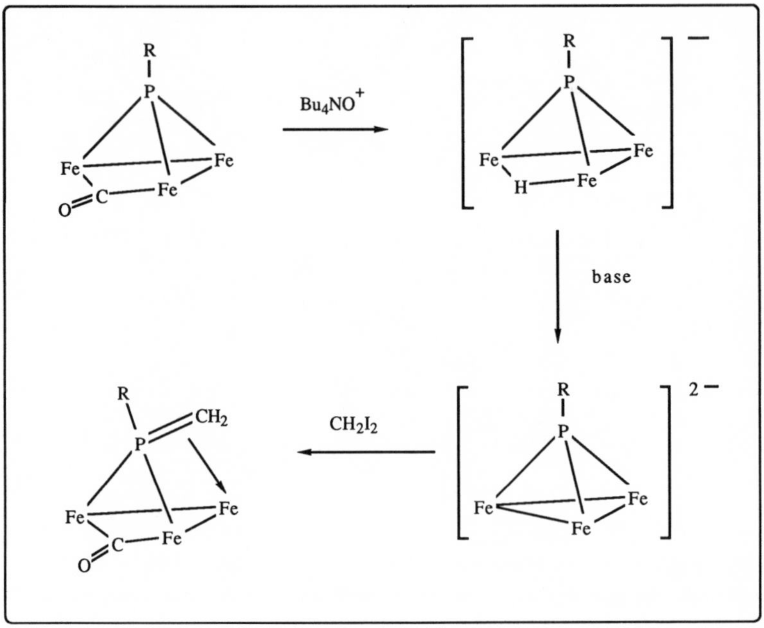

(a)

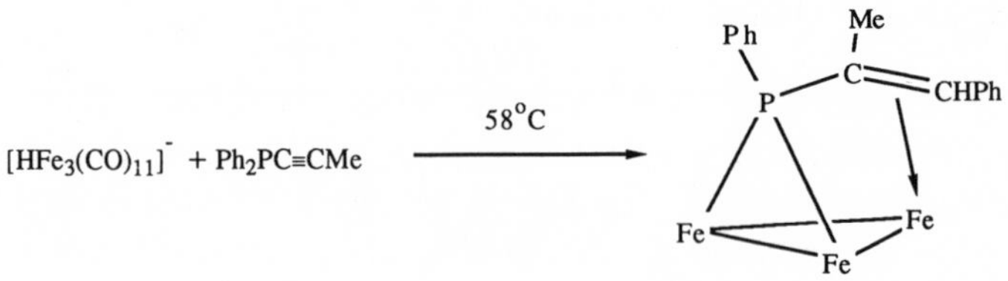

(b)

Figure 57. (a) Formation of $Fe_3(CO)_{10}P(R){=}CH_2$ and (b) the related cluster anion $[Fe_3(CO)_9PC(Me){=}CPhH]^-$.

several products are accessible (Figure 58). Isolation of a given cluster depends critically upon conditions.[249] Figure 59 summarizes a recent development in which carbene functionalized clusters synthesized from the type **28** system $Fe_3(CO)_{10}P^tBu$ have been characterized.[250]

Doubly capped M_3-triangles generate cluster type **31**. Additional cluster

R
P
Fe
Fe
Fe
O=C
PhC≡CPh
hν
−CO
R
P
Ph
Ph
Fe
Fe
Fe
hν
CO
CO
R
P
Fe
Fe
Fe
Ph
Ph
Fe(CO)3
Ph
Fe(CO)3
Ph
P
R
Fe(CO)3

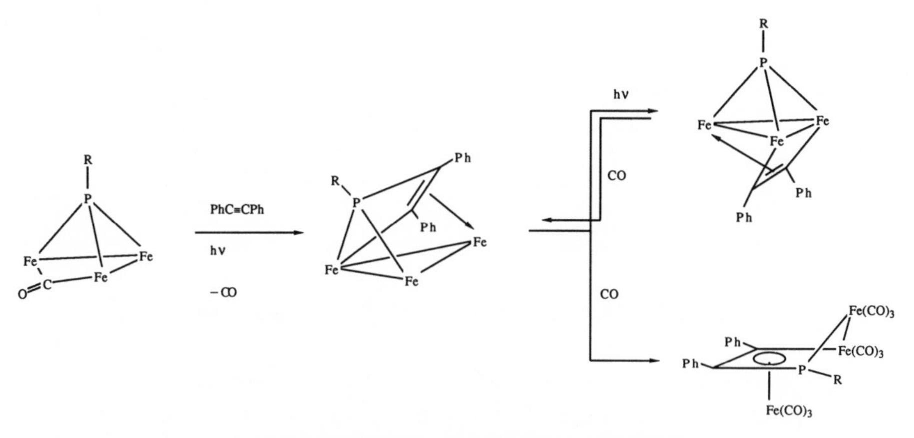

Figure 58. Reaction of a type **28** cluster with alkynes.

electrons would open the cluster to type **32**, the latter being an analog of the imido cluster **19**. However, note that **31** is only observed when each phosphorus atom donates a pair of electrons to an electron-deficient metal fragment while type **32** occurs when the P-substituent is a one-electron substituent (e.g., alkyl or aryl group); i.e., the extra two electrons required to go from **31** to **32** come not from the addition of a ligand, but by a change of P-substituent which effectively releases one valence electron per phosphorus atom for cluster bonding. Treatment of $Fe_2(CO)_9$ with L_nMPX_3 [ML_n = $CpMn(CO)_2$, $Cr(CO)_5$, $Mo(CO)_5$, $W(CO)_5$; X = halide] gives $Fe_3(CO)_9\{ML_n\}_2$, which is also a product of the reaction of $Fe_2(CO)_6\{PML_n\}_2$ with $Fe_2(CO)_9$, or of $[Fe(CO)_4]^{2-}$ with L_nMPX.(251)

Several clusters of type **32** have been characterized as well as related species in which one of the μ_3-PR ligands is replaced by a second *p*-block element or fragment. Examples include $Fe_3(CO)_9(PPh)_2$,(252)

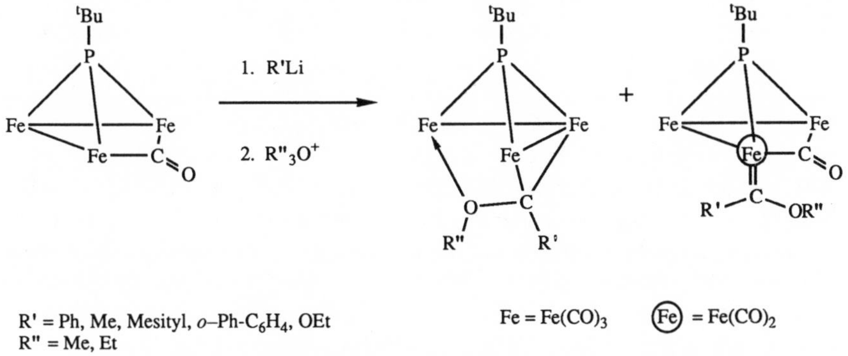

ᵗBu
P
Fe
Fe
Fe
C
O
1. R'Li
2. R"3O+
ᵗBu
P
Fe
Fe
Fe
O
C
R"
R'
+
ᵗBu
P
Fe
Fe
Fe
C
O
C
R'
OR"
R' = Ph, Me, Mesityl, o–Ph-C6H4, OEt
R" = Me, Et
Fe = Fe(CO)3
Fe = Fe(CO)2

Figure 59. Fischer-type carbene synthesis involving $Fe_3(CO)_{10}P^tBu$.

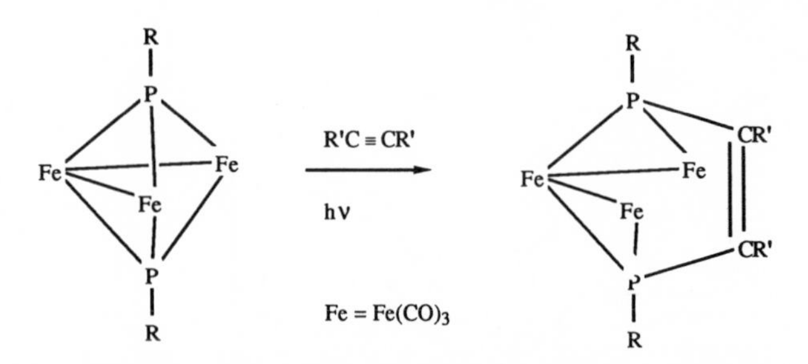

Figure 60. Opening of cluster type **32** by reaction with alkyne.

$Fe_3(CO)_9(PR)(S)$,[253] and $Ru_3(CO)_9(PPh)(S)$.[254] Pathways for the substitution of *exo*-carbonyl ligands in triiron clusters of this type have recently been investigated.[255] With alkynes, symmetric (with respect to the M_3P_2-core) cleavage of two Fe—P bonds occurs as shown in Figure 60; similar reactions between the isoelectronic (with respect to cluster electrons) compounds $Fe_3(CO)_9Se_2$ and $Fe_3(CO)_9Te_2$ give more open products.[256]

The mode of attachment of the PR unit in clusters of type **32** is mimicked in class **33**, since, although an M_4-butterfly framework is present, the PR fragment interacts with only two wing-tip and one of the hinge metal atoms. Thus, a real distinction is observed between the M_4E clusters of types **22** (E = N) and **32** (E = P); steric effects and persistence of the phosphorus atom to retain a terminal substituent are responsible for this difference. The formation of $Ru_4(CO)_{13}(PPh)$ (Figure 61) proceeds via an interesting intermediate cluster, $HRu_3(CO)_9(\mu\text{-}PPh_2)$, which represents a point along the pathway from a μ-PPh_2 bridge to an orthometalated species of the type illustrated in Figure 54. The interaction of one phenyl substituent with a metal atom must be driven by the otherwise electron-deficient (46-electron) Ru_3-triangle. With dihydrogen, the latter eliminates benzene to give $H_2Ru_3(CO)_9PPh$, but upon pyrolysis it undergoes cluster expansion to $Ru_4(CO)_{13}PPh$.[257] The related clusters $H_2Os_3M(CO)_{12}PR$ (M = Os, R = Ph or c-C_6H_{11}; M = Ru, R = c-C_6H_{11}) are products of the reaction of $H_2Os_3(CO)_9PR$ (type **27**) with $M_3(CO)_{12}$ in refluxing nonane. [$Os_3M_2(CO)_{15}PR$ and $Os_3M_3(CO)_{17}PR$ are also formed.] A crystallographic study of $H_2Os_4(CO)_{12}P(c\text{-}C_6H_{11})$ confirms the cage geometry; in solution, the mixed metal cluster $H_2Os_3Ru(CO)_{12}P(c\text{-}C_6H_{11})$ exists as three isomers with the Ru atom occupying the three unique sites, the symmetry of the butterfly having been reduced by the presence of the μ_3-PR fragment.[258] Expansion of the M_4-butterfly to an M_5-skeleton comprising three edge fused triangles (an open square based pyramid) provides the phosphorus with an opportunity to adopt a μ_4-bonding mode. This type **34** category of cluster is represented by $Cp_2Ni_2Ru_3(CO)_9PPh$, which is prepared by a route related to that shown in Figure 61; the unsaturated $HRu_3(CO)_9(\mu\text{-}PPh_2)$ reacts with

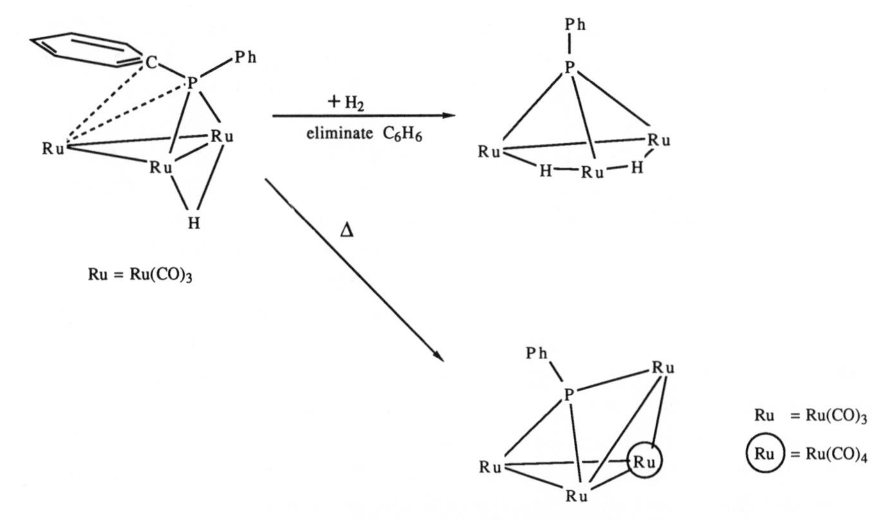

Figure 61. Reactions of the unsaturated cluster $HRu_3(CO)_9(\mu\text{-}PPh_2)$.

$Cp_2Ni_2(CO)_2$ with reductive elimination of benzene. The Ru_3-triangle is retained in the product as the central portion of the type **34** cluster.[259] Closing the open M---M edge of cluster **34** leads nicely to class **38**, although this is a formal pathway rather than one which is experimentally established.

Before considering the M_5P-clusters **38**, there are two further cage types in which the phosphorus atom is in a μ_3-bridging mode, namely, **35** and **36**, both of which are relatively unusual. In the former, the phosphorus atom carries a terminal substituent while in the latter it is bare. Reaction of $CpCo(CO)_2$ with P_4 in refluxing toluene gives the distorted cubane-like cluster $Cp_4Co_4P_4$ (type **36**). The structure consists of two Co_2P_2-butterfly units, joined through wing-tip (P) to hinge (Co) bonds.[260] The distorted trigonal prismatic structure of $H_2Ru_4(CO)_{12}(PPh)_2$ constitutes a novel cluster formed by the addition of electrons to the type **37** cluster, $Ru_4(CO)_{11}(PPh)_2$.[261] The transformation of $Ru_4(CO)_{11}(PPh)_2$ to $H_2Ru_4(CO)_{12}(PPh)_2$ formally involves the cleavage of two Ru—P bonds and two Ru—Ru bonds, and the twisting of the Ru_4P_2-framework to convert an octahedron into a trigonal prism. The octahedral M_4P_2-cluster (**37**) has been confirmed for $Co_4(CO)_{10}(PPh)_2$,[262,263] $Fe_4(CO)_{11}(PPh)_2$,[266–268] $Ru_4(CO)_{11}(PPh)_2$,[261,264,265] as well as for mixed Fe/Ru analogs.[269,270] The latter heterometallic clusters are prepared by irradiating a benzene solution of $Fe_2(CO)_6(\mu\text{-}PHPh)_2$ and $Ru_3(CO)_{12}$. This reaction mixture provides routes to $Fe_4(CO)_{11}(PPh)_2$ (3%), $Fe_3Ru(CO)_{11}(PPh)_2$ (9%), $Fe_2Ru_2(CO)_{11}(PPh)_2$ (9%), $FeRu_3(CO)_{11}(PPh)_2$ (4%), and $Ru_4(CO)_{11}(PPh)_2$ (4%).[269] These systems are of particular interest because they are electronically unsaturated; electrochemical studies show that two electrochemically reversible one-electron reductions may be accessed.[270]

Significantly, carbon monoxide (a source of two more cluster electrons) reacts differently with $Fe_4(CO)_{11}(PPh)_2$ than with $Ru_4(CO)_{11}(PPh)_2$ or any of the heterometallic analogs as Figure 62 shows.[264,266,269] It is surprising that even the presence of a *single* ruthenium atom in the $M_4(CO)_{11}(PPh)_2$ cage has such a marked effect upon its reactivity, and although factors such as relative M—P bond strengths, steric and kinetic factors have been considered qualitatively, the experimental observations have not yet been completely rationalized.[269] Clusters of type **37** have been linked together by use of bidentate phosphino ligands; the nonchelating ligand p-$(MeO)_2P—C_6H_4—P(MeO)_2$ has provided successful results as indicated by the synthesis in 58% yield and full structural characterization of $\{Fe_4(CO)_{10}(PPh)_2\}_2\{p$-$(MeO)_2P—C_6H_4—P(MeO)_2\}$.[269] One cluster related to the type **37** system is $Co_4(CO)_{10}(\mu_4\text{-}S)(\mu_4\text{-}PR)$ (R = Me, C_6H_4-p-OMe).[271]

Cluster nitrides of type **24** appear quite similar structurally to the phosphido clusters in class **38** since both possess a core formula of M_5E. However, significant differences arise from the different steric requirements of the two group 15 elements: (i) while the nitrogen atom interacts with all five metal atoms, the larger phosphorus atom resides well below the square plane of the

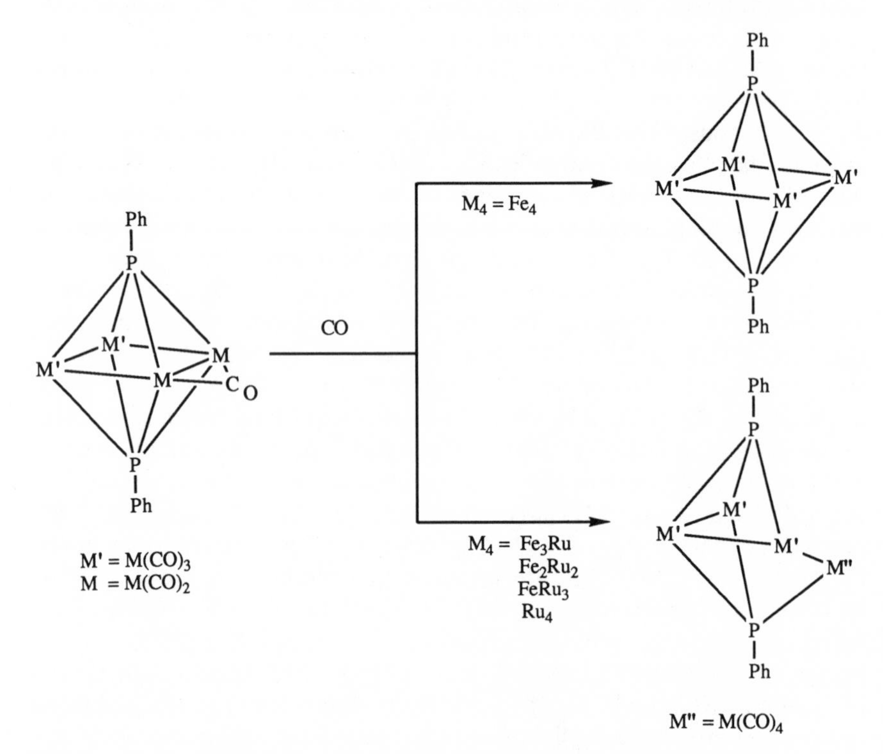

Figure 62. Dependence of the reaction of type **37** cluster with CO upon the metal(s) present.

pyramidal array of metal atoms and may therefore only participate in four P–M interactions; (ii) by virtue of carrying a terminal substituent in cluster type **38**, the phosphorus atom only provides four valence electrons to the cage as opposed to the five contributed by a nitrogen atom in cluster type **24**. Thus, pairs of formally isoelectronic compounds with the M_5E-core (E = N or P) do not exist. The simplest clusters which are truly type **38** include $Os_5(CO)_{15}POMe$,[272,273] $Ru_5(CO)_{15}PR$ (R = Et, Ph, CH_2Ph),[238,274] $Os_5(CO)_{15}PPh$,[237] and $Os_3Ru_2(CO)_{15}PPh$.[237] In the latter cluster, NMR spectroscopic data imply that the two ruthenium atoms occupy basal positions in the square pyramidal cage.[237] The first hepatruthenium cluster to be characterized comprises two type **38** clusters which are fused via a common Ru_3-face (Figure 63); $Ru_7(CO)_{18}(PPh)_2$ is produced in 2% yield by heating a toluene solution of $HRu_3(CO)_{10}(\mu\text{-}PPh_2)$ at 120 °C for 2 h.[275] The more major products from this reaction mixture are $Ru_4(CO)_{13}(\mu_3\text{-}PPh)$ (37%; type **33**), $Ru_5(CO)_{15}(PPh)$ (7%; type **38**), and $HRu_5(CO)_{13}(\mu_4\text{-}PPh)(\mu_2\text{-}PPh_2)$ (7%; derived from type **38**).

Rhenium chemistry figures only marginally in this chapter on transition metal–main group clusters but two compounds which incorporate one or more $Re_4(\mu_4\text{-}PR)$ fragments have been reported and, particularly since they appear to represent the first transition metal phosphinidene clusters *not* to be based upon group 8 metals,[276] their rather novel structures certainly deserve inclusion here. The pyrolysis of $Re_4Cl_2(CO)_{15}\{MePP(Me)PMe\}$ with $Re_2(CO)_{10}$ at 230–250 °C generates $Re_6(CO)_{18}(\mu_4\text{-}PMe)_3$ and $Re_5(CO)_{14}(\mu_4\text{-}PMe)(\mu_2\text{-}PMe_2)\{\mu_3\text{-}PRe(CO)_5\}$ (Figure 64).[276] Perhaps one important feature here is the use of the phosphidene "clamp" as a method of stabilizing high nuclearity clusters of transition metal atoms which may not otherwise

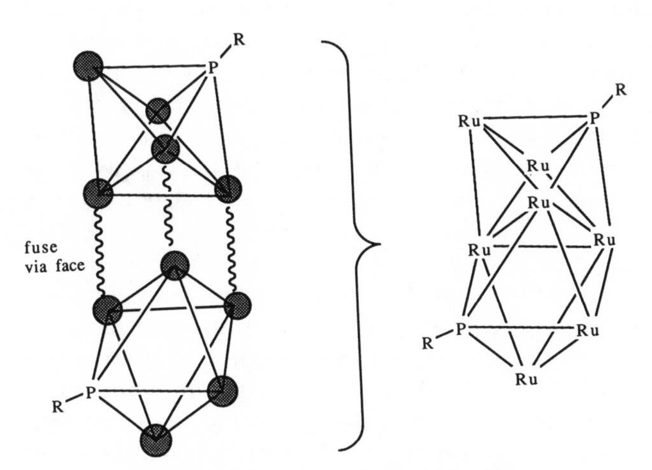

Figure 63. Cage structure of $Ru_7(CO)_{18}(PPh)_2$.

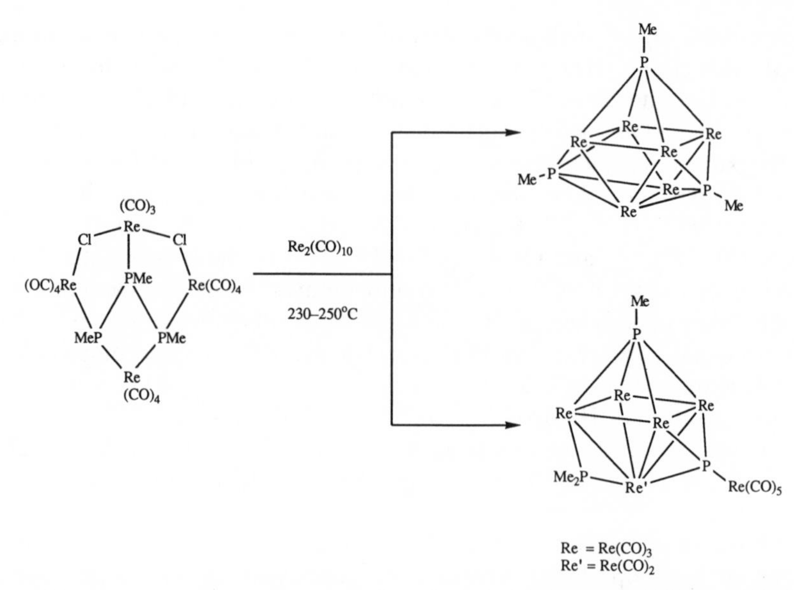

Figure 64. Synthesis and the core structures of $Re_6(CO)_{18}(\mu_4\text{-}PMe)_3$ and $Re_5(CO)_{14}(\mu_4\text{-}PMe)(\mu_2\text{-}PMe_2)\{\mu_3\text{-}PRe(CO)_5\}$; Re = $Re(CO)_3$.

tend to aggregate. The isolation of several high nuclearity nickel clusters also supports this premise. Reaction of $PhP(SiMe_3)_2$ with $NiCl_2(PPh_3)_2$ gives $Ni_8Cl_4(PPh_3)_4(\mu_4\text{-}PPh)_6$ in high yield; the cluster has a cubic Ni_8-core, each face being capped by a phosphinidene ligand.(245) The cluster $Ni_8(CO)_8(\mu_4\text{-}PPh)_6$ has an analogous structure.(277) The chlorine ligands in $Ni_8Cl_4(PPh_3)_4(\mu_4\text{-}PPh)_6$ are subject to substitution reactions, but one significant feature is the removal of the chlorine atoms to give a coordinatively unsaturated cluster $Ni_8(PPh_3)_4(\mu_4\text{-}PPh)_6$. Structural characterization illustrates the presence of a distorted Ni_8-core and reactivity studies show that, as expected, addition reactions occur readily at the unsaturated nickel sites.(278)

Cluster types **39** to **41** form a neat series in which the phosphorus atom is progressively encapsulated within an M_6-cage. The class **39** cluster has been described as part of an icosahedral framework and has one member, viz. $[Co_6(CO)_{16}P]^-$.(279) Prepared by reacting $[Co(CO)_4]^-$ with PCl_3, the $[Co_6(CO)_{16}P]^-$ anion possesses 92 valence electrons, six too many for an octahedron and two too many for a trigonal prism. Inspection of the type **39** cluster geometry (Figure 51) suggests that one suitable description, in keeping with the electron count, is that of an octahedron with three broken edges. The phosphorus atom is too large to occupy the interstitial site in a closed octahedral or trigonal prismatic cage comprising *first-row* transition metal atoms. The structure of $[Co_6(CO)_{16}P]^-$ contrasts with the regular trigonal prism found for $[Co_6(CO)_{15}N]^-$ or the octahedron adopted in $[Co_6(CO)_{13}N]^-$

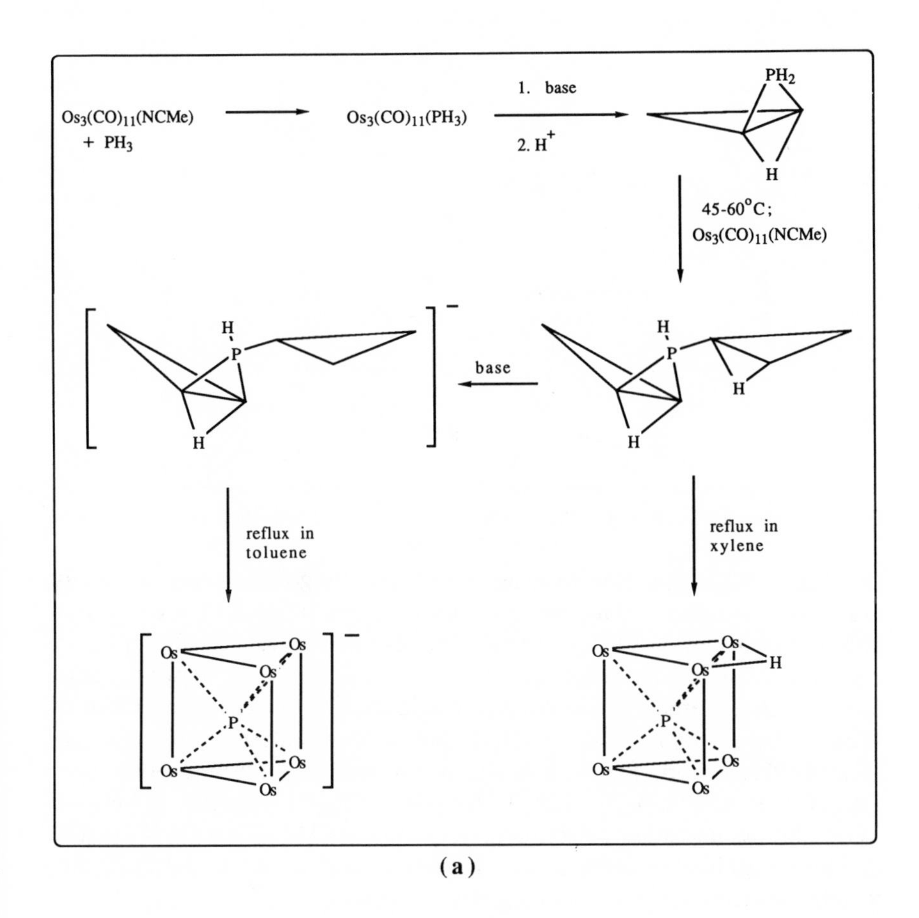

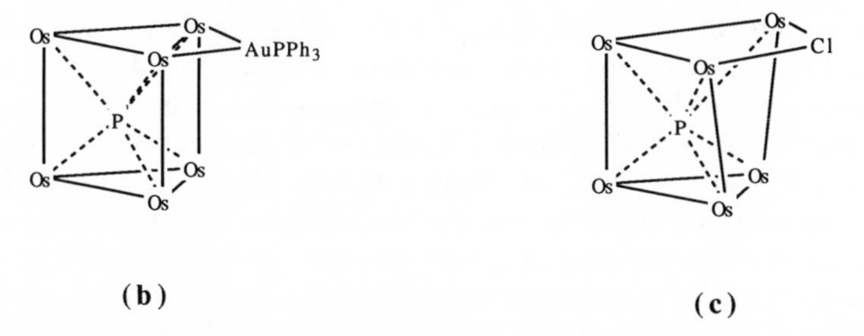

Figure 65. (a) Syntheses of $HOs_6(CO)_{18}P$ and $[Os_6(CO)_{18}P]^-$; (b) structure of $Os_6(CO)_{18}PAuPPh_3$; (c) structure of $Os_6(CO)_{18}PCl$.

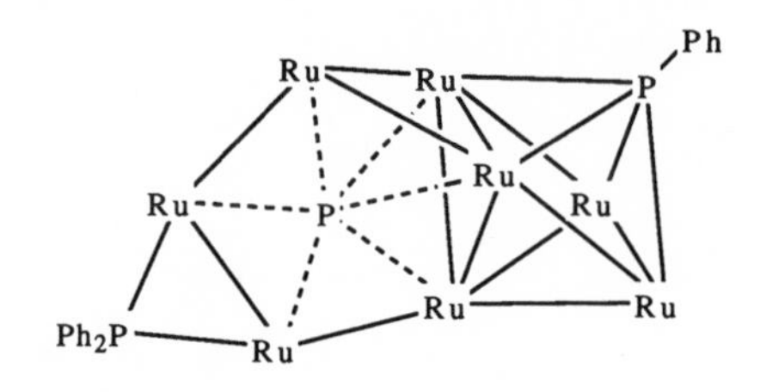

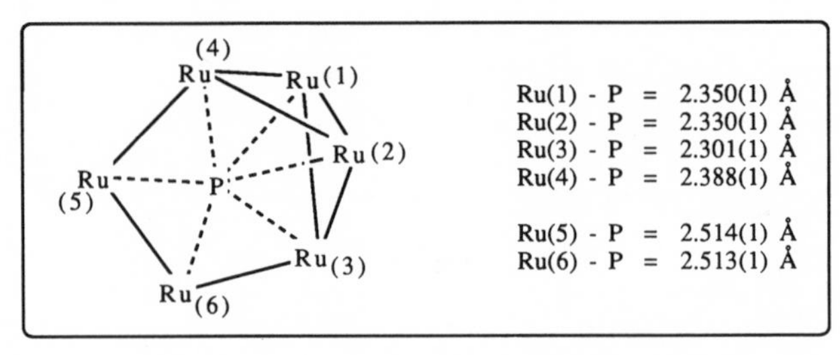

Figure 66. Structure of $Ru_8(CO)_{21}P(\mu_4\text{-}PPh)(\mu_2\text{-}PPh_2)$ and detail of the type **40** subunit present.

(see Figure 49). Once the M_6-cage is enlarged by going to the second- or third-row transition metals, the interstice becomes suitable in size to host a phosphorus atom; thus $[Os_6(CO)_{18}P]^-$ (a 90 valence electron cluster) has a regular trigonal prismatic cage structure.[280,281] The syntheses, shown in Figure 65a, of both this anion and its conjugate acid are indeed elegant and illustrate the successive activation of the P—H bonds of an initially terminally substituted PH_3 ligand. Figure 65b shows the gold(I) phosphine derivative $Os_6(CO)_{18}P(AuPPh_3)$.[281] The cluster $[Os_6(CO)_{18}P]^-$ reacts with $FeCl_3$ in THF solution, undergoing electrophilic addition of Cl^+ to form $Os_6(CO)_{18}PCl$ in which the chlorine atom acts as a three-electron donor to the Os_6P-cage thereby inducing Os—Os bond cleavage (Figure 65c).[282]

The cluster $Ru_8(CO)_{21}P(\mu_4\text{-}PPh)(\mu_2\text{-}PPh_2)$ has recently been prepared and structurally characterized.[283] As in $Ru_7(CO)_{18}(PPh)_2$ described above, the cluster core of $Ru_8(CO)_{21}P(\mu_4\text{-}PPh)(\mu_2\text{-}PPh_2)$ comprises condensed polyhedral units (Figure 66). The two subunits, a square-based pyramid and an unusual type **40** framework, share a common face. The Ru—P bond lengths within this type **40** subunit fall into two groups as shown in Figure 66 and imply that the phosphido atom interacts principally with an Ru_4-butterfly framework and has two weaker interactions to the ruthenium atoms which bridge between the wing-tips of the butterfly. This M_6-framework bears a striking resemblance to that observed in $HRu_4(CO)_{12}B\{AuPPh_3\}_2$ (Figure 17c)[81] and in $HFe_4(CO)_{12}B\{AuPEt_3\}_2$ (Figure 19)[89,104] except for the fact that in these two clusters the Au—Au vector is skewed with respect to the M_4-butterfly while in the type **40** framework the two bridging and two wing-tip ruthenium atoms lie within one plane.[283]

The reaction of $Ru_3(CO)_{12}$ with Ph_2PH has already been described as a means of preparing the type **28** cluster $H_2Ru_3(CO)_9PPh$. The reaction pathway is nonspecific but may be controlled to some extent by altering reaction conditions.(239) When equimolar quantities of the reagents are combined, one product is the type **42** phosphido cluster $Ru_8(CO)_{19}P(\mu_2\text{-}\eta^1,\eta^6\text{-}CH_2Ph)$. The phosphorus atom resides within a square antiprismatic array of metal atoms and Ru–P distances all lie within the range 2.31(1) to 2.43(1) Å.(239) The same antiprismatic cavity is present in $[Rh_9(CO)_{21}P]^{2-}$ (type **43** cluster) and $[Rh_{10}(CO)_{22}P]^{3-}$ (type **44** cluster)(284–286) and preparative routes to these anions are shown in Figure 67. Structural data for each anion illustrate that the phosphorus atom retains a μ_8-mode of attachment (e.g., in $[Rh_9(CO)_{21}P]^{2-}$, Ru_{apical}–P = 3.057(3) Å compared to an average $Ru_{antiprism}$–P = 2.425(3) Å).(284,285) When treated with carbon monoxide, $[Rh_{10}(CO)_{22}P]^{3-}$ undergoes reversible decapping to form $[Rh_9(CO)_{21}P]^{2-}$ and $[Rh(CO)_4]^-$.

Finally, phosphorus, like boron and nitrogen, possesses an NMR active nucleus, although unlike the available NMR active nuclei of boron and nitrogen, ^{31}P is present in 100% abundance and has a spin of $\frac{1}{2}$. Thus, for example, following the patterns observed for borido, carbido, and nitrido clusters, the degree of encapsulation experienced by a phosphorus atom within a transition metal cage is reflected in progressive shift of the respective resonance to lower field.

4.3. Clusters Containing Metal–Arsenic or Antimony Bonds

Progressing down group 15 from phosphorus to arsenic and antimony, two significant factors will be expected to influence the stability and structure of their derivative transition metal clusters: the size and the Lewis basicity of

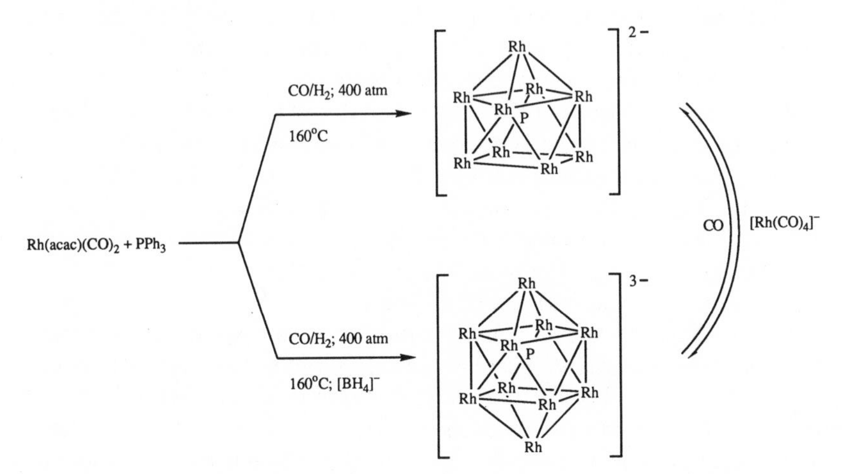

Figure 67. Preparations and interconversion of $[Rh_9(CO)_{21}P]^{2-}$ and $[Rh_{10}(CO)_{22}P]^{3-}$.

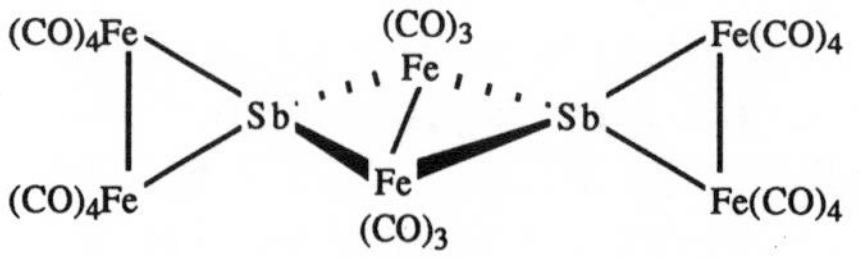

$Sb_2Fe_6(CO)_{24}$

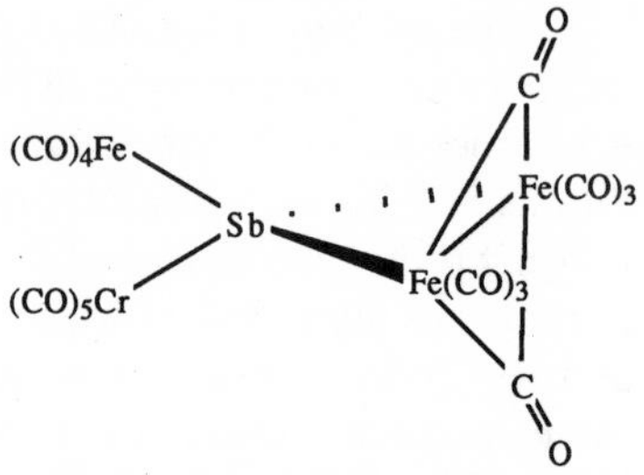

$SbFe_2(CO)_8\{Fe(CO)_4\}\{Cr(CO)_5\}$

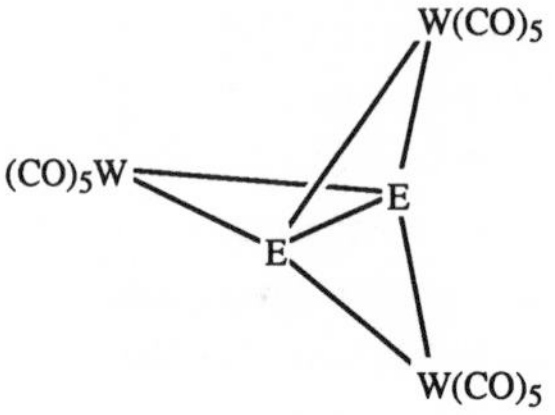

$E_2\{W(CO)_5\}_3$

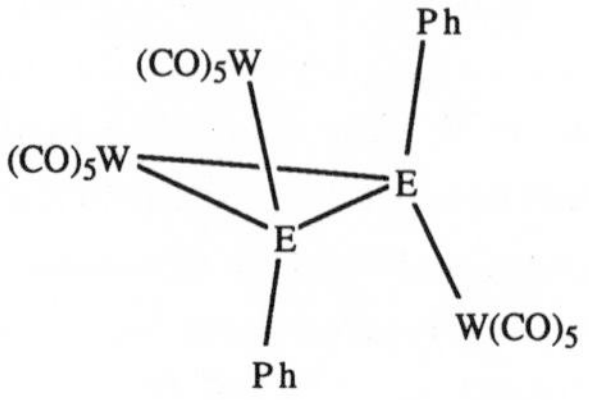

$Sb_2Ph_2\{W(CO)_5\}_3$

E = As; Sb

Figure 68. Structures of open M_nE_m (E = As, Sb) "clusters."

the main group atom. Before considering closed M_nE_m (E = As, Sb) clusters, it is worth noting examples (Figure 68) of compounds which possess open structures and comparing some of their geometries with those in Figures 22 and 23. While the spirocyclic structures of $As_2Fe_6(CO)_{22}$[228] and $Sb_2Fe_6(CO)_{22}$[287–289] compare favorably with that of $[Tl_2Fe_6(CO)_{24}]^{2-}$ (Figure 22), and the coordination sphere around antimony in $SbFe_2(CO)_8\{Fe(CO)_4\}\{Cr(CO)_5\}$ resembles that around a tin or lead atom in $[EFe_2(CO)_8\{Fe(CO)_4\}_2]^{2-}$ (Figure 30), the nature of $As_2\{W(CO)_5\}_3$[290] and $Sb_2\{W(CO)_5\}_3$[291] is not mimicked by compounds incorporating group 13 or 14 elements. In each of the latter as well as in $Ph_2Sb_2\{W(CO)_5\}_3$,[291] the E—E bond is short, implying multiple bond character.

Cluster types for arsenic and antimony containing clusters are shown schematically in Figure 69. There are clearly some similarities with some of those illustrated for the phosphorus derivatives in Figure 51, although the number of classes is less as is the extent to which each type is exemplified in practice. Cluster type **45** was not represented in the phosphorus-containing clusters, although a parallel μ_3-mode of bonding is certainly a feature en-

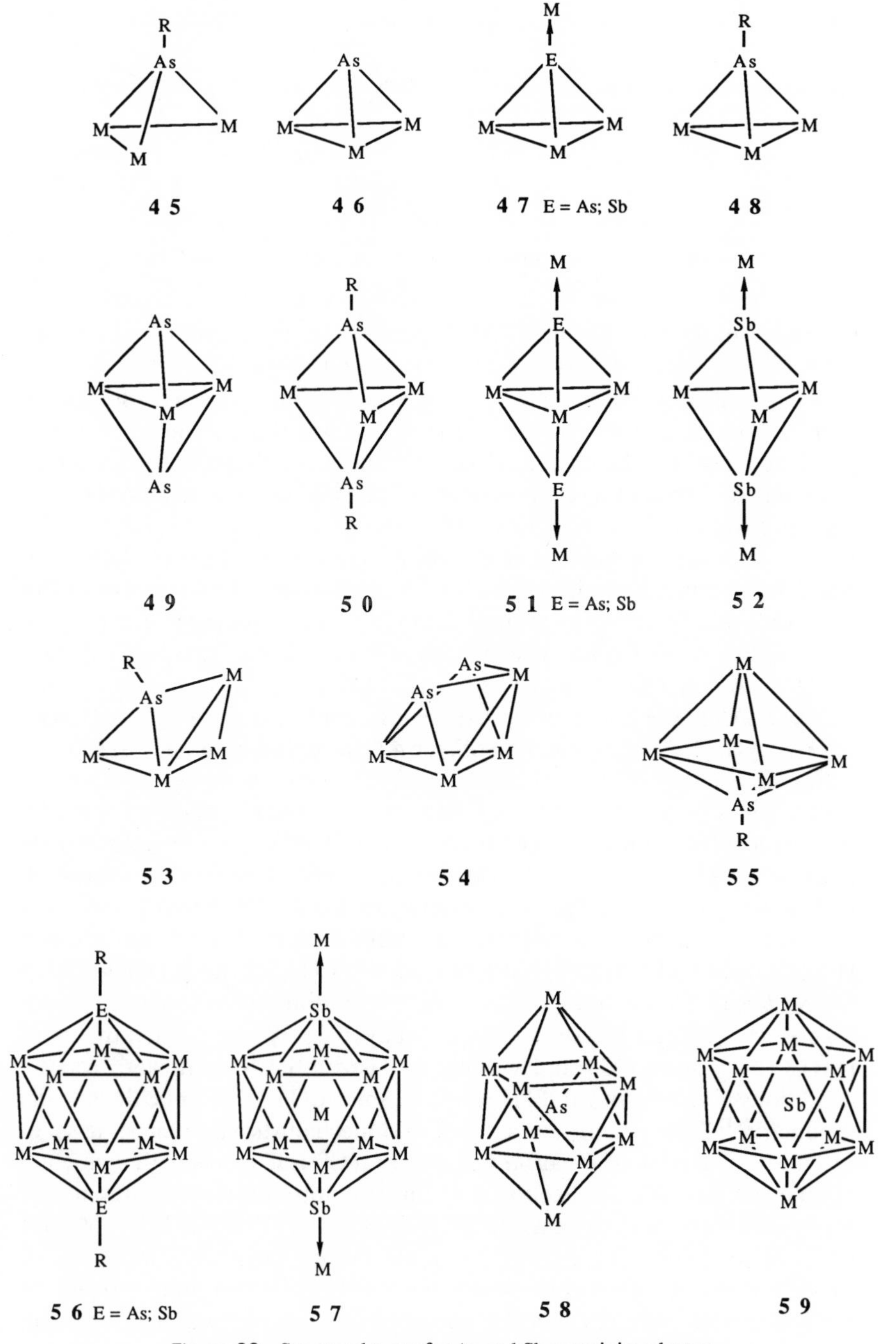

Figure 69. Structural types for As and Sb containing clusters.

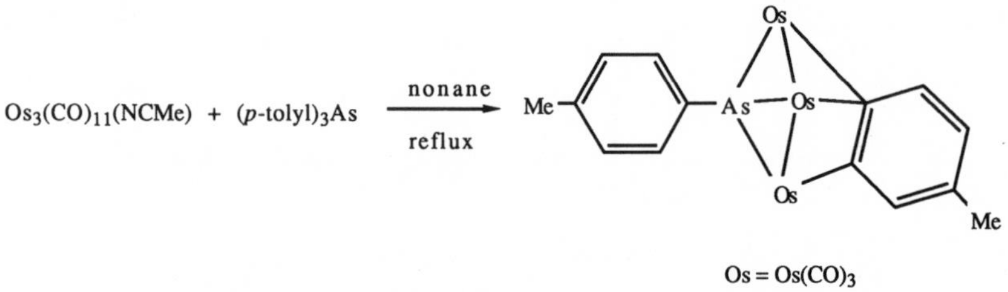

$Os_3(CO)_{11}(NCMe)$ + $(p\text{-tolyl})_3As$ $\xrightarrow[\text{reflux}]{\text{nonane}}$

Os = $Os(CO)_3$

Figure 70. Synthesis and structure of $Os_3(CO)_9\{As(p\text{-tolyl})\}(\mu_3\text{-}C_6H_3Me)$.

countered in the type **32** and **33** M_nP species. The representative of class **45** is $Os_3(CO)_9\{As(p\text{-tolyl})\}(\mu_3\text{-}C_6H_3Me)$ shown in Figure 70.(292)

Capping of a metal triangle by a bare arsenic atom has been demonstrated both in type **46** and **49** clusters. $Co_3(CO)_9As$ was the first member of this group of clusters to be characterized,(227) but it is only stable under an atmosphere of carbon monoxide. As with the phosphide congener, spontaneous trimerization occurs with concomitant loss of CO to give $\{Co_3(CO)_8As\}_3$, the structure (analogous to that shown in Figure 52 for $\{Co_3(CO)_8P\}_3$) of which has recently been determined.(228) Cobalt–arsenic bond lengths within the tetrahedral framework [average 2.285(5) Å] are comparable with those of the dative As → Co bonds which support the trimeric structure [average 2.292(5) Å]. A related cluster is $Co_3(CO)_9As\{Co_4(CO)_{11}\}$ in which the $Co_3(CO)_9As$-cluster functions as an *exo*-substituent replacing a carbonyl ligand in $Co_4(CO)_{12}$.(230) The formation of type **47** clusters relies upon $Co_3(CO)_9As$ acting as a Lewis base to electron-deficient metal fragments; compounds $Co_3(CO)_9As \rightarrow M(CO)_5$ (M = Cr, Mo, W) have been synthesized either by the direct interaction of $Co_3(CO)_9As$ with $(CO)_5M(THF)$ or indirectly by reacting $\{(CO)_5M\}_2AsCl$ with $[Co(CO)_4]^-$.(229,230) Less specific routes use $Co(CO)_3As_3$ or $Co_2(CO)_6As_2$ as starting materials.(229) Parallel antimony chemistry appears to be restricted to three closely related iron clusters $[Fe_3(CO)_{10}SbFe(CO)_4]^-$, $[H_2Fe_3(CO)_9SbFe(CO)_4]^-$, and $[HFe_3(CO)_9\text{-}SbFe(CO)_4]^{2-}$ shown in Figure 71.(293,294) In addition to the route shown in Figure 71, $[HFe_3(CO)_9SbFe(CO)_4]^{2-}$ may be prepared by treatment of $[Fe(CO)_4]^{2-}$ or $Fe(CO)_5$ in alcoholic KOH with $SbCl_5$. However, reaction conditions here are critical if $[HFe_3(CO)_9SbFe(CO)_4]^{2-}$ is to be the specific product.(294) Figure 72 depicts a novel, and mechanistically interesting set of reactions. The pentairon carbido cluster $Fe_5(CO)_{15}C$ reacts with elemental arsenic (or bismuth, see Section 4.4) but not antimony in the presence of water or sulfuric acid to generate $Fe_3(CO)_9CH(E)$ (E = As; Bi).(295) Thus (i) a carbido center is converted to an alkylidyne fragment and (ii) a μ_3-E cap is formed; the analogous cluster $Fe_3(CO)_9CH(Sb)$ is produced only if $SbCl_5$ is used in place of elemental antimony. In terms of bonding, the three-electron donor atom E replaces three μ-H atoms in the isoelectronic (valence

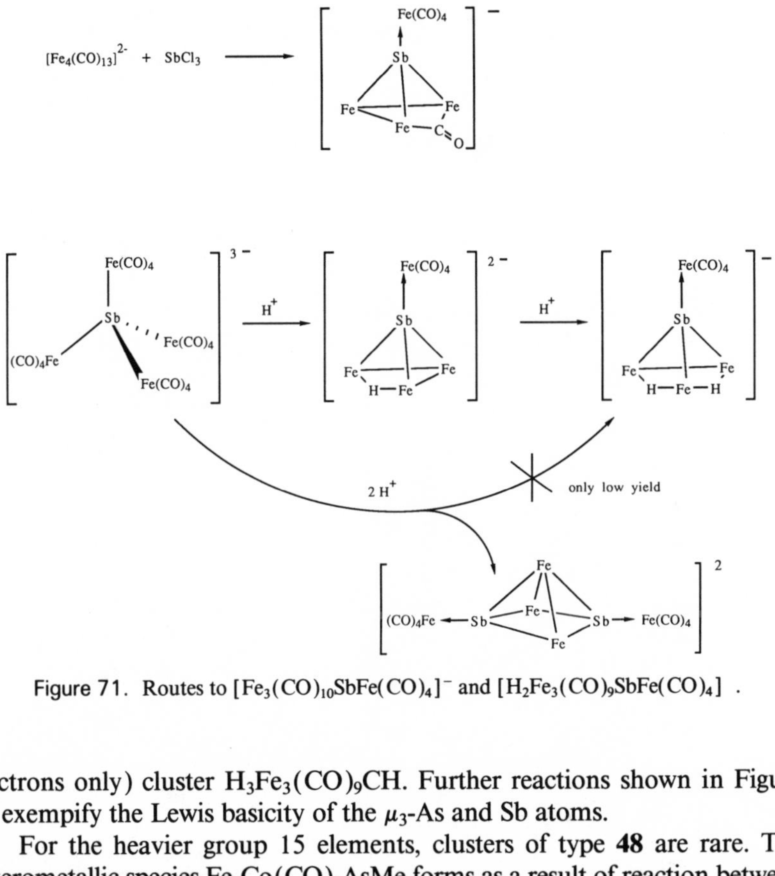

Figure 71. Routes to $[Fe_3(CO)_{10}SbFe(CO)_4]^-$ and $[H_2Fe_3(CO)_9SbFe(CO)_4]$.

electrons only) cluster $H_3Fe_3(CO)_9CH$. Further reactions shown in Figure 72 exempify the Lewis basicity of the μ_3-As and Sb atoms.

For the heavier group 15 elements, clusters of type **48** are rare. The heterometallic species $Fe_2Co(CO)_9AsMe$ forms as a result of reaction between $Co_2(CO)_8$ and $Fe(CO)_4(AsMeH_2)$.[296] Clearly, as in phosphido/phosphinidene chemistry, activation of the E—H bond is a rational strategy to the capping ER group. By use of metal exchange reactions (see parallel reactions in Section 2.2) $Fe_2Co(CO)_9AsMe$ is a precursor to other mixed metal clusters such as $CpMoFeCo(CO)_8AsMe$ also of type **48**.[297] More recently, $H_2Os_3(CO)_9AsPh$ has been reported (Figure 73.)[298]

In contrast to the phosphorus atom, naked μ_3-capping arsenic atoms may be stable as illustrated by examples given above. The type **49** cluster $Fe_3(CO)_9(As)_2$ has been crystallographically characterized.[299] The Fe—Fe bond distances are 2.626(6), 2.612(7), and 2.630(7) Å, and a closed metal triangle (i.e., as compared to type **50** and **52** clusters in which nonbonded Fe–Fe separations typically exceed 3.5 Å) is consistent with $Fe_3(CO)_9(As)_2$ possessing a Wadean count of six pairs of skeletal electrons. Attaching a terminal substituent to the capping group 15 atom *increases* the number of electrons available for cluster bonding from three to four; i.e., in

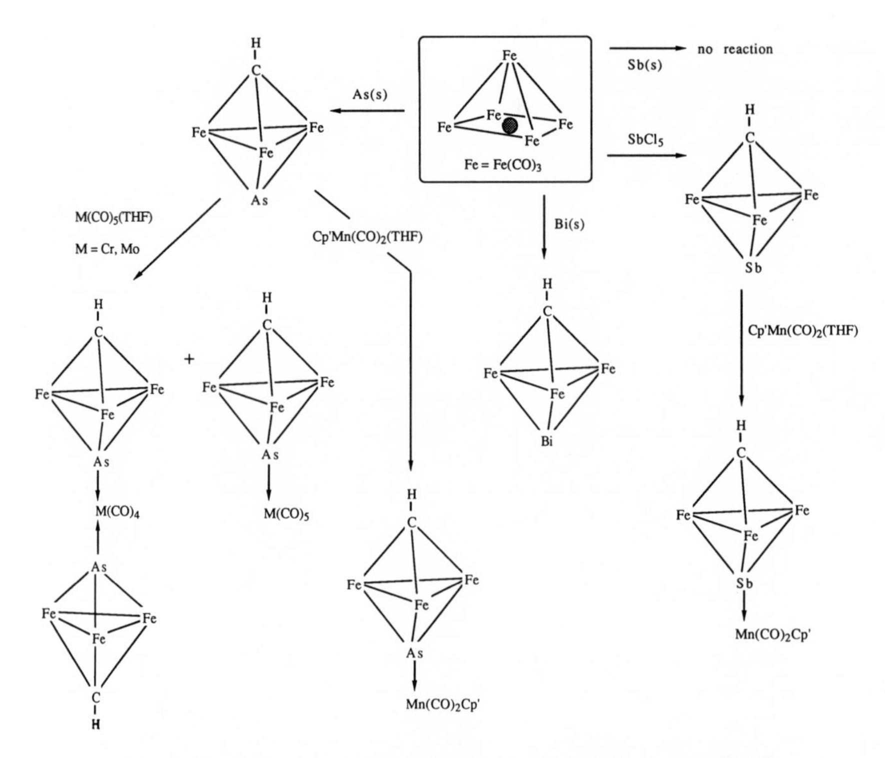

Figure 72. Formation and reactivity of $Fe_3(CO)_9CH(E)$, E = As; Sb; Bi.

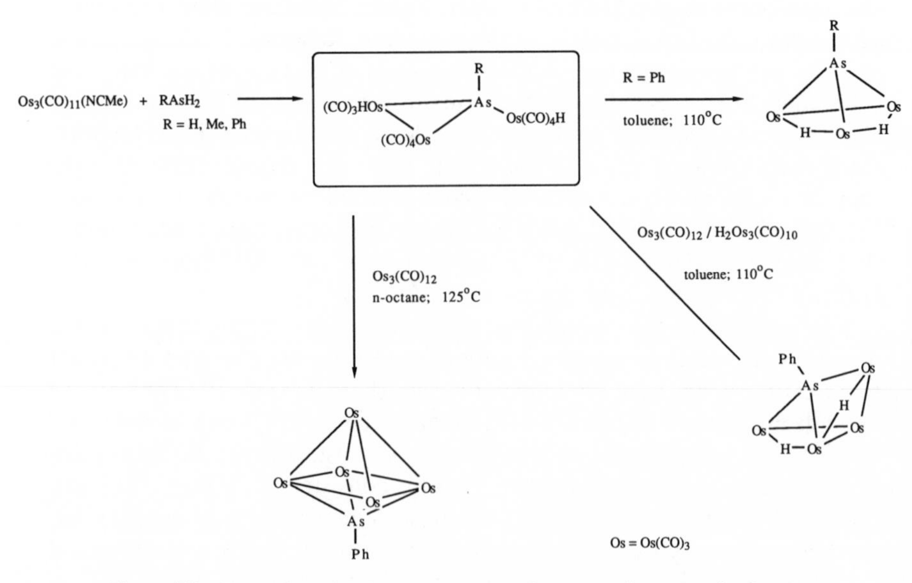

Figure 73. Formation of and core structures of some osmium–arsenic clusters.

$Fe_3(CO)_9(As)_2$ each arsenic atom exhibits a lone pair of electrons but in $Fe_3(CO)_9(AsPh)_2$,[300–302] each As—C_{Ph} bond is a localized two-center–electron bond, thus leaving four valence electrons in the frontier obitals of each PhAs fragment. Thus, $Fe_3(CO)_9(AsPh)_2$ and the isoelectronic (valence electrons only) compound $Fe_3(CO)_9(AsPh)S$[303] are type **50** clusters. $Fe_3(CO)_9(AsPh)_2$ has been prepared by the reaction of $Cp(CO)_2MnAsPhCl_2$ with $Fe_2(CO)_9$,[300,301] and by treatment of $[Fe(CO)_4]^{2-}$ with $PhAsCl_2$.[302]

If the lone pair of electrons localized on the group 15 atom in cluster type **48** is donated to a 16-electron metal center, then one would predict that the M_3E_2-core would *not* undergo structural change. Thus, cluster type **51** is verified in the cluster $Fe_3(CO)_9\{SbCr(CO)_5\}_2$ and in the arsenic analog. A further example in which this type of coordinate interaction occurs is in $[Fe_3(CO)_9\{SbFe(CO)_4\}_2]^{2-}$ but, with the extra pair of electrons contributed by the dinegative charge, an open cluster of type **52** is observed instead of **51**.[293] Formal oxidation and reduction should interconvert **51** and **52** but, as yet, such pairs of clusters have not been characterized for a single formulation.

An M_4-butterfly framework is certainly too small to support an *interstitial* arsenic or antimony atom, and, as with phosphorus, cluster type **53** replaces the $M_4(\mu_4\text{-}E)$-core which is characteristic of first-row main group atoms. $H_2Os_4(CO)_{12}AsPh$ (Figure 73) is the sole member of class **53**,[298] and is structurally analogous to $Ru_4(CO)_{13}PPh$.[257] Since the M_4-butterfly is asymmetrically "occupied" by the main group fragment in type **53** clusters, it is perhaps no surprise to find that a cluster with a type **54** core has also been prepared. Note, however, that in **53** the arsenic atom bears a terminal substituent while in **54** it has an associated lone pair of electrons. The compound (shown in Figure 74) which exhibits this unit is a condensed cluster,

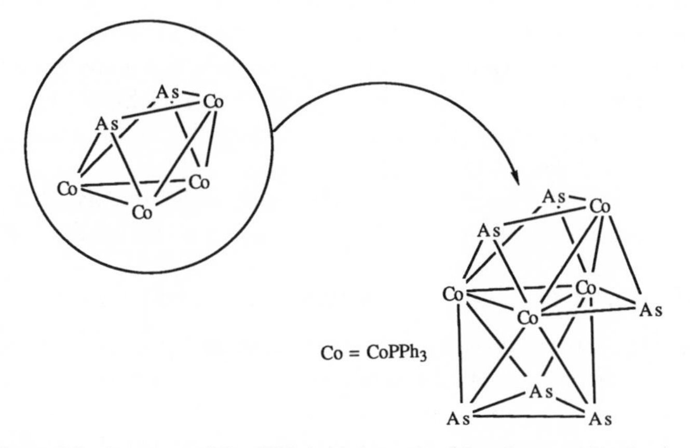

Figure 74. Structure of $Co_4(PPh_3)_4(As)_6$ emphasizing the type **54** subunit.

$Co_4(PPh_3)_4(As)_6$, and also incorporates a type **46** subunit. Presumably in response to the Lewis basicity of the bare arsenic atoms, $Co_4(PPh_3)_4(As)_6$ reacts with $Co_2(CO)_8$, $Ni(CO)_4$, and $M(CO)_5(THF)$ (M = Cr, Mo) but details of the products are not yet available.[(304)]

Although the chemistry of $M_4(\mu_4\text{-}PR)$ or $M_5(\mu_4\text{-}PR)$ clusters is now quite well developed, that of their arsenic or antimony analogs is not. Figure 73 shows the formation of $Os_5(CO)_{15}AsPh$,[(298)] an example of a type **55** cluster which is directly analogous to $Os_5(CO)_{15}PPh$.[(237)] For phosphorus, the "clamping" effect of the PPh group aids the stabilization of such molecules as $Ni_8Cl_4(PPh_3)_4(\mu_4\text{-}PPh)_6$.[(245)] Similarly $PhAs(SiMe_3)_2$ has been used as a source of a main group fragment which helps to stabilize a body-centered cubic array of nickel atoms in $Ni_9Cl_3(PPh_3)_5(\mu_4\text{-}As)_6$ and $Ni_9Cl_2(PPh_3)_6(\mu_4\text{-}As)_6$.[(305)] Significantly though, each capping arsenic atom, in contrast to phosphorus, is devoid of a terminal substituent.

Both arsenic and antimony atoms are capable of spanning five (first row to date) metal atoms and such a μ_5-mode of bonding is exemplified in the type **56** cluster $[Ni_{10}(CO)_{18}(\mu_5\text{-}ER)_2]^{2-}$ (E = As, Sb; R = Me, Ph),[(306,307)] in the related anion $[Ni_9(CO)_{15}(\mu_5\text{-}AsPh)_3]^{2-}$,[(306)] and in the unusual nickel centered type **57** cluster anions $[Ni_{11}(CO)_{18}Sb\{SbNi(CO)_3\}_2]^{n-}$ (n = 2,3,4).[(308)] While reaction of $[Ni_6(CO)_{12}]^{2-}$ with Ph_2SbCl or $PhSbCl_2$ leads to $[Ni_{10}(CO)_{18}(\mu_5\text{-}SbPh)_2]^{2-}$, use of $SbCl_3$ in place of the phenyl derivatives generates $[Ni_{11}(CO)_{18}Sb\{SbNi(CO)_3\}_2]^{n-}$. The trianion, along with the tetra-anion (not structurally characterized), are isolated from the reaction residue after treatment with carbon monoxide. Conversion between tetra-, tri-, and di-anionic clusters is achieved by reversible oxidation using Na/I_2. It has been suggested that the interstitial nickel atom in $[Ni_{11}(CO)_{18}Sb\{SbNi(CO)_3\}_2]^{n-}$ contributes only valence shell s and p orbitals, and may be considered to have an inert $3d^{10}$ configuration.[(308)]

Interstitial arsenic and antimony atoms are scarce, and one reason for this must surely be due to the steric requirements of these heavy group 15 atoms. The bicapped Archimedean antiprism provides a cavity of size compatible with that of the arsenic atom (type **58** cluster, $[Rh_{10}(CO)_{22}As]^{3-}$)[(309)] while the icosahedron is able to accommodate the antimony atom giving cluster type **58** ($[Rh_{12}(CO)_{27}Sb]^{3-}$).[(310)]

Little information is available so far regarding the reactivity of transition metal–arsenic and antimony bonds; clearly this area is ripe for exploration.

4.4. Clusters Containing Metal–Bismuth Bonds

The chemistry of bismuth-containing clusters has only recently been developed.[(311)] Several open "clusters" have been characterized, for example, $Bi_2\{W(CO)_5\}_3$[312] which is isostructural with $As_2\{W(CO)_5\}_3$ and $Sb_2\{W(CO)_5\}_3$ (Figure 68). The anion $[Bi_2Fe_2Co(CO)_{10}]^-$ exhibits a structure closely related to that of $Bi_2\{W(CO)_5\}_3$ and is produced by metal substitution (with concomitant Bi—Bi bond formation) starting from the type **61** cluster $Fe_3(CO)_9Bi_2$ (Figure 75).[313] The open framework of $Fe_2(CO)_8Bi_2Me_2$ is derived from the Zintl ion type cluster anion $[Bi_4Fe_4(CO)_{13}]^{2-}$, which is in turn synthesized either by halide displacement from $BiCl_3$ or from the type **60** anion, $[Fe_3(CO)_{10}Bi]^-$ (Figure 75).[314–316] Earlier sections in this chapter have illustrated that spirocyclic structures were a common feature for open clusters containing Tl, Si, Ge, Sn, Pb, and Sb. A spirocyclic environment for a bismuth atom is restricted to that in $HRu_5(CO)_{18}Bi$ (Figure 76) in which the group 15 element acts as a two-electron donor to the $\{Ru_2(CO)_8\}$ unit and as a three-electron donor to the $\{HRu_3(CO)_{10}\}$ unit, thereby rendering each metal atom saturated according to the 18-electron rule.[317]

Cluster types for bismuth-containing systems are summarized in Figure 77. The type **60** cluster exhibits a naked bismuth atom supported upon a trimetal framework. In comparison with elements higher than bismuth in group 15, there is a reduced tendency for the bismuth atom to function as a Lewis base toward electron-deficient transition metal fragments. Thus the clusters $Co_3(CO)_9Bi$,[154,155] $Ir_3(CO)_9Bi$,[318] $[Fe_3(CO)_{10}Bi]^-$,[319] $Fe_3(CO)_9(\mu_3\text{-COMe})Bi$,[320] $H_3Fe_3(CO)_9Bi$,[320] $Cp_3Fe_3(CO)_3Bi$,[321]

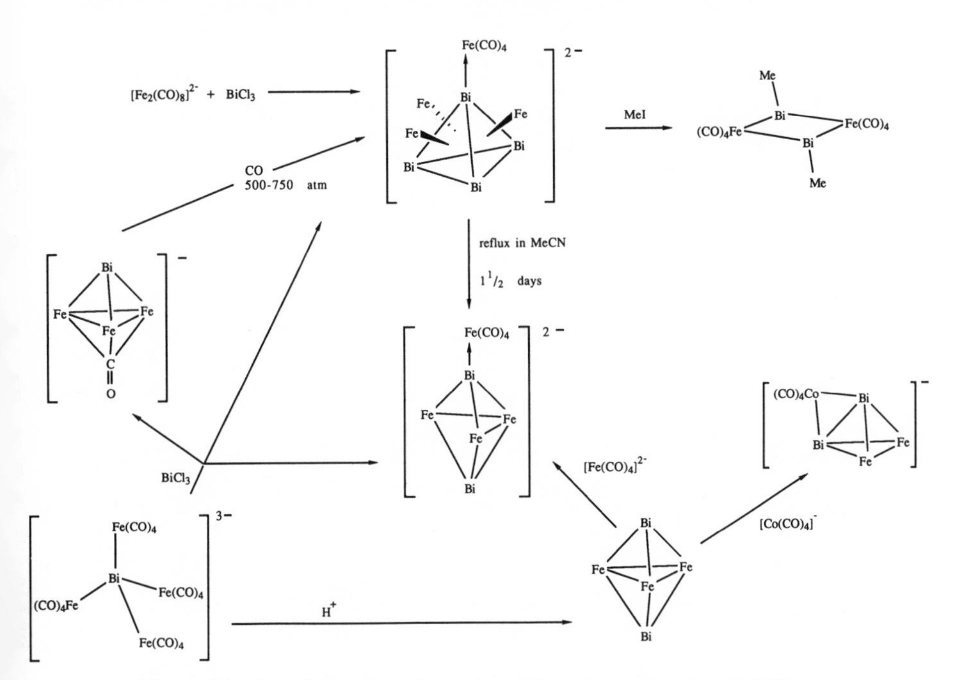

Figure 75. Transformations of some iron–bismuth clusters; Fe = $Fe(CO)_3$.

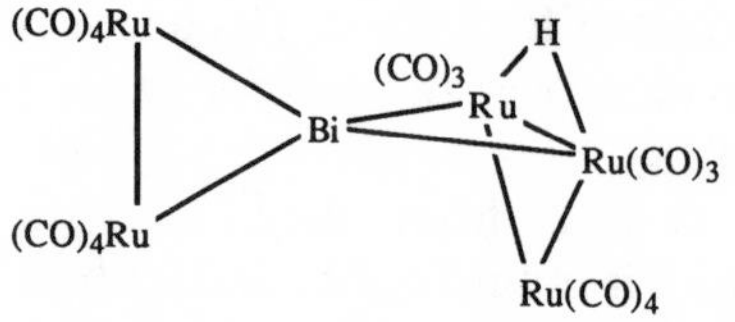

$(CO)_4Ru$ H
$(CO)_3$
Bi Ru
$Ru(CO)_3$
$(CO)_4Ru$
$Ru(CO)_4$

Figure 76. Structure of $HRu_5(CO)_{18}Bi$.

$H_3Ru_3(CO)_9Bi$,[322] and $H_3Os_3(CO)_9Bi$[322] are all stable as monomers in contrast to the spontaneous trimerization suffered by $Co_3(CO)_9P$ and $Co_3(CO)_9As$ (Sections 4.2 and 4.3, respectively).[226–228] Synthetic routes to clusters of type **61** generally fall into three groups: (i) closure of an open cluster of the type $Bi\{ML_n\}_m$ (m = 3 or 4), as shown in Figure 29 for $Co_3(CO)_9Bi$ and in Figure 75 for $[Fe_3(CO)_{10}Bi]^-$ and also observed in the photolysis of $Bi\{Fe(CO)_3Cp\}_3$ to form $Cp_3Fe_3(CO)_3Bi$[321]; (ii) the reaction of a transition metal carbonyl with bismuthate ion as illustrated in Figure 78; (iii) reaction of bismuth(III) nitrate with metal carbonyls (Figure 79). As with other transition metal–main group clusters, it is the chemistry of the group 8 metals that has received the most attention. Reaction conditions are

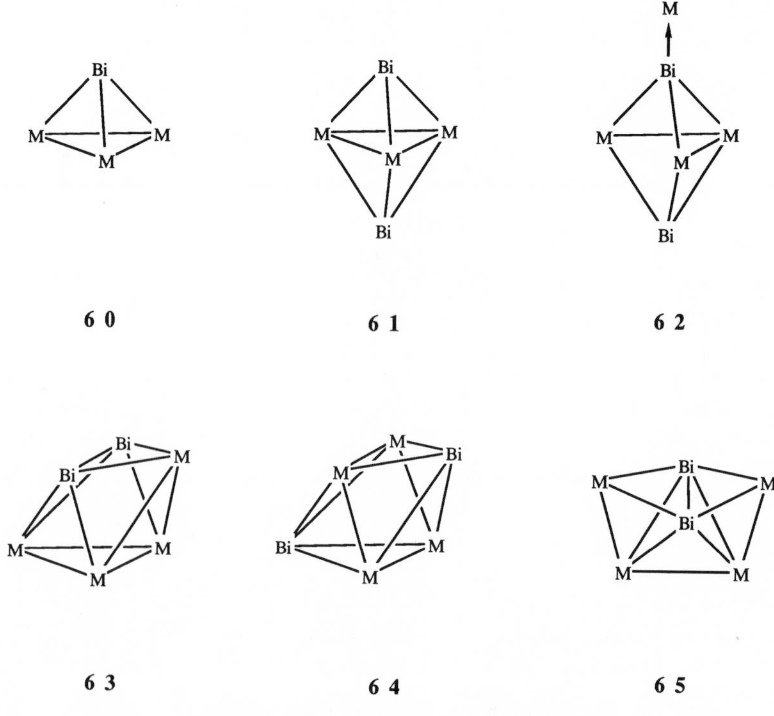

Figure 77. Structural types for M_nBi_m clusters.

critical and choice of starting cluster [e.g., $Os_3(CO)_{12}$, $Os_5(CO)_{16}$, $Os_6(CO)_{18}$, or $[HOs_3(CO)_{11}]^-$] controls the product distribution. For both the ruthenium and osmium based clusters, production of the type **60** system may be accompanied by the formation of the related type **61** cluster and either the type **63** (when M = Os) *or* type **64** (M = Ru). The relationship between clusters **60** and **61** is simply that of a *nido* and *closo* pairing; each is a six-electron pair cluster by PSEPT. For example, in the case of $H_3Ru_3(CO)_9Bi$ and $Ru_3(CO)_9Bi_2$, three valence electrons are provided either by three *endo*-hydrogen atoms or by a bismuth atom in a vertex position. Cluster types **63** and **64** are isomers of one another and, by using isolobal fragment replacements, correspond to the compounds 1,2-$C_2B_4H_6$ and 1,6-$C_2B_4H_6$ [BH is isolobal with $Ru(CO)_3$ or $Os(CO)_3$, and CH is isolobal with Bi]. In this pair of carbaboranes, the latter is thermodynamically favored over the former.[323] It is reasonable to propose that the persistence of *1,2*-Bi_2*Os_4*$(CO)_{12}$ in contrast to *1,6*-Bi_2*Ru_4*$(CO)_{12}$ is due to the relative Os—Os and Os—Bi versus Ru—Ru and Ru—Bi bond strengths.

Returning briefly to the closure of Bi$\{ML_n\}_m$ systems, one recent report emphasizes that clean elimination of CO with accompanying M—M bond formation does not always occur. In refluxing THF solution, Bi$\{Co(CO)_4\}_3$

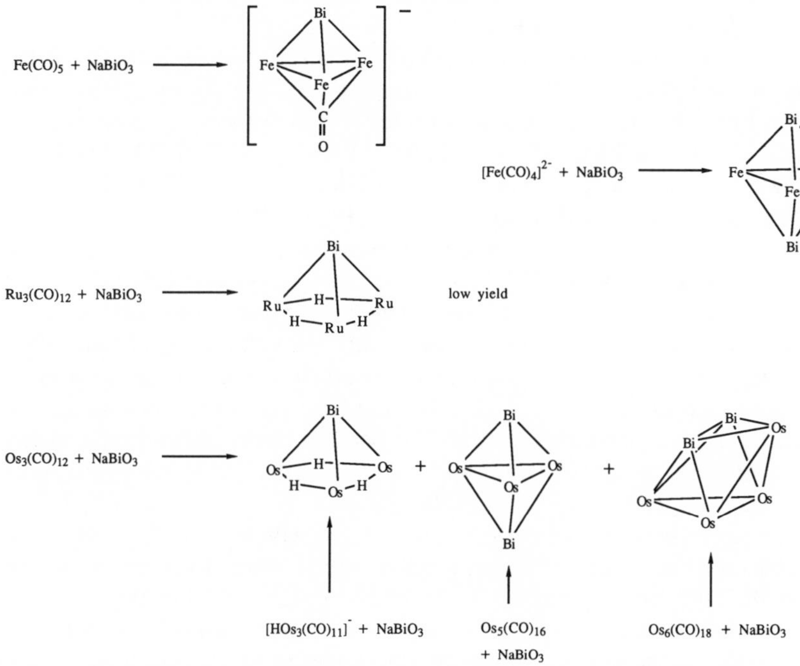

Figure 78. Use of $NaBO_3$ as a source of cluster bound bismuth atoms.

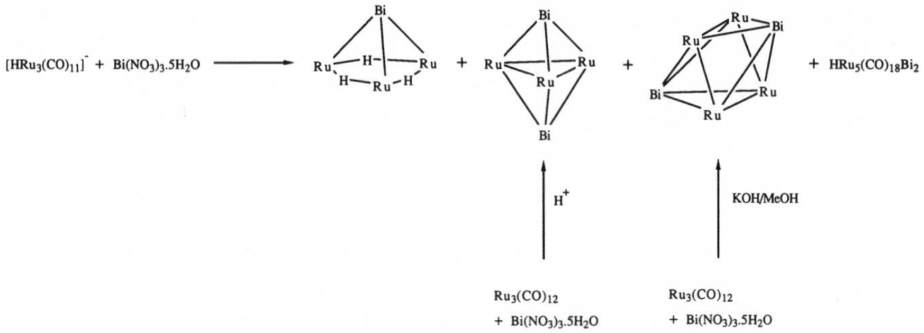

Figure 79. Synthetic routes employing $Bi(NO_3)_3 \cdot 5H_2O$.

forms the cubane-like cluster $[Co(CO)_3(\mu_3\text{-}Bi)]_4$.[324] An antimony analog of this cluster has also been characterized.[325] For the early transition metal complex $Bi\{MoCp'(CO)_3)\}_3$, photolysis yields $Cp'_2Mo_2(CO)_4Bi_2$ rather than a cluster which retains the original $BiMo_3$-core.[326]

A cluster which is related to type **61** is $Fe_3(CO)_9Bi(CH)$ illustrated in Figure 72. Once again, the lone pair located on the μ_3-Bi atom appears quite resistant to donation to 16-electron metal fragments.[295] Representative clusters of type **61** are restricted to $M_3(CO)_9Bi_2$ (M = Fe,[327] Ru,[322] Os[322]). In keeping with the expectations of PSEPT, reduction of *closo*-$Fe_3(CO)_9Bi_2$ leads to *nido*-$[Fe_3(CO)_9Bi_2]^{2-}$ and, interestingly, the paramagnetic monoanion intermediate is produced en route.[314] The process is reversible, oxidation of $[Fe_3(CO)_9Bi_2]^{2-} \rightarrow [Fe_3(CO)_9Bi_2]^{-} \rightarrow Fe_3(CO)_9Bi_2$ being achieved by using copper(I) ions. $[Fe_3(CO)_9Bi_2]^{2-}$ and $[Fe_3(CO)_9Bi_2]^{-}$ both decompose on standing in solution to give $[Bi_4Fe_4(CO)_{13}]^{2-}$.[314] As Figure 75 shows, the latter Zintl ion based complex may be converted into a type **62** cluster by thermolysis.[315] The product $[Fe_3(CO)_9Bi_2\{Fe(CO)_4\}]^{2-}$ is produced in 78% yield and its structure closely parallels that of $[Fe_3(CO)_9\{SbFe(CO)_4\}_2]^{2-}$ (Figures 71 and 74) with one significant difference: only *one* of the bismuth atoms (in contrast to both antimony atoms) functions as a Lewis base.[313–316] That the lone pair of the bismuth atom functions as an active Lewis base at all is perhaps surprising and presumably reflects changes in hybridization and charge distribution as the Fe_3Bi_2-core opens up from the type **61** to **62** cluster.

Cluster types **63** and **65** both possess an M_4Bi_2-core with the bismuth atoms bonded together. In **65**, however, the four metal atoms form an open unit rather than the more commonly exemplified butterfly. To date the paramagnetic cluster $[Co_4(CO)_{11}Bi_2]^{-}$ is the only member of group **65** (Figure 80).[155]

Preparative routes are still being developed to M_nBi_m clusters, so new chemistry of these molecules can be anticipated in the near future.

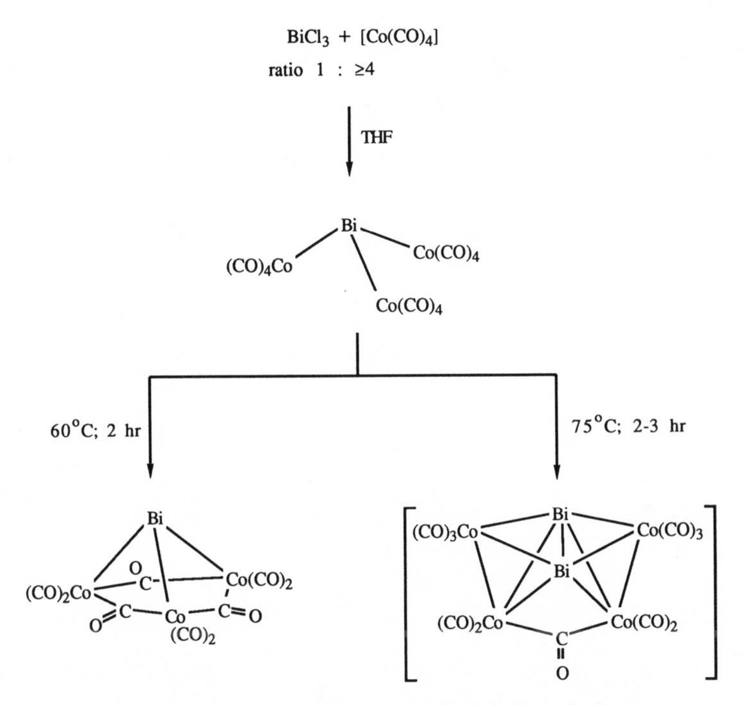

Figure 80. Synthesis and structures of cobalt bismuth clusters.

5. GROUP 16

5.1. Clusters Containing Metal–Oxygen Bonds

Discrete clusters with M_nO_m core structures are less common than one might imagine considering the wealth of metal alkoxide species[328] and the precedence for interstitial oxygen atoms in, for example, alkali metal suboxides such as Rb_9O_2 and $Cs_{11}O_3$.[329] Polyoxometallates and their organometallic derivatives are also a well-defined group of compounds.[330] The structural types represented for transition metal oxide clusters are shown in Figure 81. As for other main group element discussed in this chapter, the M_3E core is a well-documented cluster type; on the other hand, unlike other *first-row* main group elements, oxygen appears as a naked atom supported upon the M_3-framework. One cluster which very nicely illustrates the cluster/surface analogy and contains a type **66** unit is $Os_6(CO)_{18}(\mu_3\text{-}CO)(\mu_3\text{-}O)$ (Figure 2).[4]

Looking back at M_3E clusters in which E is a first-row element, a cluster such as $Ru_3(CO)_9BH_5$ needs four *endo*-hydrogen atoms in order to provide sufficient electrons to fill all the cluster bonding MOs. Progressing across the

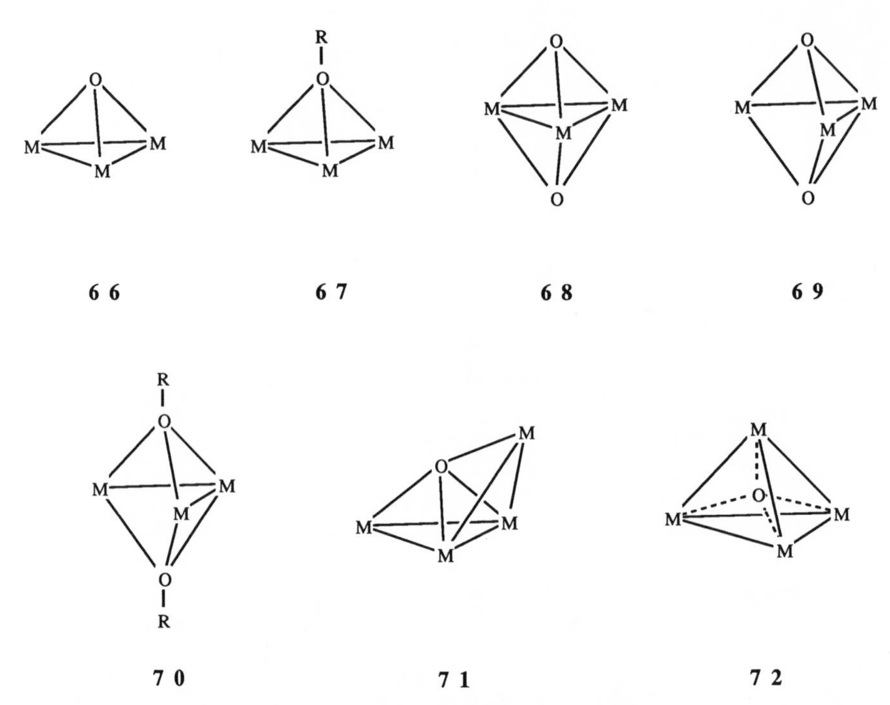

Figure 81. Structural types for M_nO_m clusters.

Periodic Table, $H_3Ru_3(CO)_9CH$ and $H_2Ru_3(CO)_9NH$ require one and two less *endo*-hydrogen atom(s), respectively. The most closely related neutral oxygen-containing compound is $H_2Ru_3(CO)_5(dppm)_2O$.[331] The preference for a naked μ_3-O and associated Ru—H—Ru bridge is significant, but in this particular compound may be attributed (at least in part) to the presence of the phosphine donors. Indeed, protonation occurs at an Ru—Ru site rather than at the oxygen atom as might be expected.[331] The isoelectronic fragments BH_3, CH_2, NH, and O each provide four cluster bonding electrons; the borane and methylene groups do so only by releasing two and one hydrogen atom(s), respectively, to the cluster (i.e., *exo*- become *endo*-hydrogen atoms). In the case of the borane, the M_3B-core is saturated with hydrogen atoms when M is a group 8 transition metal. Consequently, an M_3B-cluster of type **1** (Figure 7) which possesses a closed M_3-core comprising *early* transition metal atoms is an unlikely synthetic target. This is not the case if the capping atom is an electron-rich group 16 element; since the oxygen atom itself can contribute four cluster electrons (leaving one *exo*-lone pair), it is not surprising that, in addition to species such as $[Fe_3(CO)_9O]^{2-}$,[332] $Co_3Cp_3(\mu_3\text{-}CO)O$,[333] and $[H_3Rh_3Cp^*_3O]^{+}$[334] (Figure 82), clusters such as $[H_3Re_3(CO)_9O]^{2-}$ and $[Cp_3Mo_3(CO)_6O]^{+}$ have been characterized.[335,336] In this trirhenium system, protonation *does* (formally) occur as expected at the capping oxygen

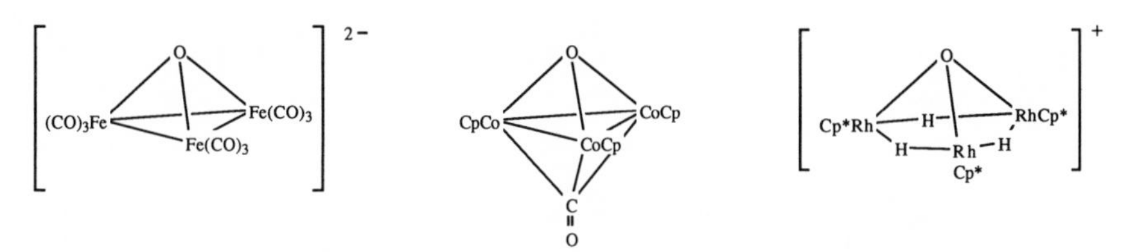

Figure 82. The type **66** clusters $[Fe_3(CO)_9O]^{2-}$, $Cp_3Co_3(\mu_3\text{-}CO)O$, and $[H_3Rh_3Cp^*_3O]^+$.

atom; the monoanion $[H_3Re_3(CO)_9OH]^-$ has been prepared from $[H_4Re_3(CO)_{10}]^-$ as illustrated in Figure 83. Whether or not $[H_3Re_3(CO)_9OH]^-$ classes as a type **66** or **67** cluster is unclear; in a crystallographic determination, the hydrogen atom was not located and so the structure is considered to lie somewhere between the two limiting forms shown in Figure 83.

In the $[Cp_3Mo_3(CO)_6O]^+$ cation[336] the bonding mode of the Mo_3O-unit is clearly related to that observed in systems such as the $[Mo_3O_4(H_2O)_9]^{4+}$ and corresponding sulfido derivatives; these in turn link through to studies on M_4O_4-cubanes.[337] In the preparation of the mixed metal cluster $Cp^*_2Mo_2Fe(CO)_7O$, the source of the capping oxygen atom is dioxygen. This type **66** cluster is formed on photolysis of $Fe_2(CO)_9$ with $Cp^*_2Mo_2(CO)_4$ in toluene which has been saturated with O_2 or air.[338] Again, this system exemplifies the ability of the capping oxygen atom to stabilize early transition metal M_3E-clusters.

In the presence of other good donor ligands, capping oxygen atoms are capable of stabilizing transition metals in group 5. The anion $[Nb_3(SO_4)_6O_2(H_2O)_3]^{5-}$ is obtained on the reduction of Nb_2O_5 in sulfuric acid,[339] or by treating $Nb_2Cl_6(THF)_3$ with H_2SO_4 (40–70% aqueous).[340] A similar type **68** cluster, $[Nb_3O_2(O_2CMe)_6(THF)_3]^+$, may also be synthesized from the chloro precursor.[340] Treating Cp^*TaCl_4 with water produces an oxo-bridged dimer which, on thermolysis, undergoes cluster aggregation

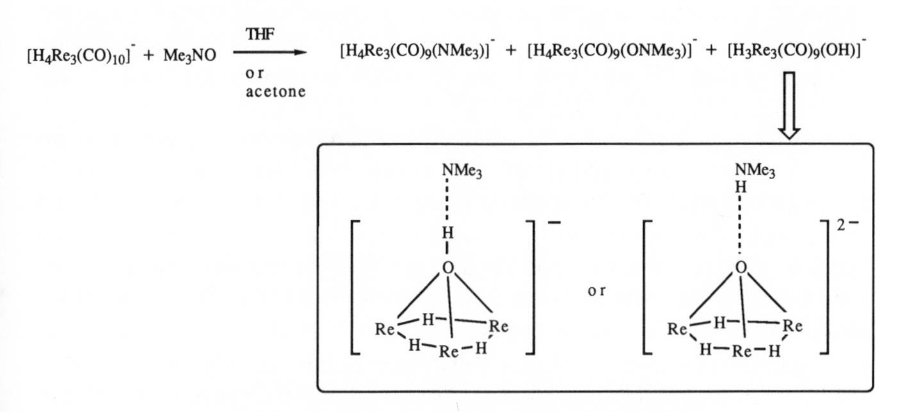

Figure 83. Synthesis and structure of $[H_3Re_3(CO)_9OH]^-$.

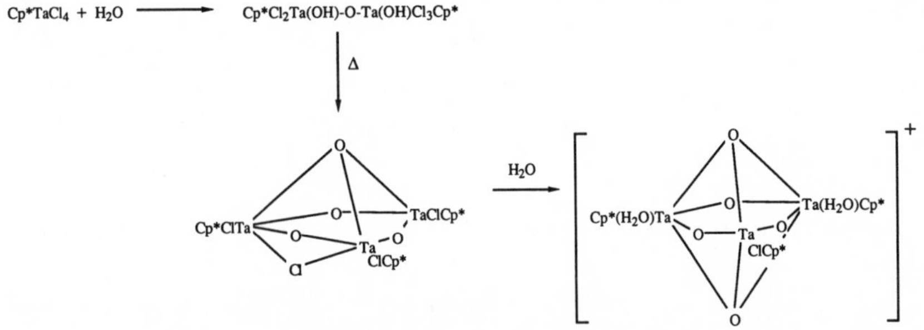

$Cp^*TaCl_4 + H_2O \longrightarrow Cp^*Cl_2Ta(OH)\text{-}O\text{-}Ta(OH)Cl_3Cp^*$

Figure 84. Formation and structures of $Cp^*_3Ta_3Cl_3(\mu\text{-}Cl)(\mu\text{-}O)_3O$ and $[Cp^*_3Ta_3Cl(H_2O)_2(\mu\text{-}O)_3O_2]^+$.

to form the type **66** compound $Cp^*_3Ta_3Cl_3(\mu\text{-}Cl)(\mu\text{-}O)_3O$ and the type **68** cluster $[Cp^*_3Ta_3Cl(H_2O)_2(\mu\text{-}O)_3O_2]^{pl}$ (Figure 84).[341,342]

As noted above, first-row elements in groups 13 through 15 exhibit a preference for the retention of an *exo*-substituent and, although cluster type **66** is well characterized for oxygen, compounds in which the trimetal framework interacts with an alkoxide or similar substituent are also known (type **67**). The anions $[H_3Re_3(CO)_9OEt]^-$[343] and $[Mo_3(CO)_6(NO)_3(\mu\text{-}OMe)_3OMe]^-$[344] fall into this category.

Capping both sides of the M_3-unit by either O or OR occurs, but may do so at the expense of opening one edge of the triangle to give the type **69** or **70** clusters, respectively. The compounds $Mn_3(CO)_9(\mu\text{-}X)(\mu_3\text{-}OEt)_2$ (X = F,I), $Mn_3(CO)_8(PMe_2Ph)\mu\text{-}OEt)(\mu_3\text{-}OEt)_2$, and $Ir_3(\eta^4\text{-}cod)_3(\mu\text{-}I)(O)_2$ have been fully characterized and a common feature is the presence of a donor ligand supporting the open edge.[345–347] A related cluster is $Ru_3(CO)_6(\mu\text{-}CO)(\mu\text{-}CO)(\mu\text{-}Cl)_2(O\text{-}c\text{-}C_6H_{11})_2$ in which two metal–metal edges have been broken but remain supported by bridging chloro-ligands.[348] Since a capping OR group is isoelectronic (valence electrons only) with a group 15 element, then the structure of $Fe_3(CO)_9(CMe)(OMe)$[349] is related to that of $Fe_3(CO)_9(CMe)E$ (E = As, Sb, Bi) (Figure 72) by the addition of one pair of cluster bonding electrons; $Fe_3(CO)_9(CMe)(OMe)$ is a *nido*-cage while each of $Fe_3(CO)_9(CMe)E$ (E = As, Sb, Bi) is a *closo*-species.

Expansion from an M_3E to M_4E cluster core via the rational capping of one M_2E-face was illustrated in Figure 20 for the synthesis of $[HFe_4(CO)_{12}BH]^-$[80,91]; similarly ketenylidene (M_3CCO-core) clusters are precursors to M_4C systems.[38,39] A parallel route allows the expansion of the type **66** to **71** cluster as shown in Figure 85a.[350] Despite residing in an exposed environment, the oxygen atom in $[Fe_3Mn(CO)_{12}O]^-$ is fairly resistant to protonation; this contrasts with the reactivity of exposed borides (type **7**) and corresponding carbides. The only other type **71** cluster to be reported so far is $Cp^*_4Ta_4O_7(OH)_2$ (Figure 85b).[351] As with the early transition metal clusters of type **66**, the tantalum aggregate is stabilized by electron-rich ligands,

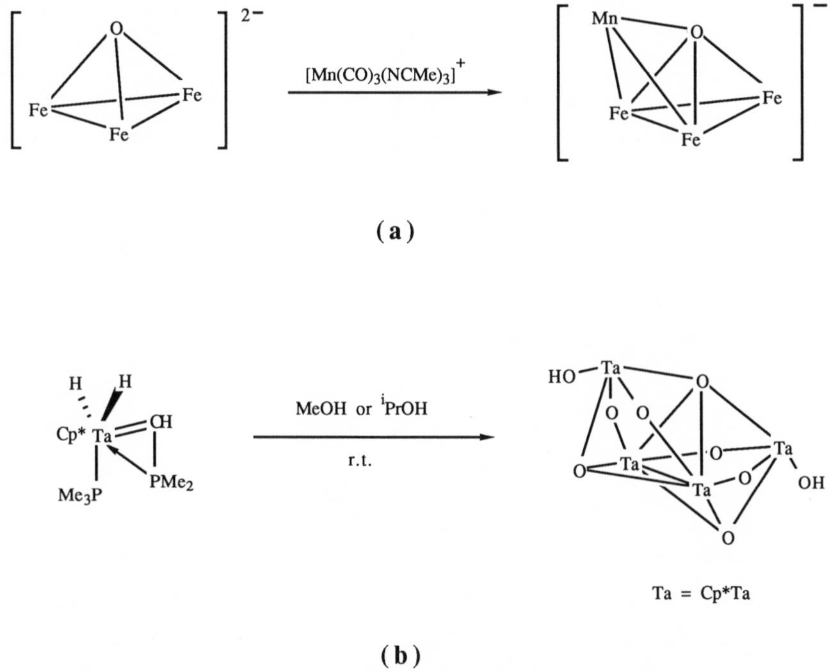

(a)

(b)

Figure 85. Synthesis and structure of the type **71** clusters (a) $[Fe_3Mn(CO)_{12}O]^-$ and (b) $Cp^*_4Ta_4O_7(OH)_2$.

one of which resides in the semi-interstitial μ_4-site. Throughout this chapter, the smallest interstice observed to *completely* encapsulate a main group element has been the octahedron. In $Ti_4(O)(S_2)_4Cl_6$, the four titanium atoms form a tetrahedral cage surrounding an oxygen atom; admittedly the disulfide bridges support a somewhat expanded Ti_4-cluster unit but nonetheless $Ti_4(O)(S_2)_4Cl_6$ establishes a new class of oxo-cluster, viz. type **72.**[352]

5.2. Clusters Containing Metal–Sulfur Bonds

On moving from oxygen to sulfur, there is a positive explosion in the number of fully characterized discrete M_nE_m clusters. The structural types for M_nS_m systems are shown in Figure 86; it is apparent that μ_3- and μ_4-modes predominate and, significantly, there are few examples of interstitial sulfur atoms. In addition, like oxygen, a cluster bound ($\geq\mu_3$) sulfur atom tends to exhibit one free lone pair of electrons; M_nS_m are far more common than $M_n(SR)_m$ clusters. Apart from these restrictions, the structures in Figure 86 do parallel many of the modes illustrated for M_nP_m clusters in Figure 51 or M_nAs_m and M_nSb_m species in Figure 69. Many of the compounds reported possess composite structures and a review of Os_nS_m clusters[353] will assist the reader in going further than space permits here.

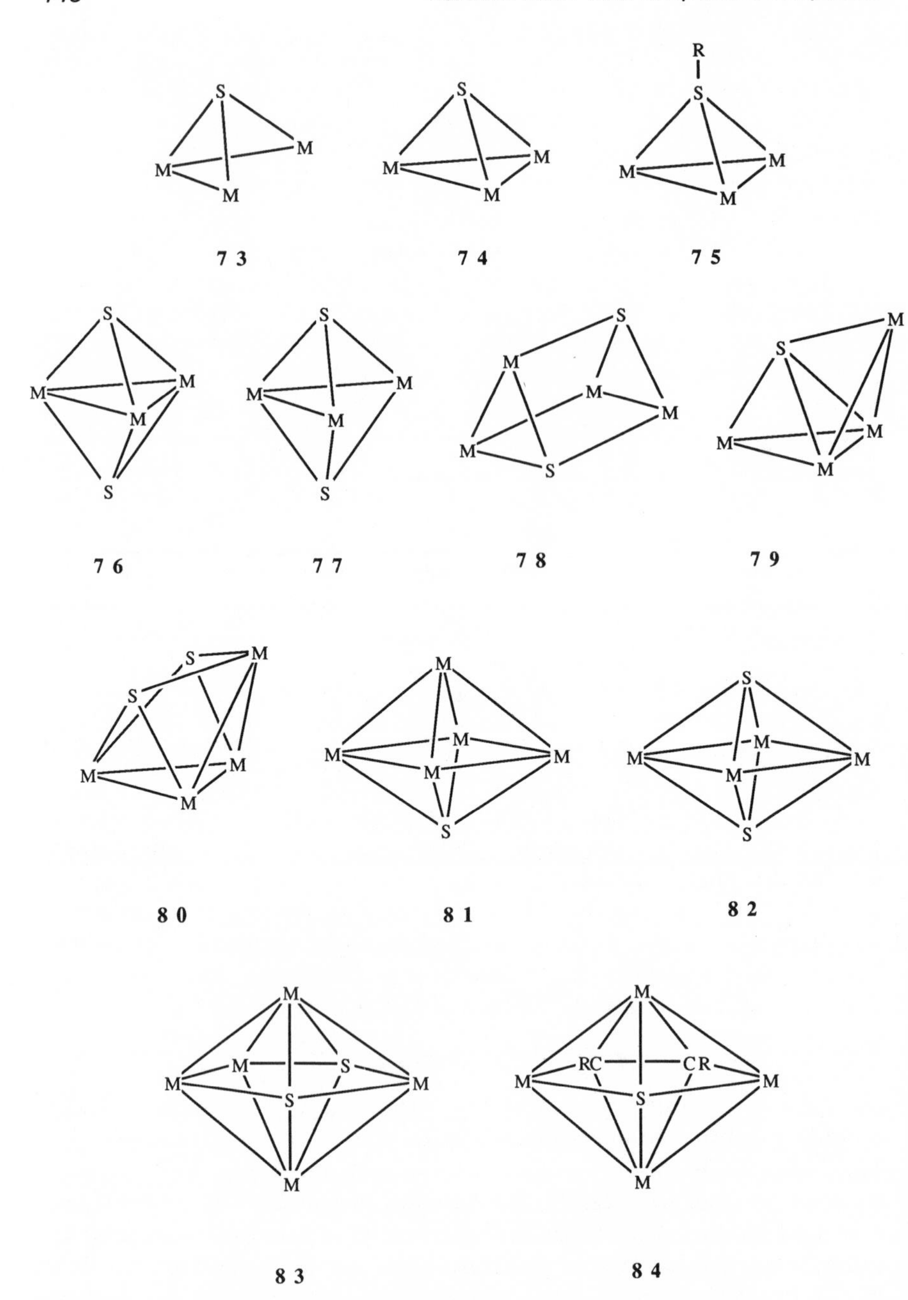

Figure 86. Structural types for transition metal–sulfido clusters.

The open type **73** cluster has been produced either directly [e.g., reaction of $Os_3(CO)_{12}$ with $S(SiMe_3)_2$ gives $Os_3(CO)_9(\mu_3\text{-}NSiMe_3)$][354] or through the addition of a pair of electrons, perhaps in the form of a halide ligand to a type **74** cluster, e.g., reaction of $[Fe_3(CO)_9S^tBu]^-$ with RX (R = alkyl; X = halide) gives *nido*-$Fe_3(CO_9S^tBu(\mu\text{-}X)$ or with XY (X = halide; Y = Ph_2P, Ph_2As, Ph_2Sb, PhSe, PhS) gives *nido*-$Fe_3(CO)_9S(\mu\text{-}Y)$.[355–357]

Clusters incorporating a trimetal platform supporting a bare sulfur atom are numerous. As long ago as 1964, $FeCo_2(CO)_9S$ was synthesized from $Fe_2(CO)_6S_2$ and $Co_2(CO)_8$ and a type **74** cluster core was proposed.[358] The structure of the related, but paramagnetic, cluster $Co_3(CO)_9S$ confirms the geometry.[359] The nonspecificity of cluster syntheses has featured repeatedly in this chapter; reactions to produce cobalt sulfide clusters are no exception[360–366] and some of the more exotic clusters[365,366] produced by treating cobalt octacarbonyl with CS_2 are shown in Figure 87. The Co_3S- and Co_4S-subunits feature as basic building blocks in these products. Related to $Co_3(CO)_9S$ are $Co_3(CO)_7(\mu\text{-}S_2COMe)S$[367] and $Co_3(CO)_7(\mu\text{-}N(H)C(Me)S)S$.[368] Since each bidentate ligand is a three-electron donor and replaces two carbonyl ligands (each a two-electron donor), these last two clusters, unlike $Co_3(CO)_9S$, are diamagnetic, having an electron count in accordance with PSEPT. When the source of the cluster sulfur atom is elemental sulfur or H_2S rather than CS_2, $Co_2(CO)_8$ reacts to form the linked cluster $[SCo_3(CO)_7]_2(S_2)$ shown in Figure 88.[369]

Once the M_3S cluster core is formed, metal exchange reactions can be carried out. For example, $Co_2Fe(CO)_9S$ reacts with $[Fe(CO)_4]^{2-}$ to form the isoelectronic type **74** cluster $HCoFe_2(CO)_9S$.[370] Similarly, $H_2Fe_3(CO)_9S$ reacts with Cp_2Ni to give $HFe_2NiCp(CO)_6S$ and $FeNi_2Cp_2(CO)_3S$.[370] As these reactions take place, the capping sulfur atom acts (as do other main group atoms) as a "clamp" holding the metal framework in place and allowing bond breakage and formation without excessive cluster degradation. The process may be continued to give type **74** clusters continuing three different transition metal fragments as illustrated by the reaction of $RuCo_2(CO)_9S$ with $[Fe(CO)_4]^{2-}$ which, after acidification, gives $HFeRuCo(CO)_9S$.[371] If the cluster possesses *endo*-hydrogen atoms, the strategy of $AuPR_3^+$ for H^+ replacement outlined in Section 2.1 may be employed to expand the metal framework. Figure 89 shows two examples in which the original M_3S-unit remains unperturbed throughout the expansion sequence.[371,372] $(Ph_3P)_2$-$Au_2Ru_3(CO)_9S$ is isostructural with its iron analog shown in Figure 89.[373]

Type **74** clusters of the group 8 metal triad are well documented. Structural characterizations of $[Fe_3(CO)_9S]^{2-}$,[374] $H_2Ru_3(CO)_9S$,[375] $[HOs_3(CO)_9S]^-$,[376] and $H_2Os_3(CO)_9S$[377] confirm the simple tetrahedral core geometry. In line with observations for the type **66** oxo clusters and with the reduced Lewis basicity of sulfur relative to oxygen, protonation of $H_2Ru_3(CO)_9S$ to $[H_3Ru_3(CO)_9S]^+$ occurs at a metal–metal bond.[378] How-

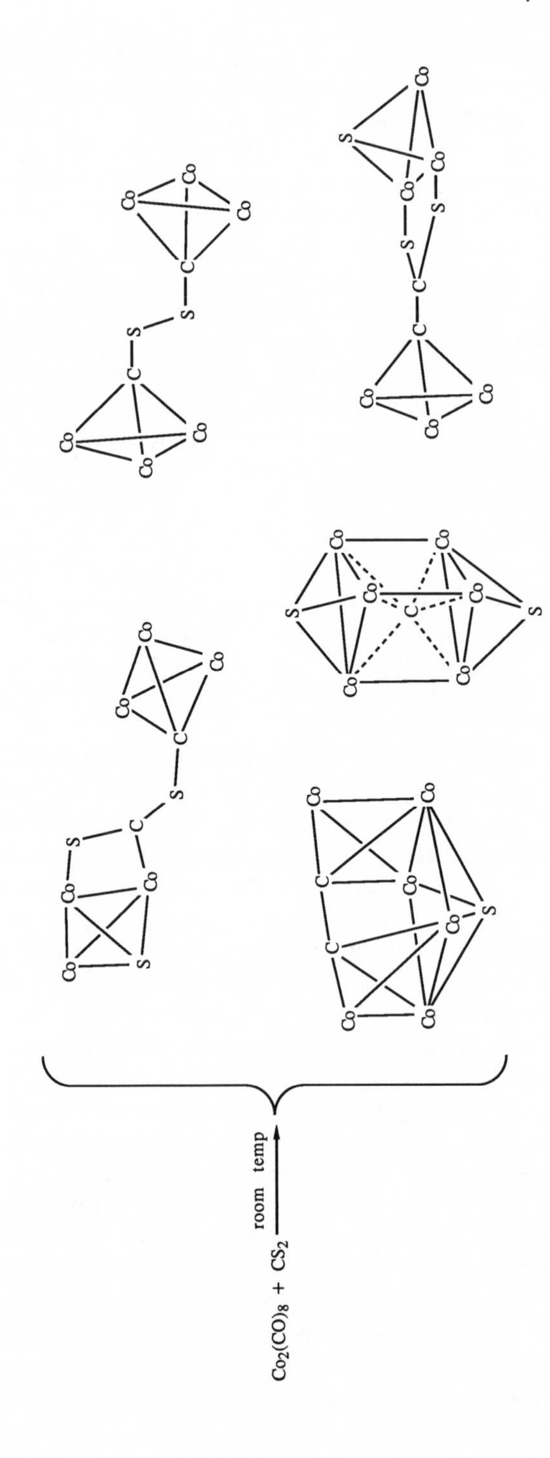

Figure 87. Selected products of the reaction of $Co_2(CO)_8$ with CS_2.

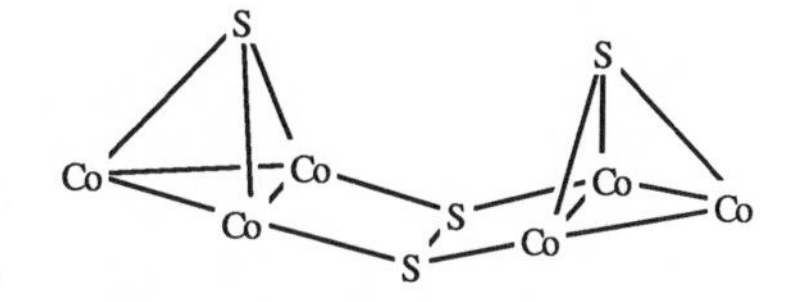

Figure 88. Structure of $[SCo_3(CO)_7]_2(S_2)$.

ever, despite the implication that the lone pair of the capping sulfur atom is not readily available, trimerization of $H_2Ru_3(CO)_9S$ to form $\{H_2Ru_3(CO)_8S\}_3$ may be achieved by refluxing in heptane or by photolysis in octane.[379] The cyclization (Figure 90) parallels that observed for $Co_3(CO)_9P$[226] (Figure 52) and for $Co_3(CO)_9As$[227,228] but in the case of sulfur is not spontaneous and is, indeed, reversible.

The stabilization by a capping sulfur atom of both early and late transition metal clusters of type **74** has been achieved. The dianion $[M_3(CO)_{12}S]^{2-}$ (M = Cr, Mo, W) is rationally constructed from $[(CO)_5MSH]^-$ via $[\{(CO)_5M\}_2SH]^-$, $[\{(CO)_5M\}S]^{2-}$, and $\{(CO)_5M\}_3S$ each of which has an open structure.[380] The final steps, shown for M = Cr in Figure 91, are analogous to the formation of the $Co_3(CO)_9Bi$ (Figure 29) or $Cp_3Fe_3(CO)_3Bi$ from $Bi\{Fe(CO)_3Cp\}_3$.[321] Clusters incorporating late transition metals are rare and, in Figure 28, the formation of Pt_3Sn-core was illustrated. The precursor for this reaction was $[(\mu\text{-dppm})_3Pt_3(\mu_3\text{-CO})]^{2+}$ and this again features in the formation of the related $[(\mu\text{-dppm})_3Pt_3(\mu_3\text{-S}]^+$ via reaction of the dication with H_2S; an analogous palladium cluster is also reported. Going from the $\{M_3(\mu_3\text{-CO})\}^{2+}$ to $\{M_3(\mu_3\text{-S})\}^+$ core (M = Pd, Pt) increases the cluster electron count and so, as expected, metal–metal bond cleavage occurs although pairs of metal centers remain supported by the dppm ligands.[381]

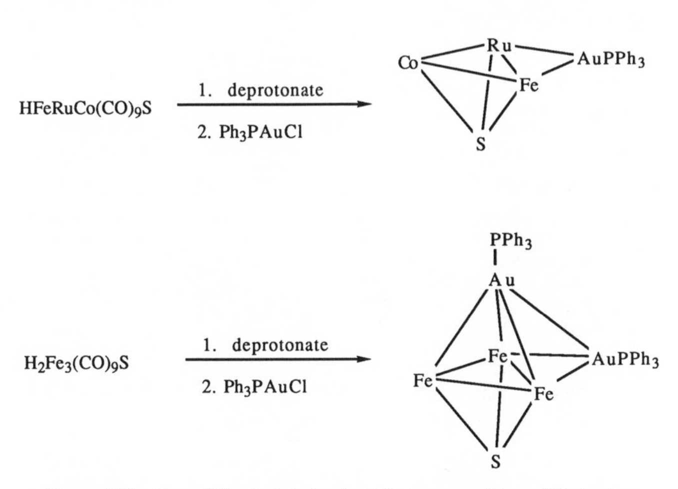

Figure 89. Use of the isolobal principle to expand type **74** clusters.

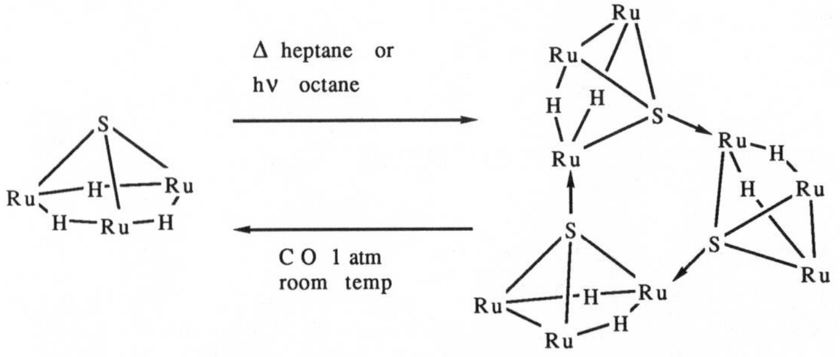

Figure 90. Reversible trimerization (with loss of CO) of $H_2Ru_3(CO)_9S$.

Before leaving the type **74** cluster, mixed metal systems involving group 6 and 8 metals are worthy of close scrutiny. Reaction of $Cp_2Mo_2(CO)_4$ or $Cp'_2Mo_2(CO)_4$ with $Fe_2(CO)_6S_2$ leads to $Cp_2Mo_2Fe_2(CO)_8S_2$ or $Cp'_2Mo_2Fe_2(CO)_8S_2$, respectively.[382–384] Two structural studies[383,384] confirm that the metal framework of $Cp_2Mo_2Fe_2(CO)_8S_2$ is an Fe_2Mo_2-butterfly (Figure 92a) with *cis*-μ_3-S caps; the compound is isostructural with its chromium analog.[385]

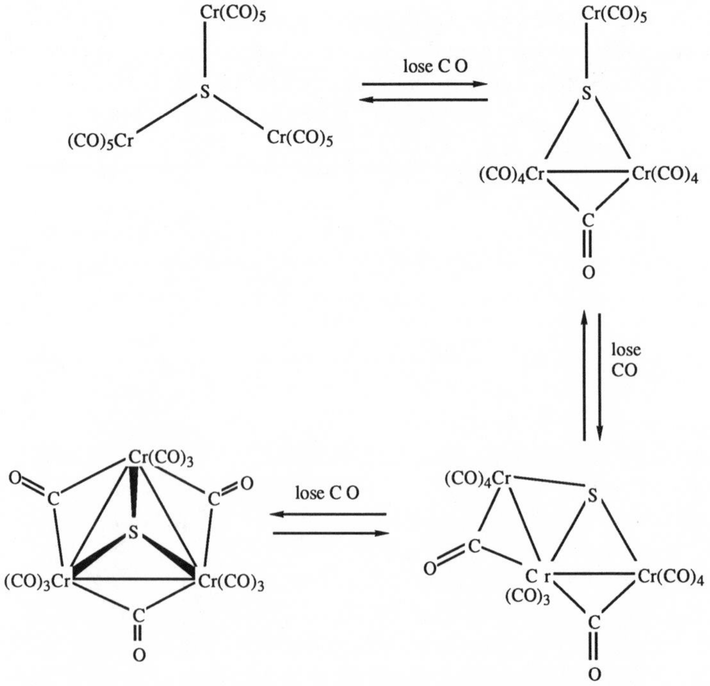

Figure 91. Stepwise closure of $\{(CO)_5Cr\}_3S$ to a type **74** cluster.

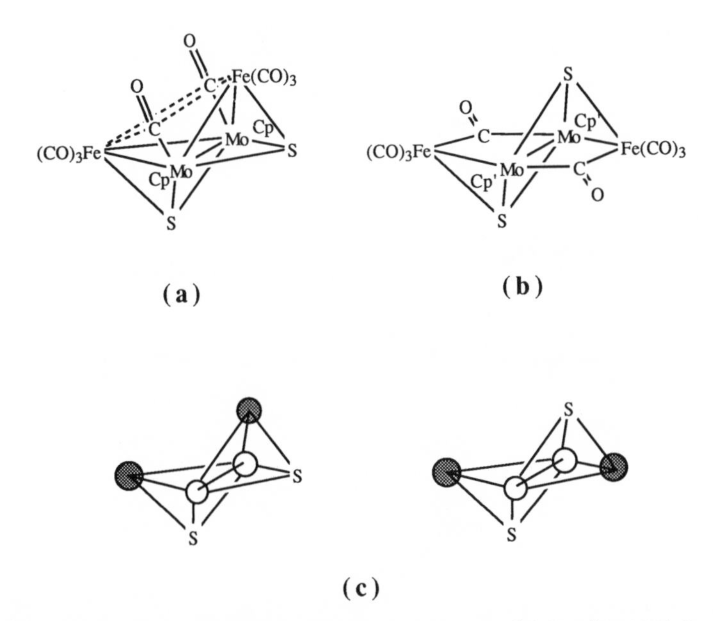

Figure 92. (a) Structure of $Cp_2Mo_2Fe_2(CO)_8S_2$, (b) structure of $Cp'_2Mo^2Fe_2(CO)_8S_2$, and (c) the isomeric relationship between them.

On the other hand, elucidation of the structure of $Cp'_2Mo_2Fe_2(CO)_8S_2$ reveals a planar arrangement of metal atoms and a *trans*-arrangement of capping sulfur atoms (Figure 92b).[382] The structural variation can be accounted for by considering the compound as a fused cluster, the subunits being two Mo_2FeS-tetrahedra.[386] Fusion to give a shared Mo–Mo edge naturally provides two isomers (*cis*- or *trans*-sulfur atoms, Figure 92c) and leads to the prediction of the two experimentally determined structures. A compound which is related to $Cp_2Mo_2Fe_2(CO)_8S_2$ is $Cp_2Mo_2Fe_2(CO)_6S_4$, which is prepared by the reaction of $Cp_2Mo_2(\mu\text{-}S)_2(\mu\text{-}SH)_2$ with $Fe_2(CO)_9$.[387] The effect of changing the number of cluster bonding electrons provided by the metal fragment is demonstrated in the set of reactions shown in Figure 93.[387]

The reactivity of the type **74** clusters includes cluster expansions. One pathway which, in terms of starting materials, might at first sight resemble those illustrated for boron in Figure 20 and for oxygen in Figure 85 appears to be affected by the steric requirement of the sulfur atom compared to that of the first-row elements; $M_3(CO)_{10}S$ (M = Fe, Ru, Os) reacts with $W(CO)_5(PMe_2Ph)$ to give $Os_3W(CO)_{11}(PMe_2Ph)_2S$ in which the sulfur retains a μ_3-bonding mode and a tetrahedral M_3W-core is formed.[388] This cluster expansion competes with metal exchange to give $M_2W(CO)_{10}L$ (L = CO or PMe_2Ph). Heterometallic clusters in particular are of interest in terms of potential catalytic properties. One intriguing example is the oligomerization

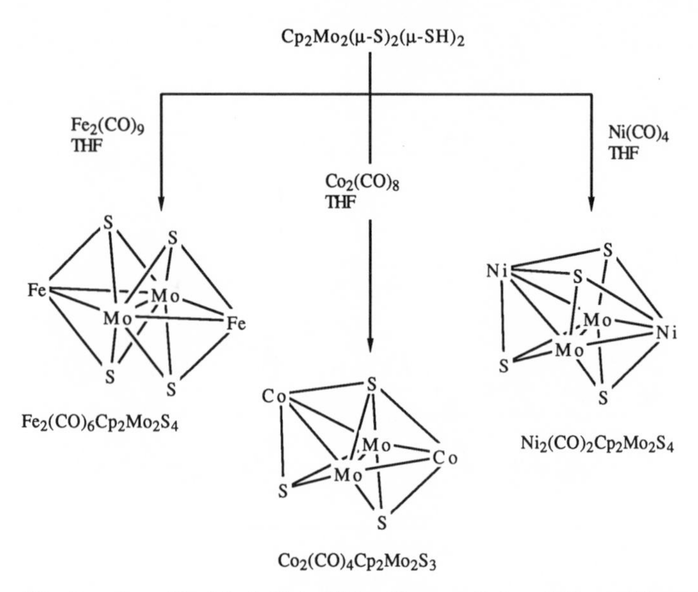

Figure 93. Reaction of $Cp_2Mo_2(\mu S)_2(\mu\text{-}SH)_2$ with group 8, 9, and 10 metal fragments.

of phenyl acetylene on the trimetal frame of the type **74** cluster $Cp_2Mo_2Ru(CO)_7S$.[389,390] The first stage is to form a metal supported $C_6H_3Ph_3$ fragment which, on heating to 125 °C, cleaves to give two allyl ligands.[390]

A difference in reactivity between clusters containing electron-rich or electron-deficient main group atoms is to be expected, and is nicely represented by reactions with Lewis bases. Competition between the metal and boron atoms in $[HFe_3(CO)_9BH_3]^-$ for Lewis bases was discussed in Section 2.1.[46] No such competition is expected when the main group cap in the M_3E-cluster carries a lone pair of electrons. An associative pathway which causes metal–metal bond cleavage is followed in the reaction of either CO or Me_2NH with $Os_4(CO)_{12}S$ (Figure 94); a contrasting addition product is formed after the oxidative addition of H_2.[391]

The type **76** cluster is quite well exemplified, although a bulky S-substituent appears to be necessary in most cases. As with oxygen, clusters with early transition metals can be stabilized, e.g., reaction of $[H_4Re_3(CO)_{10}]^-$ with ^{t}BuSH yields $[H_3Re_3(CO)_9S^tBu]^-$[392], which is isostructural with $[H_3Re_3(CO)_9OH]^-$ (Figure 83). But again, it is the group 8 metals that have received most attention. Sources of sulfur tend to be thiols, e.g., the synthesis of $HFe_3(CO)_9SR$ (R = ^{t}Bu; c-C_6H_{11}) from $Fe_3(CO)_{12}$ and RSH.[393] Deprotonation occurs easily to form the conjugate base $[Fe_3(CO)_9SR]^-$[302,393,394] and the cluster is able to withstand ligand substitution using PPh_3, $AsPh_3$, $SbPh_3$ (each mono-, di-, or tri-substitution)[393] or cycloheptatriene (replacing

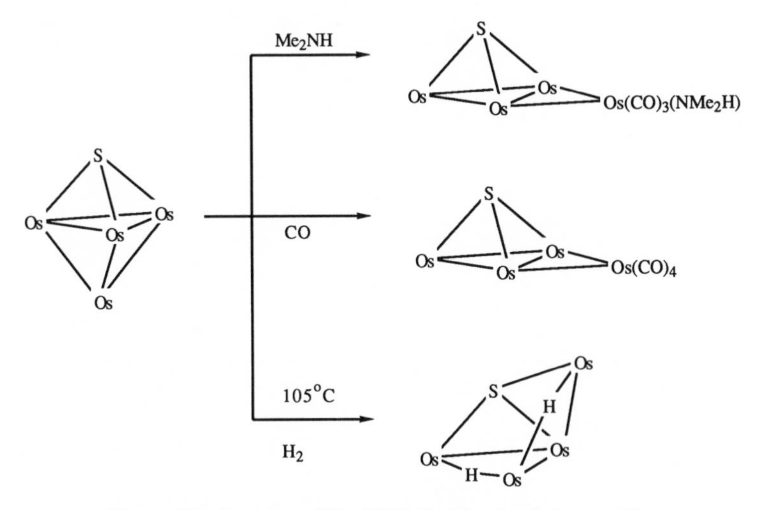

Figure 94. Reaction of $Os_4(CO)_{12}S$ with a Lewis base or H_2.

three carbonyl ligands)[395] without degradation (contrast the metalloborane species in Section 2.1).

Several examples of the type **76** cluster have been characterized. Aggregation of group 10 metal fragments with sulfur atoms occurs as Et_3P and $Ni(BF_4)_2 \cdot 6H_2O$ react together in the presence of H_2S.[396] The product $[Ni_3(PEt_3)_6S_2]^{2+}$ is isostructural with its platinum counterpart.[397] Oxidative addition of CS_2 (free or ligated) to a metal cluster provides both a sulfur atom and a thiocarbonyl ligand. This is illustrated by the reaction of $Cp(PMe_3)Co(\eta^2\text{-}CS_2)$ with $Cp(PMe_3)Co(CO)_3MnCp'$ to give $Cp_3Co_3(\mu_3\text{-}CS)S$. The relatively low capacity of the sulfur capping atom to act as a Lewis base is now neatly demonstrated as it competes with the thiocarbonyl sulfur atom for incoming electrophiles (Figure 95).[398] Returning to the true type **76** cluster, two parallel syntheses, one for rhodium and one for iridium, demonstrate the formation of $[M_3(CO)_6S_2]^-$ from $M_4(CO)_{12}$ or $M_6(CO)_{16}$ (M = Rh; Ir) and SCN^- or polysulfides.[399,400] More recently, the first arene metal sulfido complex has been produced (Figure 96) and is found to undergo reversible electrochemical reduction to the type **77** cluster (p-cymene)$_3Ru_3S_2$.[401]

The need to control reaction conditions in synthetic approaches is well exemplified in the osmium chemistry summarized in Figure 97.[402,403] As with nitrogen and phosphorus containing clusters, use is made of a trinuclear precursor; $HM_3(CO)_{10}(\mu\text{-NO})$, $HM_3(CO)_{10}(\mu\text{-NHR})$, and $HM_3(CO)_{10}(\mu\text{-PHR})$ were used widely (M = group 8; see Sections 3.1 and 3.2) and, in the case of sulfur, $HM_3(CO)_{10}(\mu\text{-SH})$ is a useful synthon. Clusters shown in Figure 97 are of types **74** [$H_2Os_3(CO)_9S$, $Os_3(CO)_{10}S$, $Os_4(CO)_{12}S$,

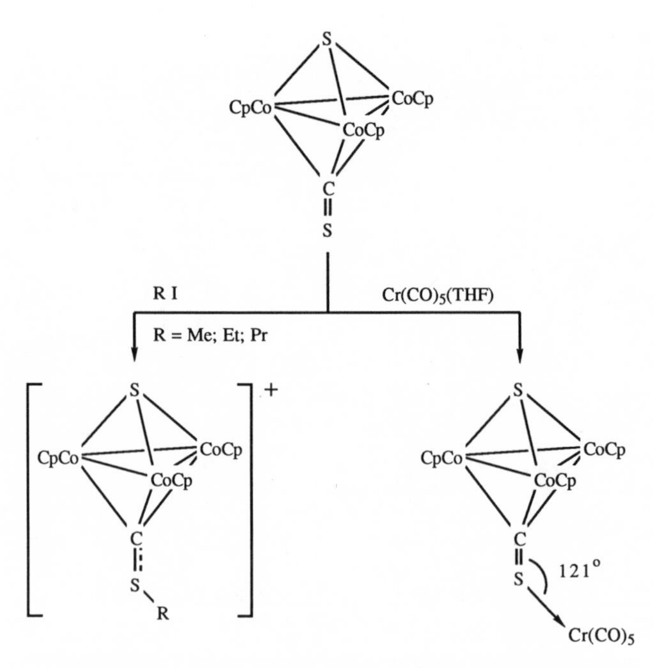

Figure 95. Test of the relative Lewis basicity of a μ_3-S atom.

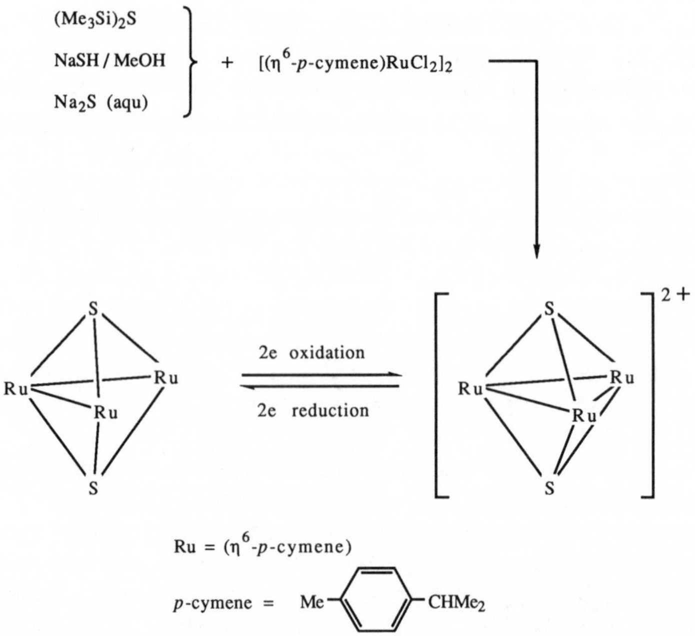

$(Me_3Si)_2S$

NaSH / MeOH

Na_2S (aqu)

+ $[(\eta^6\text{-}p\text{-cymene})RuCl_2]_2$

Ru_3S_2 (S, Ru, Ru, Ru, S)

2e oxidation

2e reduction

$[Ru_3S_2]^{2+}$ (S, Ru, Ru, Ru, S)

Ru = (η^6-*p*-cymene)

p-cymene = Me–C_6H_4–$CHMe_2$

Figure 96. Preparation and electrochemical redox behavior of $[(p\text{-cymene})_3Ru_3S_2]^{2+}$.

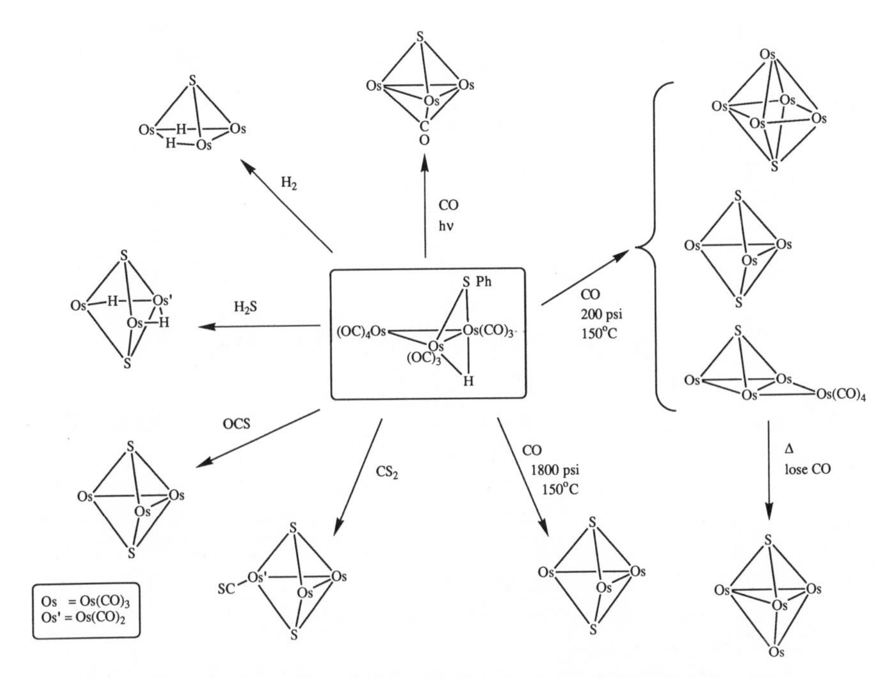

Figure 97. Routes to osmium sulfide clusters from $HOs_3(CO)_{10}(\mu\text{-}SH)$.

$Os_4(CO)_{13}S$], **77** [$Os_3(CO)_9S_2$, $H_2Os_3(CO)_8S_2$], and **82** [$Os_5(CO)_{15}S$]. Similar chemistry may be carried out for ruthenium systems.[404]

The type **77** cluster has been characterized for each of $M_3(CO)_9S_2$ (M = Fe,[405] Ru,[406] Os[407], structurally so for M = Fe and Os.[405,407] A *nido*-description is unambiguously assigned, since, for example, one M–M separation [3.662(1) Å in $Os_3(CO)_9S_2$] is significantly longer than the other two [av. 2.813(1) Å in $Os_3(CO)_9S_2$].[407] A bonding analysis of type **77** and related species has been carried out.[408] An interesting series of rhodium and iridium clusters has been characterized in which the balance between types **76** and **77** appears to be sterically controlled.[409,410] Sterically demanding ligands on the metal atoms force cluster closure, since loss of one ligand (to relieve ligand–ligand interactions) results in the loss of two cluster electrons. In the presence of a terminal phosphine substituent, a novel cluster coupling may be induced as shown in Figure 98.[411] Fusing of two type **77** clusters is also observed in [Cp′MoFe$(CO)_3S_2$]$_2$ although this is an indirect result[412] rather than a direct coupling.

In the series of group 8 clusters of type **77,** reactions with Lewis bases such as Me_2NH have been found to be metal dependent. Thus, $Fe_3(CO)_9S_2$ undergoes simple carbonyl for amine substitution while reaction of Me_2NH with $Os_3(CO)_9S_2$ leads to oxidative addition of the N—H bond and the formation of $HOs_3(CO)_8\{\mu\text{-}\eta^2\text{-}C(NMe_2)O\}S_2$.[413]

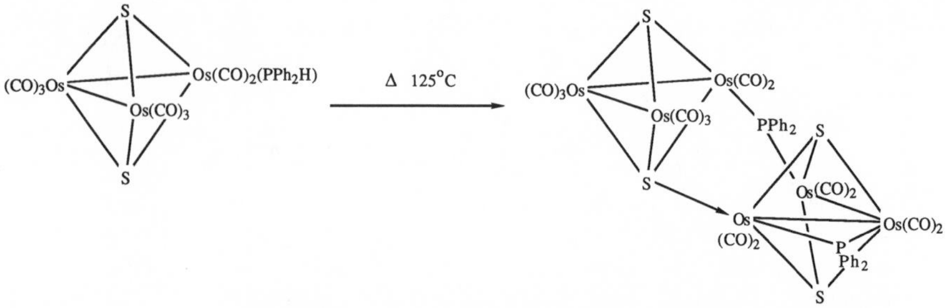

Figure 98. Coupling of $Os_3(CO)_8(PPh_2H)$ cages.

Figure 99 summarizes routes to clusters which contain an open M_3SE-core, specifically exemplified for E = PR.[302,414,415]

The μ_3-mode of bonding is obviously versatile in terms of total cluster structure. Cluster type **78** (which is analogous to the phosphido cluster **35,** since a PR group is isoelectronic with an S atom an S vertex atom) displays triply bonded sulfur atoms occupying two vertices of a trigonal prismatic skeleton. The sole representative of this group is $H_2Os_4(CO)_{12}S_2$[416] (which has a selenium analog; see Section 5.3).

As with phosphorus, incorporation of the sterically demanding sulfur atom *into* the cavity of an M_4-butterfly fragment is not observed. Cluster type **79** [e.g., observed in $Co_2(CO)_4Cp_2Mo_2(\mu_4\text{-}S)(\mu_3\text{-}S)_2$ shown in Figure 93][387] illustrates the fact that when a butterfly framework interacts with a μ_4-sulfur atom, the latter is forced into a vertex rather than semi-interstitial position. The butterfly framework also supports triply bridging sulfur atoms just as it

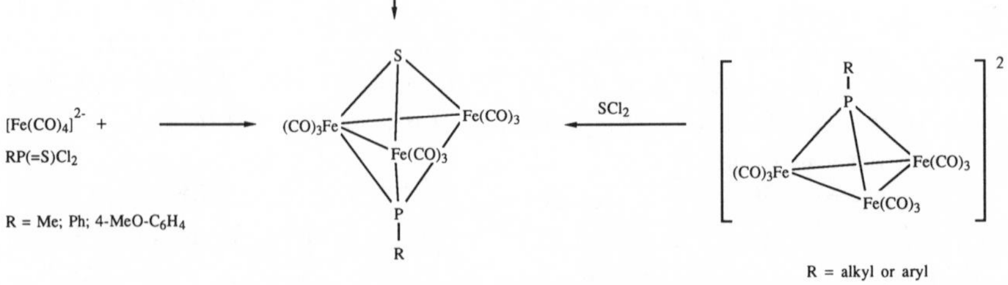

Figure 99. Routes to $Fe_3(CO)_9S(PR)$.

did μ_3-PR groups (type **33**, Figure 51), μ_3-AsR groups (type **53**, Figure 69), or μ_3-As atoms (type **54**, Figure 69). In the type **80** clusters $Os_4(CO)_{12}S_2$[417] and $Os_3W(CO)_{12}(PMe_2Ph)S_2$[418] the metal butterfly exhibits distortion with two long $M_{wingtip}—M_{hinge}$ bonds. The same M_4S_2-cluster core is present in $Ni_2(CO)_2Cp_2Mo_2S_4$ (Figure 93).[387] Reaction of $Os_4(CO)_{12}S_2$ with Lewis bases[417,419] can proceed in either an addition or substitution pathway as Figure 100 shows. Note that the site of substitution (wingtip or hinge atom) is ligand-dependent.

In each of the remaining cluster types, the sulfur atom caps four metal atoms. The type **81** and **82** clusters are quite well represented, although predominantly through compounds of osmium. By virtue of the molecular formula, $Os_5(CO)_{15}S$ may at first sight appear analogous to $Os_5(CO)_{15}C$[216] but the extra two electrons which the sulfur atom provides for cluster bonding means that $Os_5(CO)_{15}S$ is a *closo*-octahedral cluster[402] while $Os_5(CO)_{15}C$ is a *nido*-species with an interstitial carbon atom. The same applies to a comparison of the ruthenium systems $Ru_5(CO)_{15}S$[420] with $[Ru_5(CO)_{14}N]^-$.[201] The choice of a vertex site for the sulfur atom is also consistent with its steric requirements. One route to $Os_5(CO)_{15}S$ was shown in Figure 97[402]; $Ru_5(CO)_{15}S$ may be prepared along with related clusters (Figure 101) by reacting $Ru_3(CO)_{10}S$ with $Ru(CO)_5$ at 68 °C.[420] The structure of $Ru_5(CO)_{15}S$ is interesting in that two isomeric forms differing in the arrangement of carbonyl ligands only have been characterized.[420] The osmium sulfido cluster

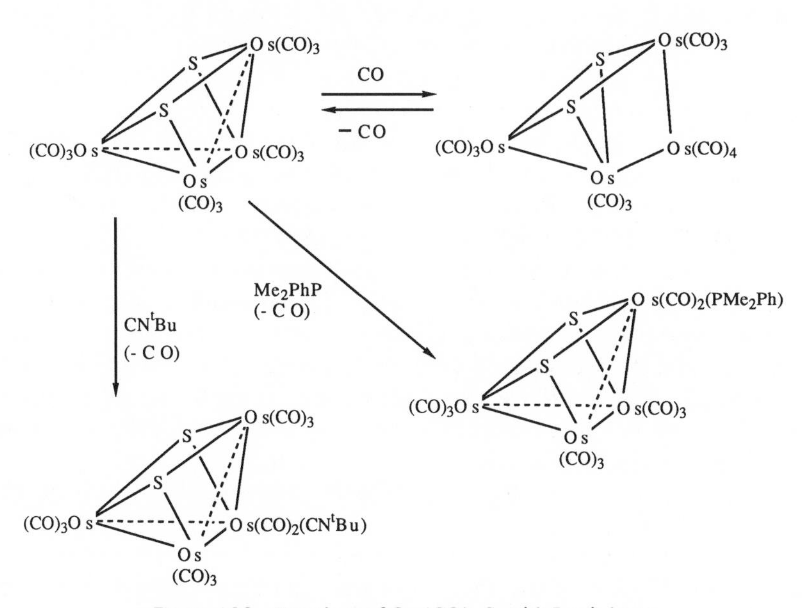

Figure 100. Reactions of $Os_4(CO)_{12}S_2$ with Lewis bases.

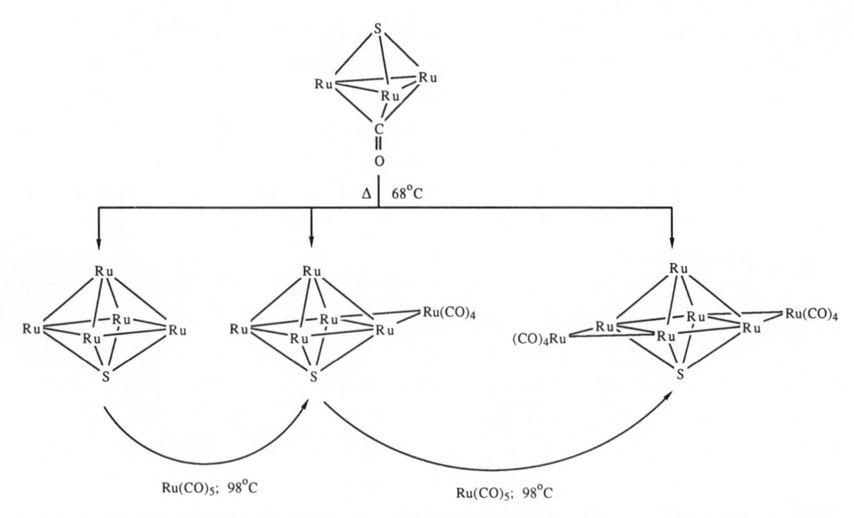

Figure 101. $Ru_5(CO)_{15}S$, $Ru_6(CO)_{18}S$, and $Ru_7(CO)_{21}S$ all of which exhibit type **81** cluster cores.

$Os_5(CO)_{15}S$ undergoes metal substitution to form $Os_4(CO)_{12}Pt(CO)(PPh_3)S$ when treated with $(Ph_3P)_2Pt(C_2H_4)$. A second product of the reaction appears from the molecular formula to be a result of simple addition of a $Pt(PPh_3)_2$ unit, but in fact the structure reveals reorganization of the Os_5Pt atoms to give a type **81** Os_4PtS-core supporting a μ-$Os(CO)_3(PPh_3)$ unit.[421]

Previous discussion has illustrated that the Lewis basicity of a transition-metal-cluster-bound group 16 atom is significantly lower than that of earlier groups. One example in which the lone pair of the sulfur atom is utilized in a coordinate bond is the formation of the adduct $Os_5(CO)_{15}S \rightarrow W(CO)_4(PMe_2Ph)$.[422] Related complexes also form when the type **82** cluster $Ru_4(CO)_9(PMe_2Ph)_2S_2$ reacts with $W(CO)_5(PMe_2Ph)$.[422]

The oxidative addition of two H—S bonds occurs when $Os_5(CO)_{15}S$ reacts with H_2S as shown in Figure 102.[423] The addition of extra cluster electrons necessitates a change of cluster geometry and the observed change from the 74 valence electron square based pyramid (counting S as a four-electron *ligand*) to the 78 electron vertex-shared fused triangles (again each S is considered as a four-electron ligand) is in keeping with the loss of one mole of CO as H_2S cleaves and adds to $Os_5(CO)_{15}S$.

A range of clusters with fused polyhedral structures, but which retain a type **81** core, are known. For example, $Os_6(CO)_{17}S$ exhibits an Os_5S-core with one Os_3-face capped by an osmium fragment.[423] Further capping produces $Os_7(CO)_{19}S$; note that the "sequential" capping is formal rather than synthetic since $Os_7(CO)_{19}S$ is derived from $Os_4(CO)_{12}S$ rather than from

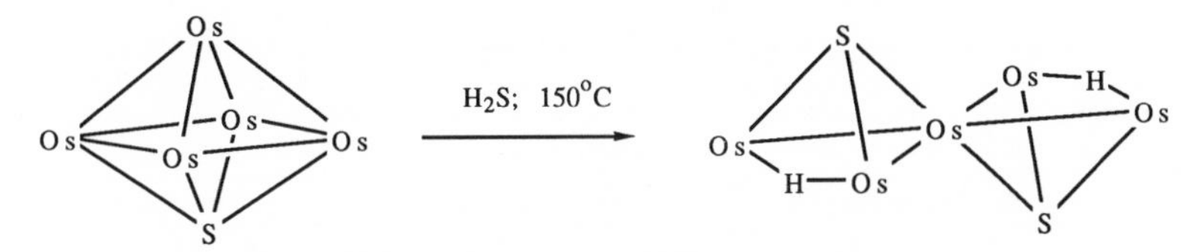

Figure 102. Oxidative addition of H_2S to $Os_5(CO)_{15}S$.

$Os_6(CO)_{17}S$.[424] A heterometallic cluster constructed upon a type **81** core is $Cp_2Mo_2Ru_5(CO)_{14}(\mu_4\text{-}\eta^2\text{-}CO)_2S$.[425]

The type **82** cluster is exemplified by $Co_4(CO)_{10}S_2$,[426] which is isostructural with $Co_4(CO)_{10}(\mu_4\text{-}S)(\mu_4\text{-}PR)$ (R = Me, C_6H_4-*p*-OMe) mentioned in Section 4.2.[271] A μ_4-bonding mode for sulfur appears to be particularly favorable and it is further observed in the pentagonal bipyramidal clusters of types **83** and **84.** Figure 103 summarizes the formations of two related molecules, each of which possesses a type **83** cluster core. The donor ability of each μ_3-sulfur atom in the precursors $Os_3(CO)_9S_2$ and $Os_4(CO)_{12}S_2$ is clearly fundamental to the success of the expansion reactions.[427,428]

The type **74** cluster $Ru_3(CO)_{10}S$ reacts with phenyl acetylene to give the insertion product shown in Figure 104.[429] Utilization of the lone pair of electrons borne by the sulfur atom allows cluster expansion, e.g., in the formation of $Ru_4(CO)_{11}SC(H)C(Ph)$ (Figure 104).[429] Further reaction with PhC≡CH has been reported for $Ru_4(CO)_{11}SC(R)C(R')$ (R = H or CO_2Me; R′ = CO_2Me) and, as Figure 104 shows, the products depend upon the substituents.[430]

As the references witness, the explosion of group 8 sulfido chemistry is largely due to the work of Adams and coworkers. In describing the work here,

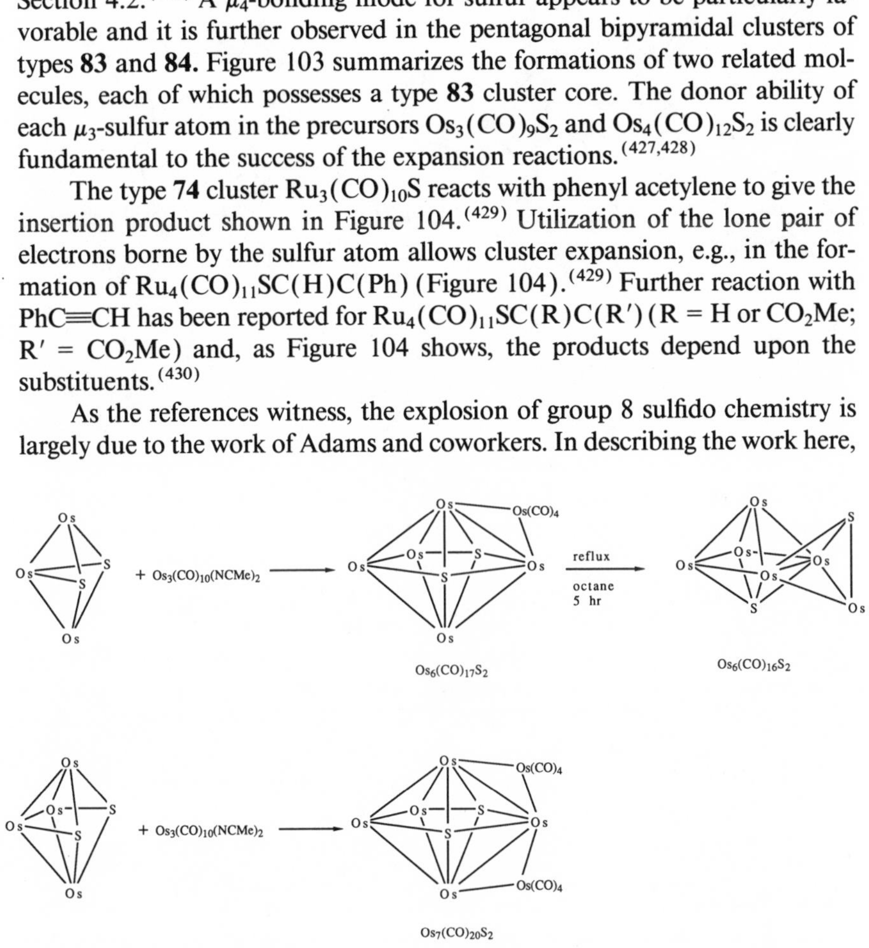

Figure 103. Conversion of μ_3- to μ_4-S in syntheses of $Os_6(CO)_{17}S_2$ and $Os_7(CO)_{20}S_2$.

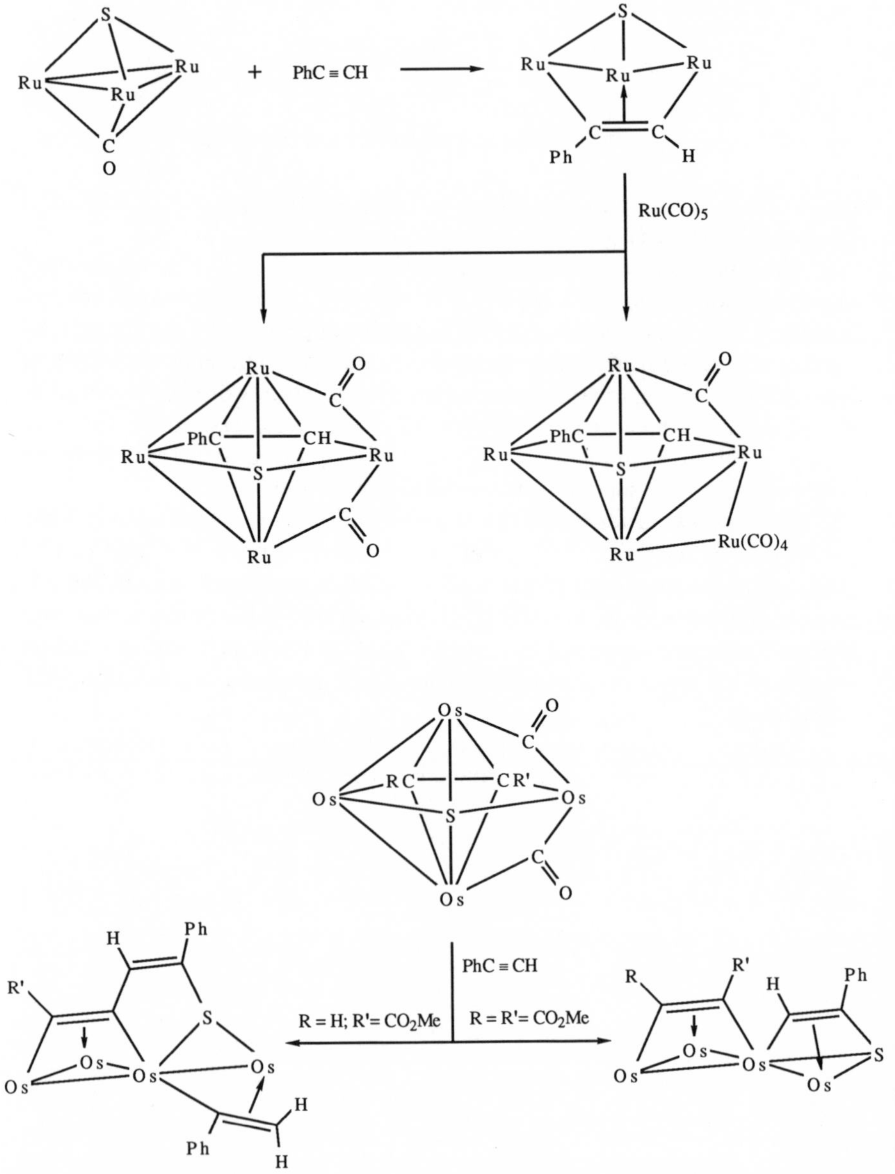

Figure 104. Formation and reactions of some type **84** clusters.

selectivity has been necessary and the reader is directed to the original literature for further examples of the basic cluster types depicted in Figure 86.

5.3. Clusters Containing Metal–Selenium or Tellurium Bonds

Transition metal clusters of core formulation M_nSe_m or M_nTe_n are not yet well exemplified. Just as with the heavier members of groups 13 through 15, open structures for selenium and, particularly, tellurium containing species are quite common, for example, $MeTe\{Mn(CO)_2Cp\}_3$,[431] $Fe_2(CO)_6$-$Te_2Fe(CO)_3(PPh_3)$,[432] $Cp(PPh_3)NiTe(\mu\text{-}NiCp)_2TeNi(PPh_3)Cp$,[433] and $Fe_2Ru_3(CO)_{17}Te_2$ (shown in Figure 105),[434] and $Fe_2Os_3(CO)_{17}Te_2$.[435]

In considering the structural classes of M_nE_m (E = Se or Te) clusters, it is striking that each structure has an analogous sulfur cluster type (compare Figures 106 and 86). Type **87** clusters are by far the most common and well studied of the clusters shown in Figure 106.

In direct parallel to the scheme shown in Figure 97, use is made of $HM_3(CO)_{10}(\mu\text{-}SePh)$ (M = group 8 metal) as a precursor to several selenium containing clusters with the driving force being elimination of benzene. Figure 107 demonstrates that four osmium–selenium cluster types may be accessed in this way.[436] As described for most of the main group elements in this chapter, the oxidative addition of an E—H bond to a transition metal center or cluster framework is an important strategy to M—E bond formation. Similarly, the reaction of H_2E (E = Se or Te) with $Os_3(CO)_{10}(NCMe)_2$ leads to the type **85** clusters $H_2Os_3(CO)_9Se$ and $H_2Os_3(CO)_9Te$, respectively.[437] Ruthenium analogs have also been characterized.[437]

Clusters of type **86** with a *closo*-M_3Se_2 skeleton exist for the group 9 metals; $[Rh_3(CO)_6Se_2]^-$ and $[Ir_3(CO)_6Se_2]^-$ are isostructural with their sulfido counterparts described in Section 5.2 and are prepared by analogous routes by using $SeCN^-$ with $M_4(CO)_{12}$ or $M_6(CO)_{16}$ (M = Rh, Ir).[399,400] A rather more open structure, rationalized in terms of the additional electrons present, is observed for $Ni_3(PPh_3)_6Se_2$.[438] The Ni–Ni distances are 3.11(1), 3.19(1), and 3.19(1) Å implying weak metal–metal interactions. By PSEPT, the Ni_3Se_2-core has two too many electrons to be a *closo*-cage, and yet the cluster does not exhibit one broken edge as might be predicted; $Ni_3(PPh_3)_6Se_2$ lies inter-

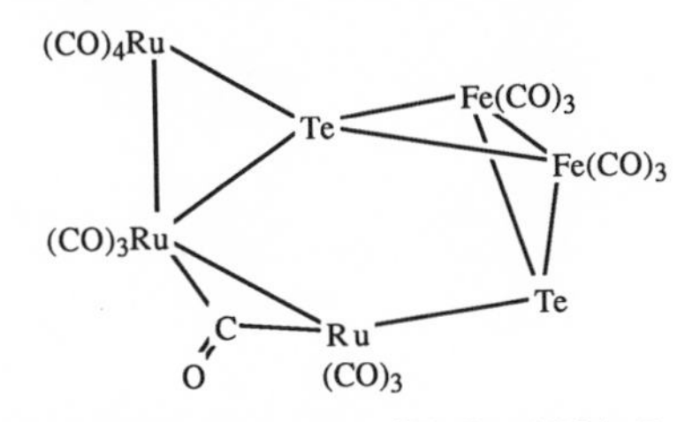

Figure 105. Structure of $Fe_2Ru_3(CO)_{17}Te_2$.

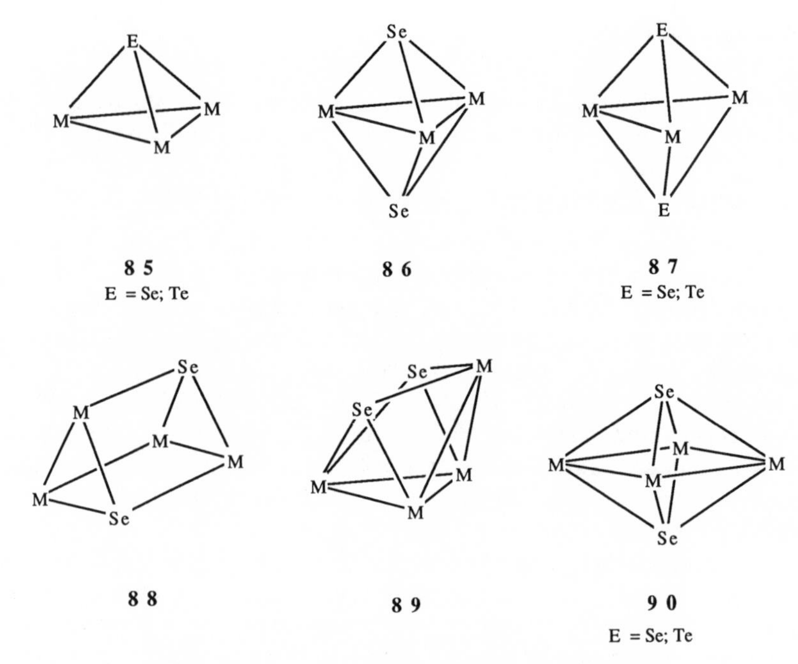

Figure 106. Structural types for M_nSe_m and M_nTe_m clusters.

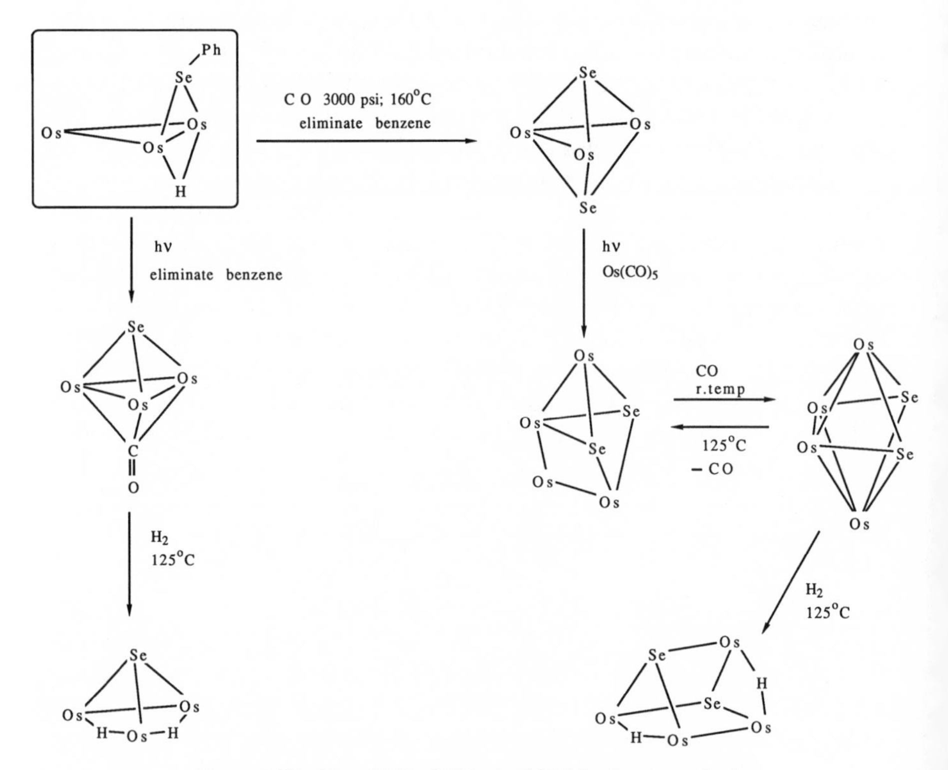

Figure 107. Use of $HOs_3(CO)_{10}(\mu\text{-}SePh)$ in cluster synthesis.

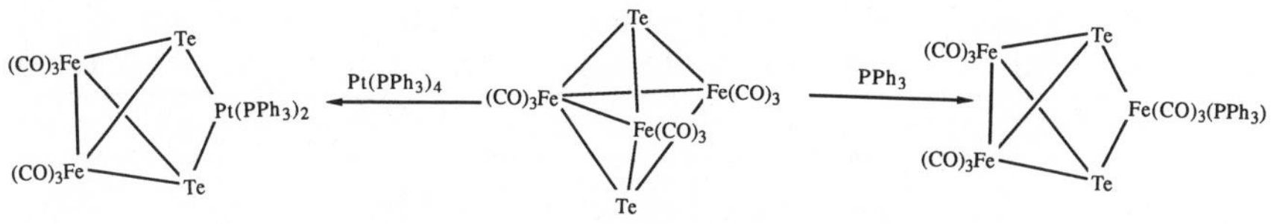

Figure 108. Some reaction of $Fe_3(CO)_9Te_2$.

mediate between cluster types **86** and **87.** A true type **87** member is $Fe_3(CO)_9Se_2$ and this compound, along with its tellurium analog, has been known since 1958[439] and was structurally characterized several years later.[440] $Fe_3(CO)_9Te_2$ was originally prepared by the reaction of $[TeO_3]^{2-}$ with $Fe(CO)_5$ in basic solution.[439] The high-temperature reaction between $Os_3(CO)_{12}$ and elemental tellurium gives $Os_3(CO)_9Te_2$[407] and a more recent synthesis from $M_3(CO)_{10}(NCMe)_2$ has been developed under ambient conditions.[435] The formation of $Os_3(CO)_9Se_2$ from $Os_3(CO)_{12}$ and elemental selenium is accompanied by that of $H_2Os_4(CO)_{12}Se_2$ (a type **88** cluster, see below).[407] Selected reactions of $Fe_3(CO)_9Te_2$ are summarized in Figure 108.[433,441] Lewis bases tend to add to the cluster with concomitant metal–metal bond cleavage.

Following in the footsteps of the osmium sulfide clusters, those containing selenium and which are based upon an M_4-butterfly framework exhibit μ_3-Se in preference to μ_4-atoms. The type **89** cluster $Os_4(CO)_{12}Se_2$ is shown in Figure 107. The compound is isostructural and $Os_4(CO)_{12}S_2$ and reactivity patterns are similar. For example, addition of CO is reversible and leads to cluster opening.[436] The distorted cage geometry of $H_2Os_4(CO)_{12}Se_2$ reflects the fact that this cluster has 66 valence electrons (four too many for a perfect M_4-butterfly).[407] In terms of electron counting, the only difference between the type **88** [e.g., $H_2Os_4(CO)_{12}Se_2$] and **89** [e.g., $Os_4(CO)_{12}Se_2$] clusters is a pair of valence electrons and, therefore, a distortion applied to cage type **89** will give **88.**

The final class of cluster (type **90**) in this section is represented by $Cp_4Ni_4E_2$ (E = Se or Te).[433] Reaction of $CpNi(PPh_3)Cl$ with $E(SiMe_3)_2$ in THF leads to both $Cp_4Ni_4E_2$ and $Cp_4Ni_4(EPPh_3)_2$ (E = Se or Te). Structural characterization of $Cp_4Ni_4Se_2$ defines the octahedral M_4E_2-core.[433] The μ_4-mode, often in conjunction with μ_3-bonding mode for selenium, is an important factor in the stabilization of several high nuclearity clusters, such as $Ni_6(PPh_3)_6(\mu_4\text{-Se})_3(\mu_3\text{-Se})_2$ and $[Co_8(PPh_3)_6(\mu_4\text{-Se})_2(\mu_3\text{-Se})_6]^+$.[278]

6. CONCLUDING REMARKS

The aim of this chapter has been to survey cluster compounds in which a group 13–16 atom interacts with three or more transition metal atoms. The scope of the chemistry is clearly large and, judging from the reference list, the reader will appreciate that it is very much an area of current interest. Structural relationships exist between many of the smaller main group containing clusters and organometallic clusters (e.g., the series M_3E where E = BR, CR, NR, O) but differences in electronegativities and numbers of, e.g., *endo*-hydrogen atoms (needed to keep constant the number of cluster bonding electrons as we progress across the Periodic Table) influence the reactivity. Progressing down a group, lone pair accessibility influences Lewis basicity, and thus reactivity and the ability of a discrete cluster to bond coordinatively to another. Steric effects are also predominant in determining the environment in which a main group atom can comfortably reside, e.g., the number of clusters containing fully interstitial atoms decreases as a function of main group atomic radius. In terms of creating models for C—H and C—C bond activation (an introductory aim of this chapter), then certainly the chemistry of transition metal main group clusters provides significant information. Again, the closer to carbon in the Periodic Table an element is, the closer will the related cluster chemistry mimic that of an organometallic analog. However, the aim should not be simply to *mimic* the organometallic compounds but to provide a perturbation to the cluster which is sufficient to *alter* the pathway preferably in a designed direction. It is this latter statement which may provide the reader with food for thought (and action).

ACKNOWLEDGMENTS

I am grateful to the Royal Society for a 1983 University Research Fellowship, and the Petroleum Research Fund and S.E.R.C. for supporting my own investigations in the area of metalloborane cluster chemistry.

REFERENCES

1. Gates, B. C.; Guczi, L.; Knözinger, H., Eds. "Metal Clusters in Catalysis"; Elsevier: Amsterdam, 1986.
2. See, for example: Lauher, J. W. *J. Am. Chem. Soc.* 1978, **100,** 5305; McPartlin, M.; Mingos, D. M. P. *Polyhedron* 1984, **3,** 1321.
3. See, for example: Gavin, R. M., Jr.; Ruett, J.; Muetterties, E. L. *Proc. Natl. Acad. Sci. USA* 1981, **78,** 3981; Muetterties, E. L. *Chem. Soc. Rev.* 1982, **11,** 283; Muetterties, E. L. *Pure Appl. Chem.* 1982, **54,** 83; Drezdon, M. A.; Shriver, D. F. *J. Mol. Catal.* 1983, **21,** 81; Somorjai, G. A. *Chem. Soc. Rev.* 1984, **13,** 321; Somorjai, G. A. *Pure Appl. Chem.* 1988, **60,** 1499.
4. Goudsmit, R. J.; Johnson, B. F. G.; Lewis, J.; Raithby, P. R.; Whitmire, K. H. *J. Chem. Soc., Chem. Commun.* 1983, 246.
5. Johnson, B. F. G.; Lewis, J.; McPartlin, M.; Nelson, W. J. H.; Raithby, P. R.; Sironi, A.; Vargas, M. D. *J. Chem. Soc., Chem. Commun.* 1983, 1476.
6. See, for example: Deeming, A. J. *Adv. Organomet. Chem.* 1986, **26,** 1; Adams, R. D.; Horvath, I. T. *Prog. Inorg. Chem.* 1985, **33,** 127 and references cited therein.
7. Schultz, A. J.; Williams, J. M.; Calvert, R. B.; Shapley, J. R.; Stucky, G. D. *Inorg. Chem.* 1979, **18,** 319.

8. Grimes, R. N. Ed. "Metal Interactions with Boron Clusters," Plenum Press: New York, 1982; Boocock, S.; Gilbert, K. B.; Shore, S. G. In "Comprehensive Organometallic Chemistry"; Abel, E.; Stone, F. G. A.; Wilkinson, G., Eds. Pergamon Press: Oxford, 1982; Vol. 6, p. 879; Grimes, R. N. *Ibid,* 1982; Vol. 1, p. 459; Kennedy, J. D. *Prog. Inorg. Chem.* 1984, **32,** 519; Kennedy, J. D. *Prog. Inorg. Chem.* 1986, **34,** 211; Greenwood, N. N. *Chem. Soc. Rev.* 1984, **13,** 353.
9. Housecroft, C. E. *Polyhedron* 1987, **6,** 1935.
10. Fehlner, T. P. *New J. Chem.* 1988, **12,** 307.
11. Housecroft, C. E.; Fehlner, T. P. *Adv. Organomet. Chem.* 1982, **21,** 57.
12. Hong, F.-E.; Coffy, T. J.; McCarthy, D. A.; Shore, S. G. *Inorg. Chem.* 1989, **28,** 3284.
13. Draper, S. M.; Housecroft, C. E.; Keep, A. K.; Matthews, D. M.; Song, X.; Rheingold, A. L. *J. Organomet. Chem.* 1992, **423,** 241.
14. Ting, C.; Messerle, L. *J. Am. Chem. Soc.* 1989, **111,** 3449.
15. Chipperfield, A. K.; Housecroft, C. E.; Matthews, D. M. *J. Organomet. Chem.* 1990, **384,** C38.
16. Feilong, J.; Fehlner, T. P.; Rheingold, A. L. *J. Am. Chem. Soc.* 1987, **109,** 1860.
17. Jacobsen, G. B.; Anderson, E. L.; Housecroft, C. E.; Hong, F.-E.; Buhl, M. L.; Long, G. J.; Fehlner, T. P. *Inorg. Chem.* 1987, **26,** 4040.
18. Vites, J. C.; Housecroft, C. E.; Eigenbrot, C.; Buhl, M. L.; Long, G. J.; Fehlner, T. P. *J. Am. Chem. Soc.* 1986, **108,** 3304.
19. Housecroft, C. E.; Buhl, M. L.; Long, G. J.; Fehlner, T. P. *J. Am. Chem. Soc.* 1987, **109,** 3323.
20. Wong, K. S.; Fehlner, T. P. *J. Am. Chem. Soc.* 1981, **103,** 966.
21. Wong, K. S.; Haller, K. J.; Dutta, T. K.; Chipman, D. C.; Fehlner, T. P. *Inorg. Chem.* 1982, **21,** 3197.
22. DeKock, R. L.; Wong, K. S.; Fehlner, T. P. *Inorg. Chem.* 1982, **21,** 3203.
23. Lynam, M. M.; Chipman, D. M.; Barreto, R. D.; Fehlner, T. P. *Organometallics* 1987, **6,** 2405.
24. Dutta, T. K.; Vites, J. C.; Jacobsen, G. B.; Fehlner, T. P. *Organometallics* 1987, **6,** 842.
25. Chipperfield, A. K.; Housecroft, C. E. *J. Organomet. Chem.* 1988, **349,** C17.
26. Vites, J. C.; Eigenbrot, C.; Fehlner, T. P. *J. Am. Chem. Soc.* 1984, **106,** 4633.
27. Shore, S. G.; Jan, D.-Y.; Hsu, L.-Y.; Hsu, W.-L. *J. Am. Chem. Soc.* 1983, **105,** 5923.
28. Shore, S. G.; Jan, D.-Y.; Hsu, W.-L.; Hsu, L.-Y.; Kennedy, S.; Huffman, J. C.; Wang, T.-C. L.; Lin Wang, T.-C.; Marshall, A. G. *J. Chem. Soc., Chem. Commun.* 1984, 392.
29. Barreto, R. D.; Fehlner, T. P.; Hsu, L.-Y.; Jan, D.-Y.; Shore, S. G. *Inorg. Chem.* 1986, **25,** 3572.
30. Shapley, J. R.; Strickland, D. S.; St. George, G. M.; Churchill, M. R.; Bueno, C. *Organometallics* 1983, **2,** 185.
31. Kolis, J. W.; Holt, E. M.; Shriver, D. F. *J. Am. Chem. Soc.* 1983, **105,** 7307.
32. Sailor, M. J.; Shriver, D. F. *Organometallics* 1985, **4,** 1476.
33. Went, M. J.; Sailor, M. J.; Bogdan, P. L.; Brock, C. P.; Shriver, D. F. *J. Am. Chem. Soc.* 1987, **109,** 6023.
34. Kolis, J. W.; Holt, E. M.; Drezdon, M.; Whitmire, K. H.; Shriver, D. F. *J. Am. Chem. Soc.* 1982, **104,** 6134.
35. Sailor, M. J.; Brock, C. P.; Shriver, D. F. *J. Am. Chem. Soc.* 1987, **109,** 6015.
36. Holmgren, J. S.; Shapley, J. R. *Organometallics* 1984, **3,** 1322.
37. Sievert, A. C.; Strickland, D. S.; Shapley, J. R.; Steinmetz, G. R.; Geoffroy, G. L. *Organometallics* 1982, **1,** 214.
38. Hriljac, J. A.; Swepston, P. N.; Shriver, D. F. *Organometallics* 1985, **4,** 158.
39. Hriljac, J. A.; Holt, E. M.; Shriver, D. F. *Inorg. Chem.* 1987, **26,** 2943.
40. Geoffroy, G. L.; Bassner, S. L. *Adv. Organomet. Chem.* 1988, **28,** 1.

41. Hriljac, J. A.; Shriver, D. F. *J. Am. Chem. Soc.* 1987, **109,** 6010.
42. Seyferth, D.; Hallgren, J. E.; Eschbach, C. S. *J. Am. Chem. Soc.* 1974, **96,** 1730.
43. Seyferth, D.; Williams, G. H.; Nivert, C. L. *Inorg. Chem.* 1977, **16,** 758.
44. Schmid, G.; Bätzel, V.; Etzrodt, G.; Pfeil, R. *J. Organomet. Chem.* 1975, **86,** 257.
45. Jan, D.-Y.; Hsu, L.-Y.; Shore, S. G. 188th ACS National Meeting, 1984; INORG 180.
46. Housecroft, C. E.; Fehlner, T. P. *J. Am. Chem. Soc.* 1986, **108,** 4867.
47. Jan, D.-Y.; Hsu, L.-Y.; Workman, D. P.; Shore, S. G. *Organometallics* 1987, **6,** 1984.
48. Jan, D.-Y.; Shore, S. G. *Organometallics* 1987, **6,** 428.
49. Shore, S. G.; Coffy, T.; Jan, D.-Y.; Krause, J.; Workman, D. 3rd Chemical Congress of the North American Continent, 1988; INORG 705.
50. Workman, D. P.; Deng, H.-B.; Shore, S. G. *Angew. Chem., Int. Ed. Engl.* 1990, **29,** 309.
51. Deeming, A. J.; Underhill, M. *J. Chem. Soc., Chem. Commun.* 1973, 277; Ref. 8 in Deeming, A. J.; Underhill, M. *J. Chem. Soc., Dalton Trans.* 1974, 1415.
52. Lewis, J.; Johnson, B. F. G. *Pure Appl. Chem.* 1975, **44,** 43.
53. See, for example: Bruce, M. I.; Swincer, A. G. *Adv. Organomet. Chem.* 1983, **22,** 59.
54. Eady, C. R.; Johnson, B. F. G.; Lewis, J. *J. Chem. Soc., Dalton Trans.* 1977, 477.
55. Greenwood, N. N.; Savory, C. G.; Grimes, R. N.; Sneddon, L. G.; Davison, A.; Wreford, S. S. *J. Chem. Soc., Chem. Commun.* 1974, 718.
56. Andersen, E. L.; Haller, K. J.; Fehlner, T. P. *J. Am. Chem. Soc.* 1979, **101,** 4390.
57. Haller, K. J.; Andersen, E. L.; Fehlner, T. P. *Inorg. Chem.* 1981, **20,** 309.
58. Housecroft, C. E. *Inorg. Chem.* 1986, **25,** 3108.
59. Fehlner, T. P. In "Boron Chemistry"; Parry, R. W.; Kodama, G., Eds. Pergamon Press; Oxford, 1980; p. 95.
60. DeKock, R. L.; Fehlner, T. P. *Polyhedron* 1982, **1,** 521.
61. Seddon, E. A.; Seddon, K. S. "Chemistry of Ruthenium"; Elsevier, Amsterdam, 1984; p. 1053.
62. Sappa, E.; Tirippichio, A.; Braunstein, P. *Chem. Rev.* 1983, **83,** 203.
63. Raithby, P. R.; Rosales, M. J. *Adv. Inorg. Chem. Radiochem.* 1985, **29,** 169.
64. Meng, X.; Rath, N. P.; Fehlner, T. P. *J. Am. Chem. Soc.* 1989, **111,** 3422.
65. Meng, X.; Fehlner, T. P.; Rheingold, A. L. *Organometallics* 1990, **9,** 534.
66. Pipal, J. R.; Grimes, R. N. *Inorg. Chem.* 1977, **16,** 3255.
67. Venable, T. L.; Grimes, R. N. *Inorg. Chem.* 1982, **21,** 887.
68. Miller, V. R.; Weiss, R.; Grimes, R. N. *J. Am. Chem. Soc.* 1977, **99,** 5646.
69. Zimmerman, G. J.; Hall, L. W.; Sneddon, L. G. *Inorg. Chem.* 1980, **19,** 3642.
70. Gromek, J. M.; Donohue, *J. Cryst. Struct. Commun.* 1981, **10,** 849.
71. Weiss, R.; Bowser, J. B.; Grimes, R. N. *Inorg. Chem.* 1978, **17,** 1522.
72. See, for example, Hughes, R. P.; Lambert, J. M. J.; Whitman, D. W.; Hubbard, J. L.; Henry, W. P.; Rheingold, A. L. *Organometallics* 1986, **5,** 789, and references cited therein.
73. Wade, K., *Adv. Inorg. Chem. Radiochem.* 1976, **18,** 1; Mingos, D. M. P. *Nature, Phys. Sci.* 1972, **236,** 99; Mingos, D. M. P. *Acc. Chem. Res.* 1984, **17,** 311.
74. Feilong, J.; Fehlner, T. P.; Rheingold, A. L. *Angew. Chem., Int. Ed. Engl.* 1988, **27,** 424.
75. Feilong, J.; Fehlner, T. P.; Rheingold, A. L. *J. Chem. Soc., Chem. Commun.* 1987, 1395.
76. Feilong, J.; Fehlner, T. P.; Rheingold, A. L. *J. Organomet. Chem.* 1988, **348,** C22.
77. Lunniss, J.; MacLaughlin, S. A.; Taylor, N. J.; Carty, A. J.; Sappa, E. *Organometallics* 1985, **4,** 2066.
78. See, for example, Pardy, R. B. A.; Smith, G. W.; Vickers, M. E. *J. Organomet. Chem.* 1983, **252,** 341.
79. Wong, K. S.; Scheidt, W. R.; Fehlner, T. P. *J. Am. Chem. Soc.* 1982, **104,** 1111.
80. Fehlner, T. P.; Housecroft, C. E.; Scheidt, W. R.; Wong, K. S. *Organometallics* 1983, **2,** 825.
81. Chipperfield, A. K.; Housecroft, C. E.; Rheingold, A. L. *Organometallics* 1990, **9,** 681.

82. Lauher, J. W.; Wald, K. *J. Am. Chem. Soc.* 1981, **103,** 7648.
83. Hall, K. P.; Mingos, D. M. P. *Prog. Inorg. Chem.* 1984, **32,** 237 and references cited therein.
84. Horowitz, C. P.; Holt, E. M.; Brock, C. P.; Shriver, D. F. *J. Am. Chem. Soc.* 1985, **107,** 8136 and references cited therein.
85. Salter, I. D. *Adv. Organomet. Chem.* 1989, **29,** 249.
86. Housecroft, C. E.; Rheingold, A. L. *J. Am. Chem. Soc.* 1986, **108,** 6420.
87. Housecroft, C. E.; Rheingold, A. L. *Organometallics* 1987, **6,** 1332.
88. References cited in Ref. 87.
89. Housecroft, C. E.; Rheingold, A. L.; Shongwe, M. S. *Organometallics* 1989, **8,** 2651.
90. Housecroft, C. E.; Buhl, M. L.; Long, G. J.; Fehlner, T. P. *J. Am. Chem. Soc.* 1987, **109,** 3323.
91. Housecroft, C. E.; Fehlner, T. P. *Organometallics* 1986, **5,** 379.
92. Wijeyesekera, S. D.; Hoffmann, R.; Wilker, C. N. *Organometallics,* 1984, **3,** 962.
93. Harris, S.; Bradley, J. S. *Organometallics,* 1984, **3,** 1086.
94. Rath, N. P.; Fehlner, T. P. *J. Am. Chem. Soc.* 1987, **109,** 5273.
95. Tachikawa, M.; Muetterties, E. L. *J. Am. Chem. Soc.* 1980, **102,** 4541.
96. Boehme, R. F.; Coppens, P. *Acta Crystallogr., Ser. B* 1981, **37B,** 1914.
97. Davis, J. H.; Beno, M. A.; Williams, J. M.; Zimmie, J.; Tachikawa, M.; Muetterties, E. L. *Proc. Natl. Acad. Sci. U.S.A.* 1981, **78,** 668.
98. Housecroft, C. E.; Fehlner, T. P. *Organometallics* 1986, **5,** 1279.
99. Khattar, R.; Puga, J.; Fehlner, T. P.; Rheingold, A. L. *J. Am. Chem. Soc.* 1989, **111,** 1877.
100. Churchill, M. R.; Wormald, J.; Knight, J.; Mays, M. J. *J. Am. Chem. Soc.* 1971, **93,** 3073.
101. Alami, M. K.; Dahan, F.; Mathieu, *J. Chem. Soc., Dalton Trans.* 1987, 1983.
102. Chipperfield, A. K.; Housecroft, C. E.; Raithby, P. R. *Organometallics* 1990, **9,** 479.
103. Sirigu, A.; Bianchi, M.; Beneditti, E. *J. Chem. Soc., Chem. Commun.* 1969, 596.
104. Housecroft, C. E.; Rheingold, A. L.; Shongwe, M. S. *Organometallics* 1988, **7,** 1885.
105. Harpp, K. S.; Housecroft, C. E.; Rheingold, A. L.; Shongwe, M. S. *J. Chem. Soc., Chem. Commun.* 1988, 965.
106. Harpp, K. S.; Housecroft, C. E. *J. Organomet. Chem.* 1988, **340,** 389.
107. Housecroft, C. E.; Rheingold, A. L.; Shongwe, M. S. *J. Chem. Soc., Chem. Commun.* 1988, 1630.
108. Bradley, J. S. *Adv. Organomet. Chem.* 1983, **22,** 1.
109. Tachikawa, M.; Muetterties, E. L. *Prog. Inorg. Chem.* 1981, **28,** 203.
110. Holt, E. M.; Whitmire, K. H.; Shriver, D. F. *J. Am. Chem. Soc.,* 1982, **104,** 5621.
111. Hriljac, J. A.; Swepston, P. N.; Shriver, D. F. *Organometallics,* 1985, **4,** 158.
112. Hriljac, J. A.; Harris, S.; Shriver, D. F. *Inorg. Chem.,* 1988, **27,** 816 and references cited therein.
113. Sappa, E.; Tiripicchio, A.; Carty, A. J.; Toogood, G. E. *Prog. Inorg. Chem.,* 1987, **35,** 437.
114. Chipperfield, A. K.; Housecroft, C. E.; Haggerty, B. S.; Rheingold, A. L. *J. Chem. Soc., Chem. Commun.* 1990, 1174.
115. Dahl, L. F.; Smith, D. L. *J. Am. Chem. Soc.* 1962, **84,** 2450.
116. Johnson, B. F. G.; Lewis, J.; Reichert, B. E.; Schorpp, K. T.; Sheldrick, G. M. *J. Chem. Soc., Dalton Trans.* 1977, 1417.
117. Fox, J. R.; Gladfelter, W. L.; Geoffroy, G. L.; Tavanaiepour, I.; Abdel-Mequid, S.; Day, V. W. *Inorg. Chem.* 1981, **20,** 3230.
118. Cazanoue, M.; Lugan, N.; Bonnet, J.-J.; Mathieu, R. *Organometallics* 1988, **7,** 2480.
119. Fehlner, T. P.; Rath, N. P. *J. Am. Chem. Soc.* 1988, **110,** 5345.
120. Fehlner, T. P.; Czech, P. T.; Fenske, R. F. *Inorg. Chem.* 1990, **29,** 3103.
121. Whitmire, K. H. *J. Coord. Chem.* 1988, **17,** 95.
122. Grimes, R. N.; Rademaker, W. J.; Denniston, M. L.; Bryan, R. F.; Greene, P. T. *J. Am. Chem. Soc.* 1972, **94,** 1865.

123. Young, D. A. T.; Wiersema, R. J.; Hawthorne, M. F. *J. Am. Chem. Soc.* 1971, **93,** 5687.
124. Young, D. A. T.; Willey, G. R.; Hawthorne, M. F.; Churchill, M. R.; Reis, A. H., Jr. *J. Am. Chem. Soc.* 1970, **92,** 6663.
125. Churchill, M. R.; Reis, A. H. *J. Chem. Soc., Dalton Trans.* 1972, 1317.
126. Beachley, O. T., Jr.; Churchill, M. R.; Fettinger, J. C.; Pazik, J. C.; Victoriano, L. *J. Am. Chem. Soc.* 1986, **108,** 4666.
127. Schwarzhans, K. E.; Steiger, H. *Angew. Chem. Int. Ed. Engl.* 1972, **11,** 535.
128. Kalbfus, W.; Kiefer, J.; Schwarzhans, K. E. *Z. Naturforsch, Ser. B* 1973, **28B,** 503.
129. Haupt, H. J.; Neumann, F.; Preut, H. *J. Organomet. Chem.* 1975, **99,** 439.
130. Haupt, H. J.; Balsaa, P.; Schwab, B. *Z. Anorg. Allg. Chem.* 1985, **521,** 15.
131. Wilson, R. D.; Bau, R. *J. Am. Chem. Soc.* 1976, **98,** 4687.
132. See, for example, Cassidy, J. M.; Whitmire, K. H. *Inorg. Chem.* 1989, **28,** 1432; Cassidy, J. M.; Whitmire, K. H. *Inorg. Chem.* 1989, **28,** 1435, and references cited therein.
133. Anema, S. G.; MacKay, K. M.; Nicholson, B. K. *Inorg. Chem.* 1989, **28,** 3158 and references cited therein.
134. Anema, S. G.; MacKay, K. M.; Nicholson, B. K. *J. Organomet. Chem.* 1989, **372,** 25.
135. Anema, S. G.; Barris, G. C.; MacKay, K. M.; Nicholson, B. K. *J. Organomet. Chem.* 1988, **350,** 207.
136. Gusbeth, P.; Vahrenkamp, H. *Chem. Ber.* 1985, **118,** 1746.
137. Gusbeth, P.; Vahrenkamp, H. *Chem. Ber.* 1985, **118,** 1770.
138. Gusbeth, P.; Vahrenkamp, H. *Chem. Ber.* 1985, **118,** 1758.
139. Anema, S. G.; MacKay, K. M.; Nicholson, B. K. *J. Organomet. Chem.* 1989, **371,** 233.
140. Schmid, G.; Bätzel, V.; Etzrodt, G. *J. Organomet. Chem.* 1976, **112,** 345.
141. Whitmire, K. H.; Lagrone, C. B.; Churchill, M. R.; Fettinger, J. C.; Robinson, B. H. *Inorg. Chem.* 1987, **26,** 3491.
142. Schmid, G.; Etzrodt, G. *J. Organomet. Chem.* 1977, **137,** 367.
143. Christie, J. A.; Duffy, D. N.; MacKay, K. M.; Nicholson, B. K. *J. Organomet. Chem.* 1982, **226,** 165.
144. McNeese, T. J.; Wreford, S. S.; Tipton, D. L.; Bau, R. *J. Chem. Soc., Chem. Commun.* 1977, 390.
145. Vahrenkamp, H. *Adv. Organomet. Chem.* 1983, **22,** 169.
146. Kolis, J. W.; Holt, E. M.; Hriljac, J. A.; Shriver, D. F. *Organometallics* 1984, **3,** 496.
147. Foster, S. P.; MacKay, K. M.; Nicholson, B. K. *Inorg. Chem.* 1985, **24,** 909.
148. Van Tiel, M.; MacKay, K. M.; Nicholson, B. K. *J. Organomet. Chem.* 1987, **326,** C101.
149. Croft, R. A.; Duffy, D. N.; Nicholson, B. K. *J. Chem. Soc., Dalton Trans.* 1982, 1023.
150. Douglas, G.; Jennings, M. C.; Manojlovic-Muir, L.; Muir, K. W.; Puddephatt, R. J. *J. Chem. Soc., Chem. Commun.* 1989, 159.
151. Cassidy, J. M.; Whitmire, K. H. *Inorg. Chem.* 1989, **28,** 2495.
152. Whitmire, K. H.; Lagrone, C. B.; Churchill, M. R.; Fettinger, J. C.; Biondi, L. V. *Inorg. Chem.* 1984, **23,** 4227.
153. Etzrodt, G.; Boese, R.; Schmid, G. *Chem. Ber.* 1979, **112,** 2574.
154. Whitmire, K. H.; Leigh, J. S.; Gross, M. E. *J. Chem. Soc., Chem. Commun.* 1987, 926.
155. Martinengo, S.; Ciani, G. *J. Chem. Soc., Chem. Commun.* 1987, 1589.
156. Lagrone, C. B.; Whitmire, K. H.; Churchill, M. R.; Fettinger, J. C. *Inorg. Chem.* 1986, **25,** 2080.
157. MacKay, K. M.; Nicholson, B. K.; Robinson, W. T.; Sims, A. W. *J. Chem. Soc., Chem. Commun.* 1984, 1276.
158. Vidal, J. L.; Walker, W. E.; Pruett, R. L.; Schoening, R. C. *Inorg. Chem.* 1979, **18,** 129.
159. Ceriotti, A.; Demartin, F.; Heaton, B. T.; Ingallina, P.; Longoni, G.; Manassero, M.; Marchionna, M.; Masciocchi, N. *J. Chem. Soc., Chem. Commun.* 1989, 786.
160. Gladfelter, W. L. *Adv. Organomet. Chem.* 1985, **24,** 41.
161. Eisenberg, R.; Hendricksen, D. E. *Adv. Catal.* 1979, **28,** 79.

162. Fjare, D. E.; Gladfelter, W. L. *Inorg. Chem.* 1981, **20,** 3533.
163. Stevens, R. E.; Gladfelter, W. L. *J. Am. Chem. Soc.* 1982, **104,** 6454.
164. Chisholm, M. H.; Folting, K.; Huffmann, J. C.; Leonelli, J.; Marchant, N. S.; Smith, C. A.; Taylor, L. C. E. *J. Am. Chem. Soc.* 1985, **107,** 3722.
165. Feasey, N. D.; Knox, S. A. R.; Orpen, A. G. *J. Chem. Soc., Chem. Commun.* 1982, 75.
166. Feasey, N. D.; Knox, S. A. R. *J. Chem. Soc., Chem. Commun.* 1982, 1062.
167. Johnson, B. F. G.; Lewis, J.; Mace, J. M. *J. Chem. Soc., Chem. Commun.* 1984, 186.
168. Smieja, J. A.; Stevens, R. E.; Fjare, D. E.; Gladfelter, W. L. *Inorg. Chem.* 1985, **24,** 3206.
169. Fjare, D. E.; Keyes, D. G.; Gladfelter, W. L. *J. Organomet. Chem.* 1983, **250,** 283.
170. Legzdins, P.; Nurse, C. R.; Rettig, S. J. *J. Am. Chem. Soc.* 1983, **105,** 3727.
171. Elder, R. C.; Cotton, F. A.; Schunn, R. A. *J. Am. Chem. Soc.* 1967, **89,** 3645.
172. Whitmire, K. H.; Shriver, D. F. *J. Am. Chem. Soc.* 1980, **102,** 1456.
173. Holt, E. M.; Whitmire, K. H.; Shriver, D. F. *J. Organomet. Chem.* 1981, **213,** 125.
174. Smieja, J. A.; Gladfelter, W. L. *Inorg. Chem.* 1986, **25,** 2667.
175. Sappa, E.; Milone, L. *J. Organomet. Chem.* 1973, **61,** 383.
176. Bhaduri, S.; Gopalkrishnan, K. S.; Sheldrick, G. M.; Clegg, W.; Stalke, D. *J. Chem. Soc., Dalton Trans.* 1983, 2339.
177. Bruce, M. I.; Humphrey, M. G.; Shawkataly, O. B.; Snow, M. R.; Tiekink, E. R. T. *J. Organomet. Chem.* 1986, **315,** C51.
178. Doedens, R. J. *Inorg. Chem.* 1969, **8,** 570.
179. Clegg, W.; Sheldrick, G. M.; Stalke, D.; Bhaduri, S.; Gopalkrishnan, K. S. *Acta Crystallogr., Sect. C* 1984, **C40,** 927.
180. Smieja, J. A.; Gozum, J. E.; Gladfelter, W. L. *Organometallics* 1987, **6,** 1311.
181. Barnett, B. L.; Krüger, C. *Angew. Chem., Int. Ed. Engl.* 1971, **10,** 910.
182. Basu, A.; Bhaduri, S.; Khwaja, H.; Jones, P. G.; Meyer-Bäse, K.; Sheldrick, G. M. *J. Chem. Soc., Dalton Trans.* 1986, 2501.
183. Burgess, K.; Johnson, B. F. G.; Lewis, J.; Raithby, P. R. *J. Chem. Soc., Dalton Trans.* 1982, 2085.
184. Chi, Y.; Hwang, D.-K.; Chen, S.-F.; Liu, L.-K. *J. Chem. Soc., Chem. Commun.* 1989, 1540.
185. Lin, Y. C.; Knobler, C. B.; Kaesz, H. D. *J. Organomet. Chem.* 1981, **213,** C41.
186. Pizzotti, M.; Porta, F.; Cenini, S.; Demartin, F. *J. Organomet. Chem.* 1988, **356,** 105.
187. Pizzotti, M.; Cenini, S.; Crotti, C.; Demartin, F. *J. Organomet. Chem.* 1989, **375,** 123.
188. Lourdichi, M., Jr.; Mathieu, R. *Nouv. J. Chem.* 1982, **6,** 231.
189. Dutta, T. K.; Fehlner, T. P.; Vites, J. C. *Organometallics* 1986, **5,** 385.
190. Nametkin, N. S.; Tyurin, V. D.; Trusov, V. V.; Nekhaev, A. I.; Batsanov, A. S.; Struchkov, Yu.T. *J. Organomet. Chem.* 1986, **302,** 243.
191. Lourdichi, M.; Mathieu, R. *Nouv. J. Chim.* 1982, **6,** 231.
192. Wucherer, E. J.; Tasi, M.; Hansert, B.; Powell, A. K.; Garland, M.-T.; Halet, J.-F.; Saillard, J.-Y.; Vahrenkamp, H. *Inorg. Chem.* 1989, **28,** 3564.
193. Beno, M. A.; Williams, J. M.; Tachikawa, M.; Muetterties, E. L. *J. Am. Chem. Soc.* 1980, **102,** 4542.
194. Beno, M. A.; Williams, J. M.; Tachikawa, M.; Muetterties, E. L. *J. Am. Chem. Soc.* 1981, **103,** 1485.
195. Tachikawa, M.; Stein, J.; Muetterties, E. L.; Teller, R. G.; Beno, M. A.; Gebert, E.; Wiliams, J. M. *J. Am. Chem. Soc.* 1980, **102,** 6648.
196. Fjare, D. E.; Keyes, D. G.; Gladfelter, W. L. *J. Organomet. Chem.* 1983, **250,** 383.
197. Braga, D.; Johnson, B. F. G.; Lewis, J.; Mace, J. M.; McPartlin, M.; Puga, J.; Nelson, W. J. H.; Raithby, P. R.; Whitmire, K. H. *J. Chem. Soc., Chem. Commun.* 1982, 1081.
198. Blohm, M. L.; Fjare, D. E.; Gladfelter, W. L. *J. Am. Chem. Soc.* 1986, **108,** 2301.
199. Fjare, D. E.; Gladfelter, W. L. *Inorg. Chem.* 1981, **20,** 3533.
200. Fjare, D. E.; Gladfelter, W. L. *J. Am. Chem. Soc.* 1981, **103,** 1572.
201. Blohm, M. L.; Gladfelter, W. L. *Organometallics* 1985, **4,** 45.

202. Collins, M. A.; Johnson, B. F. G.; Lewis, J.; Mace, J. M.; Morris, J. M.; McPartlin, M.; Nelson, W. J. H.; Puga, J.; Raithby, P. R. *J. Chem. Soc., Chem. Commun.* 1983, 689.
203. Fjare, D. E.; Gladfelter, W. L. *J. Am. Chem. Soc.* 1984, **106,** 4799.
204. Draper, S. M.; Housecroft, C. E.; Keep, A. K.; Matthews, D. M.; Song, X.; Rheingold, A. L. *J. Organomet. Chem.* 1992, **423,** 241.
205. Blohm, M. L.; Gladfelter, W. L. *Inorg. Chem.* 1987, **26,** 459.
206. Attard, J. P.; Johnson, B. F. G.; Lewis, J.; Mace, J. M.; Raithby, P. R. *J. Chem. Soc., Chem. Commun.* 1985, 1526.
207. Cowie, A. G.; Johnson, B. F. G.; Lewis, J.; Raithby, P. R. *J. Chem. Soc., Chem. Commun.* 1984, 1710.
208. Johnson, B. F. G.; Kaner, D. A.; Lewis, J.; Raithby, P. R.; Rosales, M. J. *J. Organomet. Chem.* 1982, **231,** C59.
209. Shongwe, M. S. Ph.D. Thesis, University of Cambridge, 1989.
210. Harpp, K. S.; Housecroft, C. E. *J. Organomet. Chem.* 1988, **340,** 389.
211. Blohm, M.; Gladfelter, W. L. *Organometallics* 1986, **5,** 1049.
212. Dutton, T.; Johnson, B. F. G.; Lewis, J.; Owen, S. M.; Raithby, P. R. *J. Chem. Soc., Chem. Commun.* 1988, 1423.
213. Gourdon, A.; Jeanin, R. *C. R. Acad. Sci. Paris* 1982, **295,** 1101.
214. Braye, E. H.; Dahl, L. F.; Hübel, W.; Wampler, D. L. *J. Am. Chem. Soc.* 1962, **84,** 4633.
215. Farrar, D. H.; Jackson, P. F.; Johnson, B. F. G.; Lewis, J.; Nicholls, J. N. *J. Chem. Soc., Chem. Commun.* 1981, 415.
216. Eady, C. R.; Johnson, B. F. G.; Lewis, J. *J. Chem. Soc., Dalton Trans.* 1975, 2606.
217. Blohm, M. L.; Fjare, D. E.; Gladfelter, W. L. *Inorg. Chem.* 1983, **22,** 1004.
218. Moule, A. Ph.D. Thesis, University of Cambridge, 1989.
219. Martinengo, S.; Ciani, G.; Sironi, A.; Heaton, B. T.; Mason, J. *J. Am. Chem. Soc.* 1979, **101,** 7095.
220. Stevens, R. E.; Liu, P. C. C.; Gladfelter, W. L. *J. Organomet. Chem.* 1985, **287,** 133.
221. Ciani, G.; Martinengo, S. *J. Organomet. Chem.* 1986, **306,** C49.
222. Albano, V. G.; Braga, D.; Martinengo, S. *J. Chem. Soc., Dalton Trans.* 1986, 981.
223. Martinengo, S.; Ciani, G.; Sironi, A. *J. Organomet. Chem.* 1988, **358,** C23.
224. Mason, J. In "Multinuclear NMR Spectroscopy"; Mason, J., Ed; Plenum Press: New York, 1987; Chapter 12.
225. Brint, P.; Spalding, T. R.; O'Cuill, K. *Polyhedron* 1986, **5,** 1791.
226. Vizi-Orosz, A. *J. Organomet. Chem.* 1976, **111,** 61.
227. Vizi-Orosz, A.; Galomb, V.; Pàlyi, G.; Markò, L.; Bor, G.; Natile, G. *J. Organomet. Chem.* 1976, **107,** 235.
228. Arnold, L. J.; Mackay, K. M.; Nicholson, B. K. *J. Organomet. Chem.* 1990, **387,** 197.
229. Vizi-Orosz, A.; Galomb, V.; Pàlyi, G.; Markò, L. *J. Organomet. Chem.* 1981, **216,** 105.
230. Lang, H.; Huttner, G.; Sigwarth, B.; Jibril, I.; Zsolnai, L.; Orama, O. *J. Organomet. Chem.* 1986, **304,** 137.
231. Becker, G.; Hermann, W. A.; Kalcher, W.; Kriechbaum, G. W.; Pahl, C.; Wagner, C. T.; Zeigler, L. M. *Angew. Chem., Int. Ed. Engl.* 1983, **22,** 413.
232. Lang, H.; Zsolnai, L.; Huttner, G. *Angew. Chem., Int. Ed. Engl.* 1983, **22,** 976.
233. Vizi-Orosz, A.; Pàlyi, G.; Markò, L.; Boese, R.; Schmid, G. *J. Organomet. Chem.* 1985, **288,** 179.
234. Carty, A. J. *Pure Appl. Chem.* 1982, **54,** 113.
235. Iwasaki, F.; Mays, M. J.; Raithby, P. R.; Taylor, P. L.; Wheatley, P. J. *J. Organomet. Chem.* 1981, **213,** 185.
236. Sheldrick, G. M.; Yesinowski, J. P. *J. Chem. Soc., Dalton Trans.* 1975, 873.
237. Colbran, S. B.; Johnson, B. F. G.; Lewis, J.; Sorrell, R. M. *J. Chem. Soc., Chem. Commun.* 1986, 525.
238. Haines, R. J.; Field, J. S.; Smit, D. N. *J. Organomet. Chem.* 1982, **224,** C49.

239. Bullock, L. M.; Filed, J. S.; Haines, R. J.; Minshall, E.; Moore, M. H.; Mulla, F.; Smit, D. N.; Steer, L. M. *J. Organomet. Chem.* 1990, **381,** 429.
240. Colbran, S. B.; Irele, P. T.; Johnson, B. F. G.; Kaye, P. T.; Lewis, J.; Raithby, P. R. *J. Chem. Soc., Dalton Trans.* 1989, 2033.
241. Colbran, S. B.; Irele, P. T.; Johnson, B. F. G.; Lahoz, F. J.; Lewis, J.; Raithby, P. R. *J. Chem. Soc., Dalton Trans.* 1989, 2023.
242. Natarajan, K.; Zsolnai, L.; Huttner, G. *J. Organomet. Chem.* 1981, **220,** 365.
243. Bruce, M. I.; Horn, E.; Shawkataly, O. B.; Snow, M. R.; Tiekink, E. R. T.; Williams, M. L. *J. Organomet. Chem.* 1986, **316,** 187.
244. Hay, C. M.; Jeffrey, J. G.; Johnson, B. F. G.; Lewis, J.; Raithby, P. R. *J. Organomet. Chem.* 1989, **359,** 87.
245. Fenske, D.; Merzweiler, K. *Angew. Chem., Int. Ed. Engl.* 1984, **23,** 160.
246. Knoll, K.; Huttner, G.; Zsolnai, L.; Orama, O.; Wasiucionek, M. *J. Organomet. Chem.* 1986, **310,** 225.
247. Montllo, D.; Suades, J.; Torres, M. R.; Perales, A.; Mathieu, R. *J. Chem. Soc., Chem. Commun.* 1989, 97.
248. References cited in Ref. 249.
249. Knoll, K.; Huttner, G.; Zsolnai, L.; Orama, O. *Angew. Chem., Int. Ed. Engl.* 1986, **25,** 1119.
250. Buchholz, D.; Huttner, G.; Zsolnai, L. *J. Organomet. Chem.* 1990, **381,** 97.
251. Lang, H.; Huttner, G.; Zsolnai, L.; Mohr, G.; Sigwarth, B.; Weber, U.; Orama, O.; Jibril, I. *J. Organomet. Chem.* 1986, **304,** 157.
252. Cook, S. L.; Evans, J.; Gray, L. R.; Webster, M. *J. Organomet. Chem.* 1982, **236,** 367.
253. Winter, A.; Zsolnai, L.; Huttner, G. *J. Organomet. Chem.* 1982, **234,** 337.
254. Linder, E.; Rothfuss, H. *J. Organomet. Chem.* 1988, **353,** C2.
255. Bockman, T. M.; Wang, Y.; Kochi, J. K. *New J. Chem.* 1988, **12,** 387.
256. Fässler, Th.; Buchholz, D.; Huttner, G.; Zsolnai, L. *J. Organomet. Chem.* 1989, **369,** 297.
257. Maclaughlin, S. A.; Carty, A. J.; Taylor, N. J. *Can. J. Chem.* 1982, **60,** 87.
258. Colbran, S. B.; Johnson, B. F. G.; Lahoz, F. J.; Lewis, J.; Raithby, P. R. *J. Chem. Soc., Dalton Trans.* 1988, 1199.
259. Lanfranchi, M.; Tiripicchio, A.; Sappa, E.; Carty, A. J. *J. Chem. Soc., Dalton Trans.* 1986, 2737.
260. Simon, G. L.; Dahl, L. F. *J. Am. Chem. Soc.* 1973, **95,** 2175.
261. Field, J. S.; Haines, R. J.; Honrath, U.; Smit, D. N. *J. Organomet. Chem.* 1987, **329,** C25.
262. Ryan, R. C.; Pittman, C. U., Jr.; O'Connor, J. P.; Dahl, L. F. *J. Organomet. Chem.* 1980, **193,** 247.
263. Pittman, C. U., Jr.; Wilemon, G. M.; Wilson, W. D.; Ryan, R. C. *Angew. Chem., Int. Ed. Engl.* 1980, **19,** 478.
264. Field, J. S.; Haines, R. J.; Smit, D. N.; Natarajan, K.; Scheidsteger, O.; Huttner, G. *J. Organomet. Chem.* 1982, **240,** C23.
265. Field, J. S.; Haines, R. J.; Smit, D. N. *J. Organomet. Chem.* 1982, **224,** C49.
266. Vahrenkamp, H.; Wucherer, E. J.; Wolters, D. *Chem. Ber.* 1983, **116,** 1219.
267. Jaeger, T.; Aime, S.; Vahrenkamp, H. *Organometallics* 1986, **5,** 245.
268. Jaeger, T.; Vahrenkamp, H. *Z. Naturforsch. B* 1986, **41B,** 789.
269. Jaeger, J. T.; Vahrenkamp, H. *Organometallics* 1988, **7,** 1746.
270. Jaeger, J. T.; Field, J. S.; Collison, D.; Speck, G. P.; Peake, B. M.; Hähnle, J.; Vahrenkamp, H. *Organometallics* 1988, **7,** 1753.
271. Lindner, E.; Weiss, G. A.; Hiller, W.; Fawzi, R. *J. Organomet. Chem.* 1986, **312,** 365.
272. Eady, C. R.; Johnson, B. F. G.; Lewis, J. *J. Chem. Soc., Dalton Trans.* 1977, 477.
273. Fernandez, J. M.; Johnson, B. F. G.; Lewis, J.; Raithby, P. R. *J. Chem. Soc., Chem. Commun.* 1978, 1015.
274. Natarjan, K.; Zsolnai, L.; Huttner, G. *J. Organomet. Chem.* 1981, **209,** 85.

275. Gastel, F. V.; Taylor, N. J.; Carty, A. J. *J. Chem. Soc., Chem. Commun.* 1987, 1049.
276. Taylor, N. J. *J. Chem. Soc., Chem. Commun.* 1985, 478.
277. Lower, L. D.; Dahl, L. F. *J. Am. Chem. Soc.* 1976, **98,** 5046.
278. Fenske, D.; Ohmer, J.; Hachgenei, J.; Merzweiler, K. *Angew. Chem., Int. Ed. Engl.* 1988, **27,** 1277 and references cited therein.
279. Chini, P.; Ciani, G.; Martinengo, S.; Sironi, A.; Longetti, L.; Heaton, B. T. *J. Chem. Soc., Chem. Commun.* 1979, 188.
280. Colbran, S. B.; Lahoz, F. J.; Raithby, P. R.; Lewis, J.; Johnson, B. F. G.; Cardin, C. J. *J. Chem. Soc., Dalton Trans.* 1988, 173.
281. Colbran, S. B.; Hay, C. M.; Johnson, B. F. G.; Lahoz, F. J.; Lewis, J.; Raithby, P. R. *J. Chem. Soc., Chem. Commun.* 1986, 1766.
282. Colbran, S. B.; Housecroft, C. E.; Johnson, B. F. G.; Lewis, J.; Raithby, P. R. *Polyhedron* 1988, **7,** 1759.
283. Gastel, F. V.; Taylor, N. J.; Carty, A. J. *Inorg. Chem.* 1989, **28,** 384.
284. Vidal, J. L.; Fiato, R. A.; Cosby, L. A.; Pruett, R. L. *Inorg. Chem.* 1978, **17,** 2574.
285. Vidal, J. L.; Walker, W. E.; Pruett, R. L.; Schoening, R. C. *Inorg. Chem.* 1979, **18,** 129.
286. Vidal, J. L.; Walker, W. E.; Pruett, R. L.; Schoening, R. C. *Inorg. Chem.* 1981, **20,** 238.
287. Rheingold, A. L.; Geib, S. J.; Shieh, M.; Whitmire, K. H. *Inorg. Chem.* 1987, **26,** 463.
288. Ferrer, M.; Rossell, O.; Seco, M.; Braunstein, P. *J. Organomet. Chem.* 1989, **364,** C5.
289. Arif, A. M.; Cowley, A. H.; Pakulski, M. *J. Chem. Soc., Chem. Commun.* 1987, 622.
290. Sigarth, B.; Zsolnai, L.; Berke, H.; Huttner, G. *J. Organomet. Chem.* 1982, **226,** C5.
291. Huttner, G.; Weber, U.; Sigarth, B.; Scheidsteger, O. *Angew. Chem., Int. Ed. Engl.* 1982, **21,** 215.
292. Johnson, B. F. G.; Lewis, J.; Massey, A. D.; Braga, D.; Grepioni, F. *J. Organomet. Chem.* 1989, **369,** C43.
293. Whitmire, K. H.; Leigh, J. S.; Luo, S.; Shieh, M.; Fabiano, M. D.; Rheingold, A. L. *New J. Chem.* 1988, **12,** 397.
294. Luo, S.; Whitmire, K. H. *Inorg. Chem.* 1989, **28,** 1424.
295. Caballero, C.; Nuber, B.; Ziegler, M. L. *J. Organomet. Chem.* 1990, **386,** 209.
296. Röttinger, E.; Vahrenkamp, H. *J. Organomet. Chem.* 1981, **213,** 1.
297. Dietrich, W.-R.; Vahrenkamp, H. *J. Chem. Res. (S)* 1985, 200.
298. Guldner, K.; Johnson, B. F. G.; Lewis, J. *J. Organomet. Chem.* 1988, **355,** 419.
299. Delbraere, L. T. J.; Kruczynski, L. T.; McBride, D. W. *J. Chem. Soc., Dalton Trans.* 1973, 307.
300. Huttner, G.; Mohr, G.; Frank, A.; Schubert, U. *J. Organomet. Chem.* 1976, **118,** C73.
301. Huttner, G.; Mohr, G.; Friedrich, P.; Schmid, H. G. *J. Organomet. Chem.* 1978, **160,** 59.
302. Jacob, M.; Weiss, E. *J. Organomet. Chem.* 1977, **131,** 263.
303. Winter, A.; Zsolnai, L.; Huttner, G. *J. Organomet. Chem.* 1982, **234,** 337.
304. Fenske, D.; Hachgenei, J. *Angew. Chem., Int. Ed. Engl.* 1986, **25,** 175.
305. Fenske, D.; Merzweiler, K.; Ohmer, J. *Angew. Chem., Int. Ed. Engl.* 1988, **27,** 1512.
306. Rieck, D. F.; Montag, R. A.; McKechnie, T. S.; Dahl, L. F. *J. Am. Chem. Soc.* 1986, **108,** 1330.
307. DesEnfants, R. E.; Gavney, J. A., Jr.; Hayashi, R. K.; Rae, A. D.; Dahl, L. F.; Bjarnason, A. *J. Organomet. Chem.* 1990, **383,** 543.
308. Albano, V. G.; Demartin, F.; Iapalucci, M. C.; Longoni, G.; Sironi, A.; Zanotti, V. *J. Chem. Soc., Chem. Commun.* 1990, 547.
309. Vidal, J. *Inorg. Chem.* 1981, **20,** 243.
310. Vidal, J.; Troup, J. M. *J. Organomet. Chem.* 1981, **213,** 351.
311. Norman, N. C. *Chem. Soc. Rev.* 1988, **17,** 269.
312. Huttner, G.; Weber, U.; Zsolnai, L. *Z. Naturforsch. B* 1982, **37B,** 707.
313. Whitmire, K. H.; Raghuveer, K. S.; Churchill, M. R.; Fettinger, J. C.; See, R. F. *J. Am. Chem. Soc.* 1986, **108,** 2778.

314. Whitmire, K. H.; Shieh, M.; Lagrone, C. B.; Robinson, B. H.; Churchill, M. R.; Fettinger, J. C.; See, R. F. *Inorg. Chem.* 1987, **26,** 2798.
315. Whitmire, K. H.; Shieh, M.; Cassidy, J. *Inorg. Chem.* 1989, **28,** 3164.
316. Whitmire, K. H.; Churchill, M. R.; Fettinger, J. C. *J. Am. Chem. Soc.* 1985, **107,** 1057.
317. Johnson, B. F. G.; Lewis, J.; Raithby, P. R.; Whitton, A. J. *J. Chem. Soc., Chem. Commun.* 1988, 401.
318. Kruppa, W.; Bläser, D.; Boese, R.; Schmid, G. *Z. Naturforsch. B* 1982, **37B,** 209.
319. Whitmire, K. H.; Lagrone, C. B.; Churchill, M. R.; Fettinger, J. C.; Biondi, L. V. *Inorg. Chem.* 1984, **23,** 4227.
320. Whitmire, K. H.; Lagrone, C. B.; Rheingold, A. L. *Inorg. Chem.* 1986, **25,** 2472.
321. Wallis, J. M.; Müller, G.; Schmidbauer, H. *J. Organomet. Chem.* 1987, **325,** 159.
322. Hay, C. M.; Johnson, B. F. G.; Lewis, J.; Raithby, P. R.; Whitton, A. J. *J. Organomet. Chem.* 1987, **330,** C5.
323. See, for example: Fehlner, T. P.; Housecroft, C. E. In "Molecular Structure and Energetics"; Liebman, J. F.; Greenberg, A., Eds.; VCH Publishers; Florida, 1986; Vol. 1, Chapter 6.
324. Ciani, G.; Moret, M.; Fumagalli, A.; Martinengo, S. *J. Organomet. Chem.* 1989, **362,** 291.
325. Foust, A. S.; Dahl, L. F. *J. Am. Chem. Soc.* 1970, **92,** 7337.
326. Clegg, W.; Compton, N. A.; Errington, R. J.; Norman, N. C. *Polyhedron,* 1988, **7,** 2239.
327. Churchill, M. R.; Fettinger, J. C.; Whitmire, K. H. *J. Organomet. Chem.* 1985, **284,** 13.
328. Chisholm, M. H. In "Inorganic Chemistry Toward the 21st Century"; Chisholm, M. H., Ed.; *ACS Symp. Ser.* **211,** 1983; Chapter 16.
329. See, for example: Simon, A. *Angew. Chem., Int. Ed. Engl.* 1988, **27,** 159.
330. Bottomly, F.; Sutin, L. *Adv. Organomet. Chem.* 1988, **28,** 339.
331. Colombié, A.; Bonnet, J.-J.; Fompeyrine, P.; Lavigne, G.; Sunshine, S. *Organometallics* 1986, **5,** 1154.
332. Ceriotti, A.; Resconi, L.; Demartin, F.; Longoni, G.; Manassero, M.; Sansoni, M. *J. Organomet. Chem.* 1983, **249,** C35.
333. Uchtman, V. A.; Dahl, L. F. *J. Am. Chem. Soc.* 1969, **91,** 3763.
334. Nutton, A.; Bailey, P. M.; Maitlas, P. M. *J. Organomet. Chem.* 1981, **213,** 313.
335. Bertolucci, A.; Freni, M.; Romiti, P.; Ciani, G.; Sironi, A.; Albano, V. G. *J. Organomet. Chem.* 1976, **113,** C61.
336. Schloter, K.; Nagel, U.; Beck, W. *Chem. Ber.* 1980, **113** 3775.
337. Kathirgamanathan, P.; Martinez, M.; Sykes, A. G. *J. Chem. Soc., Chem. Commun.* 1985, 1437.
338. Gibson, C. P.; Huang, J.-S.; Dahl, L. F. *Organometallics* 1986, **5,** 1676.
339. Bino, A. *Inorg. Chem.* 1982, **21,** 1917.
340. Cotton, F. A.; Diebold, M. P.; Llusar, R.; Roth, W. J. *J. Chem. Soc., Chem. Commun.* 1986, 1276.
341. Jernakoff, P.; de Bellefon, C. de M.; Geoffroy, G. L.; Rheingold, A. L.; Gelb, S. J. *Organometallics* 1987, **6,** 1362.
342. Jernakoff, P.; de Bellefon, C. de M.; Geoffroy, G. L.; Rheingold, A. L.; Gelb, S. J. *New J. Chem.* 1988, **12,** 329.
343. Ciani, G.; D'Alfonso, G.; Freni, M.; Romiti, P.; Sironi, A. *J. Organomet. Chem.* 1981, **219,** C23.
344. Kirtley, S. W.; Chanton, J. P.; Love, R. A.; Tipton, D. L.; Sorrell, T. N.; Bau, R. *J. Am. Chem. Soc.* 1980, **102,** 3451.
345. Abel, E. W.; Towle, I. D. H.; Cameron, T. S.; Cordes, R. E. *J. Chem. Soc., Dalton Trans.* 1979, 1943.
346. Cotton, F. A.; Lahuerta, P.; Sanau, M.; Schwotzer, W. *J. Am. Chem. Soc.* 1985, **107,** 8284.
347. Abel, E. W.; Towle, I. D. H.; Cameron, T. S.; Cordes, R. E. *J. Chem. Soc., Chem. Commun.* 1977, 285.
348. Bhardi, S.; Sapre, N. Y.; Sharma, K. R.; Jones, P. G. *J. Organomet. Chem.* 1989, **364,** C8.

349. Wong, W.-K.; Chiu, K. W.; Wilkinson, G.; Galas, A. M. R.; Thornton-Pett, M. T.; Hursthouse, M. B. *J. Chem. Soc., Dalton Trans.* 1983, 1557.
350. Schauer, C. K.; Shriver, D. F. *Angew. Chem., Int. Ed. Engl.* 1987, **26,** 255.
351. Gibson, V. C.; Kee, T. P.; Clegg, W. *J. Chem. Soc., Chem. Commun.* 1990, 29.
352. Cotton, F. A.; Feng, X.; Kibala, P. A.; Sandor, R. B. W. *J. Am. Chem. Soc.* 1989, **111,** 2148.
353. Adams, R. D. *Polyhedron* 1985, **4,** 2003.
354. Süss-Fink, G.; Thewalt, U.; Klein, H.-P. *J. Organomet. Chem.* 1982, **224,** 59.
355. Winter, A.; Zsolnai, L.; Huttner, G. *J. Organomet. Chem.* 1982, **232,** 47.
356. Adams, R. D.; Katahira, D. A. *Organometallics* 1982, **1,** 53.
357. Winter, A.; Zsolnai, L.; Huttner, G. *J. Organomet. Chem.* 1983, **250,** 409.
358. Khattab, S. A.; Markó, L.; Bor, G.; Markó, B. *J. Organomet. Chem.* 1964, **1,** 373.
359. Wei, C. H.; Dahl, L. F. *Inorg. Chem.* 1967, **6,** 1229.
360. Bor, G.; Gervasio, G.; Rossetti, R.; Stanghellini, P. L. *J. Chem. Soc., Chem. Commun.* 1978, 841.
361. Stanghellini, P. L.; Gervasio, G.; Rossetti, R.; Bor, G. *J. Organomet. Chem.* 1980, **187,** C37.
362. Bor, G.; Stanghellini, P. L. *J. Chem. Soc., Chem. Commun.* 1979, 886.
363. Bor, G.; Dietler, U. K.; Stanghellini, P. L.; Sbrignadello, G.; Battison, G. A. *J. Organomet. Chem.* 1981, **213,** 277.
364. Gervasio, G.; Rossetti, R.; Stanghellini, P. L.; Bor, G. *Inorg. Chem.* 1982, **21,** 3781.
365. Gervasio, G.; Rossetti, R.; Stanghellini, P. L.; Bor, G. *Inorg. Chem.* 1984, **23,** 2073.
366. Wei, C. H. *Inorg. Chem.* 1984, **23,** 2973.
367. Markó, L.; Gervasio, G.; Stanghellini, P. L.; Bor, G. *Transition Metal Chem.* 1985, **10,** 344.
368. Benoit, A.; Darchen, A.; Le Marouille, J.-Y.; Mahé, C.; Patin, H. *Organometallics* 1983, **2,** 555.
369. Stevenson, D. L.; Magnuson, V. R.; Dahl, L. F. *J. Am. Chem. Soc.* 1967, **89,** 3727.
370. Fischer, K.; Deck, W.; Schwanz, M.; Vahrenkamp, H. *Chem. Ber.* 1985, **118,** 4946.
371. Fischer, K.; Müller, M.; Vahrenkamp, H. *Angew. Chem., Int. Ed. Engl.* 1984, **23,** 140.
372. Roland, E.; Vahrenkamp, H. *Angew. Chem., Int. Ed. Engl.* 1983, **22,** 326.
373. Farrugia, L. J.; Freeman, M. J.; Green, M.; Orpen, A. G.; Stone, F. G. A.; Salter, I. D. *J. Organomet. Chem.* 1983, **249,** 273.
374. Al-Ani, F. T.; Hughes, D. L.; Pickett, C. J. *J. Organomet. Chem.* 1986, **307,** C31.
375. Adams, R. D.; Katahira, D. A. *Organometallics* 1982, **1,** 53.
376. Johnson, B. F. G.; Lewis, J.; Pippard, D.; Raithby, P. R. *Acta Crystallogr., Ser. B* 1978, **34B,** 3767.
377. Johnson, B. F. G.; Lewis, J.; Pippard, D.; Raithby, P. R.; Sheldrick, G. M.; Rouse, K. D. *J. Chem. Soc., Dalton Trans.* 1979, 616.
378. Deeming, A. J.; Ettorre, R.; Johnson, B. F. G.; Lewis, J. *J. Chem. Soc. (A)* 1971, 1797.
379. Adams, R. D.; Männing, D.; Segmüller, B. E. *Organometallics* 1983, **2,** 149.
380. Darensbourg, D. J.; Zalewski, D. J.; Sanchez, K. M.; Delford, T. *Inorg. Chem.* 1988, **27,** 821.
381. Jennings, M. C.; Payne, N. C.; Puddephatt, R. J. *Inorg. Chem.* 1987, **26,** 3776.
382. Williams, P. D.; Curtis, M. D.; Duffy, D. N.; Butler, W. M. *Organometallics* 1983, **2,** 165.
383. Braunstein, P.; Jud, J.-M.; Tiripicchio, A.; Tiripicchio-Camellini, M.; Sappa, E. *Angew. Chem., Int. Ed. Engl.* 1982, **21,** 307.
384. Curtis, M. D.; Williams, P. D.; Butler, W. M. *Inorg. Chem.* 1988, **27,** 2853.
385. Braunstein, P.; Jud, J.-M.; Tiripicchio, A.; Tiripicchio-Camellini, M.; Sappa, E. *Inorg. Chem.* 1981, **20,** 3586.
386. Fehlner, T. P.; Housecroft, C. E.; Wade, K. *Organometallics* 1983, **2,** 1426.
387. Curtis, M. D.; Williams, P. D. *Inorg. Chem.* 1983, **22,** 2661.

388. Adams, R. D.; Babin, J. E.; Mathur, P.; Natarjan, K.; Wang, J.-G. *Inorg. Chem.* 1989, **28,** 1441.
389. Adams, R. D.; Babin, J. E.; Tasi, M. *Organometallics* 1988, **7,** 219.
390. Adams, R. D.; Babin, J. E.; Tasi, M. *Organometallics* 1987, **6,** 2247.
391. Adams, R. D.; Wang, S. *Inorg. Chem.* 1986, **25,** 2534.
392. Bonfichi, R.; Ciani, G.; Al'fonso, G. D.; Romiti, P.; Sironi, A. *J. Organomet. Chem.* 1982, **231,** C35.
393. Winter, A.; Zsolnai, L.; Huttner, G. *Chem. Ber.* 1982, **115,** 1286.
394. Frank, L.-R.; Winter, A.; Huttner, G. *J. Organomet. Chem.* 1987, **335,** 249.
395. Cresswell, T. A.; Howard, J. A. K.; Kennedy, F. G.; Knox, S. A. R.; Wadepohl, H. *J. Chem. Soc., Dalton Trans.* 1981, 2220.
396. Ghilardi, C. A.; Midollini, S.; Sacconi, L. *Inorg. Chim. Acta* 1978, **31,** L431.
397. Chatt, J.; Mingos, D. M. P. *J. Chem. Soc. (A)* 1970, 1243.
398. Werner, H. H.; Leonhard, K.; Kolb, O.; Rottinger, E.; Vahrenkamp, H. *Chem. Ber.* 1980, **113,** 1654.
399. Galli, D.; Garlaschelli, L.; Ciani, G.; Fumagalli, A.; Martinengo, S.; Sironi, A. *J. Chem. Soc., Dalton Trans.* 1984, 55.
400. Pergola, R. D.; Garlaschelli, L.; Martinengo, S.; Demartin, E.; Manassero, M.; Sironi, A. *J. Chem. Soc., Dalton Trans.* 1986, 2463.
401. Lockemeyer, J. R.; Rauchfuss, T. B.; Rheingold, A. L. *J. Am. Chem. Soc.* 1989, **111,** 5733.
402. Adams, R. D.; Horváth, I. T.; Segmüller, B. E.; Yang, L.-W. *Organometallics* 1983, **2,** 1301.
403. Adams, R. D.; Horváth, I. T.; Kim, H.-S. *Organometallics* 1984, **3,** 548.
404. Adams, R. D.; Babin, J. E.; Tasi, M. *Inorg. Chem.* 1986, **25,** 4514.
405. Wei, C. H.; Dahl, L. F. *Inorg. Chem.* 1965, **4,** 493.
406. Johnson, B. F. G.; Lewis, J.; Raithby, P. R.; Henrick, K.; McPartlin, M. *J. Chem. Soc., Chem. Commun.* 1979, 719.
407. Adams, R. D.; Horváth, I. T.; Segmüller, B. E.; Yang, L.-W. *Organometallics* 1983, **2,** 144.
408. Rives, A. B.; Xiao-Zeng, Y.; Fenske, R. F. *Inorg. Chem.* 1982, **21,** 2286.
409. Bright, T. A.; Jones, R. A.; Koshmieder, S. U.; Nunn, C. M. *Inorg. Chem.* 1988, **27,** 3819.
410. Arif, A. M.; Hefner, J. G.; Jones, R. A.; Koshmieder, S. U. *Polyhedron* 1988, **7,** 561.
411. Adams, R. D.; Horváth, I. T.; Segmüller, B. E. *J. Organomet. Chem.* 1984, **262,** 243.
412. Cowans, B.; Noordvik, J.; DuBois, M. R. *Organometallics* 1983, **2,** 931.
413. Adams, R. D.; Babin, J. E. *Inorg. Chem.* 1986, **25,** 3418.
414. Fackler, J. P.; Mazany, A. M.; Seyferth, D.; Withers, H. P.; Wood, T. G.; Campana, C. F. *Inorg. Chim. Acta* 1984, **82,** 31.
415. Lindner, E.; Weiss, G. A.; Hiller, W.; Fawzi, R. *J. Organomet. Chem.* 1983, **25,** 245.
416. Adams, R. D.; Horváth, I. T.; Yang, L.-W. *Organometallics* 1983, **2,** 1257.
417. Adams, R. D.; Yang, L.-W. *J. Am. Chem. Soc.* 1983, **105,** 235.
418. Adams, R. D.; Horváth, I. T.; Mathur, P. *J. Am. Chem. Soc.* 1984, **106,** 6296.
419. Adams, R. D.; Horváth, I. T.; Natarajan, K. *Organometallics* 1984, **3,** 1540.
420. Adams, R. D.; Babin, J. E.; Tasi, M. *Organometallics* 1988, **7,** 503.
421. Adams, R. D.; Babin, J. E.; Mathab, R.; Wang, S. *Inorg. Chem.* 1986, **25,** 1623.
422. Adams, R. D.; Babin, J. E.; Natarajan, K.; Tasi, M.; Wang, J.-G. *Inorg. Chem.* 1987, **26,** 3708.
423. Adams, R. D.; Horváth, I. T.; Mathur, P. *Organometallics* 1984, **3,** 623.
424. Adams, R. D.; Foust, D. F.; Mathur, P. *Organometallics* 1983, **2,** 990.
425. Adams, R. D.; Babin, J. E.; Tasi, M. *Inorg. Chem.* 1988, **27,** 2618.
426. Wei, C. H.; Dahl, L. F. *Inorg. Chem.* 1975, **4,** 583.
427. Adams, R. D.; Horváth, I. T.; Yang, L.-W. *J. Am. Chem. Soc.* 1983, **105,** 1533.
428. Adams, R. D.; Horváth, I. T.; Mathur, P.; Segmüller, B. E.; Yang, L.-W. *Organometallics* 1983, **2,** 1078.

429. Adams, R. D.; Babin, J. E.; Tasi, M.; Wolfe, T. A. *Organometallics* 1987, **6,** 2228.
430. Adams, R. D.; Wang, S. *Organometallics* 1987, **6,** 739.
431. Herrmann, W. A.; Rohrmann, J.; Hecht, C. *J. Organomet. Chem.* 1985, **290,** 53.
432. Lesch, D. A.; Rauschfuss, T. B. *Organometallics* 1982, **1,** 499.
433. Fenske, D.; Hollnagel, A.; Merzweiler, K. *Angew. Chem., Int. Ed. Engl.* 1988, **27,** 965.
434. Mathur, P.; Mavunkal, I. J.; Rheingold, A. L. *J. Chem. Soc., Chem. Commun.* 1989, 382.
435. Mathur, P.; Mavunkal, I. J.; Rugmini, V. *Inorg. Chem.* 1989, **28,** 3616.
436. Adams, R. D.; Horváth, I. T. *Inorg. Chem.* 1984, **23,** 4718.
437. Schacht, H.-T.; Powell, A. K.; Vahrenkamp, H.; Koike, M.; Kneuper, H.-J.; Shapley, J. R. *J. Organomet. Chem.* 1989, **368,** 269.
438. Cecconi, F.; Ghilardi, C. A.; Midollini, S. *Inorg. Chem.* 1983, **22,** 3802.
439. Hieber, W.; Gruber, J. *Z. Anorg. Allg. Chem.* 1958, **296,** 91.
440. Dahl, L. F.; Sutton, P. W. *Inorg. Chem.* 1963, **2,** 1067.
441. Mathur, P.; Mavunkal, I. J. *J. Organomet. Chem.* 1988, **350,** 251.

4

Bonding Connections and Interrelationships

D. M. P. Mingos

1. INTRODUCTION

The historical development of the theoretical ideas which have brought the areas of transition metal and main group chemistry closer together have been described in some detail elsewhere[1] and will not be recounted here. The major developments and key references are, however, listed in chronological order in Table 1.[2–4] In the 1950s the contributions of Longuet-Higgins proved decisive, since he provided not only the three-center two-electron bonding concept for B_2H_6, which was elegantly generalized in Lipscomb's styx formalism, but also pioneered the use of molecular orbital theory to describe the skeletal bonding in deltahedral boranes. This approach led, for example, to the successful prediction of the closed-shell requirements for the icosahedral $B_{12}H_{12}^{2-}$ ion, several years before it was isolated.

In the 1960s Cotton and Haas pioneered the development of these molecular orbital ideas to transition metal clusters such as $[Mo_6Cl_8]^{4+}$, $[Ta_6Cl_{12}]^{2+}$, and $[Re_3Cl_{12}]^{3-}$ where the *d*-orbital interactions between the metals made an important bonding contribution. An alternative localized bonding description related to the styx methodology was proposed for these metal clusters by Kettle. These theoretical developments tended to emphasize the differences between main group and transition metal chemistry at a time when synthetic

D. M. P. Mingos • Department of Chemistry, Imperial College of Science Technology and Medicine, London SW7 2AY, England.

Inorganometallic Chemistry, edited by Thomas P. Fehlner. Plenum Press, New York, 1992.

Table 1. Summary of Important Theoretical Developments

Year	Essential conceptual development	Reference
1949	Three-center two-electron molecular orbital description of bonding in B_2H_6	Longuet-Higgins, H. C. *J. Chim. Phys.*, 1949, **46,** 275
1954	Styx semilocalized description of the bonding in boranes B_nH_{n+4} and B_nH_{n+6}	Lipscomb, W. N. *et al. J. Chem. Phys.*, 1954, **22,** 989
1955	Delocalized description of the bonding in icosahedral boride anions $B_{12}{}^{2-}$ and prediction of $B_{12}H_{12}{}^{2-}$	Longuet-Higgins, H. C. *et al. Proc. Roy. Soc.* 1955, **A230,** 110
1962	General molecular orbital analysis of borane anions $B_nH_n{}^{2-}$	Hoffmann, R.; Lipscomb, W. N. *J. Chem. Phys.* 1962, **36,** 2179
1963	Zintl concept popularized for solid state main group infinite clusters containing clusters moieties	Klemm, W. *et al. Z. Anorg. Allg. Chem.* 1963, **319,** 217
1964	Molecular orbital description of the bonding in metal clusters, $[Mo_6Cl_8]^{4+}$, $[Ta_6Cl_{12}]^{2+}$, and $[Re_3Cl_{12}]^{3-}$	Cotton, F.A.; Hass, T.E. *Inorg. Chem.* 1964, **3,** 10
1965	Localized description of the bonding in $[Mo_6Cl_8]^{4+}$ and $[Ta_6Cl_{12}]^{2+}$	Kettle, S. F. A. *Theor. Chim. Acta* 1965, **3,** 211
1965–1970	Discussion of analogies between bonding capabilities of main group and transition metal fragments, e.g., $Co(CO)_3$ and P and between d^8 square-planar complexes and carbenes	Halpern, J. *Discuss. Faraday Soc.* 1968, **46,** 7 Dahl, L. F. *et al. J. Am. Chem. Soc.* 1969, **91,** 5631
1971	Structural relationships between *closo-*, *nido-*, and *arachno*-boranes overcome the general view that these structures are based on icosahedra	Williams, R. E. *Inorg. Chem.* 1971, **10,** 210
1971	Electronic relationship between octahedral boranes and transition metal clusters, e.g. $[B_6H_6]^{2-}$ and $[Ru_6(CO)_{18}]^{2-}$ proposed. Electronic bases of *closo-*, *nido-*, *arachno-* relationships defined	Wade, K. *J. Chem. Soc., Chem. Commun.* 1971, 792

1972	Classification of electron-deficient, electron-precise, and electron-rich polyhedra and the importance of bond-breaking processes in electron-rich polyhedra. Capping principle illustrated	Mingos, D.M.P. *Nature Phys. Sci.* 1972, **236,** 99
1972	The role of lone pairs in polyhedral hetero-boranes; molecular orbital analysis of *closo-, nido-, arachno-*transformations	Rudolph, R. E. *Inorg. Chem.* 1972, **11,** 1974
1976	The definition of the *isolobal* relationships in polyhedral main group and transition metal clusters	Hoffmann, R.; Mingos, D. M. P. *et al. Inorg. Chem.* 1976, **15,** 1148
1977	Quantum mechanical description of capping principle	Mingos, D. M. P. *J. Chem. Soc., Dalton Trans.* 1977, 610
1979–1980	Molecular orbital calculations based on the extended Hückel approximation underpin the closed-shell requirements of metal clusters	Lauher, J. E. *J. Am. Chem. Soc.* 1979, **101,** 2604 Ciani, G.; Sironi, A. *J. Organomet. Chem.* 1980, **197,** 233
1977	Graph theoretical description of polyhedral molecules	King, R. B.; Rouvray, D. H. *J. Am. Chem. Soc.* 1977, **99,** 7814
1980	Tensor surface harmonic theory developed	Stone, A. J. *Mol. Phys.* 1980, **41,** 1339
1983	Principle for defining the closed-shell requirements of condensed clusters	Mingos, D. M. P. *J. Chem. Soc., Chem. Commun.* 1983, 206
1984	Topological electron-counting model	Teo, B. K. *Inorg. Chem.* 1984, **23,** 1251
1985	Group theoretical consequences of Tensor Surface Harmonic Theory developed	Ceulemans, A. *Mol. Phys.* 1985, **54,** 161
1985	Polyhedral inclusion principle proposed for defining the closed-shell requirements of high nuclearity clusters	Mingos, D. M. P. *J. Chem. Soc., Chem. Commun.* 1985, 1352

organometallic chemists were generating compounds whose structures could not readily be described in localized terms. The complex products resulting from the reactions of acetylenes with metal-carbonyls[5] and the metallocarboranes synthesized by Hawthorne's group[6] provide particularly memorable examples of structures which did not fit comfortably into the main group and transition metal conceptual straightjackets.

The structural and electronic relationships connecting main group and transition metal polyhedral molecules were defined in the 1970s largely as a result of the contributions of Williams, Wade, Rudolph, and Mingos. The ideas introduced by these chemists effectively broke down the conceptual barriers separating large areas of main group and transition metal chemistry and suggested the synthesis of many compounds containing both main group and transition metal fragments. The *isolobal* relationships connecting main group and transition metal fragments with similar bonding capabilities found particular favor among synthetic chemists, because they could directly relate the bonding abilities of particular fragments to molecules which could be readily generated in the laboratory.[7,8] The *capping* and *debor* principles also proved to be particularly useful for making structural relationships between molecules whose structures were formed either by the addition of capping atoms to polyhedral faces or the loss of a vertex atom from a complete polyhedron.

During the 1980s the theoretical basis of the analogies between main group and metal polyhedral molecules has been underpinned by Stone's Tensor Surface Harmonic Theory. The group theoretical implications of this model have also been developed and used to account for apparent exceptions to the model. In addition, this model has been developed to account for the closed-shell requirements of fused clusters and high nuclearity metal carbonyl clusters.

The main purpose of this chapter is to introduce the reader to those key bonding concepts which are necessary to understand the structures of molecules that fall within the "inorganometallic" classification. In addition, illustrations of the applications of these principles will be given and, where appropriate, some guidance to those situations where exceptions to the generalizations are likely to occur.

2. ALTERNATIVE MODES OF DEFINING THE STRUCTURAL RELATIONSHIPS IN POLYHEDRAL MOLECULES

2.1. General Considerations

The structural relationships which have unified main group and metal cluster chemistry are based on the definition of specific closed-shell electronic relationships for particular three-dimensional structural types. These closed-shell requirements are dictated by the spectrum of molecular orbitals generated for a particular skeletal geometry. For three-connected clusters the number of bonding and antibonding skeletal molecular orbitals is equal. As the connectivity of the cluster is increased the number of bonding skeletal MOs decreases and the number of antibonding skeletal molecular orbitals correspondingly increases. Although the structure of a cluster and the total number of valence electrons are well defined, it is sometimes convenient to partition the total number of electrons in a manner which more conveniently expresses the structural relationships. Since unfamiliarity with the field can sometimes cause confusion when faced with alternative partitioning schemes, an attempt will be made to illustrate and define these alternative partitioning schemes below.

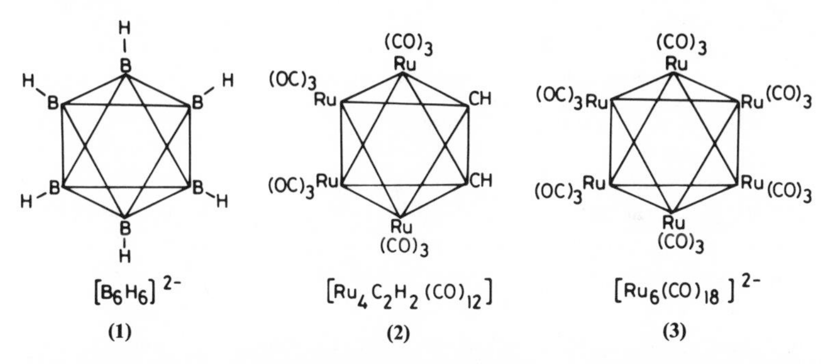

The isostructural octahedral clusters $B_6H_6^{2-}$ **(1)**, $[Ru_4C_2H_2(CO)_{12}]$ **(2)** and $[Ru_6(CO)_{18}]^{2-}$ **(3)** will be used to illustrate the important differences since they can be viewed as representative of main group, inorganometallic, and transition-metal cluster compounds. The alternative description of $[Ru_4C_2H_2(CO)_{12}]$ **(2)** as a butterfly Ru_4 cluster with a bridging alkyne ligand is also valid, but suppressed in the discussion below in order to emphasize the common octahedral structural unit present in **(1)**, **(2)**, and **(3)**. These ambiguities associated with defining the relevant polyhedral skeleton have been discussed in some detail in Chapters 2 and 3 of this book.

2.2. Polyhedral Skeletal Electron Pair Theory

According to the *polyhedral skeletal electron pair* theory the polyhedral geometry is related directly to the total number of valence electrons in the molecule, i.e., the electrons in skeletal, nonbonding, and ligand–metal MOs. In this review this number of valence electrons is described as the *polyhedral electron count* (pec). The relevant relationships are summarized for main group and transition metal polyhedral molecules in Table 2. The relationships are expressed in terms of the total number of vertices, *n*, associated with the polyhedral structure. For rings and three-connected polyhedral molecules the results are exactly the same as those anticipated from the noble gas rule, since

Table 2. Summary of Closed-Shell Requirements for Main Group and Transition Metal Cluster Molecules

Polyhedral geometry	Total number of valence electrons (pec[a])		Total number of unavailable skeletal MOs
	Main group	Transition metal	
Rings	$6n$	$16n$	n
Three-connected	$5n$	$15n$	$3n/2$
Four-connected	$4n+2$	$14n+2$	$2n-1$
Deltahedral[b]			
closo-	$4n+2$	$14n+2$	$2n-1$
nido-	$4n+4$	$14n+4$	$2n-2$
arachno-	$4n+6$	$14n+6$	$2n-3$

[a] pec = polyhedral electron count.
[b] A deltahedral molecule with triangular faces exclusively is described as *closo-*. The related structure which is derived from the *closo*-molecule by the loss of a single vertex is described as *nido-*. The loss of two vertices from the *closo*-structure results in an *arachno*-structure.

for each element–element bond the total number of valence electrons is reduced from that anticipated by the noble gas rule by two.

For example:

CH_4 8 valence electrons; $H_3C—CH_3$ 14 valence electrons ($8n–2$)

$$\begin{array}{c} CH_2 \\ / \quad \backslash \\ CH_2—CH_2 \end{array}$$ 18 valence electrons ($8n–6$)

$Mo(CO)_6$ 18 valence electrons; $OC_5\,Mn—Mn(CO)_5$ 34 valence electrons ($18n–2$)
$Fe_3(CO)_{12}$ 48 valence electrons ($18n–6$)

This connection is lost for four-connected and deltahedral polyhedral molecules because the main group and metal–carbonyl fragments use only three frontier orbitals for skeletal bonding. The one-to-one relationship between the number of edges radiating from the vertex and the number of orbitals available is lost, and the bonding can no longer be described in terms of localized two-center two-electron bonds along all of the polyhedral edges.[9] In deltahedral and four-connected clusters the three frontier orbitals associated with each vertex atom generate $n+1$ bonding and $2n-1$ antibonding skeletal molecular orbitals. The partitioning of skeletal molecular orbitals into $(n+1)$ bonding–$(2n-1)$ antibonding may be proved by MO calculations on specific molecules and, more, generally by Stone's Tensor Surface Harmonic Theory. The latter is a free electron particle on a sphere model which utilizes scalar and vector harmonics to describe the radial and tangential bonding interactions in a cluster. The main group and transition metal polyhedral clusters therefore share a common set of unavailable skeletal molecular orbitals which depend only on the polyhedral geometry. These molecular orbitals are described as unavailable because they are antibonding as far as the skeletal atoms are concerned and hybridized toward the center of the cluster and therefore unsuitable for accepting electron pairs from ligands. Therefore, in both main group and transition metal clusters a closed shell is achieved when all the molecular orbitals except the unavailable molecular orbitals are filled. The relevant closed-shell requirements for main group and transition metal clusters may be defined as in Table 3. Filling all the available MOs leads to the polyhedral electron counts (pec) summarized in Table 2.

The common set of unavailable molecular orbitals in isostructural main group and transition metal polyhedral molecules suggests that inorganometallic molecules with the same structure also have the same number of unavailable molecular orbitals. Therefore, the total number of valence electrons increments by 10 for every transition metal vertex introduced into the polyhedral main group molecule. For example, octahedral $B_6H_6^{2-}$ (**1**), $[Ru_4C_2H_2(CO)_{12}]$ (**2**), and $[Ru_6(CO)_{18}]^{2-}$ (**3**) have 26, 66, and 86 valence electrons, respectively.

Figures 1 and 2 provide more extensive compilations of isostructural three-connected and deltahedral inorganometallic molecules showing these incremental relationships. Figure 1 illustrates series of trigonal bipyramidal, octahedral, and pentagonal bipyramidal clusters, where the main group atoms are successively replaced by transition metal fragments. The series are almost complete for the trigonal bipyramid and octahedron. For the pentagonal bipyramid there are few examples of metal-rich clusters. In Figure 2 related series of molecules based on the square-pyramid (*nido*-octahedron) and but-

Table 3. Closed-Shell Requirements for Main Group and Transition Metal Clusters

Total number of valence orbitals		Number of unavailable MOs	Total number of available MOs
Main group	$4n$	$3n/2$ for 3-connected	$5n/2$
		$2n-1$ for *closo*-deltahedra	$2n+1$
		$2n-2$ for *nido*-deltahedra	$2n+2$
		$2n-3$ for *arachno*-deltahedra	$2n+3$
Transition metals	$9n$	$3n/2$ for 3-connected	$15n/2$
		$2n-1$ for *closo*-deltahedra	$7n+1$
		$2n-2$ for *nido*-deltahedra	$7n+2$
		$2n-3$ for *arachno*-deltahedra	$7n+3$

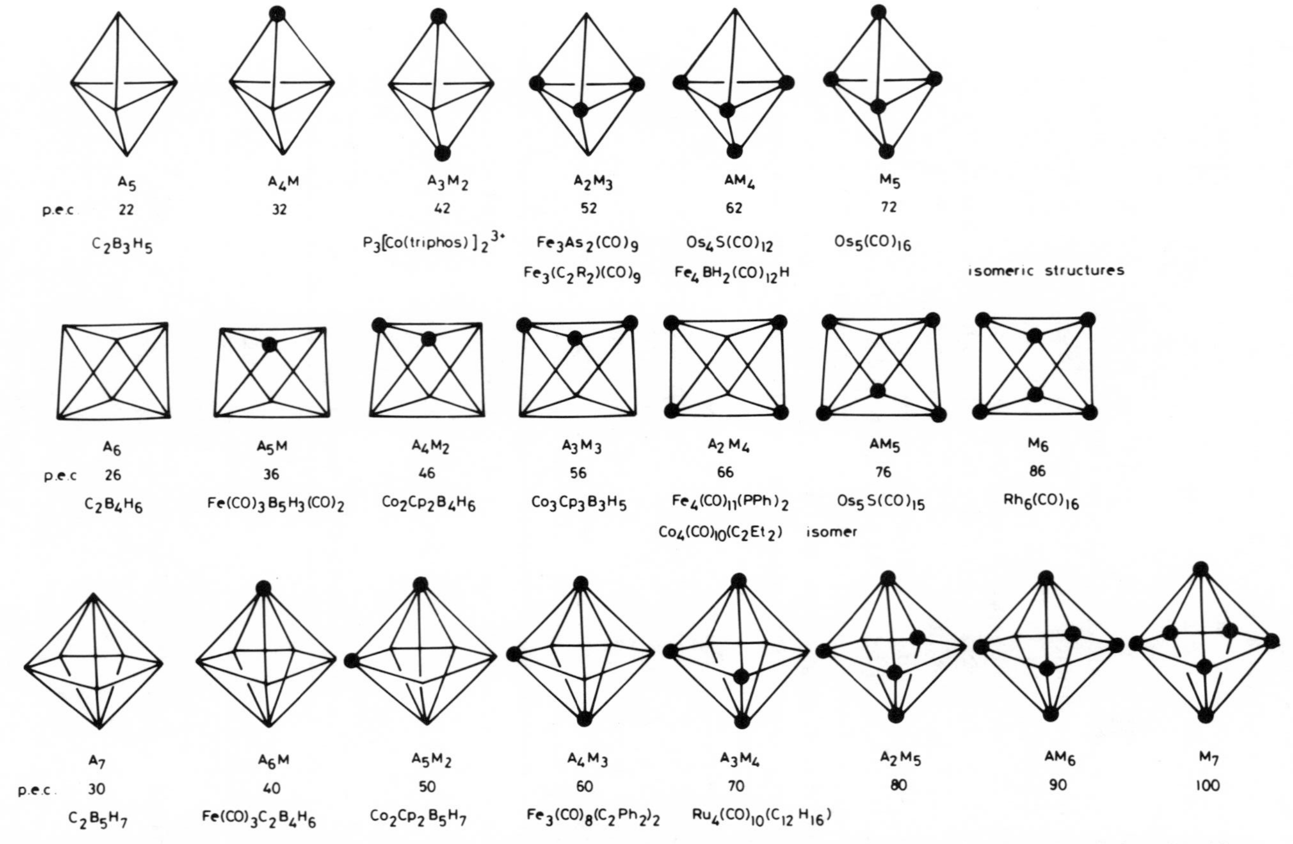

Figure 1. Isostructural series of *closo*-transition metal and main group clusters E_nM_m. Each time a transition metal atom is replaced by a main group atom the total valence electron count falls by 10. pec denotes polyhedral electron count. The *closo*-structures at the extremes are characterized by $4n+$ *and* $14n+2$ electrons, respectively.

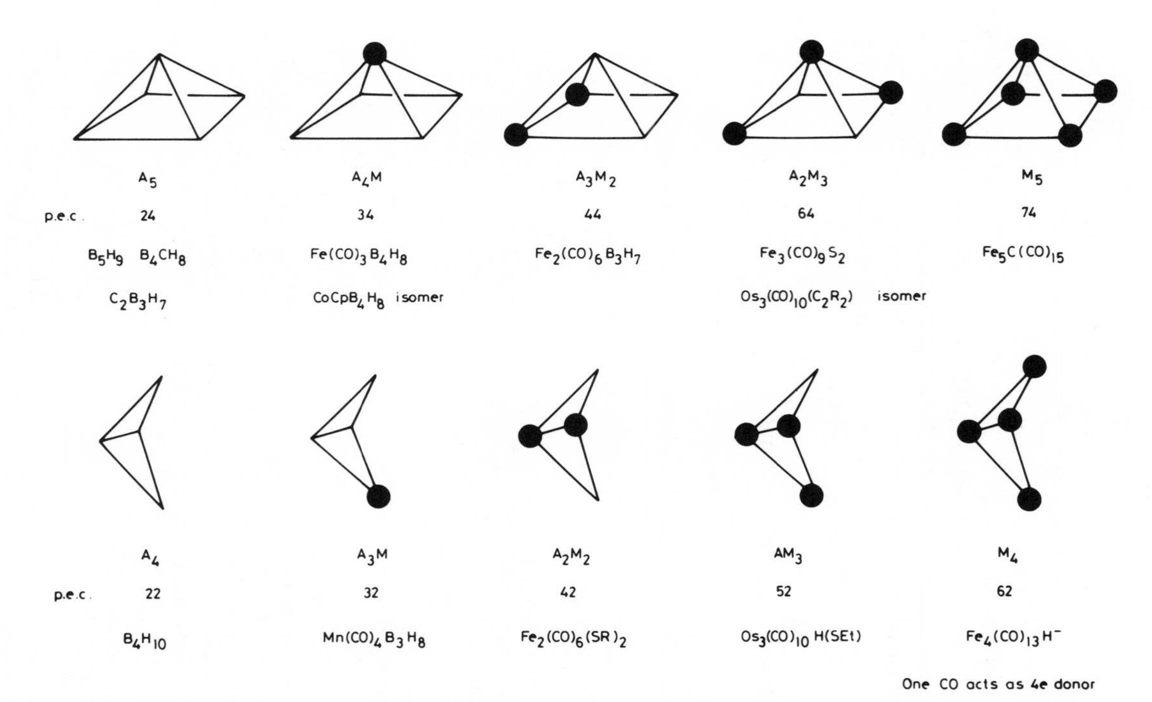

Figure 2. Series of isostructural clusters *nido*- and *arachno*-E_nM_m with different contributions of main group and transition metal atoms. The *nido*- and *arachno*-clusters containing main group atoms exclusively are characterized by $4n+4$ and $4n+6$ valence electrons, respectively.

terfly (*arachno*-octahedron) are illustrated. These compilations emphasize the generality of the relationships derived from the *polyhedral skeletal electron pair theory* and provide examples of as yet unknown molecules which are reasonable targets for the synthetic chemist.[2]

In summary, the *polyhedral skeletal electron pair* approach has its basis in two important factors. First, the metal–ligand combinations are those which are commonly associated with the adherence to the noble gas rule, i.e., π-acid ligands and metals of the later transition metals in low oxidation states.[10] Second, the polyhedral geometry defines a characteristic set of unavailable skeletal molecular orbitals which dictates the closed-shell requirements of main group, inorganometallic, and transition metal polyhedral molecules. It is significant that this approach requires no detailed knowledge of the number of ligands at particular metal centers and their geometries and the distribution of ligands between bridging and terminal sites, as long as the number of electrons donated by the ligands remains constant. Table 4 summarizes the donating characteristics of commonly used terminal and bridging ligands.

Table 4. Donor Characteristics of Some Common Ligands

Terminal ligands	
One-electron donors	H, Cl, CN, CH_3, SiR_3
Two-electron donors	CO, SO_2, CNR, PR_3
Three-electron donors	η-C_3R_3, η-allyl,
Four-electron donors	$PR_2CH_2PR_2$, η-C_4H_4
Five-electron donors	η-C_5H_5
Edge bridging ligands (μ_2)	
One-electron donors	H, CH_3, Ph, SiR_3
Two-electron donors	CO, CS, CNR, CR_2, SO_2
Three-electron donors	PR_2, SR, OR, NO, Cl, Br, I
Face bridging ligands (μ_3 or μ_4)	
One-electron donors	H
Two-electron donors	CO, CS, $SnCl_2$
Three-electron donors	NO, CR, P, As, Bi
Four-electron donors	PR, S, O
Five-electron donors	Cl, Br, I
Interstitial ligands	
One-electron donors	H
Two-electron donors	Mg
Three-electron donors	B
Four-electron donors	C, Si, Ge
Five-electron donors	P, As, Bi, Sb
Nine-electron donors	Rh, Co
Ten-electron donors	Pt, Pd

Many transition metal clusters have interstitial atoms located at the center of the cluster. These atoms form delocalized molecular orbitals with the skeletal molecular orbitals and effectively donate their electrons for skeletal bonding. Therefore, Table 4 also summarizes the donating characteristics of commonly observed interstitial atoms. Although beyond the scope of the present review the approach may be extended to more complex inorganometallic structures derived from vertex, edge, and face sharing of the primary polyhedral types discussed in Table 2. A recent book has provided a detailed account of these applications.[2]

Multiple bonding between atoms provides the possibility of isomeric structures which do not conform to the structural generalizations presented

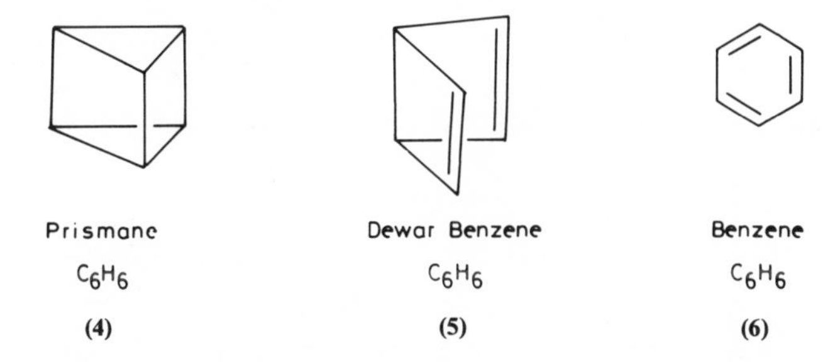

in Table 2. For example, prismane (**4**), Dewar benzene (**5**), and benzene (**6**) are isoelectronic but have quite different skeletal geometries. In inorganometallic compounds the occurrence of multiple bonds leads to a similar ambiguity and some specific examples are illustrated in Figure 3. The linear examples have 36 valence electrons and are therefore electronically related to CS_2, and the angular tellurium compound with 38 valence electrons is related to SO_2.

2.3. Fragment Molecular Orbital Methods

The fragment molecular orbital approaches focus on the specific bonding capabilities of individual metal–ligand or main group fragments. In main group chemistry the availability of *ns* and *np* valence orbitals provides a basis for tetrahedral sp^3 hybrid orbitals, which can be partially or fully used in ligand–element bonds. The bonding possibilities available for such main group fragments ER_v are summarized in Figure 4. In the limit such fragments can contribute a maximum of three frontier orbitals for skeletal bonding. The fourth hybrid is constrained by symmetry to be pointed away from the center of the cluster and can be utilized only for forming a bond either to a ligand or accommodating a lone pair of electrons.

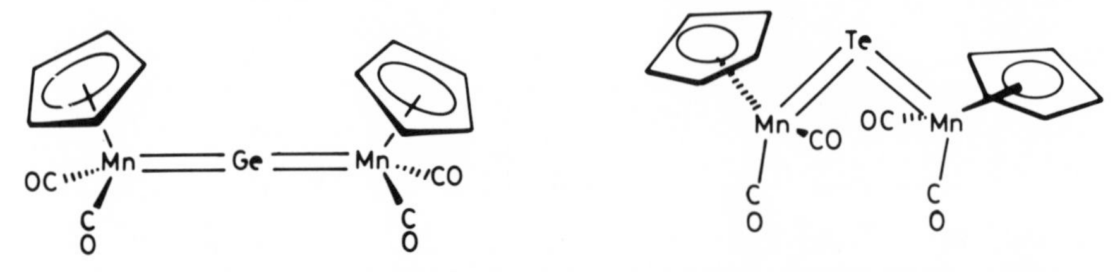

Figure 3. Examples of inorganometallic compounds exhibiting multiple bonding effects.

Metal complexes which conform to the noble gas rule have coordination numbers of 4 to 9 and the most commonly observed coordination geometries for these complexes are summarized in Table 5. Hybridization schemes for the metal–ligand bonds in these complexes have been calculated and illustrations of the relevant hybrids are given in Figure 5.[11] In contrast to the situation in main group chemistry these complexes may have additional nonbonding orbitals localized on the metal. These nonbonding orbitals are either

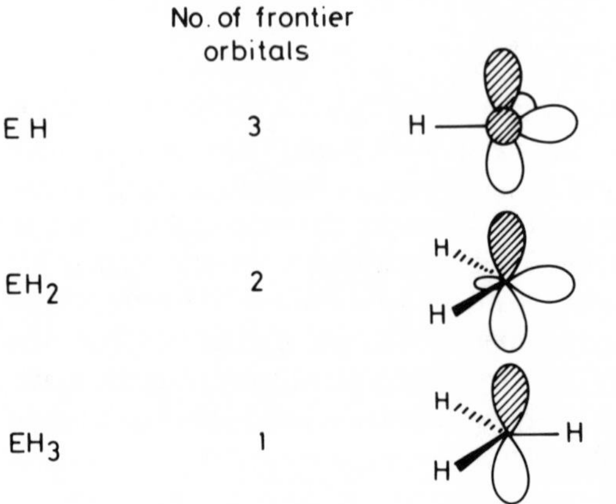

Figure 4. The bonding possibilities for main group fragments ER_v (v = 1–3).

Table 5. Valence Orbitals for a Range of Spherical or Approximately Spherical Polyhedral Complexes

Coordination number and geometry	Hybrid schemes	Hybrid nonbonding orbitals	Pure atomic nonbonding orbits
4 Tetrahedron	$s(pd)^3$	t_2 $\begin{cases} 0.790p_z + 0.612d_{xy} \\ 0.790p_z + 0.612d_{yz} \\ 0.790p_y + 0.612d_{xz} \end{cases}$	$e(d_{z^2}, d_{x^2-y^2})$
5 Trigonal bipyramid	$spd(pd)^2$	e' $\begin{cases} 0.745p_x + 0.667d_{xy} \\ 0.745p_y + 0.667d_{x-y} \end{cases}$	$e''(d_{xz},d_{yz})$
6 Octahedron	sp^3d^2	None	$t_{2g}(d_{xy},d_{xz},d_{yz})$
6 Trigonal prism	$spd^2(pd)^2$	e' $\begin{cases} 0.670p_z - 0.740d_{x^2-y^2} \\ 0.670p_y + 0.740d_{xy} \end{cases}$	$a_1'(d_{z^2})$
7 Pentagonal bipyramid	sp^3d^3	None	$e''(d_{xz},d_{yz})$
8 Square antiprism	sp^3d^4	None	$a_1(d_{z^2})$
8 Dodecahedron	sp^3d^4	None	$b_1(d_{xy})$
9 Tricapped trigonal prism	sp^3d^5	None	None

hybrids which point away from the metal–ligand bonds or pure atomic orbitals. These orbitals are summarized in columns 3 and 4 of Table 5. Although not suitable for σ-bond formation, these orbitals are ideal for metal–ligand π-bonding. The adherence to the noble gas rule in mononuclear metal complexes is dependent on the stabilization of these orbitals by interactions with π-acceptor ligands.[10]

Fragments derived from these primary coordination polyhedra by the loss of one to three adjacent ligands have out-pointing *dsp* hybrids available for skeletal bonding which are analogous to those described in Figure 4 for main group fragments. The possibilities are illustrated in Figure 6 for coordination numbers 4–7. The first column illustrates those complexes which conform to the noble gas rule and have filled "nonbonding" orbitals. Although such complexes can expand their coordination geometries by electrophilic addition reactions, they have no empty orbitals available for nucleophilic addition. The fragments in the second column all have a total of 16 valence electrons and an empty out-pointing hybrid orbital suitable for accepting an

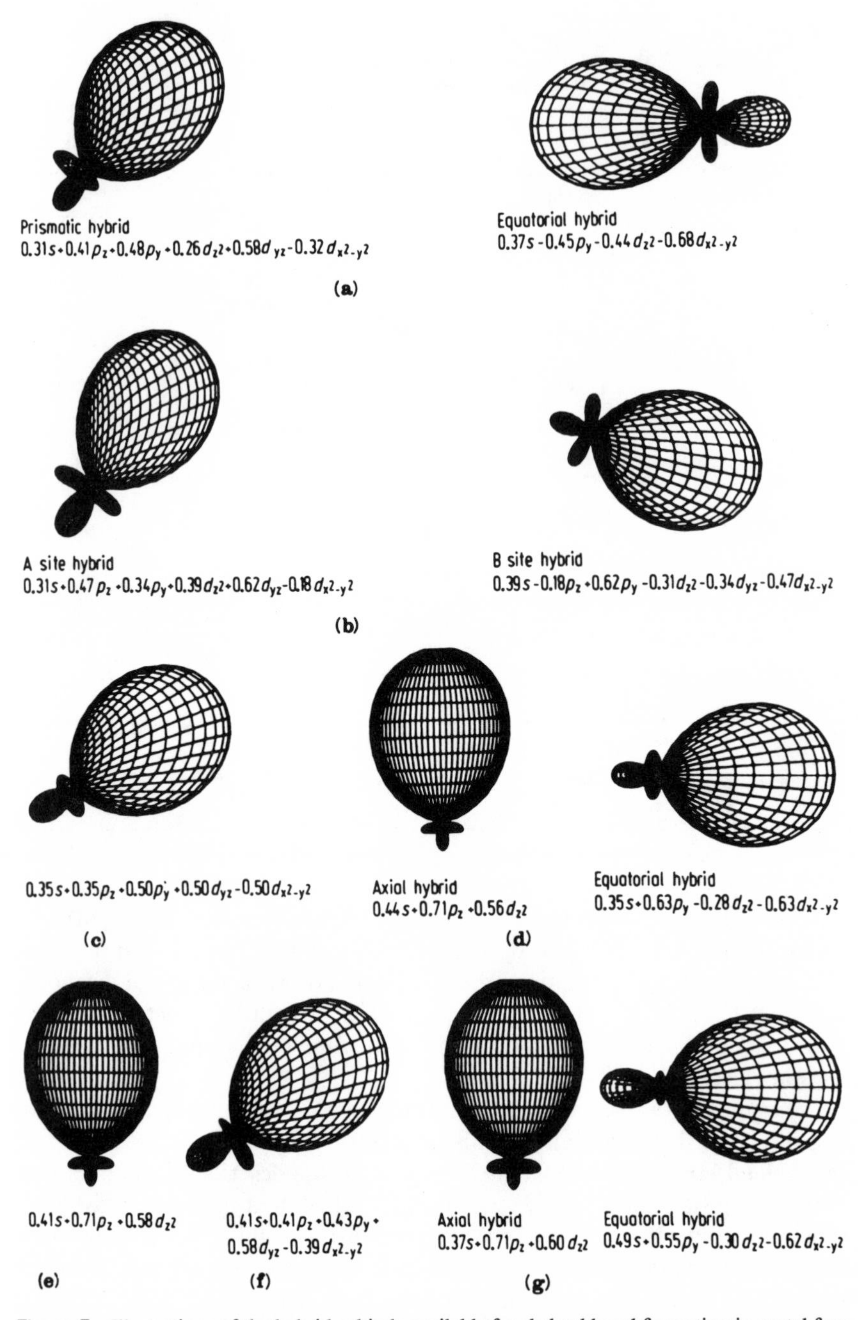

Figure 5. Illustrations of the hybrid orbitals available for skeletal bond formation in metal fragments derived from common coordination polyhedra. (a) Tricapped trigonal prism; (b) dodecahedron; (c) square antiprism; (d) pentagonal bipyramid; (e) octahedron, (f) trigonal prism; (g) trigonal bipyramid.

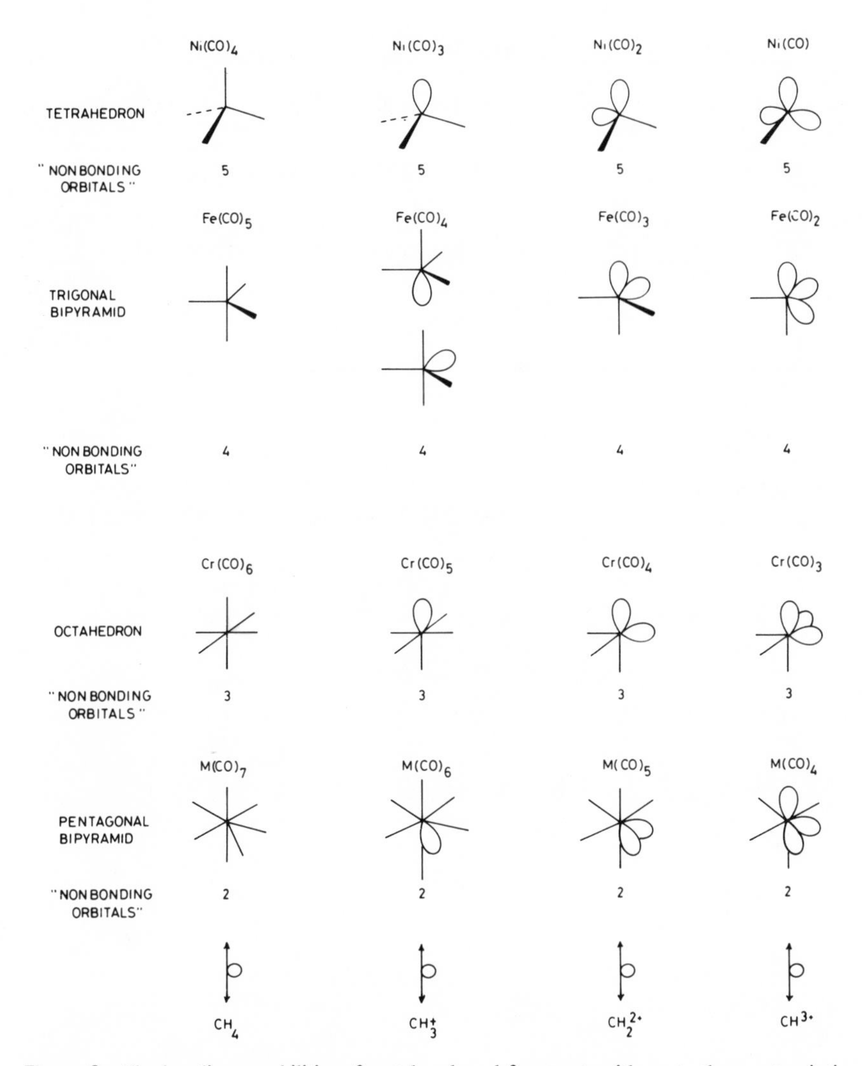

Figure 6. The bonding capabilities of metal carbonyl fragments with up to three out-pointing hybrid orbitals.

electron pair. These fragments therefore, to a first approximation, are expected to behave in an analogous fashion to CH_3^+ fragments. The term *isolobal* has been coined to describe this similarity and the following symbolism has been introduced to represent it:

$$Ni(CO)_3 \underset{\circ}{\longleftrightarrow} Fe(CO)_4 \underset{\circ}{\longleftrightarrow} Cr(CO)_5 \underset{\circ}{\longleftrightarrow} Ti(CO)_6 \underset{\circ}{\longleftrightarrow} CH_3^+$$

The presence of an additional electron in the out-pointing hybrid orbitals of these fragments leads to the following series of *isolobal* and pseudo-isoelectronic fragments:

$Cu(CO)_3 \leftrightarrow Co(CO)_4 \leftrightarrow Mn(CO)_5 \leftrightarrow V(CO)_6 \leftrightarrow CH_3$

The *isolobal* analogy suggests permutational possibilities leading to series of related molecules, e.g.,

$$H_3C—CH_3 \quad H_3C—Mn(CO)_5 \quad (OC)_5Mn—Mn(CO)_5$$

Less trivially the CH_5^+ carbonium ion and the metal dihydrogen complex $[W(CO)_3(PR_3)_2H_2]$ are clearly seen to bear an *isolobal* relationship:

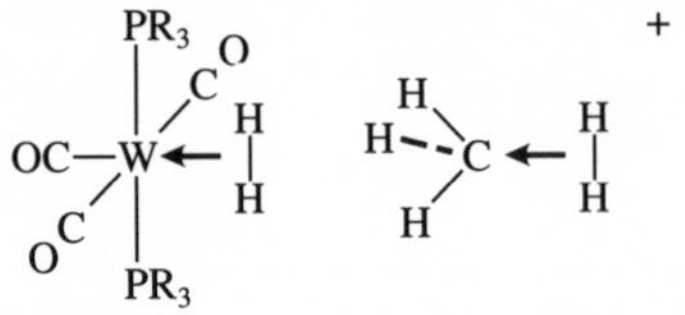

Fragments in the third column of Figure 6 each have two out-pointing dsp hybrid orbitals and may be described as isolobal with CH_2^{2+}, BH_2^+, and BeH_2.

$Ni(CO)_2 \leftrightarrow Fe(CO)_3 \leftrightarrow Cr(CO)_4 \leftrightarrow Ti(CO)_5 \leftrightarrow CH_2^{2+}$

$Cu(CO)_2 \leftrightarrow Co(CO)_3 \leftrightarrow Mn(CO)_4 \leftrightarrow V(CO)_5 \leftrightarrow CH_2^+$

$Ni(CO)_3 \leftrightarrow Fe(CO)_4 \leftrightarrow Cr(CO)_5 \leftrightarrow CH_2$

Therefore, the following compounds are related by the *isolobal* analogy:

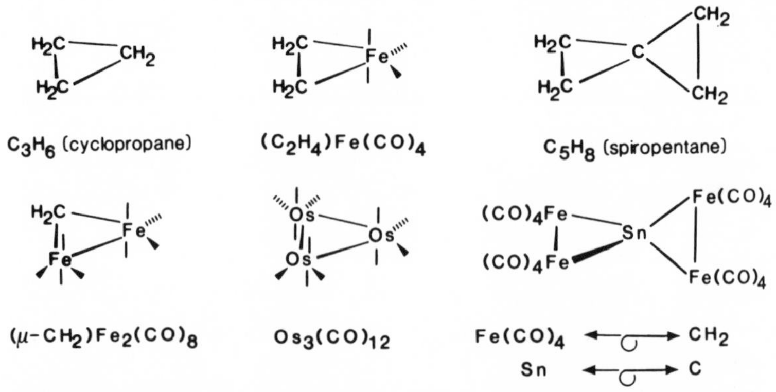

C_3H_6 (cyclopropane) $(C_2H_4)Fe(CO)_4$ C_5H_8 (spiropentane)

$(\mu\text{-}CH_2)Fe_2(CO)_8$ $Os_3(CO)_{12}$

$Fe(CO)_4 \leftrightarrow CH_2$

$Sn \leftrightarrow C$

Fragments in the fourth column of Figure 6 have three confacial dsp hybrids available for bonding and are therefore described as isolobal with CH^{3+}, BH^{2+}, and BeH^+.

$Ni(CO) \leftrightarrow Fe(CO)_2 \leftrightarrow Cr(CO)_3 \leftrightarrow Ti(CO)_4 \leftrightarrow CH^{3+}$

$Cu(CO) \longleftrightarrow Co(CO)_2 \leftrightarrow Mn(CO)_3 \leftrightarrow V(CO)_4 \leftrightarrow CH^{3+}$

$Ni(CO)_2 \leftrightarrow Fe(CO)_3 \leftrightarrow Cr(CO)_4 \leftrightarrow CH^+$

$Co(CO)_3 \leftrightarrow Mn(CO)_4 \leftrightarrow CH$

The structural consequences of these analogies may be illustrated by the following series of compounds:

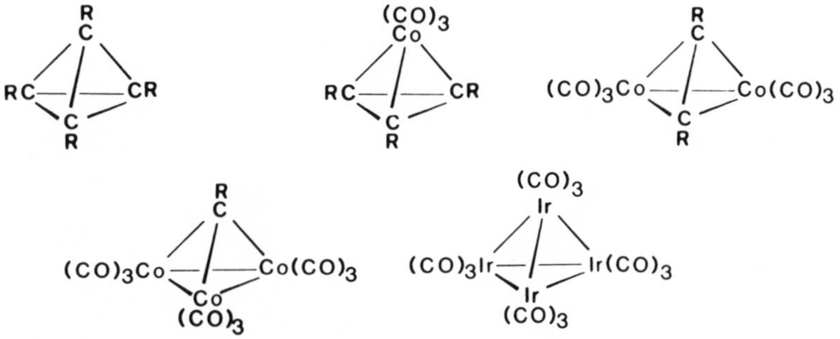

RC—C(R)—CR, C(R) (tetrahedrane C_4R_4); $(CO)_3Co$, RC, CR, C(R); $(CO)_3Co$, C(R), $Co(CO)_3$, C(R); $(CO)_3Co$, C(R), $Co(CO)_3$, $Co(CO)_3$; $(CO)_3Ir$, $Ir(CO)_3$, $Ir(CO)_3$, $Ir(CO)_3$

Clearly these *isolobal* relationships are generating series of inorganometallic compounds which are directly analogous to those generated by the *polyhedral skeletal electron pair theory* and illustrated in Figures 1 and 2. In fact they are identical because they share a common set of assumptions concerning the importance of the *noble gas rule* and the availability of one to three orbitals for skeletal bonding. Furthermore, each series of *isolobal* fragments given above involves an increment of 10 valence electrons when the main group fragment is replaced by a transition metal fragment. Therefore, the *isolobal* analogy provides an alternative representation of the results in Table 2, which defines the closed-shell requirements of main group and transition metal clusters.

Experimental chemists have often shown a preference for discussing their results in terms of the *isolobal* analogy rather than the total electron counts, because they can see a more direct relationship between the reagents which they use in their synthetic procedures and the *isolobal* fragments. For example, it is possible to design a synthetic procedure involving the incorporation of a $Cr(CO)_4$ fragment into a cluster knowing that such a fragment may be readily generated from $Cr(CO)_4(CH_3CN)_2$. Stone has been particularly effective in utilizing *isolobal* analogies for designing syntheses of heterometallic clusters.[12–16]

The *isolobal* relationships for metal fragments containing cyclic polyene and polyenyl ligands may be derived in an analogous fashion from a parent compound conforming to the noble gas rule. For example, Figure 7 illustrates some *isolobal* relationships for $M(\eta\text{-}C_5H_5)(CO)_x$ (x = 0–2) fragments. $Mn(CO)_3(\eta\text{-}C_5H_5)$ obeys the noble gas rule and therefore $Mn(CO)_2(\eta\text{-}C_5H_5)$, $Mn(CO)(\eta\text{-}C_5H_5)$, and $Mn(\eta\text{-}C_5H_5)$ are isolobal with CH_3^+, CH_2^{2+}, and CH^{3+}.[17]

Some care has to be taken when deriving isolobal relationships for bis(polyenyl) sandwich complexes, because the out-pointing hybrids associated with such molecules do not have the same symmetry and topological rela-

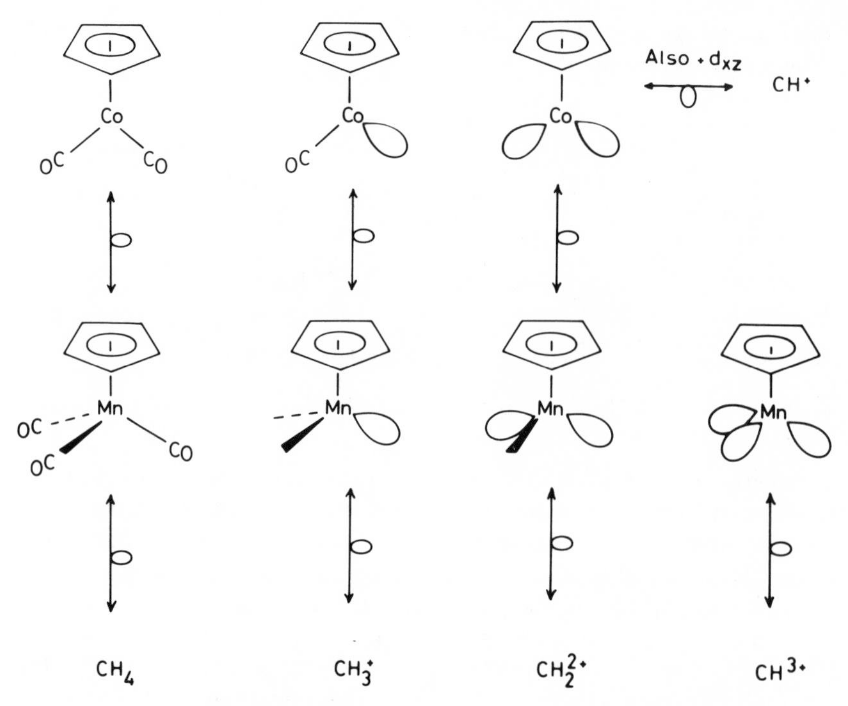

Figure 7. *Isolobal* relationships for some metal cyclopentadienyl fragments.

tionships as those discussed above. Specifically in $[Ta(\eta\text{-}C_5H_5)_2H_3]$ (**7**), which conforms to the noble gas rule, the three hydrido-ligands lie in one plane and therefore the localized hybrids associated with these bonds do not have the same conical relationship as that discussed above for main group and transition metal carbonyl fragments. Therefore a $(\eta\text{-}C_5H_5)_2Ta$ fragment is not *isolobal* with CH, because the symmetries and locations of the hybrids are quite different. The related complex $(\eta\text{-}C_5H_5)_2)WH_2$, which also conforms to the noble gas rule, can be used as a parent to develop the following *isolobal* relationships (see Figure 8)[18]:

$$(\eta\text{-}C_5H_5)_2WH \leftarrow\!\circ\!\rightarrow CH_3$$

$$(\eta\text{-}C_5H_5)_2W \leftarrow\!\circ\!\rightarrow CH_2$$

The latter relationship is important not only when accounting for the oxidative addition reactions of $(\eta\text{-}C_5H_5)_2W$, but also the metallocyclopropane analog shown in (**8**).[19]

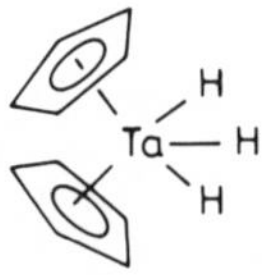

(7) 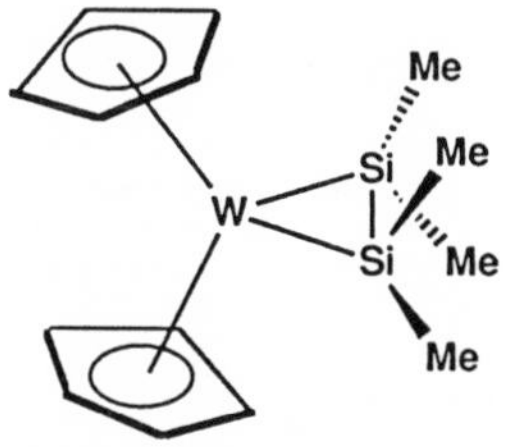 (8)

The presence of the additional *d* orbitals in transition metal fragments leads to a flexibility in bonding capabilities not available to main group fragments. The *isolobal* relationships described above have assumed that the "nonbonding" metal orbitals do not participate in bonding and the bonding capabilities of the fragments are completely dominated by the out-pointing hybrid orbitals. The overrestrictive nature of this assumption can be illustrated by reference to a specific example. The $Cr(CO)_5$ fragment derived from octahedral $Cr(CO)_6$ is *isolobal* with CH_3^+ and BH_3, but has in addition a set of t_{2g} (d_{xz}, d_{yz}, d_{xy}) "nonbonding" orbitals. A strict interpretation of the *isolobal* analogy would highlight the relationship between

	$H_3B \leftarrow NH_3$	and	$(OC)_5Cr \leftarrow NH_3$
Total no. of valence electrons	14		24

but would not reveal the following analogies arising from the participation of the "nonbonding" orbitals in multiple bonds.

	$R_2C{=}CR_2$	$(OC)_5Cr{=}CR_2$
Total no. of valence electrons	12	22
	+	+
	$RN{\equiv}CR$	$(OC)_5Cr{\equiv}CR$
	10	20

These compounds, all of which conform to the inert gas rule, underline the additional flexibility of metal–carbonyl fragments. The sixteen-electron $Cr(CO)_5$ fragment is therefore *isolobal* with CH_3^+, CH_2, and CH^-. The manner in which contributions from the t_{2g} set of orbitals contribute to these relationships is illustrated in Figure 8. An alternative way of appreciating this flexibility is to recognize that a $Cr(CO)_5$ fragment need not necessarily have been derived from an octahedron, but could also have been derived from 7 and 8 coordinate polyhedra and thereby be associated with either one or two additional frontier hybrid orbitals.[17]

The bonding capabilities of metal–carbonyl fragments can also be varied

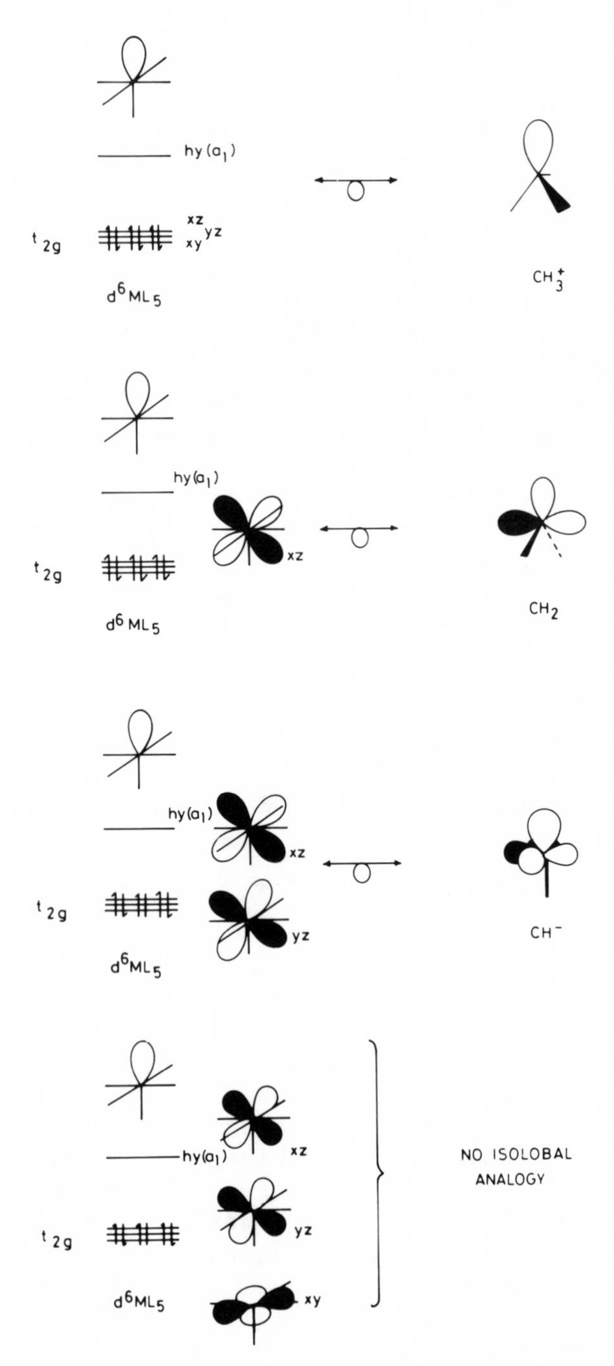

Figure 8. The utilization of the "t_{2g}" orbitals in C_{4v} $M(CO)_5$ fragments for multiple bonding.

by changing the geometric disposition of the carbonyl ligands. For example, a $M(CO)_4$ fragment may behave as a 1-orbital fragment if derived from a trigonal bipyramid, a 2-orbital fragment if derived from an octahedron, and a 3-orbital fragment if it takes up a C_{4v} conical geometry. These possibilities are illustrated in Figure 9. Therefore, the bonding capabilities of a metal–carbonyl fragment should not be viewed as immutable, and its character may change in order to maximize its bonding interactions with the substrate to which it is binding. The multifunctional character is of course unavailable to main group ER_v and E fragments. Indeed, the ability of transition metals to catalyze a wide range of reactions is intimately connected with this flexibility, which enables the metal fragment to change its bonding capabilities along a reaction coordinate and thereby reduce the activation energy barrier.

A further limitation of the *isolobal* analogy is highlighted at the bottom of Figure 8. The third component of the ML_5 t_{2g} set is d_{xy} which has δ symmetry with respect to the principal rotation axis. The utilization of this orbital in bonding has no analog in main group chemistry, because of the inaccessibility of *d* orbitals. For the earlier transition metals the presence of this orbital is crucial not only for the formation of quadruple bonds but also in cluster compounds. Examples of compounds where the fragments are using four frontier orbitals σ, 2π, and δ for multiple and skeletal bond formation are illustrated in Figure 10.[20]

In summary, the bonding capabilities of metal fragments are considerably more flexible than those of main group fragments and one particular fragment can function in related compounds as if it were *isolobal* with CH_3^+, CH_2, and CH^-. These alternative bonding modes all conform to the *inert gas rule* and reflect the increased participation of "nonbonding" orbitals in multiple bonds. The availability of orbitals with δ symmetry gives rise to bonding possibilities for transition metal fragments which cannot be mimicked by main group atoms.

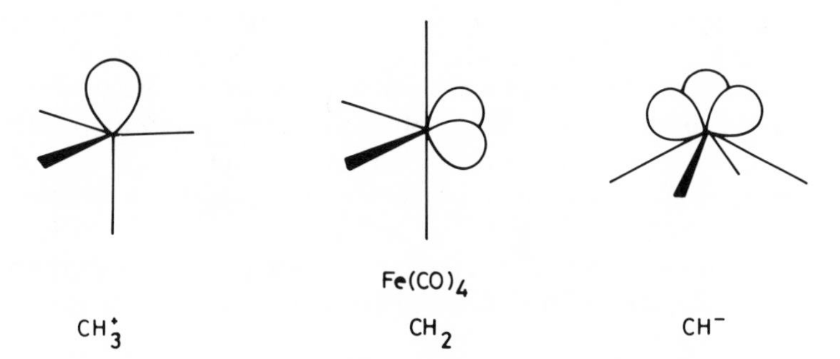

Figure 9. Alternative bonding capabilities of $M(CO)_4$ fragments.

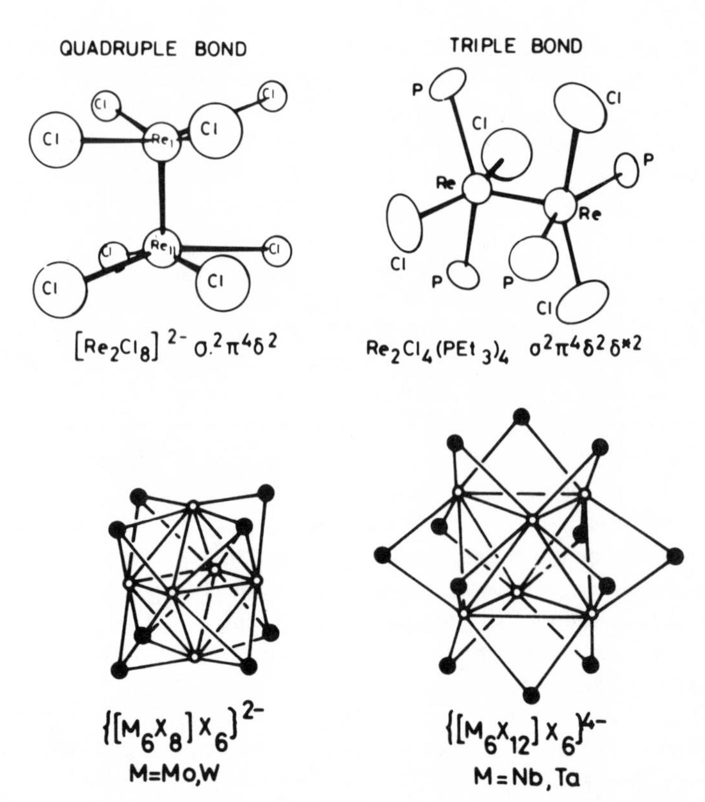

Figure 10. Examples of compounds where the metal fragment is utilizing σ, π, and δ frontier orbitals.

Isolobal Relationships for Ligands

Although isolobal relationships for metal fragments have been discussed above, the way in which they have been derived makes the procedure sufficiently flexible to be applied to ligands also. Specifically, the loss of an H^+ or $CH_3{}^+$ from a ligand, which itself has a stable closed-shell configuration, will create a filled out-pointing orbital suitable for donating to an empty fragment orbital. Similarly, the loss of a $B{-}H^{2+}$ fragment from a deltahedral cluster generates a set of filled a_1 and e orbitals on the open face which has been created. The resulting *nido*-fragment can therefore behave as a 6-electron donor toward a metal atom. It was this type of reasoning by Hawthorne which led to the development of metallocarborane chemistry.[6]

2.4. *Topological Electron Counting Theory*

It is apparent from the discussion above that both the *polyhedral skeletal electron pair approach* and the *isolobal* analogies have their origins in the *noble gas rule* and the skeletal molecular orbital patterns for different classes of polyhedra. Teo[21] has developed a *topological electron counting theory* which also utilizes the *noble gas rule* and Euler's theorem for three-dimensional polyhedra. For metal clusters this approach is based on the fundamental formula which relates the number of cluster valence molecular orbitals (CVMO) to the number of vertices (V) and faces (F) of the polyhedron:

$$\mathrm{CVMO} = 8\mathrm{V} - \mathrm{F} + 2 + \mathrm{X}$$

Each polyhedron is characterized by the parameter X, which is defined as the number of "extra" electron pairs in excess of the *noble gas rule*. For the majority of 3-connected, deltahedral, and capped polyhedra this approach yields results identical to the *polyhedral skeletal electron pair theory*, although the inherent flexibility of the parameter X may permit some useful alternative electron counts for certain polyhedra, e.g., bipyramids and antiprisms.[22]

2.5. Ambiguous Fragments

The importance of the noble gas rule has been stressed in the previous sections; however, it is well recognized that this generalization is not ubiquitously applicable for all transition metal complexes. Therefore, it is necessary to address the consequences of having fragments which have alternative bonding requirements.

The earlier transition metals, particularly when π-donor ligands are coordinated to them, frequently deviate from the noble gas rule. The heavier transition metals in their low oxidation states also exhibit many exceptions. For example, d^8 palladium(II) and platinum(II) complexes almost always form square-planar 16-electron complexes in preference to trigonal-bipyramidal and square-pyramidal 18-electron complexes. Similarly, d^{10} platinum(0) and gold(I) complexes are frequently trigonal planar and linear rather than tetrahedral. The bonding capabilities of these fragments reflect these closed-shell requirements not only in metal clusters, but also in inorganometallic compounds derived from them. To a first approximation the significant frontier orbitals of these fragments can be derived from those hybrid orbitals which are used to define the bonding in the stable parent mononuclear molecule. Specifically the T-shaped $[PtL_3]^{2+}$ and angular $[PtL_2]^{2+}$ fragments shown in Figure 11 have either one or two dsp^2 empty hybrid orbitals pointing toward the missing vertices of the square plane, i.e., the following *isolobal* relationships are relevant.[23,24]:

$$[PtL_3]^{2+} \leftrightarrow CH_3^{+}$$

$$[PtL_2]^{2+} \leftrightarrow CH_2^{2+}$$

This analysis has neglected the presence of the empty $6p_z$ orbital which lies perpendicular to the square-plane and the nonbonding d orbitals. In conven-

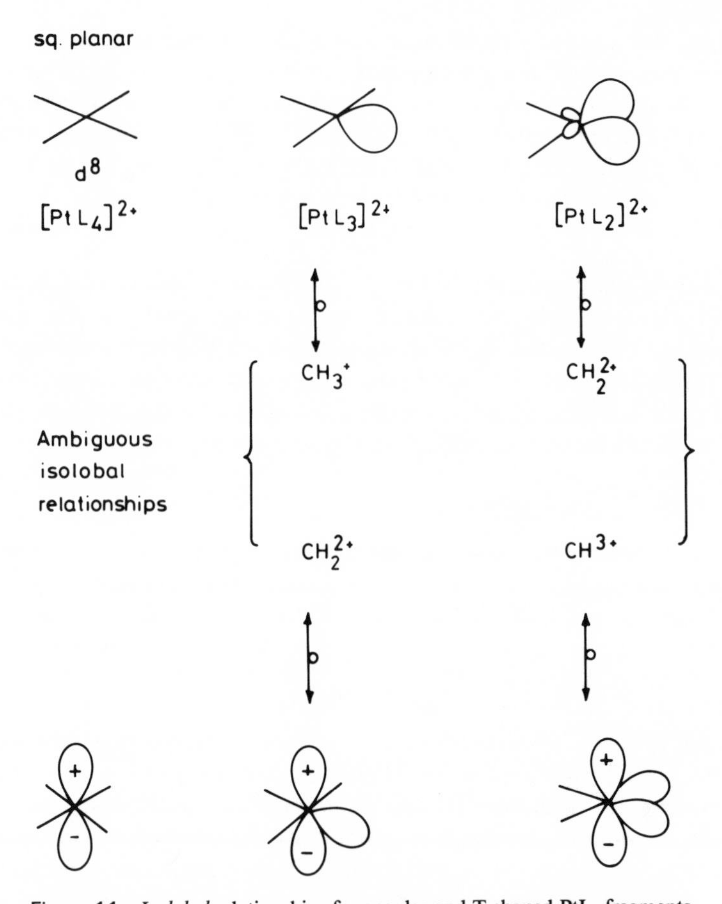

Figure 11. *Isolobal* relationships for angular and T-shaped PtL_3 fragments.

tional coordination compounds it is the nonparticipation of the p_z orbital which leads to 16-electron complexes. However, this orbital is sufficiently available to participate in complexes with good π-acid ligands, such as $[Pt(SnCl_3)_5]^{3-}$, complexes with polydentate ligands designed to stabilize trigonal-bipyramidal geometries, and transition states in nucleophilic substitution reactions.

If this orbital makes a significant contribution to bonding, then the following isolobal relationships are germane:

$$[PtL_3]^{2+} \longleftrightarrow CH^{3+}$$

$$[PtL_2]^{2+} \longleftrightarrow CH_2^{2+}$$

In addition, the square-planar platinum complexes have a filled d_π orbital perpendicular to the square plane which can also make a contribution. If this orbital makes a larger contribution than the $6p_z$ orbitals, then the following *isolobal* relationships are relevant:

$$[PtL_3]^{2+} \longleftrightarrow CH_2$$
$$[PtL_2]^{2+} \longleftrightarrow CH^+$$

Therefore, these fragments have ambiguous bonding capabilities which have interesting structural and reactivity consequences. Specifically, a ML_2 (M = Ni, Pd, or Pt) fragment may behave in a manner reminiscent of CH_2 CH^+ or CH^- depending on the relative energies of the metal *nd* and $(n+1)p$ orbitals and the π-acceptor qualities of the ligand. These orbital energies may be "fine tuned" by varying the donor properties of the ligands, the angle between the ligands and the metal.

The *isolobal* connection $PtL_2 \longleftrightarrow CH_2$ is most widely applicable and is illustrated by the following set of compounds:

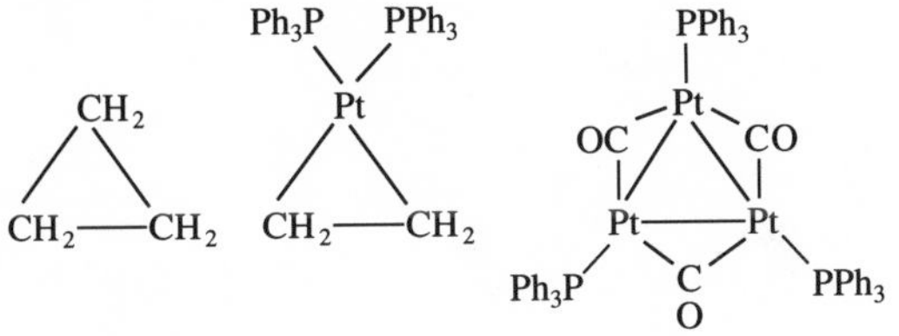

Total no. of valence electrons	18	26	42

In the last compound the bridged form of the cluster is more stable than the strictly analogous compound with terminal Ph_3P and CO ligands. It is noteworthy that the total number of valence electrons increments by 8 for each metal atom in this series. The following set of related nickel and platinum compounds indicates the ambiguous nature of the fragments.[25–28]:

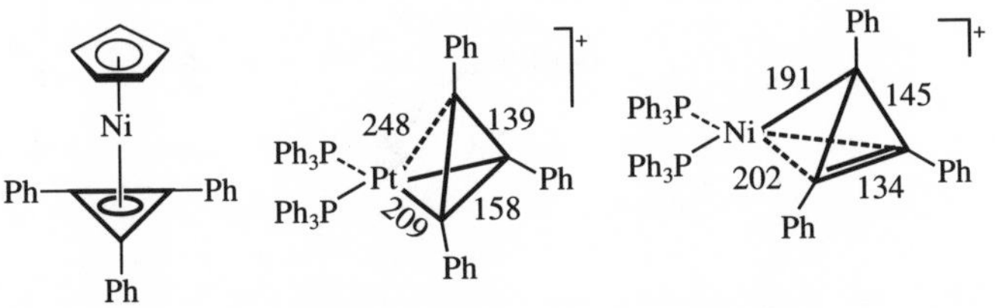

In the complex $[Ni(\eta\text{-}C_5H_5)(C_3Ph_3)]$ the $Ni(\eta\text{-}C_5H_5)^-$ fragment is behaving as if it were isolobal with CH^- and thereby generates a tetrahedrane analog. The $Pt(PPh_3)_2$ fragment is behaving in a manner intermediate between CH^- and CH_2 leading to a structure which lies between tetrahedrane and the bicyclobutane cation. The nickel complex $[Ni(PPh_3)_2(C_3Ph_3)]^+$ lies between tetrahedrane and cyclopropenylcarbinyl cation. Albright and co-workers have given a more detailed MO analysis of the bonding in these complexes and analyzed the relationship between the conformation of the ML_2 fragment and the observed bond-length variations.[25–28]

The ambiguous nature of ML_2 fragments has also been revealed in the metallocarboranes $[C_2B_9H_{11}(PtL_2)]$, which are illustrated in Figure 12.[29,30] Depending on the metal, the π-acceptor abilities of the ligands, and the locations of the carbon atoms, the molecules have either *closed*-icosahedral structures or more open structures with the ML_2 fragment located in a "slipped" position away from the center of the pentagonal face. Similarly, Stone has reported two isomers for $[Pt(PR_3)_2C_2B_6H_6]$, which are illustrated in Figure 13.[31] One isomer has a symmetrical *closo*-tricapped trigonal prismatic structure with the $Pt(PR_3)_2$ fragment behaving like a BH fragment in

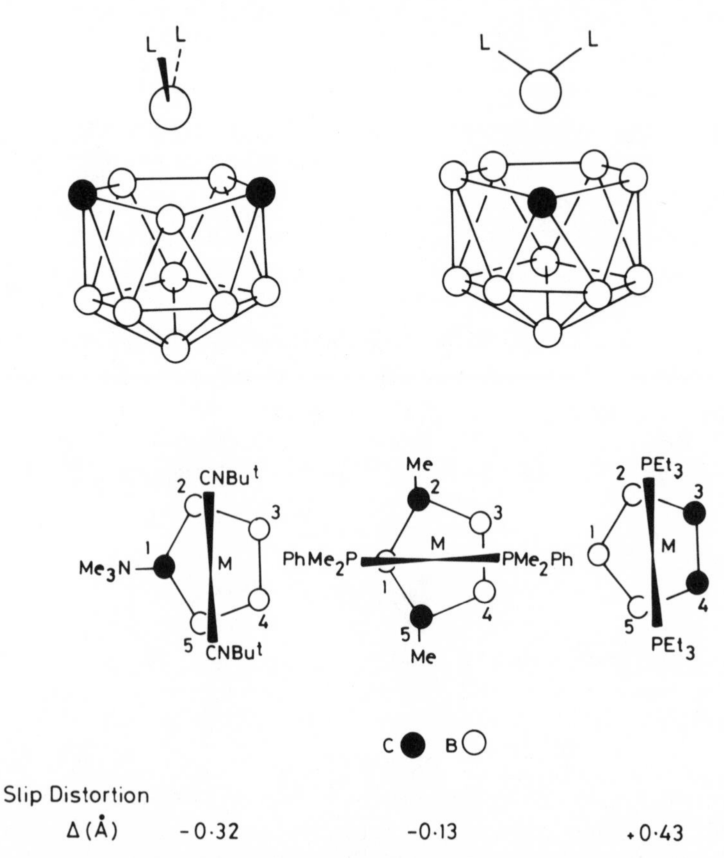

Figure 12. Illustrations of "slipped" metallocarboranes based on $[C_2B_9H_{11}(ML_2)]$ (M = Pd or Pt), for example, $B_9(CMe^{2,4})_2[Pt(PMe_2Ph)_2{}^1]H_9$, and $[CB_{10}H_{11}PtL_2]^-$ e.g., $B_{10}[C(NMe_3)^2]Pt(CNBu^t)_2{}^1]H_{10}$. The distortions away from the idealized icosahedral structures are indicated by the slip distortion parameter Δ.

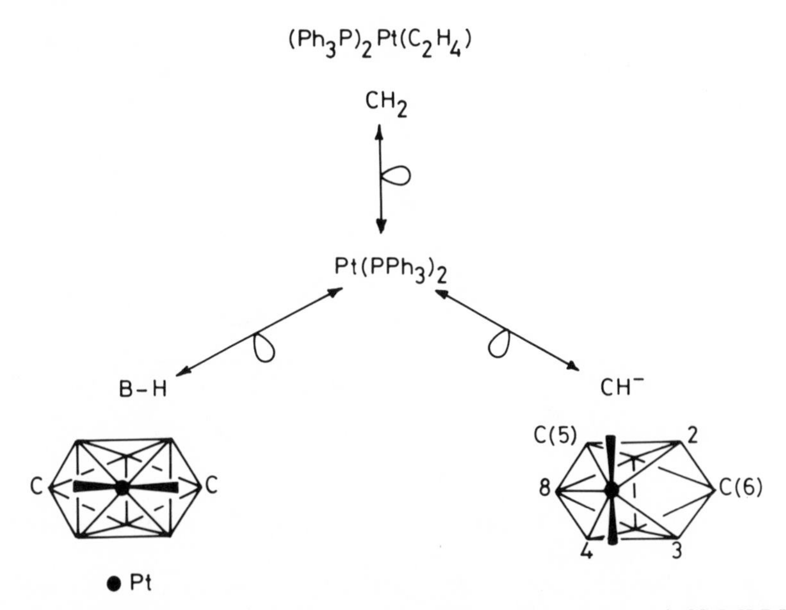

Figure 13. Illustrations of the ambiguous nature of PtL_2 fragments in isomeric $[C_2B_6H_6PtL_2]$ complexes.

a capping position. The second isomer has a *nido*-B_9 cage with the platinum atom adjacent to an open square-face, suggesting that the $Pt(PR_3)_2$ fragment is behaving like a BH^{2-} fragment.

The d^{10} $M{-}L^+$ (M = Cu, Ag, or Au) exhibits an analogous ambiguity, since it can be derived from a linear 14-electron $L{-}M{-}L^+$ molecule.[32,33] The $M{-}L^+$ fragment has a single out-pointing dsp hybrid, which provides the primary bonding interactions. Therefore, these fragments are formally isoelectronic with CH_3^+ and H^+. The *isolobal* relationship between $AuPPh_3^+$ and H^+ has proved to be a particularly important one for cluster aggregation reactions, because the $AuPPh_3^+$ can be readily generated from $Au(PPh_3)Cl$. Since the $L{-}M{-}L^+$ molecule also has a pair of empty *p* orbitals perpendicular to the metal–ligand bonds, the ML^+ fragment is also *isolobal* with CH^{3+}, BH^{2+}, and BeH^+. This ambiguity is clearly illustrated by the pairs of related compounds shown in Figure 14.

In the copper complex the $CuPR_3^+$ fragment is behaving like CH^{3+} while the $AuPR_3^+$ fragment more closely approximates CH_3^+.

3. LIGAND VARIATIONS

3.1. Variations in Frontier Orbitals of Fragments

Although the bonding capabilities of transition metal and main group fragments are sufficiently similar to form the basis of the *isolobal* analogy,

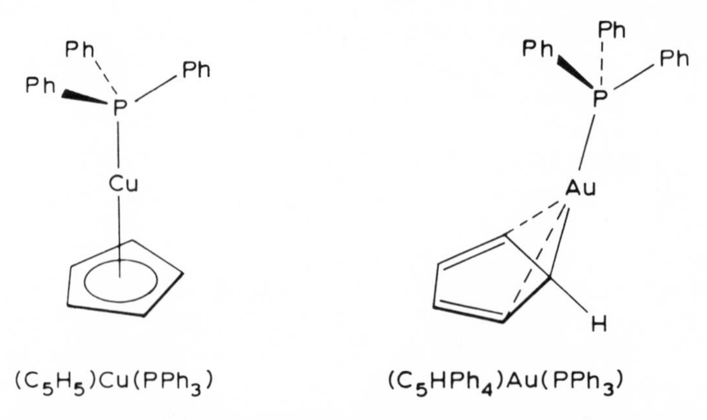

Figure 14. Illustrations of the ambiguous *isolobal* relationships for $M(PPh_3)^+$ fragments.

there are also important differences which need to be recognized. Figure 15 provides a comparison of the orbital energies of B—H, C—H, and some commonly occurring conical ML_3 fragments. The ML_3 fragments have been chosen to have d^6 metal configurations, i.e., $Cr(CO)_3$, $Cr(PH_3)_3$, $Cr(\eta\text{-}C_6H_6)$, and $Cr(OMe)_3^{3-}$. Since the conical fragments may be derived from an octahedron, the "t_{2g}" orbitals of the parent octahedron remain recognizable and

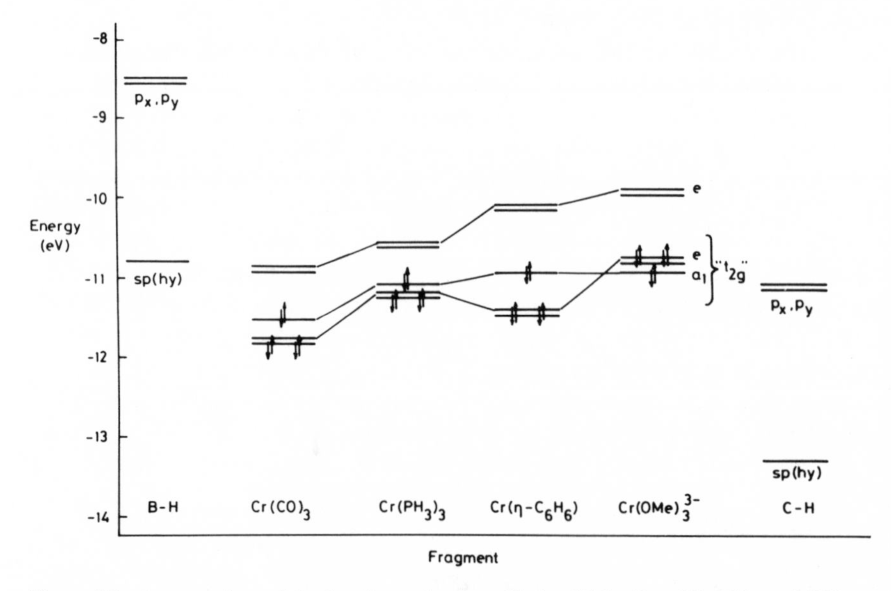

Figure 15. A comparison of the frontier molecular orbitals of CrL_3 (L = CO, PH_3, and OMe^-, $L_3 = \eta\text{-}C_6H_6$), B—H and C—H.

are indicated in the figure. In the axially symmetric point groups associated with these fragments the t_{2g} orbitals transform as a_1 and e. In the $Cr(PH_3)_3$ fragment, where the ligands are behaving essentially as σ-donors, the a_1 and e orbitals remain essentially degenerate. In contrast, the π-acceptor and donor ligands in $Cr(CO)_3$ and $Cr(OMe)_3^{3-}$ remove the degeneracy of the t_{2g} set with the e set lying below a_1 in $Cr(CO)_3$ and above a_1 in $Cr(OMe)_3^{3-}$. In $Cr(\eta$-$C_6H_6)$ the splitting is in the same sense as that observed for $Cr(CO)_3$. The average energy of the t_{2g} set decreases in the order $Cr(CO)_3 > Cr(\eta\text{-}C_6H_6) > Cr(PH_3)_3 > Cr(OMe)_3^{3-}$ suggesting that these fragments become more nucleophilic as the σ-donor and π-donor ligands are introduced. The localization of the t_{2g} orbitals are summarized in Table 6; they suggest that although the orbitals are localized almost exclusively on the metal d orbitals, there is significantly more delocalization in the fragments with π-donor and π-acceptor ligands.

The acceptor orbitals of these ligands are also illustrated in Figure 15 and their orbital compositions are summarized in Table 6. Each of the fragments has a relatively low-lying e acceptor pair of orbitals, which have a substantial amount of d orbital character but with sufficient p orbital character to hybridize the orbital away from the ligands. The $Cr(CO)_3$ fragment has only 52.5% d orbital character, but the remaining fragments have 75–80% d orbital character.

The orbital energies of these e sets shown in Figure 15 indicate that their relative electrophilic character falls in the order $Cr(CO)_3 > Cr(PH_3)_3 > Cr(\eta\text{-}C_6H_6) > Cr(OMe)_3^{3-}$. The fragments have in addition a higher-lying a_1 mo-

Table 6. Comparison of the Orbital Compositions of the Frontier Orbitals of Conical ML_3 Fragments

	$Cr(CO)_3$	%Cr s	%Cr p	%Cr d	$Cr(PH_3)_3$	%Cr s	%Cr p	%Cr d
	a_1	5.5	16.5	20.2	a_1	9.5	77.0	0
	e	0	12.7	52.5	e	0	12.3	75
t_{2g}	a_1	1.4	1.6	74.0	a_1	1.3	0	97.0
t_{2g}	e	0	0.5	70.0	e	0	0	82.0

	$Cr(OMe)_3^{3-}$	%Cr s	%Cr p	%Cr d	$Cr(\eta\text{-}C_6H_6)$	%Cr s	%Cr p	%Cr d
	a_1	26.5	68.5	1.4	a_1	51	46.5	0
	e	0	11.2	80.0	e	0	11.3	75.5
t_{2g}	e	0	0.6	81.0	a_1	0.4	0.5	86.0
t_{2g}	a_1	5.3	0.1	90.0	e	0	0	78.0

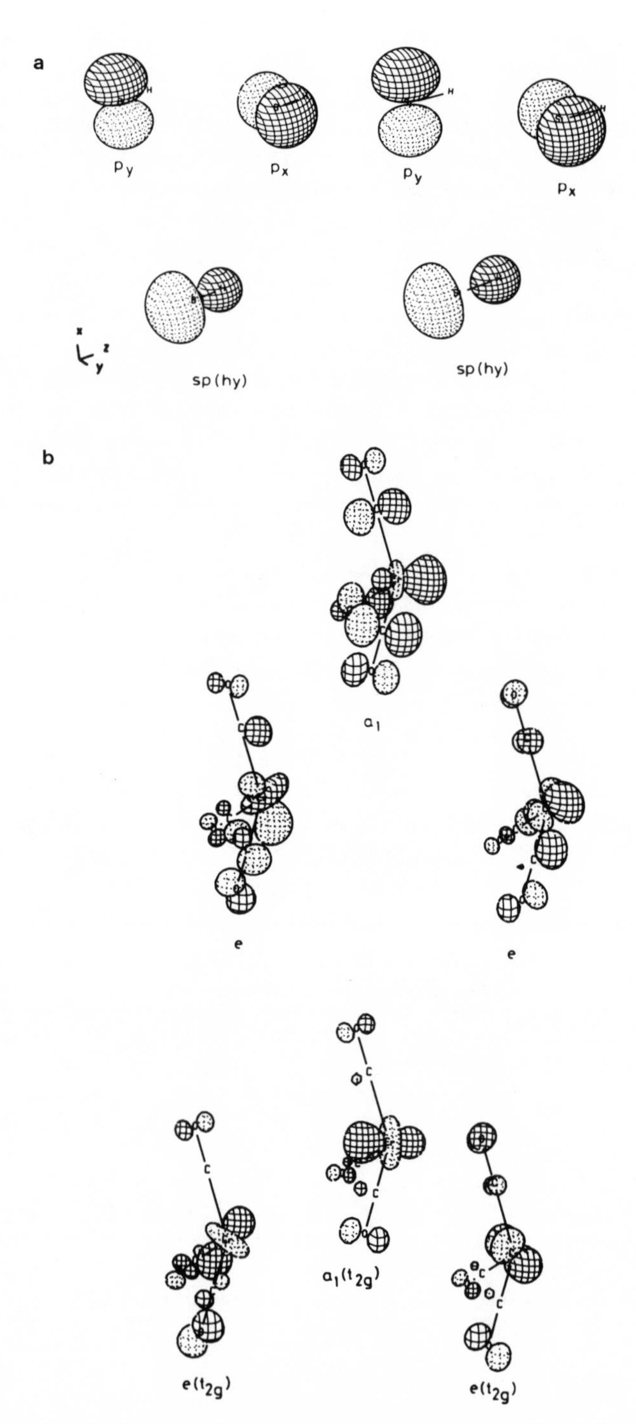

Figure 16. Electron density plots of the frontier orbitals of B—H, Si—H, and CrL_3 fragments. (a) B—H and Si—H, (b) $Cr(CO)_3$, (c) $Cr(PH_3)_3$, (d) $Cr(\eta\text{-}C_6H_6)$, and (e) $Cr(OMe)_3^{3-}$. The orbitals

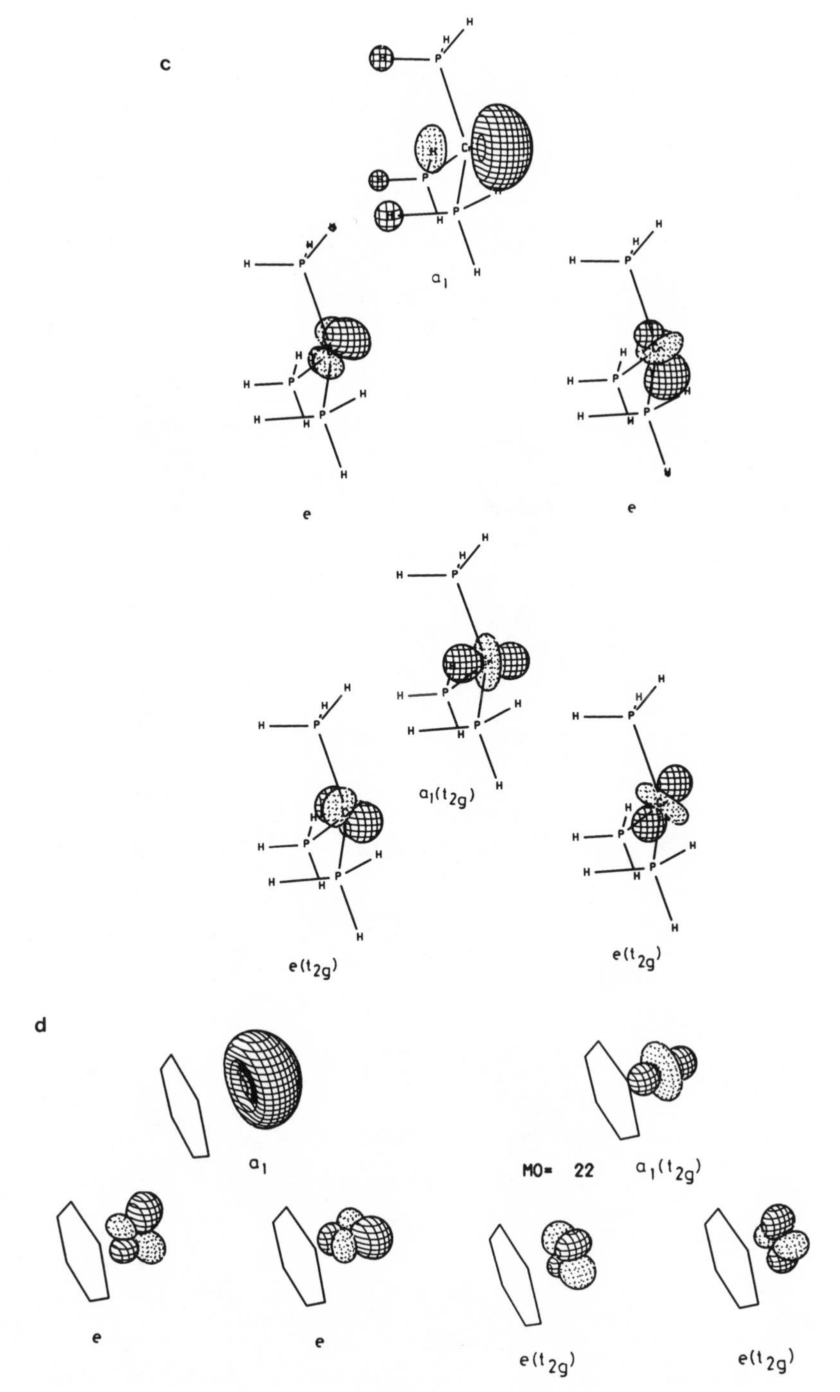

Figure 16. (*Continued*) have been plotted on a common scale using the programs developed by C. Mealli and D. M. Proserpio and described in *J. Chem. Educ.* 1990, **67,** 399.

e

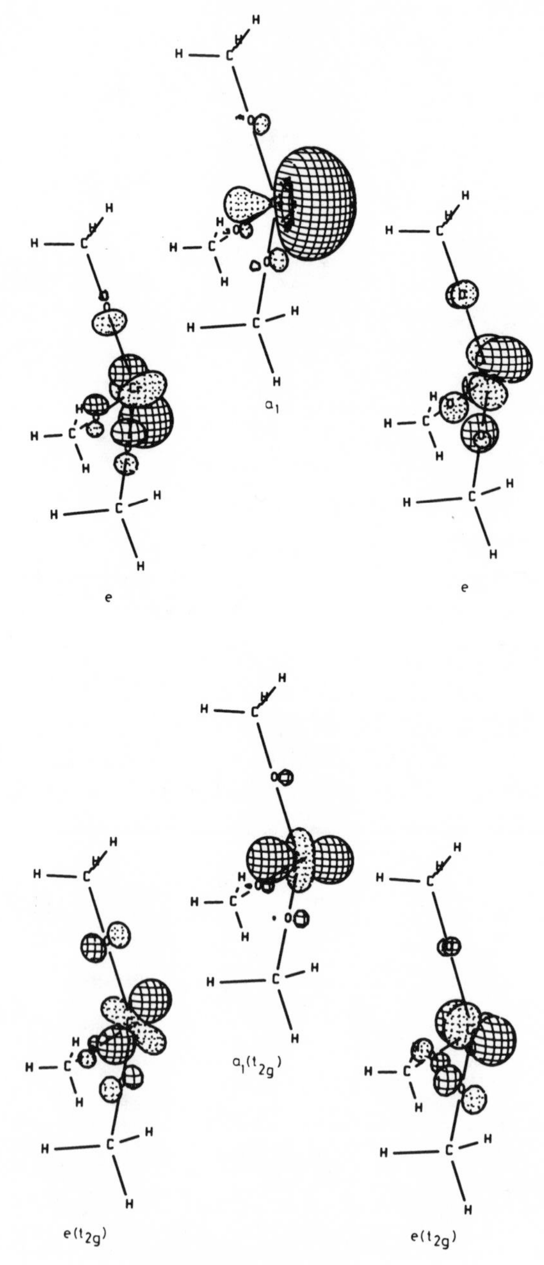

Figure 16. (*Continued*)

lecular orbital which, in the case of $Cr(CO)_3$, has only a total of 41% metal character, but for the remaining fragments this orbital is much more localized on the metal (see Table 6 and Figure 16). The significant degree of metal *s* and *p* character associated with these orbitals ensures that these orbitals are out-pointing and suitable for forming bonds to other fragments. These orbitals are, however, much higher lying than the *e* acceptor orbitals discussed above.

The frontier orbitals of B—H and Si—H are also illustrated in Figure 16. The lower-lying orbital is the out-pointing *s–p* hybrid orbital illustrated in Figure 16 and, in the case of B—H, is occupied by an electron pair. The higher-lying empty acceptor orbitals of the B—H fragment are the degenerate pair of atomic orbitals shown in Figure 16. The *isolobal* relationship between B—H and the ML_3 fragments rests on the similarity between these orbitals and the *e* and a_1 out-pointing orbitals discussed above and illustrated in Figure 16. The B—H and CrL_3 fragments have similar orbital energies, but the relative positions of a_1 and *e* are reversed. The B—H fragment has a_1 more stable than the p_x and p_y orbital pair, while the CrL_3 fragments have *e* below a_1. This reversal, which is a direct consequence of the relative energies of the atomic orbitals, does not change the bonding abilities of these fragments but can have a marked influence on those properties of molecules which depend on the precise molecular orbital energies, e.g., the photoelectron spectra.[34] It is also apparent from Figure 15 that the change in main group atom from B to C has a much greater effect on the orbital energies than those induced by ligand variations in the CrL_3 series.

In Figure 17 the relative energies of the frontier orbitals of some common main group fragments are compared with those of $Cr(CO)_3$. The electronegativity effect of the atoms is clearly discernable. If the relative energies of the main group and CrL_3 *e* orbitals are compared, it is apparent that CrL_3 corresponds most closely to B—H and C—H for first-row atoms and P—H and S—H for second-row atoms. It is also apparent from the figure that the amount of *s* character associated with the out-pointing hybrid diminishes as the electronegativity of the atom increases. It is also smaller for the lighter atom in a pair of E—H fragments from the same column of the Periodic Table.

3.2. Alternative Isolobal Analogies for $Cr(OMe)_3$

Chisholm and his co-workers[35,36] have characterized a range of alkoxide clusters of the early transition metals and noted a striking analogy between alkoxide and carbonyl clusters. For example, the following pairs of compounds:

$Co_3(\mu^3\text{-}CR)(CO)_9$	$W(\mu^3\text{-}CR)(OR)_9$
$Fe_3(\mu^3\text{-}NH)(CO)_{10}$	$W(\mu^3\text{-}NH)(OR)_{10}$

have closely related triangular metal clusters capped by either CR or NH μ^3 bridging groups. Similarly the carbido-clusters $[Fe_4C(CO)_{13}]$ and $[W_4C(\mu^2$-

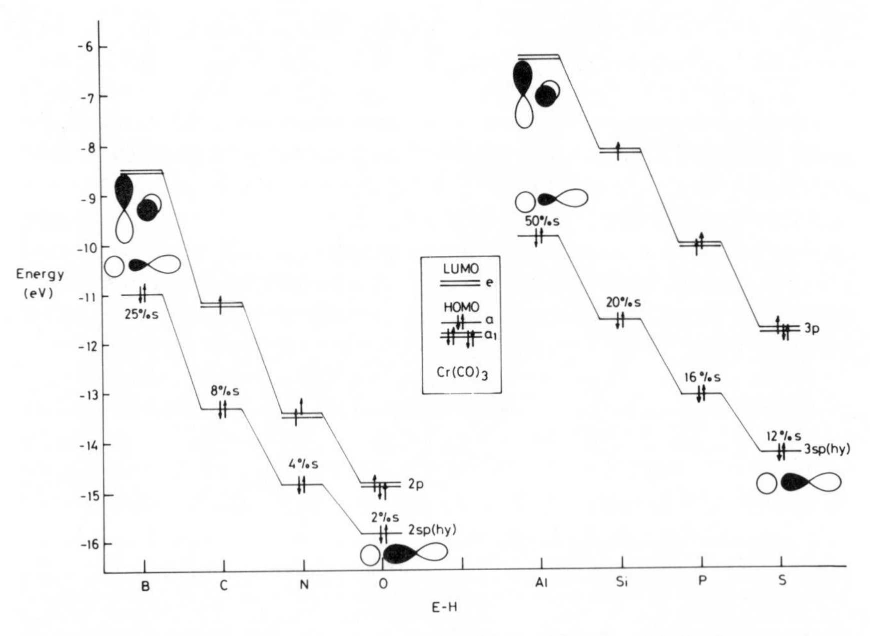

Figure 17. Comparison of the energies and compositions of frontier orbitals of commonly occurring main group fragments.

O)(OPr^i)$_{12}$] both have butterfly arrangements of the metal atoms with the carbido-ligands located between the wing-tip atoms and also bonded to the remaining two metal atoms. Chisholm has proposed that the CO and OR ligands lead to similar skeletal geometries because of the complementary nature of their bonding interactions with the metal frameworks. It is apparent from Figure 15 that ML_3 fragments have two sets of a_1 and e orbitals and the more stable a_1 and e sets can also behave as a frontier orbital set if L is a strong π-donor. Therefore a d^3 $Cr(OMe)_3$ fragment with three electrons in the t_{2g} set may behave in an analogous fashion to $Co(CO)_3$ with the t_{2g} set fully occupied in metal–ligand π-bonding and three electrons in the higher-lying e and a_1 frontier orbitals. In $Cr(OMe)_3$ the t_{2g} set is, of course, metal–ligand π-antibonding. The following fragments which differ by 6 valence electrons may exhibit *isolobal* relationships:

$$\underset{(M=Cr,Mo,W)}{M(OR)_3} \quad \longleftrightarrow\!\!\!\!\!\!\!\!\bigcirc \quad \underset{(M=Co,Rh,Ir)}{M(CO)_3}$$

4. EXCEPTIONS TO THE POLYHEDRAL SKELETAL ELECTRON PAIR THEORY

It is naive to imagine that the frontier orbitals of main group and transition metal fragments are sufficiently similar to always produce a spectrum of skeletal molecular orbitals which predict the closed-shell requirements of the whole gamut of structures containing these fragments. Differences in the radial characteristics and electronegativities of the atoms can cause sufficiently large perturbations to the molecular orbitals of the cluster to invalidate the generalizations developed above. These exceptions can be classified into the following groups.

4.1. Topological Factors

The closed-shell requirements for four connected and deltahedral clusters summarized in Table 1 depend on a spherical free electron model. Real deltahedral molecules and four connected molecules have lower symmetries and this can lead to a spectrum of molecular orbitals which can lead to alternative electron counts. These deviations can be explained using group theoretical techniques based on the symmetry consequences of Stone's Tensor Surface Harmonic Theory. The following general classes of polyhedra can be identified, which do not always conform to the simple generalizations developed above.[37,38]

4.1.1. Polar Deltahedra

Main group deltahedral clusters are generally associated with $4n+2$ valence electrons; however, if they belong to the C_{3v} and T_d point groups they are associated with the electron counts $4n$ or $4n+4$. These polyhedral molecules have an e set of frontier orbitals which is approximately nonbonding and these alternative electron counts correspond to either having these orbitals empty or fully occupied. These polar deltahedra have $3p+1$ atoms (where $p = 1, 2, \ldots$) and examples include the tetrahedron ($p = 1$), the capped octahedron ($p = 2$), and the hexadecahedron ($p = 3$). A specific example of these alternative closed-shell requirements is provided by the tetrahedron which is observed for B_4Cl_4 (16 valence electrons; $4n$) and C_4R_4 (20 valence electrons; $4n+4$). More examples of the closed-shell requirements of polar deltahedral clusters are given in Table 7.

4.1.2. Bipolar Deltahedral Clusters

Group theoretical considerations have shown that these clusters can be associated with $4n-2$, $4n+2$, and $4n+6$ valence electrons. Bipolar clusters

Table 7. Polar Clusters

Cluster	No. of skeletal electron pairs	Ref.
Tetrahedral ($p = 1$)		
B_4Cl_4	4	*a*
$Re_4H_4(CO)_{12}$	4	*b*
C_4R_4	6	*c*
$Ru_4H_4(CO)_{12}$	6	*d*
Capped octahedral ($p = 2$)		
$[\{(\eta\text{-}C_5H_5)Co\}_3B_4H_4]$	7	*e*
$Os_7(CO)_{21}$	7	*f*
Hexadecahedron ($p = 3$)		
$[H(PPh_3)ClRuB_9H_7(PPh_3)_7]$	10	*g*
$[\{(\eta^5\text{-}C_5H_5)Fe\}_2C_2B_6H_8]$	10	*h*

[a] Atoji, M.; Lipscomb, W. N. *Acta Crystallogr.* 1953, **6,** 547.
[b] Saillant, R.; Barcelo, G.; Kaesz, H. D. *J. Am. Chem. Soc.* 1970, **92,** 5739.
[c] Maier, G.; Pfriem, S.; Schafter, U.; Matusch, R. *Angew. Chem.* 1978, **90,** 652.
[d] Wilson, R. D.; Wu, S. M.; Love, R. A.; Bau, R. *Inorg. Chem.* 1978, **17,** 1271.
[e] Pipal, J. R.; Grimes, R. N. *Inorg. Chem.* 1977, **16,** 3255.
[f] Eady, C. R.; Johnson, B. F. G.; Lewis, J.; Mason, R.; Hitchcock, P. B.; Thomas K. M. *J. Chem. Soc., Chem. Commun.* 1977, 385.
[g] Crook, J. E.; Eltrington, M.; Greenwood, N. N.; Kennedy, J. D.; Thornton-Pitt, M.; Woollins, J. D. *J. Chem. Soc., Dalton Trans.* 1985, 2407.
[h] Callahan, K. P.; Evans, W. J.; Lo, F. Y.; Strouse, C. E.; Hawthorne, M. F. *J. Am. Chem. Soc.* 1975, **97,** 296.

have two atoms on the principal rotation axis and the most commonly encountered examples include the trigonal bipyramid, the pentagonal bipyramid, and the bicapped square-antiprism. Examples of these alternative closed-shell requirements are given in Table 8.

4.1.3. Nonpolar Deltahedral Clusters

Nonpolar clusters have no atoms lying on the principal axis and include the dodecahedron and the tricapped trigonal prism. These clusters may be associated with $4n$, $4n+2$, and $4n+4$ valence electrons when the vertices are occupied by main group atoms. Examples of these clusters with related inorganometallic clusters are given in Tables 9 and 10.

4.2. Electronegativity Effects

The presence of vertex atoms with very different electronegativities and orbital radial characteristics can remove the spherical symmetry and provide

Table 8. Bipyramidal Clusters

Cluster	Skeletal electron pairs	Ref.
(a) Trigonal bipyramidal		
$C_2B_3H_5$	6 $(N + 1)$	*a*
$[Os_5(CO)_{16}]$	6	*b*
$[Sn_5]^{2-}$	6	*c*
$[Ni_5(CO)_{12}]^{2-}$	8 $(N + 3)$	*d*
$[Rh_5(CO)_{14}I]^{2-}$	8	*e*
(b) Pentagonal bipyramidal		
$[B_7H_7]^{2-}$	8 $(N + 1)$	*f*
$[(CO)_3FeC_2B_4H_6]$	8	*g*
$[(\eta^5\text{-}C_5H_5)Co(MeC_2B_3H_4)\text{-}Co(\eta^5\text{-}C_5H_5)]$	8	*h*
$[(\eta^5\text{-}C_5H_5)Ni(\mu\text{-}C_5H_5)Ni(\eta^5\text{-}C_5H_5)]$	10 $(N + 3)$	*i*

[a] Williams, R. E. *Adv. Inorg. Chem. Radiochem.* 1976, **18,** 67.
[b] Eady, C. R.; Johnson, B. F. G.; Lewis, J.; Reichert, B. E.; Sheldrick, G. M. *J. Chem. Soc., Chem. Commun.* 1976, 271.
[c] Edwards, P. A.; Corbett, J. D. *Inorg. Chem.* 1977, **16,** 903.
[d] Longoni, G.; Chini, P.; Lower, L. D.; Dahl, L. F. *J. Am. Chem. Soc.* 1975, **97,** 5034.
[e] Martingengo, S.; Ciani, G.; Sironi, A. *J. Chem. Soc., Chem. Commun.* 1979, 1059.
[f] Lipscomb, W. N. "Boron Hydrides"; Benjamin: New York, 1963.
[g] Grimes, R. N. *J. Am. Chem. Soc.* 1973, **95,** 3046.
[h] Beer, D. C.; Miller, V. R.; Sneddon, L. G.; Grimes, R. N.; Mathew, M.; Palenik, G. J. *J. Am. Chem. Soc.,* 1973, **95,** 3046.
[i] Salzer, A.; Werner, H. *Angew. Chem., Int. Ed. Engl.* 1978, **17,** 869.

Table 9. Tricapped Trigonal Prismatic Clusters

Cluster	Skeletal electron pairs	Ref.
B_9Cl_9	9 (N)	*a*
$[B_9H_9]^{2-}$	10 $(N + 1)$	*b*
$[Ge_9]^{2-}$	10	*c*
$[TlSn_8]^{3-}$	10	*d*
$[(\eta^5\text{-}C_5H_5)CoC_2B_6H_8]$	10	*e*
$[Bi_9]^{5+}$	11 $(N + 2)$	*f*

[a] Hursthouse, M. B.; Kane, J.; Massey, A. G. *Nature (London)* 1970, **228,** 659.
[b] Guggenberger, L. J. *Inorg. Chem.* 1968, **7,** 2260.
[c] Belin, C. H. E.; Corbett, J. D.; Cisarm, A. *J. Am. Chem. Soc.* 1977, **99,** 7163.
[d] Burns R. C.; Corbett, J. D. *J. Am. Chem. Soc.* 1982, **104,** 2804.
[e] Dustin, D. F.; Evans, W. J.; Jones, C. J.; Wiersma, R. J.; Gong, H.; Chan, S.; Hawthorne, M. F. *J. Am. Chem. Soc.* 1973, **95,** 4565.
[f] Friedman, R. M.; Corbett, J. D. *Inorg. Chem.* 1973, **12,** 1134.

Table 10. Triangulated Dodecahedral Clusters

Cluster	Skeletal electron pairs	Ref.
B_8Cl_8	8 (N)	*a*
$[\{(\eta^5\text{-}C_5H_5)Co\}_4B_4H_4]$	8	*b*
$[B_8H_8]^{2-}$	9 ($N+1$)	*c*
$[(\eta^5\text{-}C_5H_5)CoSnC_2B_4Me_2H_4]$	9	*d*
$[\{(\eta^5\text{-}C_5H_5)Ni\}_4B_4H_4]$	10 ($N+2$)	*e*

[a] Jacobson, R. A.; Lipscomb, W. N. *J. Chem. Phys.* 1959, **31,** 605.
[b] Pipal, J. R.; and Grimes, R. N. *Inorg. Chem.* 1979, **18,** 257.
[c] Guggenberger, L. J. *Inorg. Chem.* 1969, **8,** 2771; Klanberg, F.; Eaton, D. R.; Guggenberger, L. J.; Muetterties, E. L. *Inorg. Chem.* 1967, **6,** 1271.
[d] Wong, K. S.; Grimes, R. N. *Inorg. Chem.* 1977, **16,** 2053.
[e] Bowser, J. R.; Bonny, A.; Pipal, J. R.; Grimes, R. N. *J. Am. Chem. Soc.* 1979, **101,** 6229.

the opportunity for alternative closed-shell requirements. Specific examples of this behavior include the octahedral clusters

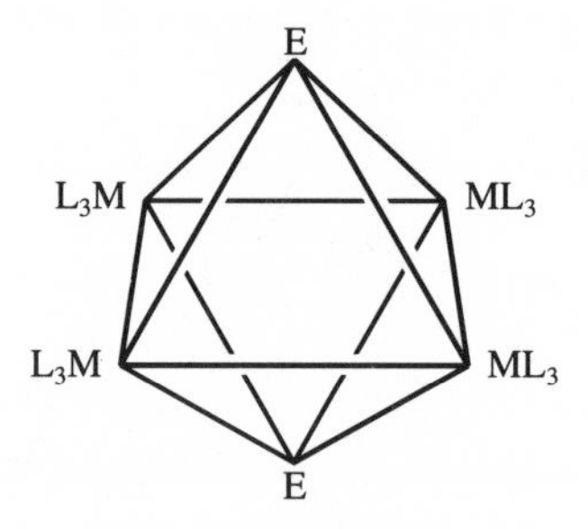

where the ligand E is PR, AsR, S, or Te and ML_3 is a local coordination entity which has bridging or terminal ligands. These clusters have been isolated with 66 valence electrons conforming to the polyhedral skeletal electron pair approach ($14n+2 \rightarrow 4n+2$), but more commonly are associated with 68 valence electrons. Molecular orbital calculations have demonstrated that the additional electron pair resides in a π-antibonding molecular orbital localized on the metals and which is illustrated below:

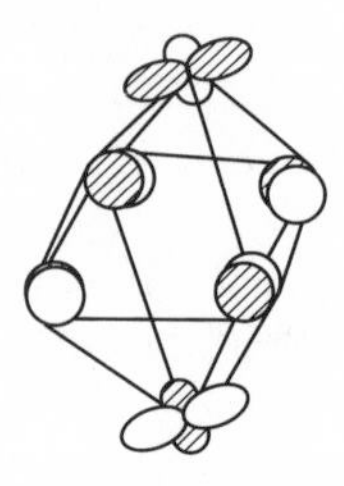

this orbital may be stabilized further by overlap with the *d* orbitals of the main group atoms.[39–41]

This orbital is reminiscent of the highest-lying π^* antibonding molecular orbital of cyclo-butadiene. Therefore, the electronegativity difference between the apical main group and the transition metal atoms has led to a localization of the skeletal MOs on the metal atoms. The two classes of molecule are therefore structurally and electronically related to cyclo-$[butadiene]^{2-}$ and cyclo-$[butadiene]^{4-}$. This localized relationship can be emphasized by viewing the molecule not as an octahedron but a square-planar metal cluster. The PR groups donate 4 electrons each. The incremental relationships associated with metal cyclobutadiene analogs and the parent molecules are evident from the total electron counts above.

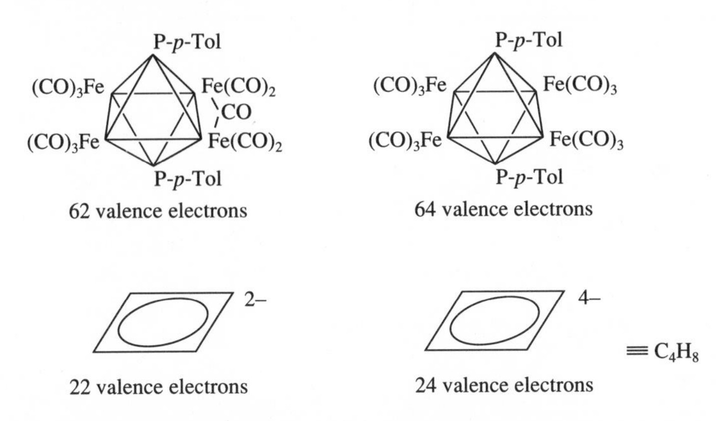

Halet and Saillard have noted that a similar anomalous electron count occurs in the *nido*-$[E_2(ML_3)_4]$ molecules when the main group atoms (E) are axial, e.g., $[Co_4(CO)_{13}[(F_2NMe)_4](\mu_4\text{-}PPh)_2]$, but not when they are equatorial, e.g., $[Os_4(CO)_{12}(\mu\text{-}S)_2]$ and $[WOs_3(CO)_{12}(PMe_2Ph)(\mu_3\text{-}S)_2]$.(40,41)

4.3. Mismatch of the Skeletal Molecular Orbital Bands(42,43)

The polyhedral skeletal electron pair and isolobal approaches depend on a continuity in the bonding skeletal molecular orbitals as one transcends the series from main group to inorganometallic to transition metal clusters. In general, the perturbations associated with electronegativity effects are sufficiently small that the basic pattern of bonding and antibonding orbitals is retained. However, the overlap integrals associated with the constituent atoms of a cluster can be very different, leading to exceptions to the structural generalizations. The total spread in energies between the most-bonding and the least-bonding skeletal molecular orbitals is much larger for a main group cluster than a transition metal cluster, because the overlaps between the main group atoms at a distance appropriate for a cluster are significantly larger than those for transition metal group atoms. In an inorganometallic cluster with approximately equal main group and transition metal atoms, this may lead to the situation illustrated in Figure 18. The relatively low-lying anti-

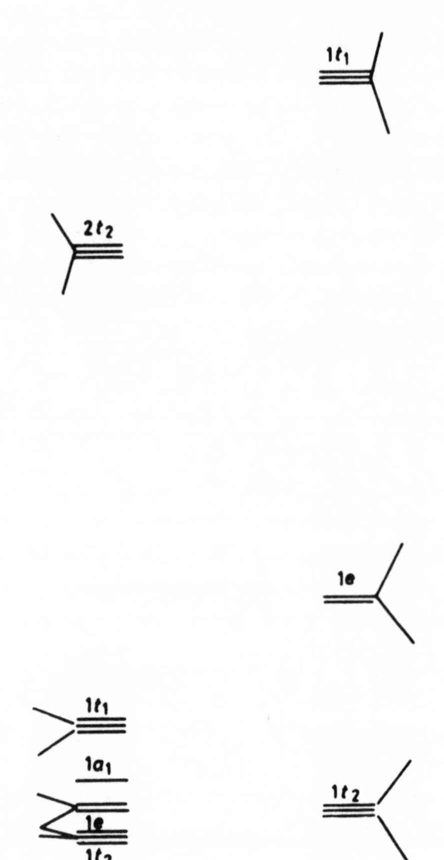

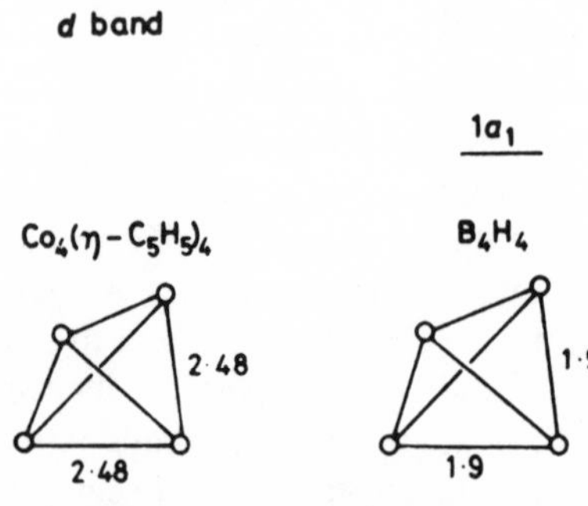

Figure 18. Schematic illustration of the consequences of the orbital mismatch in inorganometallic clusters. The larger spread in orbital energies for the tetrahedral boron cluster is a direct consequence of the larger overlap integrals for B—B relative to Co—Co. Clearly the occupation of antibonding skeletal molecular orbitals based on the metal is energetically favored.

bonding skeletal molecular orbitals of $[M(CO)_n]_y$ are stabilized by overlap with the antibonding molecular orbitals of E_x and may end up in an energy range suitable for electron occupation. Therefore, the occupied molecular orbitals no longer match those anticipated on the basis of the polyhedral skeletal electron pair approach. In these situations it may be possible to resolve the bonding problem by focusing attention only on the metal skeletal geometry. For example, the following molecules all have dodecahedral geometries:

	$B_8H_8^{2-}$	$B_4H_4(Co(C_5H_5))_4$	$B_4H_4(Ni(\eta\text{-}C_5H_5)_4)$
Total no. of electrons	34	72	76
No. of skeletal electron pairs	9	8	10

Neither of the inorganometallic examples have the predicted 74 valence electrons. It is apparent from the metal skeletal structures (see Figure 19) that in $[B_4H_4(Co(\eta\text{-}C_5H_5))_4]$ the metals define a folded rhombus while in $[B_4H_4(Ni(\eta\text{-}C_5H_5))_4]$ the metals form a pair of separated dimers. Both of these structures can be derived from a tetrahedron by breaking either two or four bonds, i.e., they are associated with either $60 + 4 = 64$ or $60 + 8 = 68$ valence electrons. Therefore, in $[B_4H_4(Co(\eta\text{-}C_5H_5))_4]$ the B_2H_2 groups act as bridging groups donating 4 electrons each and in $[B_4H_4(Ni(\eta\text{-}C_5H_5))_4]$ the rhombic bridging

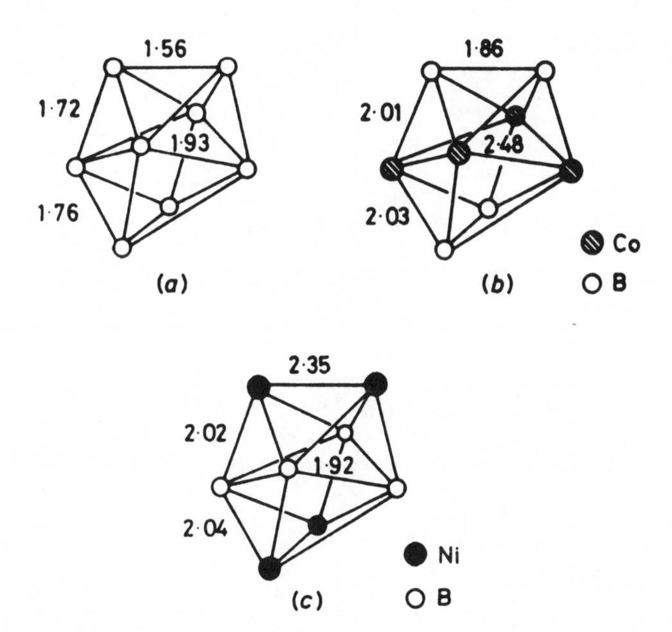

Figure 19. Skeletal geometries of $B_8H_8^{2-}$, $[B_4H_4(Co(\eta\text{-}C_5H_5))_4]$, and $[B_4H_4(Ni(\eta\text{-}C_5H_5))_4]$.

B_4H_4 fragment donates 8 electrons to both metal atoms. From the discussion above it should be apparent that these exceptions will only be observed when there are a number of one or both types of constituent atoms clustered together. A more detailed molecular orbital analysis of this class of polyhedral metalloborane has been presented elsewhere.[42]

ACKNOWLEDGMENTS

Dr. Davide M. Proserpio is thanked for supplying the electron density plots in Figure 16 and Dr. Javier Modrego for the calculations on the relative energies of metal and main group fragments. The S.E.R.C. is thanked for financial support.

REFERENCES

1. Mingos, D. M. P.; Johnston, R. L. *Struct. Bonding* 1987, **68,** 31.
2. Mingos, D. M. P.; Wales, D. J. "Introduction to Metal Cluster Chemistry"; Prentice Hall: New York, 1990.
3. Mingos, D. M. P. In "The Chemistry of Metal Cluster Complexes"; Shriver, D. F.; Kaesz, H. D.; Adams, R. D., Eds.; Verlag Chemie: New York, 1990; p. 11.
4. Roesky, H. W. "Rings, Cluster and Polymers of Main Group and Transition Elements"; Elsevier: New York, 1989.
5. Braye, E. H.; Hübel, W.; Caplier, I. *J. Am. Chem. Soc.* 1962, **83,** 4406.
6. Hawthorne, M. F. *J. Organomet. Chem.* 1975, **100,** 97.
7. Adams, R. D. In "The Chemistry of Metal Cluster Complexes"; Shriver, D. F.; Kaesz, H. D.; Adams, R. D., Eds.; Verlag Chemie: New York, 1990; 121.
8. Stone, F. G. A. *Angew. Chem., Int. Ed. Engl.* 1984, **23,** 89.
9. Johnston, R. L.; Mingos, D. M. P. *J. Organomet. Chem.* 1985, **280,** 407, 419.
10. Hawes, J. C.; Mingos, D. M. P. *Struct. Bonding* 1985, **63,** 1.
11. Lin, Z.; Mingos, D. M. P. *Struct. Bonding* 1990, **72,** 73.
12. Stone, F. G. A. *Inorg. Chim. Acta* 1981, **50,** 33.
13. Stone, F. G. A. *Pure Appl. Chem.* 1986, **58,** 529.
14. Ashworth, T. V.; Howard, J. A. K.; Laguna, M.; Stone, F. G. A. *J. Chem. Soc., Dalton Trans.* 1980, 1593.
15. Green, M.; Mills, R. M.; Pain, G. N.; Stone, F. G. A.; Woodward, P. *J. Chem. Soc., Dalton Trans.* 1982, 1309.
16. Green, M.; Jeffrey, J. C.; Porter, S. J.; Razay, H.; Stone, F. G. A. *J. Chem. Soc., Dalton Trans.* 1982, 2475.
17. Hoffmann, R. *Angew. Chem., Int. Ed. Engl.* 1982, **21,** 711.
18. Lauher, J. W.; Hoffmann, R. *J. Am. Chem. Soc.* 1976, **48,** 1729.
19. Berry, D. H.; Chey, J. H.; Zipin, H. S.; Carroll, P. J. *J. Am. Chem. Soc.* 1990, **112,** 452.
20. Cotton, F. A.; Watton, R. A. "Multiple Bonds Between Metal Atoms"; Wiley: New York, 1982.
21. Teo, B. K. *Inorg. Chem.* 1984, **23,** 1251.
22. Mingos, D. M. P. *Inorg. Chem.* 1985, **24,** 114.
23. Mingos, D. M. P.; Welch, A. J. *J. Chem. Soc., Dalton Trans.* 1980, 1674.
24. Mingos, D. M. P. *J. Chem. Soc., Dalton Trans.* 1977, 602.
25. Mealli, C.; Midollini, S.; Moneti, S.; Sacconi, L.; Silvestre, J.; Albright, T. A. *J. Am. Chem. Soc.* 1982, **104,** 45.

26. Silvestre, J.; Albright, T. A. *J. Am. Chem. Soc.* 1985, **107,** 6829.
27. Albright, T. A. *Acc. Chem. Res.* 1982, **15,** 135.
28. Albright, T. A.; Hoffmann, R.; Tse, Y.-C.; D.'Ottavio, C. *J. Am. Chem. Soc.* 1979, **107,** 3812.
29. Stone, F. G. A. *J. Organomet. Chem.* 1975, **100,** 257.
30. Forsyth, M. I.; Mingos, D. M. P.; Welch, A. J. *J. Chem. Soc., Dalton Trans.* 1978, 1363.
31. Green, M.; Spencer, J. L.; Stone, F. G. A.; Welch, A. J. *J. Chem. Soc., Chem. Commun.* 1974, 794.
32. Evans, D. G.; Mingos, D. M. P. *J. Organomet. Chem.* 1982, **232,** 171.
33. Evans, D. G.; Mingos, D. M. P. *J. Organomet. Chem.* 1985, **295,** 389.
34. Green, J. C.; Mingos, D. M. P.; Seddon, E. A. *Inorg. Chem.* 1981, **20,** 2595.
35. Chisholm, M. H.; Clark, D. L.; Hampden-Smith, M. J. *Angew. Chem., Int. Ed. Engl.* 1989, **28,** 432.
36. Chisholm, M. H.; Folting, K.; Hammond, C. E.; Hoffmann, J. C.; Makin, J. C. *Angew. Chem., Int. Ed. Engl.* 1989, **28,** 1368.
37. Johnston, R. L.; Mingos, D. M. P. *J. Chem. Soc., Dalton Trans.* 1987, 1643.
38. Johnston, R. L.; Mingos, D. M. P. *J. Chem. Soc., Dalton Trans.* 1987, 1445.
39. Halet, J.-F.; Hoffmann, R.; Saillard, J.-Y. *Inorg. Chem.* 1985, **24,** 1695.
40. Halet, J.-F.; Saillard, J.-Y. *J. Organomet. Chem.* 1987, **327,** 365.
41. Halet, J.-F.; Saillard, J.-Y. *New J. Chem.* 1987, **11,** 315.
42. Cox, D. N.; Mingos, D. M. P.; Hoffmann, R. *J. Chem. Soc., Dalton Trans.* 1981, 1788.
43. Scott-Leigh, J.; Whitmore, K. H.; Yee, K. A.; Albright, T. A. *J. Am. Chem. Soc.* 1989, **111,** 2726.

5

Experimental Comparison of the Bonding in Inorganometallic and Organometallic Complexes by Photoelectron Spectroscopy

Dennis L. Lichtenberger, Anjana Rai-Chaudhuri, and Royston H. Hogan

1. INTRODUCTION

The broad strides that have been made in recent years in the synthesis and structural characterization of organometallic and inorganometallic complexes have led to increasing interest in understanding the electronic structure and bonding in these molecules. The principles and models of bonding that have developed for organometallic complexes, as typified by the Dewar–Chatt–Duncanson model of synergistic metal–ligand bonding, have also provided the foundation for understanding the chemical and physical properties of inorganometallic complexes. These principles go beyond the basic crystal field or valence bond descriptions of *d*-orbital splitting diagrams of classical coordination compounds. The models now consider the precise electron-donor, electron-acceptor, and bond-forming capabilities of the ligands. For every

Dennis L. Lichtenberger, Anjana Rai-Chaudhuri, and Royston H. Hogan • Laboratory for Electron Spectroscopy and Surface Analysis, Department of Chemistry, University of Arizona, Tucson, Arizona 85721.
Inorganometallic Chemistry, edited by Thomas P. Fehlner. Plenum Press, New York, 1992.

interaction of an organic ligand with a metal there is a symmetry-equivalent interaction of an inorganic ligand with a metal. This has been termed the "isolobal analogy."[1,2] The differences lie in the precise spatial distribution and energy of the valence electrons of the ligands. Experimental approaches are essential to obtain a quantitative comparison of the electron distribution and bonding of organic and inorganic ligands with transition metals.

In recent years, photoelectron spectroscopy has been increasingly utilized as an experimental tool to reveal the electronic structure and bonding features of metal complexes.[3–9] Photoelectron spectra offer well-defined and detailed experimental information about electron richness, electron distributions, and the strength of bonding interactions in molecules. The general electron richness of a molecule may be correlated with its first ionization potential (IP). A high ionization potential (binding energy) indicates relatively stable and inaccessible electrons, while a low ionization potential indicates relatively unstable and available electron density. These features are directly related to Lewis acid–base and oxidation–reduction reactivity. The bonding within a molecule defines the molecular geometry and stability. Strong covalent interactions result in ionization bands with generally higher IPs, while nonbonding and antibonding orbitals result in ionizations with generally lower IPs. Recently, the direct relationship between the ionization energy and the bond energy has been placed on a more formal basis.[10] The breadth of an ionization band is also significant because it indicates the bonding character in the ionized orbital and the extent of bond distance changes with the ionization. An ionization associated with a strongly bonding orbital results in a broad ionization envelope, while ionization associated with a nonbonding orbital generally results in a narrow ionization peak. Thus the photoelectron spectrum contains a wealth of information on the electron distribution and bonding with the metal.

In the following we will explore in some depth the comparisons that are revealed in the relative bonding in organometallic and inorganometallic complexes. We will use results from photoelectron data collected in our laboratory to illustrate the similarities in orbital symmetry interactions that occur between metals with organic and inorganic ligands. In each case the ligands have similar formal electron counts and isolobal symmetries, but the different spatial and energy characteristics of the orbitals lead to clear effects in the photoelectron spectra. The study of both classes of compounds leads to a much more complete understanding of the electronic structure of each. We have previously termed this the "electronic structure perturbation" (ESP) approach to probing the electronic structure.[3] That is, by perturbing the electronic structure with well-defined chemical substitutions, such as inorganic ligands in place of organic ligands, we can learn not only about the electron distribution in a particular class of molecules, but we can also investigate the fluidity of electron charge and its ability to adapt to chemical changes.

2. THE σ-DONOR AND π-ACCEPTOR CAPABILITIES OF ORGANIC AND INORGANIC LIGANDS FROM PHOTOELECTRON SPECTROSCOPY

The role of photoelectron spectroscopy in characterizing the relative bonding capabilities of ligands will first be illustrated with an analysis of the relative σ-donor and π-acceptor abilities of simple formal two-electron donor organic and inorganic ligands. The extent of σ-donation and π-acceptance by molecules such as CO and C_2H_4 is prevalent in the discussion of the chemistry and spectroscopy of complexes with these ligands. The same properties may be identified for classically two-electron donating inorganic ligands such as N_2, amines, phosphines, and SO_2, as will be shown here.

In general, photoelectron spectroscopy provides a measure of the relative σ-donor ability of a ligand by observing the shift of the metal-based ionizations. Better electron donors make the metal more electron rich and lower the metal-based ionization energies. This measure of the change in electron richness is most clear when one of the metal orbitals has no other symmetry interactions with the ligand. This is not uncommon for d^6 "pseudo-octahedral" complexes, where one of the metal orbitals from the t_{2g} set is primarily δ symmetry with respect to the ligand. The ionizations of other ligands attached to the metal, such as the cyclopentadienyl ring in many of the systems we have investigated, also sense the increased electron richness at the transition metal center.[11]

The stabilization of the ionization of the donor orbital of the free ligand when it forms the bond to the metal is also a measure of the strength of the interaction. The extent of this stabilization is closely related to the strength of the bond.[12–14] An example is provided by the lone pairs of phosphorus and other inorganic two-electron donor ligands. The ionizations of phosphorus lone pairs are usually stabilized about 1 eV when coordinated to a metal. Metal–carbon, metal–hydrogen, and metal–oxygen σ-bond ionizations are also commonly observed in the low-valence ionization region.

The π-acceptor capability of a ligand is characterized in photoelectron spectroscopy by observing the shift of the ionizations of occupied metal donor orbitals that are stabilized by delocalization to the ligand acceptor orbitals.[3,4] The extent of delocalization is also roughly reflected in the relative intensities (cross sections) of the ionizations as a function of the energy of the ionizing photon. For instance, in comparison to the intensities of the ionizations of main group atoms, the relative intensities of metal-based ionizations generally increase when He II (40.8 eV) photons are used in place of He I (21.21 eV) photons. The intensities of ionizations based on second-row main group atoms, such as Si, P, and Cl, are exceptionally weak with He II excitation and are easily identified.[31] Other features such as vibrational fine structure[15] and spin–orbit coupling[16] provide additional information.

The most direct way to compare the relative bonding of ligands is to compare the photoelectron spectra of complexes in which the different ligands are bound to a common metal fragment.[4] For purposes of comparing the σ-donor and π-acceptor abilities of ligands, the fragment [$CpMn(CO)_2$] has proven to be extremely valuable (Cp is η^5-$C_5H_5^-$). A large number of inorganic and organic ligands form sufficiently stable molecules with this fragment that they can be studied by high resolution spectroscopy in the gas phase. The spectra of $CpMn(CO)_3$, $CpMn(CO)_2C_2H_4$, $CpMn(CO)_2N_2$, $CpMn(CO)_2NH_3$, $CpMn(CO)_2PMe_3$, and $CpMn(CO)_2SO_2$ will be compared here to reveal the relative bonding capabilities of these ligands.

The electronic structure of $CpMn(CO)_3$ and the bonding capabilities of [$CpMn(CO)_2$] toward ligands has been well documented.[11,17] $CpMn(CO)_3$ may be considered pseudo-octahedral with the Cp ring occupying three coordination sites. Figure 1 shows the photoelectron spectrum of $CpMn(CO)_3$. The first ionization band is metal-based and contains the three ionizations of the formally d^6 metal center. These "t_{2g}" metal ionizations are barely split in the relative area ratio of 2:1, reflecting the near-octahedral electronic symmetry at the metal center. The ionizations which correlate predominantly with the Cp ring e_1'' π-levels appear at a higher energy than the metal ionizations, at 9.9 and 10.3 eV.

Now consider replacing one of the CO ligands with an inorganic ligand L to form $CpMn(CO)_2L$. A convenient coordinate system is to place the L ligand along the z-axis with the x-axis bisecting the two carbonyls as shown in Figure 2. The LUMO of the $CpMn(CO)_2$ fragment is predominantly d_{z^2} in character and is the σ-acceptor orbital of the metal for a ligand on the z-axis. The occupied "t_{2g}" orbitals are comprised of the d_{xz}, d_{yz}, and $d_{x^2-y^2}$ orbitals in this coordinate system. The predominantly metal d_{xz} and d_{yz} orbitals of the $CpMn(CO)_2$ fragment have π symmetry with respect to L and can behave as π-donors. The predominantly metal $d_{x^2-y^2}$ orbital backbonds to the two carbonyls in the x–y plane and is essentially δ with respect to the ligand L. The shift in ionization energy of this $d_{x^2-y^2}$ orbital reflects the change in

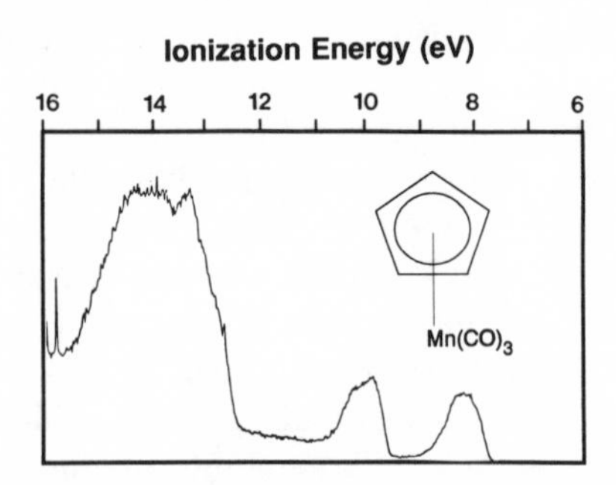

Figure 1. He I spectrum of $CpMn(CO)_3$.

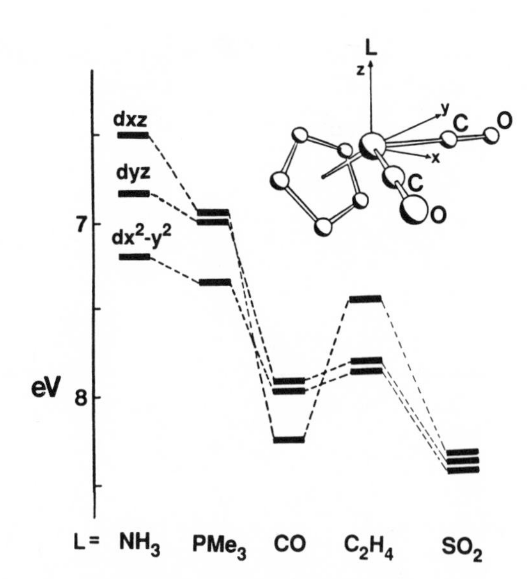

Figure 2. Correlation diagram of experimental ionization energies of $MeCpMn(CO)_2L$ complexes where L is NH_3, PMe_3, CO, C_2H_4, and SO_2.

charge potential at the metal center with substitution of the ligand L on the z-axis. Because each d_{xz} and d_{yz} orbital backbonds to one L acceptor orbital and one CO acceptor orbital, while $d_{x^2-y^2}$ backbonds to two CO acceptor orbitals, the stabilizations of d_{xz}/d_{xz} orbital ionizations relative to the $d_{x^2-y^2}$ ionization are a measure of the π-acceptor abilities of L in each plane relative to CO.

A simple example of these principles is provided by the photoelectron spectrum of the dinitrogen complex $CpMn(CO)_2N_2$,[18] shown in Figure 3. The trends in valence ionization energies from the tricarbonyl to the dinitrogen complex are clarified by the molecular orbital diagrams of $CpMn(CO)_3$ and

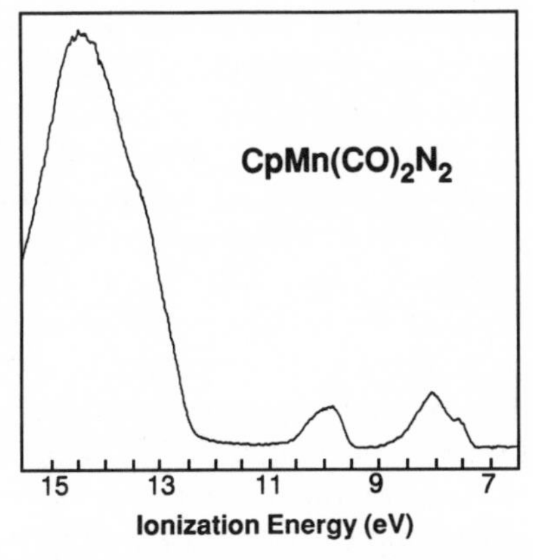

Figure 3. He I spectrum of $CpMn(CO)_2N_2$.

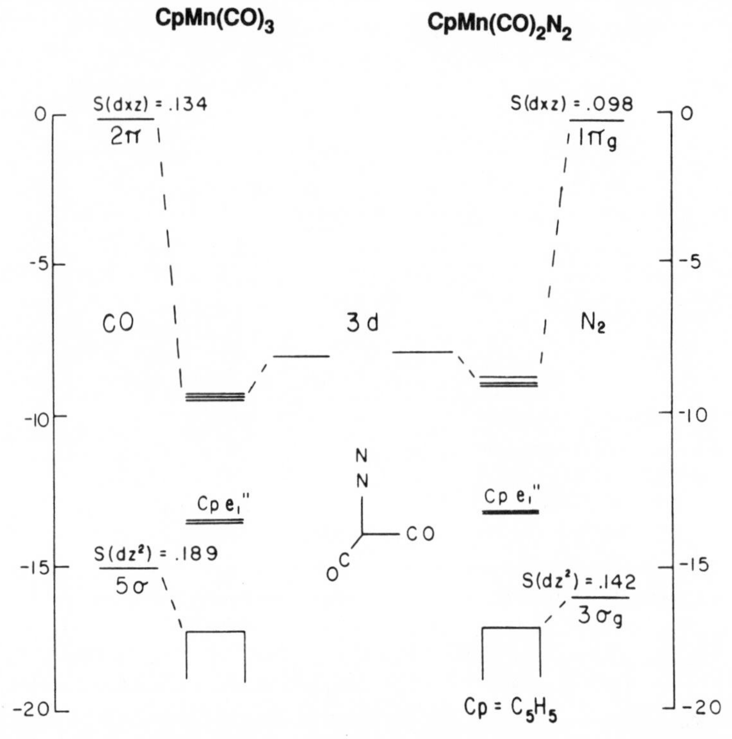

Figure 4. Comparison of molecular orbital diagrams of $CpMn(CO)_3$ and $CpMn(CO)_2N_2$.

$CpMn(CO)_2N_2$ shown in Figure 4. The energies of the σ-donor and π-acceptor orbitals of CO and N_2 are similar. However, because N_2 is a homonuclear diatomic molecule, these orbitals are distributed equivalently on the two nitrogen centers, while the corresponding σ-donor and π-acceptor orbitals of CO are more localized toward the carbon atom that is bound to the metal. This spatial distribution reduces the overlap of these N_2 orbitals in comparison to the CO orbitals with the d_{xz} and d_{yz} orbitals. The lower π-acceptor ability of N_2 compared to CO reduces the electronic symmetry at the metal center, with the d_{xz} and d_{yz} orbitals destabilized relative to the $d_{x^2-y^2}$ orbital. The lower π-acceptor ability also leaves more electron density on the metal center, which lowers the ionization energy of all the metal-based ionizations. The increased electron density at the metal center is also sensed by the Cp ring, and the Cp e_1'' ionizations occur at lower energy in the dinitrogen complex. Thus although the complexes are isoelectronic and the key ligand orbitals are nearly isoenergetic, the different spatial distributions of the ligand orbitals and their different overlaps with the metal center, as shown in Figure 4, lead to different bond strengths and electron distributions with the metal center.

These differences are clearly observed in the photoelectron spectrum. Because the changes are largely based on overlap considerations, approximate molecular orbital calculations are quite successful at reproducing the shifts of the valence ionizations of the complexes.[18]

The metal-based ionizations of $CpMn(CO)_2L$ complexes have been used to indicate the donor and acceptor strengths of a variety of ligands L. Figure 2 shows the correlation diagram of the metal IPs of $CpMn(CO)_2NH_3$, $CpMn(CO)_2PMe_3$, $CpMn(CO)_3$, $CpMn(CO)_2C_2H_4$, and $CpMn(CO)_2SO_2$. NH_3 is obviously a good σ-donor through its nitrogen lone pair and the metal ionizations are destabilized by 0.7 eV from those in $CpMn(CO)_3$. NH_3 is a poor π-acceptor and there is a considerable loss of electronic symmetry at the metal center as revealed by the splitting of the metal ionizations.[18] PMe_3 is also an effective σ-donor as seen by the destabilization of the $d_{x^2-y^2}$ orbital ionization by 0.6 eV.[19] The acceptor orbitals on the PMe_3 ligand would be the $3d$-orbitals, but this interaction is obviously small since the metal ionizations are substantially split (0.4 eV). The difference in core (Mn $^2P_{3/2}$) ionization potentials between $CpMn(CO)_3$ and $CpMn(CO)_2PMe_3$ of -1.4 eV indicates a 0.1–0.15 electron potential difference at the metal center. This is due to a combination of the loss of π-backbonding stabilization when CO is replaced by PMe_3 and the gain in electron richness from σ-donation of PMe_3 compared to CO.

In $MeCpMn(CO)_2C_2H_4$, the ethylene ligand has only one π-acceptor orbital that can interact with a metal orbital. In the coordinate system that is used, the ethylene π^*-orbital interacts with the metal d_{yz} orbital, thus stabilizing it. The d_{yz} orbital is the HOMO of the fragment and is also hybridized to be the best π-donor of the metal fragment.[17] The d_{xz} orbital is orthogonal to π-accepting orbitals on ethylene and the corresponding ionization is 0.4 eV destabilized from the $d_{x^2-y^2}$ orbital ionization as a result of the removal of the backgonding stabilization of the carbonyl ligand that was on the z-axis. The d_{yz} orbital is stabilized by backbonding to the ethylene ligand. The corresponding ionization is degenerate with the $d_{x^2-y^2}$ orbital ionization. Therefore the metal ionizations are split in a 1:2 pattern. The degeneracy of the d_{yz} metal ionization with the $d_{x^2-y^2}$ metal ionization indicates that one ethylene π-acceptor orbital is as effective as a single carbonyl π-acceptor orbital in stabilizing a metal-based ionization in this complex. Core XPS studies show that the charge potential at the metal is intermediate between CO and PMe_3.

In $CpMn(CO)_2SO_2$, the SO_2 ligand is bound to the metal in an η^1 fashion through the sulfur atom. The isolobal analogies between the π-acceptor orbitals of SO_2 and those of CO are shown in Figure 5.[20] As can be seen, the spatial distributions that the metal sees when the SO_2 is coordinated through the sulfur are quite similar. The similarity between the CO π^* and the SO_2 $3b_1$ is especially striking. The d_{yz} metal orbital aligns with and π-backbonds to

Figure 5. Isolobal analogies between the π-acceptor orbitals of CO and SO_2.

the $3b_1$ out-of-plane π^* orbital of SO_2. This is the LUMO of SO_2 and the best π-acceptor. The d_{xz} orbital and $6b_2$ of SO_2 are also of π-symmetry and mix in this orientation. Interestingly, the electronic symmetry at the metal center is not significantly perturbed when SO_2 replaces CO. The "t_{2g}" set of ionizations remains nearly degenerate. The slight stabilization of the metal-based ionizations is traced to the effective positive charge of the sulfur atom in SO_2 compared to the carbon atom in CO, rather than an overall charge shift between the metal and the ligand. This conclusion is the result of a combined XPS and UPS investigation.[21] In fact, η^1-coplanar SO_2, although in appearance much different than CO, is more like CO in its overall bonding effects in these molecules than any other ligand studied to date.

The host of studies of $CpMn(CO)_2L$ complexes have shown that the electron density in the $CpMn(CO)_2$ fragment is quite fluid and able to adapt to the bonding characteristics of the ligand L. This no doubt contributes to the stability of the wide variety of organic and inorganic ligands that can be bound to this fragment.

3. MULTIPLE BONDING TO TRANSITION METALS: THE ALKYLIDYNE AND NITRIDE LIGANDS

A major theme in the current study of transition metal systems involves multiple bonding. Just as in organic systems, this "unsaturation" leads to interesting reactivity and properties. One of the most fascinating examples from the organometallic arena is the alkylidyne ligand, CR^{3-}, which forms a formal triple bond with the metal and undergoes a variety of metathesis and polymerization reactions.[22–25] The simplest inorganometallic analog is the nitride ligand, N^{3-}. Both the alkylidyne and nitride ligands are formally six-electron donor ligands, with two electrons in a largely sp hybrid orbital donating in a σ fashion to the metal center while four electrons, residing in principally $p\pi$ orbitals, donate into empty $d\pi$ acceptor orbitals on the metal fragment. Formally, three bonds, one σ and two π, exist between the metal center and the alkylidyne or nitride ligands. The electron distribution in these bonds, the covalent or ionic character of these bonds, and the sensitivity of these bonds to their environment are important for understanding the chemistry of these species.

An extensive number of transition metal alkylidyne and nitride complexes in a number of electronic and structural environments have been synthesized. A particularly interesting series of alkylidyne and nitride complexes possess the general formula $(E)Mo(OR)_3$ [$E = CCMe_3$ or N and $R = CMe_3$, CMe_2CF_3, $CMe(CF_3)_2$ or Cl]. These stable, d^0, molybdenum alkylidyne complexes demonstrate a varying degree of catalytic activity with respect to alkyne metathesis. The catalytic activity is dependent upon the type of alkyne metathesized, i.e., an internal or external alkyne, and the solvent employed by the catalytic system.[26,27] The rate of metathesis as well as the product of the reaction between catalyst and alkyne is very sensitive to the ligand environment around the metal center. The nature of the metal–carbon triple bond in these molybdenum catalysts is therefore a source of extreme academic and commercial interest. The nitride complexes, because of their comparative simplicity, assist in understanding the bonding of this class of ligands to the metal center.

Figure 6 shows the He I valence photoelectron spectra of $(Me_3CC)Mo(CH_2CMe_3)_3$ (A), $(Me_3CC)Mo(OCMe_3)_3$ (B), $(Me_3CC)Mo(OCMe_2CF_3)_3$ (C), $(Me_3CC)Mo(OCMe(CF_3)_2)_3$ (D), and $(Me_3CC)MoCl_3$ (E).[28] The spectrum of each complex shows a broad region of overlapping ionizations from 11–16 eV due to primarily C—C, C—O, and C—H σ ionizations. The alkoxide complexes also show ionizations in this region except with a concomitant decrease in relative intensity due to the lesser number of C—H σ ionizations with fluorine substitution. The broad band at about 8–9.5 eV in the spectrum of $(Me_3CC)Mo(CH_2CMe_3)_3$ is due to metal–neopentyl carbon σ bonds. On proceeding from the alkyl to the alkoxide complexes, the metal–neopentyl carbon ionizations are replaced by a different set of ionizations in the 9–10

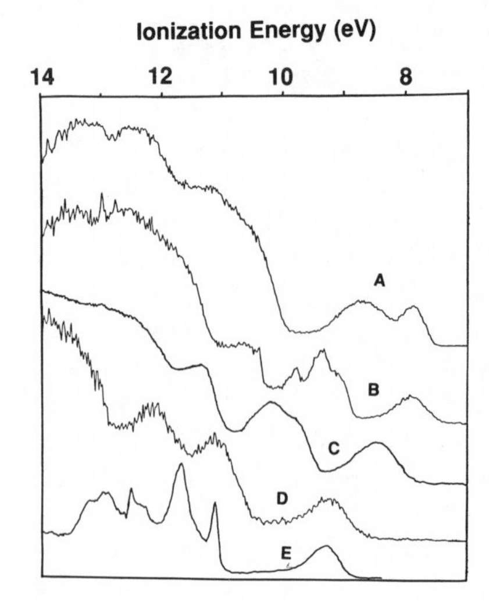

Figure 6. He I spectra of the molybdenum alkylidynes: A, $(Me_3CC)Mo(CH_2CMe_3)_3$; B, (Me_3CC)-$Mo(OCMe_3)_3$; C, $(Me_3CC)Mo(OCMe_2CF_3)_3$; D, $(Me_3CC)Mo(OCMe(CF_3)_2)_3$; E, $(Me_3CC)MoCl_3$.

eV region which are due primarily to the alkoxide oxygen $p\pi$ and Mo—O σ-bonding electrons.

The area of greatest interest in these complexes lies in the low binding region from about 7–10 eV. In all these complexes the leading ionization is due to the two degenerate metal–alkylidyne carbon π-bonds. An array of interesting chemical information regarding the electronic characteristics of these complexes is provided by studying the trends in the metal–alkylidyne carbon π ionizations in these compounds.

On proceeding from $(Me_3CC)Mo(CH_2CMe_3)_3$ to $(Me_3CC)Mo(OCMe_3)_3$, the Mo—C π-band shifts by an insignificant amount. This is an unexpected result considering the stronger σ-donating ability of alkyl compared to alkoxide ligands. This indicates that the alkoxide ligands are also acting as π-donors to the molybdenum center. The π-donation by the alkoxide ligands offsets its weaker σ-donating ability (compared to the neopentyl ligand) resulting in little change in the charge potential at the metal center in the neopentyl and non-substituted alkoxide complexes.

In the series of complexes from $(Me_3CC)Mo(OCMe_3)_3$ to $(Me_3CC)Mo$-$(OCMe_2CF_3)_3$ and $(Me_3CC)Mo(OCMe(CF_3)_2)_3$, the successive substitution of three hydrogens in the alkoxide ligands by three fluorines allows for substantial charge transfer from the metal center to the alkoxide ligands owing to the strong electron-withdrawing effect of fluorine atoms. Even though these substitutions are on the periphery of the alkoxide ligands, these effects result in a substantial 0.65 eV shift of the Mo—C π-bond ionization to higher binding energy on going from $(Me_3CC)Mo(OCMe_3)_3$ to $(Me_3CC)Mo(OCMe_2CF_3)_3$. Substitution by six fluorines to give $(Me_3CC)Mo(OCMe(CF_3)_2)_3$ leads to a total shift of 1.32 eV. As can be seen, the shifts with fluorination are essentially additive.[3,19,33]

These shifts have a large influence on the rates of the metathesis reactions.[28] The metathesis reaction is initiated by nucleophilic attack of an incoming alkyne on the metal complex. The photoelectron spectra show that fluorination of the alkoxide ligands makes the metal significantly more electron poor, and favors the nucleophilic attack by the alkyne. The rates of these reactions increase with fluorination of the alkoxide ligands.

One particularly intriguing aspect of the photoelectron spectra of these complexes concerns the location of the Mo—C σ ionization in these compounds. The total disappearance of the peak associated with the Mo—C σ-bonds in the neopentyl complex upon substitution of alkoxide for alkyl ligands indicates that the Mo–alkylidyne carbon σ-bond was not among this group of ionizations. Although the ionization cannot be located precisely in the spectral data, the available evidence indicates that it lies 1.5 eV or more to higher binding energy than the Mo—C π-band since it was not observed in the spectrum of $(Me_3CC)Mo(OCMe_3)_3$. Further evidence to support this hypothesis is found in the location of the Cr—C σ-bond in the photoelectron spectra of pentacarbonyl chromium carbene complexes.[29]

The photoelectron spectrum of $(N)Mo(Cl)_3$ offers a unique opportunity to assign each valence ionization in the photoelectron spectrum of a transition metal complex, and assists in understanding the alkylidyne complexes.[28] Figure 7 shows the full valence spectrum of $(N)Mo(Cl)_3$. Band A arises from the "*e*" chlorine lone-pair combinations while bands B and C are assigned to the Mo—Cl "*a*" and "*e*" σ-bonding combinations. The chlorine lone-pair of band A possesses appreciable metal character (as evidenced by He I/He II comparisons) indicating that a filled–unfilled interaction is taking place between the chlorine lone pairs and the unoccupied metal levels. This same

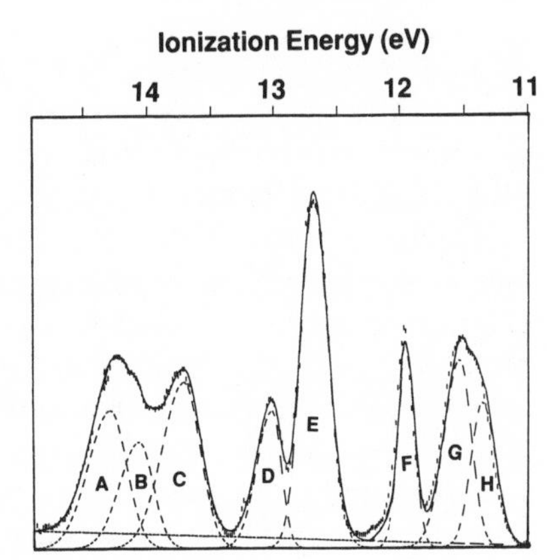

Figure 7. Fitted He I valence spectrum of $NMoCl_3$.

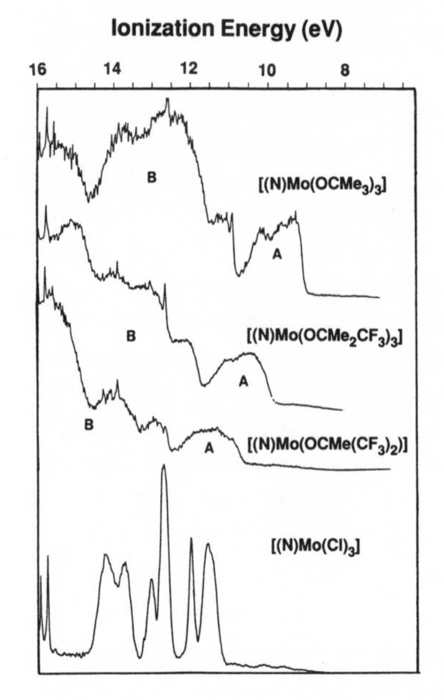

Figure 8. He I spectra of the molybdenum nitrides.

interaction was indicated above for the alkoxide complexes. Band D is assigned to the Mo—N σ-bonding orbital while bands E and F are associated with chlorine lone-pairs which have no appreciable interaction with the unfilled metal levels as indicated by the nearly total collapse of E and F in the He II spectrum. These assignments are further supported by similar assignments for the chlorine lone-pairs in chloroform[30] and trichlorosilane.[31]

The only valence electrons left unaccounted for are those from the Mo—N π and nitrogen lone-pair electrons. Accordingly, the band near 11.5 eV is attributed to these, with bands G and H being assigned to the Mo—N π and nitrogen lone-pair electrons, respectively.

For the related Mo—N alkoxide complexes, $(N)Mo(OCMe_3)_3$, $(N)Mo(OCMe_2CF_3)_3$, and $(Me_3CC)Mo(OCMe(CF_3)_2)_3$, the assignments become increasingly complex because of overlap with alkoxide ionizations. This is expected in comparison to the alkylidyne complexes because of the greater inherent stability of nitrogen atomic orbitals compared to carbon atomic orbitals. Figure 8 shows the full valence spectra of the complexes. The low-valence spectra of these complexes can be divided into regions A and B. Band B is due primarily to ionizations associated with the oxygen lone-pairs of the alkoxide ligands. The assignment for band B is well supported by similar assignments of the oxygen lone-pairs of the free alkoxide ligands.

Region A comprises the oxygen lone-pairs, the Mo—N π and nitrogen lone-pair, and probably the Mo—O σ ionizations as well. Since this band consists of a number of overlapping peaks, the location of the respective ionizations is only possible by considering additional information. Comparison of the nitride spectra to the analogous alkylidyne spectrum, along with the He I and He II spectra of region A of the nitrides, allows reasonable estimates to be made regarding the location of the Mo—N ionizations. In each case, the spectrum of the nitride complex shows an increase in intensity on the leading edge of band A in comparison to the corresponding ionization of the alkylidyne complex, indicating that Mo—N π and/or nitrogen lone-pair ionizations are the first ionizations of the complexes. This is another example where comparison of organometallic and inorganometallic analogs assists interpretation of the spectra.

4. THE CYCLOPROPENYL LIGAND AND AN INORGANOMETALLIC ANALOG, P_3^+

A major feature of organometallic chemistry from its inception is the bonding of aromatic organic rings, most commonly cyclopentadienyl, to transition metal centers. Another notable example in the history of organometallic chemistry is the stabilization of cyclobutadiene by coordination to metals. In recent years, the synthesis and characterization of a number of complexes have been described in which a transition metal fragment is bound in an η^3 fashion to a cyclopropenyl ligand, $C_3R_3^+$.[32–38] An inorganometallic (and isolobal) analog is provided by metal complexes with a P_3^+ ligand. In the following we describe the experimental characterization of the bond between the metal center of a transition metal fragment and cyclopropenyl and cyclophosphopropenyl ligands.

A useful point to begin is with a qualitative description of the metal–ligand bond and then proceed to the experimental photoelectron data available on various cyclopropenyl and cyclophosphopropenyl complexes. Consider a general metal ML_n^- fragment bonded in an η^3 fashion to either a $C_3R_3^+$ or P_3^+ ligand in the coordinate system shown in Figure 9. The choice of formal charges for the fragments is not required, but simplifies the following discussion. For convenience, only the $C_3R_3^+$ ligand interactions will be described. The same conclusions apply to the isolobal cyclophosphopropenyl ligand, where the CR bonds are replaced by phosphorus lone-pairs. The frontier orbitals of the $C_3R_3^+$ ligand comprise an "*a*" and "*e*" set of the p_π orbitals as shown in Figure 9. The filled "*a*" orbital is π-bonding between the carbons in the ring while the empty "*e*" set is antibonding with respect to the ring. The rings are formally considered to be cations to avoid occupying an antibonding orbital in the free ligand. On allowing the cyclopropenyl ligand to interact with the metal fragment, the filled "*a*" and empty "*e*" set will interact with frontier orbitals possessing similar symmetry in the metal fragment. In the coordinate system adopted in Figure 9, the frontier orbitals of the metal

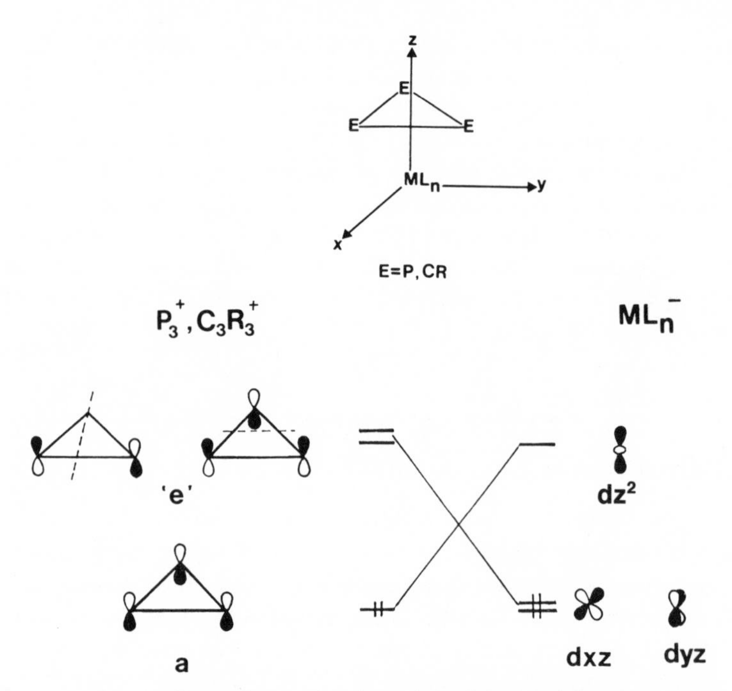

Figure 9. The frontier orbitals of the $C_3R_3^+$ ligand.

fragment of the right symmetry to interact with the frontier orbitals of the cyclopropenyl ligand would comprise an empty orbital of largely d_{z^2} character and two filled orbitals of largely metal d_{xz} and d_{yz} character. Ideally, the filled d_{xz} and d_{yz} orbitals of the metal fragment donate electron density to the acceptor "*e*" set of the cyclopropenyl ligand while the filled "*a*" orbital of the cyclopropenyl ligand donates electron density to the empty d_{z^2} orbital of the metal fragment. The cyclopropenyl ligand acts as a two-electron σ-donating ligand which back-accepts metal electron density in a synergistic bonding interaction according to the Dewar–Chatt–Duncanson model.

The low-valence regions of the He I and He II photoelectron spectra of $(\eta^3\text{-}C_3Bu^t_3)Co(CO)_3$ are shown in Figure 10.[39] Adopting the coordinate system shown in Figure 9 for $(\eta^3\text{-}C_3Bu^t_3)Co(CO)_3$, the metal center in the $Co(CO)_3^-$ fragment is formally d^{10}, with filled d_{z^2}, d_{xz}, and d_{yz} orbitals. The bonding interactions between the cyclopropenyl ligand and the $Co(CO)_3^-$ fragment are therefore expected to be very similar to those discussed with one important exception. The d_{z^2} orbital in the $Co(CO)_3^-$ fragment is filled and so no net bonding interaction results from overlap of this orbital with the filled "*a*" $p\pi$ orbital of the cyclopropenyl ring.

The valence photoelectron spectrum of this complex agrees closely with the qualitative picture of metal-ring bonding presented above. Band A in

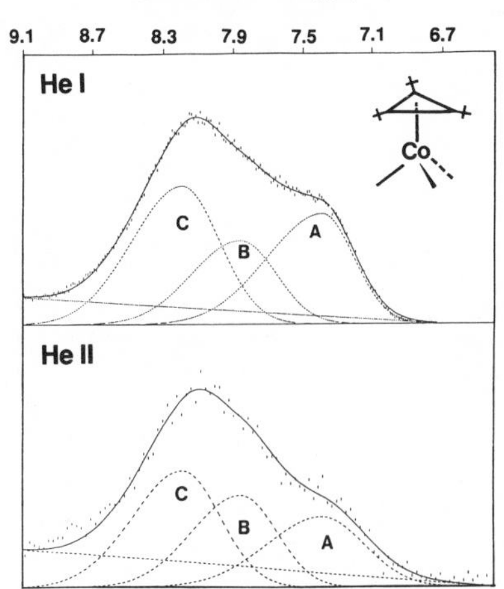

Figure 10. Fitted closeup spectra of the upper valence region for the cyclopropenyl cobalt tricarbonyl complex.

Figure 10 comprises the filled d_{xz} and d_{yz} orbitals donating into the "e" set of the cyclopropenyl ring and destabilized by σ interactions with the other ligands. This interaction forms the primary bonding interaction between the ring and the cobalt center. Band B is mostly d_{z^2}, stabilized by back-bonding to the carbonyl ligands, and somewhat destabilized by interaction with the filled "a" orbital of the cyclopropenyl ring. The remaining metal d-orbitals (degenerate $d_{x^2-y^2}$ and d_{xy} orbitals) are assigned to band C. These are stabilized by back-bonding to the carbonyl π^* orbitals and have no interaction with the ring.

Not surprisingly, the spectrum of the isoelectronic $(\eta^3\text{-}C_3Bu^t_3)Fe(CO)_2NO$ complex is very similar to that of the cobalt complex. The NO^+ ligand, however, being a better π-acceptor than CO, brings about a relative stabilization of the d-orbitals which are suitably oriented to interact with it. A consequence of replacing CO by the NO^+ ligand is a reduction in the overall symmetry of the complex to C_s which results in a splitting of the degenerate "e" sets (d_{xz}, d_{yz} and d_{xy}, $d_{x^2-y^2}$ orbitals, respectively) into four nondegenerate orbitals. The d_{xy} and $d_{x^2-y^2}$ orbitals are not sufficiently different in energy to be resolved in the photoelectron spectrum and thus they comprise the broad asymmetric Gaussian peak D in Figure 11. Bands A and B are assigned to ionizations from metal orbitals of largely d_{xz} and d_{yz} character while band C is assigned to ionization from the largely metal d_{z^2} orbital.

An inorganometallic analog of a metal–cyclopropenyl complex is provided by $(\eta^3\text{-}P_3)W(OCH_2Bu^t)_3NMe_2H$. The valence photoelectron spectrum of $(\eta^3\text{-}P_3)W(OCH_2Bu^t)_3NMe_2H$ in the 7–8 eV region (shown in Figure 12) is very similar to the metal regions of the spectra of $\eta^3\text{-}C_3Bu^t_3)Co(CO)_3$ and $(\eta^3\text{-}C_3Bu^t_3Fe(CO)_2NO$.[40] The ionization potentials at 7.57 and 7.89 eV are

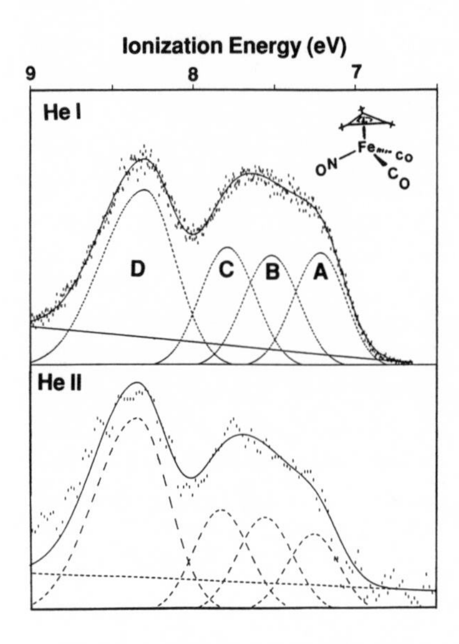

Figure 11. Fitted closeup spectra of the upper valence region for the cyclopropenyl iron dicarbonyl nitrosyl complex.

due to the metal-ring π-bonding orbitals. In this complex, unlike the cyclopropenyl complexes discussed earlier, the d_{z^2} orbital of the $W(OCH_2Bu^t)_3$-NMe_2H^- fragment is unfilled and capable of accepting electron density from the filled "*a*" orbital of the phosphocyclopropenyl ligand. Calculations indicate that this ionization is likely to be contained in band C (of Figure 12), along with the oxygen lone-pairs of the alkoxide ligands.

While the $C_3R_3^+$ and P_3^+ ligand are expected to be similar in many respects, there are some electronic characteristics of the P_3^+ ligand which render it unique. Most important are the phosphorus lone-pairs of the P_3^+ ligand, which opens the door to an interesting question, namely: are the phos-

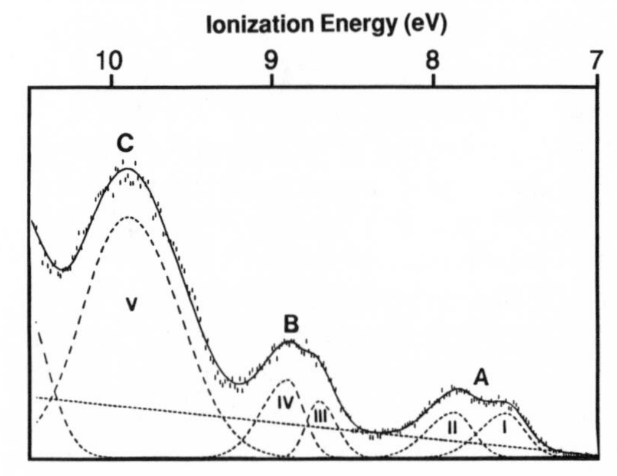

Figure 12. Closeup He I spectrum of $(\eta^3\text{-}P_3)W(OCH_2Bu^t)_3NMe_2H$.

phorus lone-pairs of the P_3^+ ligand capable of acting as two-electron donor ligands to appropriate metal fragments? In general, the lower the ionization potential of the phosphine lone-pair, the better the energy match with the acceptor orbital of a metal fragment. On these terms, the phosphorus lone-pairs of the P_3^+ ligand are seen in band B to compare quite favorably with those of the phosphine ligands. On these grounds, coordination compounds using these lone pairs are favorable provided the geometric constraints can be met.

A further point of interest regarding the P_3^+ ligand involves the nature of the ligating characteristics of the P_3^+ ligand with respect to other commonly encountered isolobal ligands such as the cyclopropenyl, carbyne, and NO^+ ligands. Unfortunately, complexes in which these ligands are bound to a common metal fragment are not yet available. Despite the evident similarity of the $C_3R_3^+$ and P_3^+ ligands, the available complexes are substantially different with regard to the ligand environment and formal oxidation state at the metal center that comparisons become complicated.

One comparison that can be considered is between $(\eta^3\text{-}P_3)W(OCH_2Bu^t)_3$-$NMe_2H$ and $(CCMe_3)Mo(Me_3CO)_3$. The ionization potentials associated with the oxygen lone-pairs in $(\eta^3\text{-}P_3)W(OCH_2Bu^t)_3NMe_2H$ are shifted $\approx$0.5 eV to higher binding energy compared to the corresponding ionizations in $(Me_3CO)_3$-$Mo(CCMe_3)$. This result is surprising for a number of reasons. First, the cyclophosphopropenyl ligand is formally considered to be a cation while the alkylidyne ligand is formally considered to be a -3 anion. This means that the tungsten is formally d^4 in the P_3 complex compared to d^0 in the alkylidyne complex. Furthermore, the "extra" NMe_2H ligand in the P_3 complex may be expected to make the W center even more electron-rich resulting in lower ionization potentials for the oxygen lone-pairs. The opposite is observed. Explanations invoking geometrical effects and/or alkoxide π-donation have been ruled out.

This is simply another example where models of formal charges and oxidation states do not reflect the charge potential at the metal center. Even if the alkylidyne pulls the full -3 electron charge from the metal center, that carbon center is still sufficiently close to the metal that the negative charge significantly destabilizes the valence ionizations of the metal and the oxygen lone-pairs. The P_3+ ligand is apparently a strong π-acceptor and is able to substantially remove electron density a greater distance from the metal center. The photoelectron experiment is able to show the formal electron counts and oxidation states at the atomic centers by the primary characters of the ionizations, and it is able to show the effects of the charge distributions by the energies of the core and valence ionizations.

Comparing the ionization potentials of the metal–ligand π levels of $(\eta^3$-$P_3)W(OCH_2Bu^t)_3NMe_2H$ and $(Me_3OC)_3Mo(OCMe)_3$ further illustrates these effects. These ionizations of the alkylidyne complex are a degenerate set with an ionization potential of 7.89 eV. These ionizations of the P_3 complex are non-degenerate due to the lower symmetry with ionization potentials of 7.57 eV and 7.89 eV. Thus the ionization potentials of these levels in the P_3 complex and the alkylidyne complex are very similar. Because the metal levels are inherently more stable in the P_3 complex as evidenced by the ionizations of the oxygen lone-pairs, the P_3 ligand is less effective than alkylidyne at overlap bonding stabilization of the highest occupied levels.

5. COMPARISON OF METAL–ALKYL BONDS AND METAL–SILYL BONDS

The nature of metal–alkyl bonding has been extensively studied in organometallic chemistry. The strengths of these bonds are important in relation to oxidative addition reactions and carbon–hydrogen bond activation, β-hydride transfer processes, and insertion reactions. Metal–silyl bonds are similarly significant to the inorganometallic chemistry of silicon. Comparison of the characteristics of metal–alkyl and metal–silyl bonds gives a better understanding of each.

A few related metal–alkyl and metal–silyl complexes have been investigated by photoelectron spectroscopy.[41,42] As an example, consider the bonding of the methyl group and different silyl groups to the common metal fragment [$CpFe(CO)_2$].[42] The He I/He II ionization intensity trends of $CpFe(CO)_2CH_3$ are shown in Figure 13. The band at 9.23 eV is assigned to the Fe—C σ ionization. Using the coordinate system described earlier for $CpM(CO)_2L$ systems, the d_{xz} and d_{yz} metal orbitals are π to the methyl ligand. The $d_{x^2-y^2}$ metal orbital has electrons which backbond to the empty carbonyl π^*-orbitals and the ionization arising from this orbital primarily tracks the charge potential at the metal center. From He I/He II intensity trends, the ionizations at 7.8 and 8.0 eV arise due to d_{xz} and d_{yz} metal electrons while

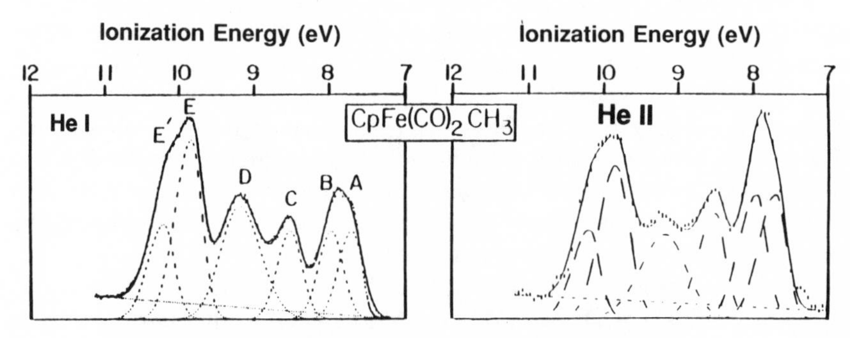

Figure 13. He I/He II intensity trends in $CpFe(CO)_2CH_3$.

the ionization at 8.6 eV is the $d_{x^2-y^2}$ orbital ionization. These metal-based ionizations are followed by the Fe—C σ bond ionization, and at higher energy still are the characteristic Cp ring ionizations.

The difference in ionization energy between the $d_{x^2-y^2}$ metal ionizations and the d_{xz} and d_{yz} metal ionizations gives a measure of the π-accepting ability of the L ligand in comparison to CO in the $CpFe(CO)_2L$ system. In the case of the methyl ligand, this split in energy is 0.6 to 0.8 eV. Thus the methyl ligand, like the NH_3 ligand discussed earlier, is not a π-acceptor ligand in comparison to CO, and may be a weak π-donor ligand.[44,45]

The related $CpFe(CO)_2SiR_3$ complexes (R = Cl, Me) have been utilized as models for catalysts to probe the mechanism of hydrosilation reactions.[46] As such, the electronic structure study of these complexes is very important. A central issue is the strength of the Fe—Si bond. Structural studies have shown that Fe—Si bonds are shorter than usual in these complexes. For instance, the metal–silicon bond distances in $Cp(CO)FeH(SiCl_3)_2$ (2.3 Å), $NEt_4[(CO)_4FeSiCl_3]$ (2.2 Å), and $CpFe(CO)_2SiCl_3$ (2.2 Å) are substantially smaller than the predicted value of an Fe—Si bond of 2.5 Å from single bond covalent radii. The short bond length has been attributed[47] to $d\pi$–$d\pi$ π-bonding involving donation of d-electron density from the metal center to empty Si d-orbitals of the required symmetry. However, it should be remembered that covalent single-bond distances depend on substituent effects, coordination number, oxidation state of the metal, and relative electronegativities in addition to bond multiplicity. Mossbauer and infrared spectroscopy on $CpFe(CO)_2SiR_3$ complexes reveal greater s-electron density at the metal center in the silyl complexes denoting the silyl ligands as good σ-donors. As shown before, the π-acceptor abilities of two ligands can be compared when these ligands are bound to a common fragment. This principle is used here where the common fragment is $CpFe(CO)_2$ and the ligands whose π-acceptor abilities are compared are $SiMe_3$ and CH_3.

The He I and He II photoelectron spectra of $CpFe(CO)_2SiMe_3$ are shown in Figure 14. The ionization at 9.7 eV is assigned to the Cp e_1'' ring ionizations by correlation with the same ionizations of $CpFe(CO)_2CH_3$ (9.9 eV). From He I/He II intensity trends, the ionization at 7.9 eV is assigned to the d^4 metal electrons of the nearly degenerate d_{xz} and d_{yz} orbitals, and the ionization at 8.7 eV is assigned to the d^2 metal electrons of the $d_{x^2-y^2}$ orbital to complete the formal d^6 configuration at the metal center. The ionization at 8.4 eV decreases in relative intensity on going from He I to He II excitation and is assigned to the Fe—Si bond ionization. The substantially lower ionization energy of the Fe—Si bond compared to the ionization energy of the Fe—C bond follows from the lower electronegativity of Si compared to C and the inherently less stable Si atomic orbitals compared to C atomic orbitals. The remaining ionizations at 10.1 and 10.6 eV are assigned to the Si—C σ ionizations.

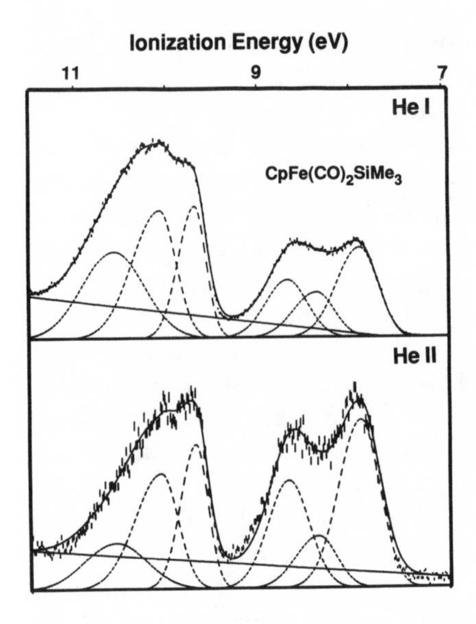

Figure 14. He I/He II intensity trends in $CpFe(CO)_2SiMe_3$.

As mentioned before, the split in ionization energy between the d_{xz}/d_{yz} metal ionizations and the $d_{x^2-y^2}$ metal ionizations reveal the π-acceptor ability of L. When L is CH_3, this separation is 0.6 to 0.8 eV and we know that the methyl ligand is a poor π-acceptor. In the case of L being $Si(CH_3)_3$, this separation is 0.8 eV. This shows that the silyl ligand is also a poor π-acceptor. The $d_{x^2-y^2}$ orbital ionization best tracks the charge potential at the metal center. This ionization occurs at the same position in the methyl and the silyl complexes denoting that the σ-donor ability of the two ligands are similar. Both CH_3 and $Si(CH_3)_3$ are good σ-donating ligands. The methyl ligand and its inorganometallic counterpart, the silyl ligand, are quite similar in their donor and acceptor effects on the valence ionizations.

6. Si—H AND Ge—H BOND ACTIVATION AS COMPARED TO C—H BOND ACTIVATION

In transition-metal chemistry, the activation of C—H bonds and the study of the interaction of the C—H bond with transition metal centers has received much attention.[48–52] In the early stages of the bond activation process, there is donation of electron density from the C—H σ-bonding orbital into the empty metal LUMOs (see Figure 15). This is termed the σ interaction of the C—H bond with the transition metal. This interaction may be accompanied by competing filled–filled orbital interactions (i.e., steric repulsions) which will also play an important role in determining whether activation of the C—H bond will actually take place. For the C—H bond to be actually

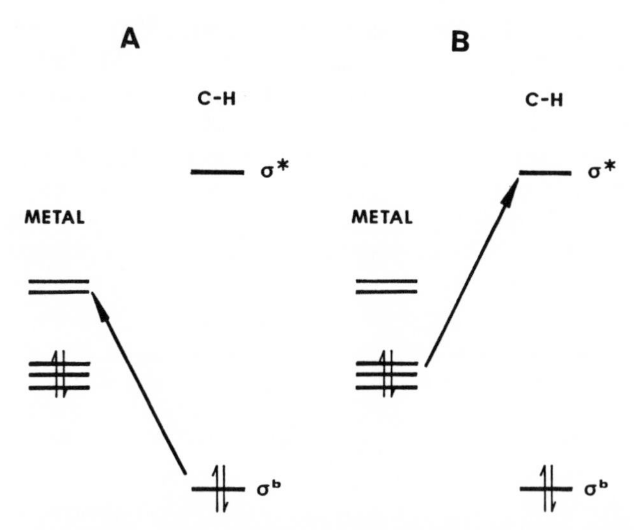

Figure 15. (A) σ activation (donation of electron density from C—H σ bond into empty metal levels). (B) σ^* activation (donation of electron density from filled metal levels into empty C—H σ^* orbital).

broken, the second stage of interaction must take place in which electron density from the highest occupied metal orbitals is donated into the empty C—H σ^*-orbital. This interaction is termed the σ^* interaction. When this interaction reaches its limit, the net result is the collapse of the C—H σ and σ^* levels and there is full oxidative addition resulting in the formation of the alkyl–metal–hydride.

Thus the ability of a particular metal complex to activate these bonds is dependent on its ability to accept electron density from the C—H bond, its ability to donate electron density to the C—H σ^* orbital, the strength of the C—H bond that is broken, and the strengths of the metal–carbon, metal–hydrogen, and other bonds that are formed or altered in the process. These electronic structure factors have been the subject of numerous theoretical investigations. The challenge remains to obtain experimental information relating to the relative energy contributions of each of these different interactions at different stages of the activation process. Photoelectron spectroscopy is directly sensitive to these electronic structure and bonding factors, and is able to provide direct *experimental* information on the relative significance of each. For example, if σ interaction is the predominant electronic structure mechanism of bond activation in a complex, the splitting pattern of the metal orbitals derived from the t_{2g} set is not perturbed (see Figure 15). If σ^* interaction is the predominant electronic structure mechanism, one of the filled t_{2g} metal orbitals will be stabilized as a result of interaction with the C—H σ^* orbital (see Figure 15). The ionization energies are dependent on the electron richness and charge potential at the metal center, and the strengths of individual bonds with the metal center. The characters of the orbital ionizations will also reveal the formal electron count at the metal center and whether oxidative addition has occurred.

Brookhart and Green have contributed much to the understanding of the interaction of carbon-hydrogen bonds with transition metals.[52] They have proposed the term "agostic" to describe molecules where the C—H bond is primarily a two-electron donor to the metal resulting in the formation

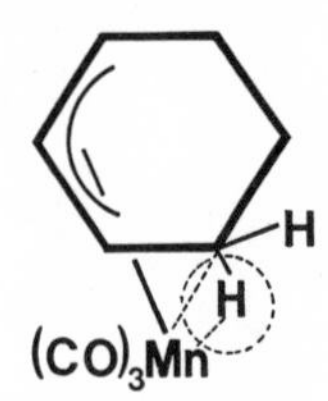

of a Mn—H—C, 3-center–2-electron bond. An example of this kind of agostic interaction is found in cyclohexenyl manganese tricarbonyl, (η^3-C_6H_9)$Mn(CO)_3$.[54] This complex shows a C—H bond length of 1.19 Å and a Mn—H bond length of 1.84 Å, both bonds being lengthened from direct bonds. The photoelectron study of the electronic structure factors of C—H bond activation in (cyclohexenyl)manganese tricarbonyl reveals some interesting information.[55] The photoelectron spectrum of cyclohexenyl manganese tricarbonyl (Figure 16) shows a metal band which is not split.[55] As mentioned above, if σ^* interaction is significant, one of the metal orbitals would be stabilized as a result of interaction with the C—H σ^* orbital. This would break the degeneracy of the metal HOMOs from the pseudo-octahedral electronic symmetry and give rise to a split metal band in the PES. Therefore, it follows that σ interaction is the predominant factor determining the structure and properties of this C—H bond in this complex. The C—H bond should be stabilized as a result of σ interaction, but unfortunately the C—H ioniza-

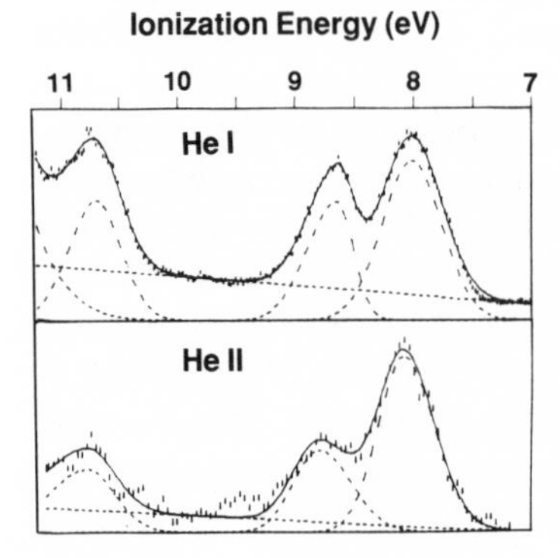

Figure 16. He I/He II intensity trends in (cyclohexenyl)manganese tricarbonyl.

tions are buried under other ionizations in the photoelectron spectrum and are not observed. It was necessary to study other related complexes to gain more information on the factors that determine the electronic interactions.

The Si—H studies are important in relation to C—H activation, because the qualitative features of Si—H activation are largely the same but the specific quantitative contributions are much different. One key factor that is different is the strength of the E—H bond (E = C, Si, Ge) that is broken. Other key factors that are quantitatively different include the strengths of the bonds with the metals that are formed, the electronegativity of silicon in comparison to carbon as seen before, and the ability of the Si—H system to donate and accept electrons.

A remarkable class of complexes which show an interaction of the Si—H, Ge—H, or Sn—H bond with a transition metal center has been discovered and well studied.(56–58) The molecular formula of the complexes are $CpMn(CO)_2HER_3$ where E is Si, Ge, or Sn and R is Ph or Cl. The Si—H, Ge—H, and Sn—H bonds were postulated to interact with the transition metal in an η^2 fashion and thus act as neutral $2e^-$ donors to the Mn center. The bonds are thought to be "activated" and as such are very important as models for catalysts in industrial processes.

The different ligands and substitutions available in this class of complexes provide for a wide range of electron donating and accepting abilities of the metal complex and the E—H bond. Depending on these substitutions, these molecules display stages of activation ranging from very weak coordination of the E—H bond to the metal to essentially complete oxidative addition, as shown in **1** through **4** below. Structures **1** and **4** are limiting descriptions of the bonding prevalent in these complexes. Structure **1** can be described as a Mn(I), d^6 center where the silane is acting as a neutral two-electron donor. The electronic structure mechanism of interaction in this complex is σ interaction of the Si—H bond with the transition metal with no σ^* interaction present. Structure **4** represents a complex which can be described as a full oxidative addition product, resulting in a Mn(III), d^4 complex with direct Mn—Si and Mn—H bonds. Structures **2** and **3** represent intermediates which differ in the degree of 3-center–2-electron bonding and σ^* interaction. Although structural and other methods indicate that the compounds are "activated," there is still controversy about the extent of Si—H bond interaction.

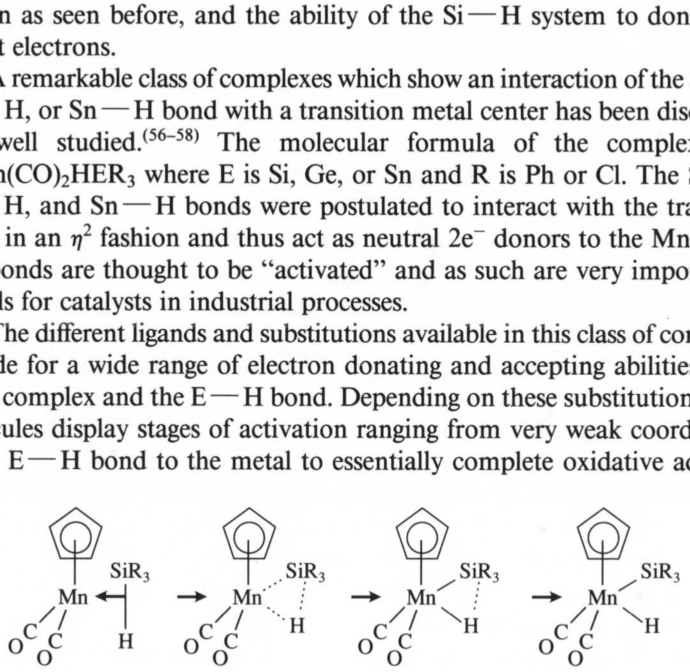

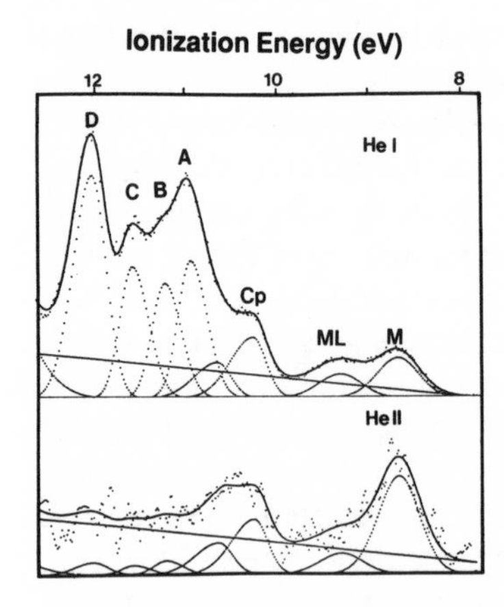

Figure 17. He I/He II comparison of the valence ionizations of $CpMn(CO)_2HSiCl_3$.

The primary question is which of the structures **1** through **4** best represents the ground state of each complex.

The observation of the metal and ligand ionizations in the photoelectron spectra of the $CpMn(CO)_2HSiR_3$ complexes reveal their stages of Si—H bond interaction with the transition metal center. The He I/He II intensity trends in $CpMn(CO)_2HSiCl_3$[31] are shown in Figure 17. The A, B, C, and D ionizations are the Cl lone-pair ionizations. These Cl lone-pair ionizations are destabilized 1 eV from their position in $HSiCl_3$. This shift is indicative of a negative charge potential in the region of $SiCl_3$ when $HSiCl_3$ is coordinated to the metal, consistent with representation of $SiCl_3$ as $SiCl_3^-$ bound to the metal. The ionizations labeled Cp are assigned to the Cp ring π ionizations and are 0.4 eV stabilized from that present in $CpMn(CO)_3$. This shift is indicative of a more positive metal center in $CpMn(CO)_2HSiCl_3$. The ionizations near 8 eV represent the predominantly metal levels of the d^6 complex split in a 2:1 ratio. The higher ionization energy band is labeled ML because it shows considerable ligand character in the He I/He II comparisons. This band is assigned to the Mn—Si bond ionization. The first ionization peak (M) is predominantly metal in character and is assigned to the d^4 electrons. This band is 0.7 eV stabilized from the metal band in $CpMn(CO)_3$, additional evidence of a more positive metal center. From all these observations, $CpMn(CO)_2HSiCl_3$ is concluded to have a formally Mn(III), d^4 metal center and represents stage IV of the bond-breaking process.

To ascertain whether Si substituents play a role in the extent of Si—H bond interaction with the Mn center, valence He I studies were also carried out on $MeCpMn(CO)_2HSiPh_3$.[59] Figure 18 shows the spectral comparison of $MeCpMn(CO)_2HSiPh_3$ with $MeCpMn(CO)_3$ and the free ligand, $HSiPh_3$. The band at 9.1 eV in $MeCpMn(CO)_2HSiPh_3$ is assigned to the phenyl π ionizations while the band at 10.4 eV is assigned to the Si—C σ ionizations. These ionizations occur at the same position in free $HSiPh_3$. This shows that

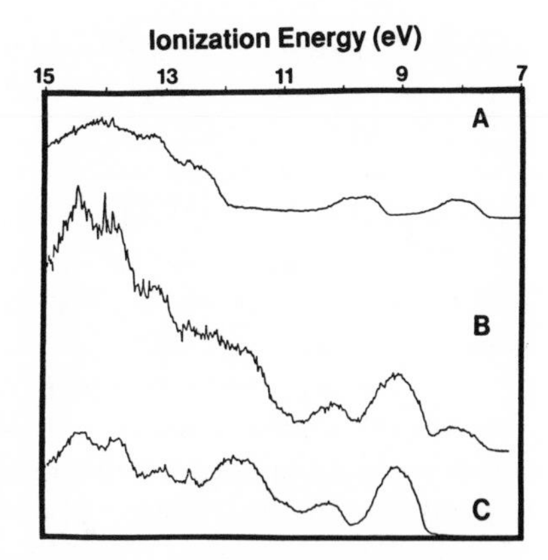

Figure 18. Comparison of the He I photoelectron spectra of (A) $MeCpMn(CO)_3$, (B) $MeCpMn(CO)_2HSiPh_3$, and (C) $HSiPh_3$.

there is not much electron charge shift from the metal to the ligand on complexation, which is different from the results on the $CpMn(CO)_2HSiCl_3$ complex. The first ionization band is the metal band. It can be analytically represented by the fit of three asymmetric Gaussian peaks consistent with the formal assignment of six electrons to the Mn(I) center (see Figure 19). This metal band is not split to a great extent and occurs at the same position as the metal band in $MeCpMn(CO)_3$, signifying that the σ-donor/π-acceptor properties of the silane balance out in comparison to CO. The metal band is not stabilized from the parent tricarbonyl like in $CpMn(CO)_2HSiCl_3$. From these observations it is concluded that $MeCpMn(CO)_2HSiPh_3$ has a formally

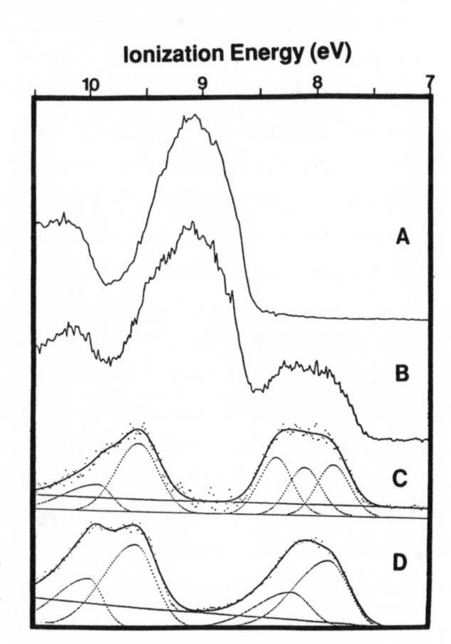

Figure 19. He I spectral comparison of 7 to 10.5 eV region of (A) $HSiPh_3$, (B) $MeCpMn(CO)_2HSiPh_3$, (C) subtracted spectra, and (D) $MeCpMn(CO)_3$.

Mn(I), d^6 center and therefore there is Si—H σ interaction with the metal center. There is some σ^* Si—H interaction with the metal also present but it is not the predominant activation mode. $MeCpMn(CO)_2HSiHPh_2$, CpMn-$(CO)_2HSiHPh_2$, $C_5Me_5Mn(CO)_2HSiHPh_2$, and $MeCpMn(CO)_2HSiFPh_2$ also show similar ionization features in their PES and also have Mn(I), d^6 centers.[60]

Thus, the nature of the Si substituents on the silyls in the $CpMn(CO)_2$-$HSiR_3$ complexes plays a predominant role in the extent of Si—H σ interaction with the Mn center. The electronegative chlorine substituents on the Si drive the complex to complete oxidative addition while the less electronegative Ph substituents leads to a 3-center–2-electron activated complex like (cyclohexenyl)manganese tricarbonyl. We had thought from our C—H activation results that since the Si—H σ^* is lower in energy than the C—H σ^*, the silyl complexes would undergo full oxidative addition. The activation process depends on the electronegativity of substituents as shown.

The $CpMn(CO)(L)HER_3$ (where L is CO or PMe_3; R is Cl or Ph, E is Si, Ge) complexes exhibit a wide range of substitutions and there are subtle changes in the magnitude of Si—H bond interaction with the transition metal with each kind of substitution. In $MeCpMn(CO)(PMe_3)HSiCl_3$, there is a Mn(III), d^4 center while in $MeCpMn(CO)(PMe_3)HSiHPh_2$ there is a Mn(I), d^6 center.[61] Again, in the phosphine-substituted complexes, like their carbonyl counterparts, the Si substituents dictate the magnitude of bond activation. However due to the more electron-rich metal center in the phosphine complexes, the extent of Si—H σ^* interaction is more in $MeCpMn(CO)(PMe_3)$-$HSiHPh_2$ compared to $MeCpMn(CO)_2HSiHPh_2$. The Ph π ionizations in the phosphine complexes are destabilized from their carbonyl analogs, showing there is more negative charge on the silane ligand and that the phosphine complex has advanced to a greater degree of oxidative addition. The greater electron density at the metal center due to σ donation of a phosphine can now be dissipated by backbonding to only one carbonyl. The silane ligand is in a position to act as a π-acceptor and backbonding to the silane increases resulting in the greater σ^* interaction. Also, one of the metal orbitals is 0.1 eV stabilized in the —PMe_3 complex compared to the dicarbonyl complex due to the σ^* interaction.

In $C_5Me_5Mn(CO)_2HSiHPh_2$, the metal is electron-rich because of the methyl groups on the Cp ring and we expected full oxidative addition. Instead, we found that due to steric effects of the bulky Cp ring, Si—H σ^* interaction in this complex is even less than in the Cp analog.[60] In this case, steric factors dominate over electronic factors of bond activation.

Bond activation studies have been extended to germyl complexes. $MeCpMn(CO)_2HGePh_3$ is found to have a Ge—H σ interaction with the transition metal from reaction chemistry. Therefore $CpMn(CO)_2HGePh_3$, $MeCpMn(CO)_2HGePh_3$, and $C_5Me_5Mn(CO)_2HGePh_3$ were studied by photoelectron spectroscopy.[62] Figure 20 shows the valence ionizations of

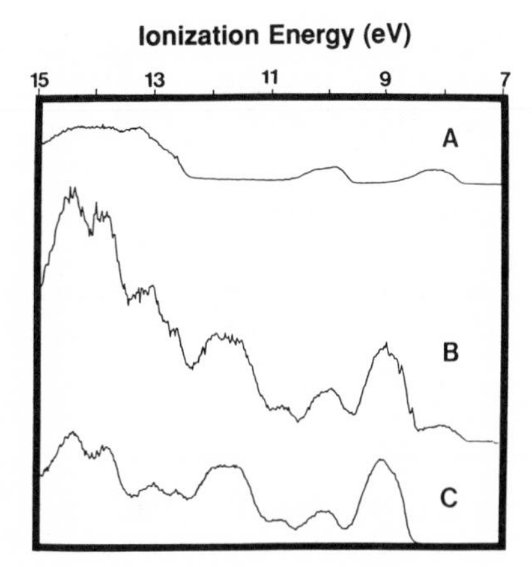

Figure 20. He I valence spectral comparison of (A) $CpMn(CO)_3$, (B) $CpMn(CO)_2HGePh_3$, and (C) $HGePh_3$.

$CpMn(CO)_2HGePh_3$ as compared to $CpMn(CO)_3$ and $HGePh_3$. The Ph π ionizations are unshifted from their position in the ligand denoting no substantial electron charge shift from the metal to the ligand. The metal ionizations are also unshifted from their position in $CpMn(CO)_3$. All these observations suggest a Mn(I), d^6 center in the complex and σ interaction of the Ge—H bond with the Mn center. The MeCp— and C_5Me_5— analogs also display predominantly Ge—H σ interaction with the transition metal. All complexes show Ge—H σ^* interaction with the Mn center but, as in the silyl cases, this is not the predominant mode of activation. The Si—H and Ge—H σ^* interaction was measured in the complexes $MeCpMn(CO)_2HSiPh_3$ and $MeCpMn(CO)_2HGePh_3$ by observing the metal bands. The highest-energy metal orbital in $MeCpMn(CO)_2HGePh_3$ is a little stabilized from that in $MeCpMn(CO)_2HSiPh_3$. This shows that for analogous compounds, Ge—H σ^* interaction is more than Si—H σ^* which is to be expected, since the Ge—H σ^* level is lower in energy than the Si—H σ^* level.

Thus photoelectron spectroscopy can be used to study different stages of C—H, Si—H, and Ge—H bond interactions with transition metals. The "inorganometallic" analogs of C—H bonds show different stages of interaction because of their different electronegativities and bond strengths. The study of all the complexes leads to a more complete understanding of bond activation processes and chemistry. The recognition of these relationships between organometallic and inorganometallic chemistry is an important benefit to both fields.

Acknowledgments

We wish to thank all the former students and associates who have contributed their time and enthusiasm to these investigations. They are listed in the references. Our continuing studies of the electron distributions and bonding of small molecules coordinated to transition metals has been generously funded by the U.S. Department of Energy (Division of Chemical Sciences, Office of Basic Energy Sciences Office of Energy Research, DE-FG02-86ER13501). The National Science Foundation (Grant CHE8519560) and the Materials Characterization Program of the University of Arizona have contributed to the purchase and construction of the equipment used in these studies. A.R.C. acknowledges the Petroleum Research Fund (PRF# 20408-AC3) for support of the studies of Si—H and Ge—H activation by metals.

REFERENCES

1. Elian, M.; Chen, M. L.; Mingos, D. M. P.; Hoffmann, R. *Inorg. Chem.* 1976, **15,** 1148.
2. Hoffmann, R. *Angew. Chem.,* Int. Ed. *Engl.* 1982, **21,** 711.
3. Lichtenberger, D. L.; Kellogg, G. E. *Acc. Chem. Res.* 1987, **20,** 379.
4. Lichtenberger, D. L.; Kellogg, G. E.; Pang, L. S. K. *ACS Symp. Ser.* 1987, 357.
5. Lichtenberger, D. L.; Kellogg, G. E. In "Gas Phase Inorganic Chemistry"; Russell, D. H., Ed; Plenum: New York, 1989.
6. Lichtenberger, D. L.; Johnston, R. L. In "Metal–Metal Bonds and Clusters in Chemistry and Catalysis"; Fackler, J. P., Jr., Ed; Plenum: New York, 1989.
7. Green, J. C. *Struct. Bonding (Berlin)* 1981, **43,** 37.
8. Cowley, A. H. *Prog. Inorg. Chem.* 1979, **26,** 45.
9. Solomon, E. I. *Comments Inorg. Chem.* 1984, **3,** 227.
10. Lichtenberger, D. L.; Copenhaver, A. S. *ACS Symp. Ser.,* in press.
11. Calabro, D. C.; Hubbard, J. L.; Blevins, C. H., II; Campbell, A. C.; Lichtenberger, D. L. *J. Am. Chem. Soc.* 1981, **103,** 6839.
12. Lichtenberger, D. L.; Darsey, G. P.; Kellogg, G. E.; Sanner, R. D.; Young, V. G.; Clark, J. R. *J. Am. Chem. Soc.* 1989, **111,** 5019.
13. Healy, M. D.; Barron, A. R.; Lichtenberger, D. L.; Hogan, R. H. *J. Am. Chem. Soc.* 1990, **112,** 3369.
14. Bancroft, G. M.; Bailey-Dignard, L.; Puddephatt, R. J. *Inorg. Chem.,* 1986, **25,** 3675.
15. Lichtenberger, D. L.; Copenhaver, A. S. *J. Electron. Spectrosc. Relat. Phenom.,* 1990, **50,** 335.
16. Hall, M. B. *J. Am. Chem. Soc.* 1975, **97,** 2057.
17. Schilling, B. E. R.; Hoffmann, R.; Lichtenberger, D. L. *J. Am. Chem. Soc.* 1979, **101,** 585.
18. Lichtenberger, D. L.; Fenske, R. F. *J. Organomet. Chem.* 1976, **117,** 253.
19. Kellogg, G. E. *Diss. Abstr. Int., B* 1986, **46,** 3838.
20. Campbell, A. C. Master of Science Thesis, University of Arizona 1979.
21. Blevins, C. H. II. *Diss. Abstr. Int., B* 1984, **45,** 1186.
22. Schrock, R. R. *Acc. Chem. Res.* 1986, **19,** 342.
23. Schrock, R. R.; Rocklage, S.; Wengrovius, J.; Rupprecht, G.; Fellmann, J. *J. Mol. Catal.* 1980, **8,** 73.
24. Wengrovius, J.; Schrock, R. R.; Churchill, M. R.; Missert, J. R.; Youngs, W. J. *J. Am. Chem. Soc.* 1980, **102,** 4515.
25. Kress, J.; Wesolek, M.; Le Ny, J. P.; Osborn, J. A. *J. Chem. Soc., Chem. Commun.* 1981, 1039.
26. Churchill, M. R.; Ziller, J. W.; Freudenberger, J. H.; Schrock, R. R. *Organometallics* 1984, **3,** 1554.
27. Freudenberger, J. H.; Schrock, R. R.; Churchill, M. R.; Rheingold, A. L.; Ziller, J. W. *Organometallics* 1984, **3,** 1563.

28. Hoppe, M. L. Ph.D. Dissertation, University of Arizona, 1988.
29. Bloch, T. F.; Fenske, R. F. *J. Am. Chem. Soc.* 1977, **99,** 4321.
30. Turner, D. W.; Baker, C.; Baker, A. D.; Brundle, C. R. "Molecular Photoelectron Spectroscopy"; Wiley: New York, 1970; p. 346.
31. Lichtenberger, D. L.; Rai-Chaudhuri, A. *J. Am. Chem. Soc.* 1989, **111,** 3583.
32. Churchill, M. R.; Ziller, J. W.; McCullough, L.; Pederson, S. F.; Schrock, R. R. *Organometallics* 1983, **2,** 2046.
33. Feltham, R. D.; Brant, P. *J. Am. Chem. Soc.* 1982, **104,** 641.
34. Hughes, R. P.; Lambert, J. M. J.; Hubbard, J. L. *Organometallics* 1986, **5,** 797.
35. Drew, M. G. B.; Brisdon, B. J.; Day, A. *J. Chem. Soc., Dalton. Trans.* 1981, 1310.
36. Rausch, M. D.; Tuggle, R. M.; Weaver, D. L. *J. Am. Chem. Soc.* 1970, **92,** 4981.
37. Olander, W. K.; Brown, T. L. *J. Am. Chem. Soc.* 1972, **94,** 2139.
38. Schneider, M.; Weiss, E. *J. Organomet. Chem.* 1976, **121,** 345.
39. We thank Prof. R. P. Hughes for supplying the samples of the cyclopropenyl complexes.
40. Hogan, R. H. Ph.D. Dissertation, University of Arizona, 1990.
41. Louwen, J. N.; Andrea, R. R.; Stufkens, D. J.; Oskam, A. Z. *Naturforsch.* 1982, **38b,** 194.
42. Cradock, S.; Ebsworth, E. A. V.; Robertson, A. *J. Chem. Soc., Dalton Trans.* 1973, **22,** 22.
43. Lichtenberger, D. L.; Rai-Chaudhuri, A. *J. Am. Chem. Soc.,* 1991, **113,** 2923.
44. Zeigler, T.; Tschinke, V.; Becke, A. *J. Am. Chem. Soc.* 1987, **109,** 1351.
45. Hall, M. B.; Williamson, R. L. *J. Am. Chem. Soc.* 1988, **110,** 4428.
46. Randolph, C. L.; Wrighton, M. S. *J. Am. Chem. Soc.* 1986, **108,** 3366.
47. Tilley, D. T. Unpublished review.
48. Crabtree, R. H. *Chem. Rev.* 1985, **85,** 245.
49. Periana, R. A.; Bergman, R. G. *J. Am. Chem. Soc.* 1986, **108,** 7346.
50. Wenzel, T. T.; Bergman, R. G. *J. Am. Chem. Soc.* 1986, **108,** 4856.
51. Halpern, J. *Inorg. Chim. Acta* 1985, **100,** 41.
52. Jones, W. D.; Maguire, J. A. *Organometallics* 1986, **5,** 590.
53. Brookhart, M.; Green, M. L. H. *J. Organomet. Chem.* 1983, **250,** 395.
54. Brookhart, M.; Lamanna, W.; Humphrey, M. B. *J. Am. Chem. Soc.* 1982, **104,** 2117.
55. Lichtenberger, D. L.; Kellogg, G. E. *J. Am. Chem. Soc.* 1986, **108,** 2560.
56. Jetz, W.; Graham, W. A. G. *Inorg. Chem.* 1971, **10,** 4.
57. Schubert, U.; Scholz, G.; Muller, J.; Ackermann, K.; Worle, B. *J. Organomet. Chem.* 1986, **306,** 303.
58. Colomer, E.; Corriu, R. J. P.; Vioux, A. *Inorg. Chem.* 1979, **18,** 695.
59. Lichtenberger, D. L.; Rai-Chaudhuri, A. *J. Am. Chem. Soc.* 1990, **112,** 2492.
60. Lichtenberger, D. L.; Rai-Chaudhuri, A. *Organometallics* 1990, **9,** 1686.
61. Lichtenberger, D. L.; Rai-Chaudhuri, A. *Inorg. Chem.* 1990, **29,** 975.
62. Lichtenberger, D. L.; Rai-Chaudhuri, A. *J. Chem. Soc., Dalton. Trans.* 1990, 2161.

6

Transition Metal-Promoted Reactions of Main Group Species and Main Group-Promoted Reactions of Transition Metal Species

Russell N. Grimes

1. TRANSITION METAL-PROMOTED REACTIONS OF MAIN GROUP SPECIES: INTRODUCTION

Given the wide application of transition metal reagents to organic chemistry, it is not surprising to find a broad range of main group element reactions which are mediated by transition metals. It is often useful to compare and contrast such processes with transition metal–hydrocarbon interactions, but there are important differences and the analogy must not be carried too far. Indeed, on surveying the range of known metal-facilitated transformations of main-group species, it is apparent that many such reactions are in fact mechanistically distinct from organometallic processes (and from each other) and frequently generate structurally dissimilar products. Not surprisingly, main group–transition metal interactions tend to be characterized by special features endemic to the specific element or compound type, defying broad generalizations.

This section will survey several types of main group element reactions

Russell N. Grimes • Department of Chemistry, University of Virginia, Charlottesville, Virginia 22901.

Inorganometallic Chemistry, edited by Thomas P. Fehlner. Plenum Press, New York, 1992.

which are promoted by transition metal compounds, in order to illustrate both the breadth and diversity of this chemistry. Certain of these reactions will be discussed in some depth, and parallels with organotransition metal chemistry will be drawn where possible; however, comprehensive coverage of this subject is clearly not possible in a chapter-length review and consequently some reaction types have been omitted which might well have been included. Boron compounds are prominently featured in the discussion, primarily because the extraordinarily rich area of boron chemistry has numerous examples of transition metal promoted reactions which have been examined in some detail.

2. PROMOTION OF X—X AND X—Y BOND FORMATION (X,Y = P-BLOCK ELEMENTS OTHER THAN CARBON)

2.1. Dimerization and Oligomerization

The ability of transition metal reagents to facilitate carbon–carbon and carbon–oxygen bond formation under mild conditions is well documented.[1] In main group chemistry, oxidative coupling is often promoted by transition metals although an M—X bonded intermediate is seldom isolated. Examples can be found throughout main group chemistry, a few of which will serve as illustrations.[2]

Formation of dinitrogen tetrafluoride:

$$2NF_2H \xrightarrow[H^+]{FeCl_3(aq)} N_2F_4$$

Formation of nitrous oxide from hydroxyammonium ion:

$$4Fe^{3+} + 2NH_3OH^+ \rightleftharpoons 4Fe^{2+} + N_2O + 6H^+ + H_2O$$

Formation of dithionate ion:

$$MnO_2 + 2SO_3^{2-} + 4H^+ \rightleftharpoons Mn^{2+} + S_2O_6^{2-} + 2H_2O$$

An example of a transition metal-facilitated dimerization, which offers a clear analogy with hydrocarbon–metal chemistry, is the conversion of $BH_3 \cdot L$ complexes to metalladiborane species containing an olefin-like $B_2H_5^-$ unit,[3] as in the reactions

$$K_2[M(CO)_4] + 3THF \cdot BH_3 \rightarrow K[M(CO)_4(\eta^2\text{-}B_2H_5)] + K[BH_4] + 3THF$$

$$M = Fe, Ru, Os$$

$$K[M'Cp(CO)_2] + 3Me_2O \cdot BH_3 \rightarrow M'Cp(CO)_2(\eta^2\text{-}B_2H_5) + K[BH_4] + 3Me_2O$$

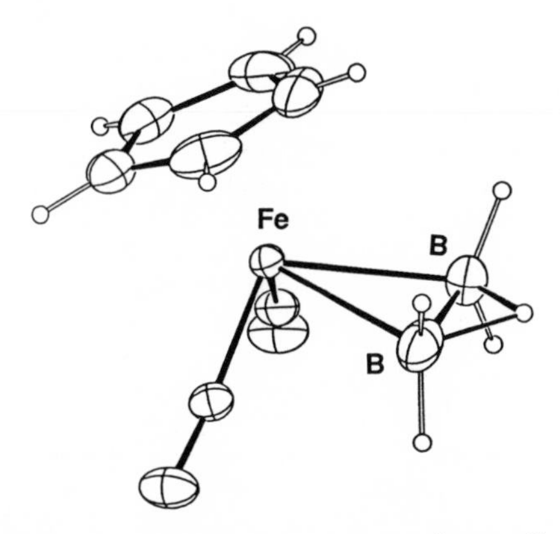

Figure 1. Structure of $FeCp(CO)_2(\eta^2\text{-}B_2H_5)$.[3]

The structure of the latter complex (M′ = Fe), shown in Figure 1, together with other evidence supports a 3-center B—Fe—B 3-center bonding interaction.[3]

A related case of metal-promoted borane dimerization is the conversion of borohydride ion to complexes incorporating the $B_2H_6^{2-}$ moiety (Figure 2).[4] The structure of complex **2** (X = Br) has been crystallographically confirmed.[4] In view of the isoelectronic relationship between BH_4^- and CH_4, this reaction has been proposed as a model for dehydrodimerization of CH_4 to C_2H_6. It is noteworthy that the $B_2H_6^{2-}$ ligand has no known existence as a free ligand, so that the tantalum in this case not only promotes B—B linkage but also stabilizes the dimerized borane structure.

The platinum-induced dehydrocoupling of boron hydrides, discovered by Sneddon and co-workers,[5] is a good example of the application of metal catalysis in borane synthesis. As shown in Figure 3, pentaborane(9) in the presence of $PtBr_2$ powder generates 1,2′-$(B_5H_8)_2$ cleanly, albeit slowly at 25 °C; neither of the other possible isomers (1,1′ or 2,2′) is formed here, although both are obtainable via other methods.[6] Tetraborane(10) can be similarly dimerized to 1,1′-$(B_4H_9)_2$, and the tetraborane and pentaborane units can be coupled to give 1,2-(B_4H_9)-(B_5H_8) (Figure 3). Again, these are the only isomers formed.

These observations are consistent with the mechanism shown in Figure 4, which has been proposed[5b] for the B_5H_9 coupling. The initial step is electrophilic attack of Pt at the apex BH to form a B_5H_8-1-$PtBr_2$ intermediate, which is suggested to undergo oxidative addition at a basal B—H bond on a second pentaborane molecule; the intermediate thus formed then undergoes reductive elimination of HPtBr to give the dimer, the structure of which has been confirmed by X-ray crystallography.[7]

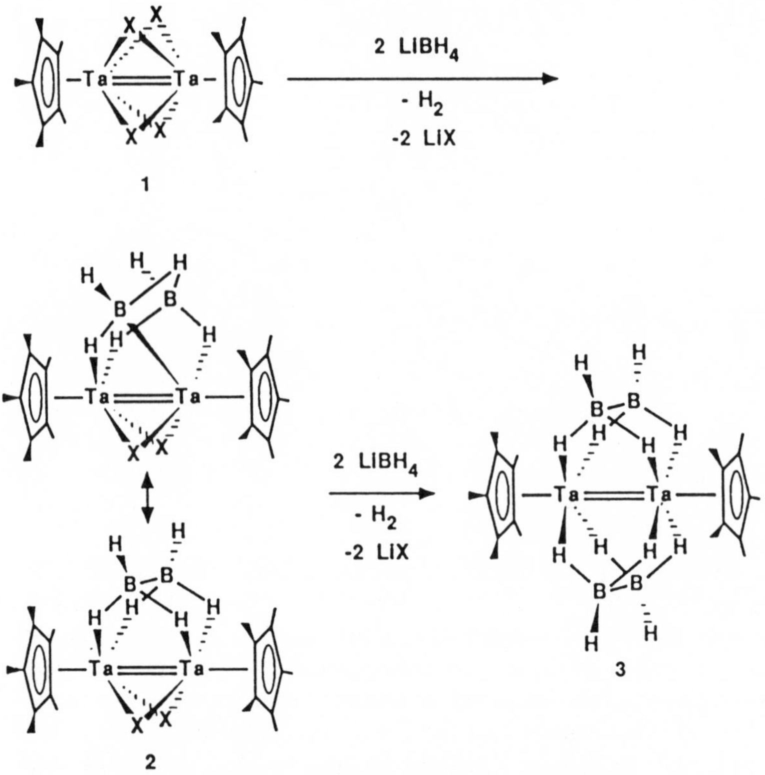

Figure 2. Synthesis of ditantalum complexes of $B_2H_6^{2-}$.[4]

Platinum(II) bromide also catalyzes the B—B coupling of small carboranes,[8] affording quantitative yields at room temperature (Figure 5). Although the borane and carborane dehydrocoupling reactions are slow (on the order of a few catalyst turnovers per day), they do appear to represent a distinct synthetic advance over other routes (e.g., thermolysis) to B—B linked boron clusters. The coupled species in some cases can serve as precursors to larger clusters, as discussed in Section 2.5.2 below.

Transition metal-promoted oxidative linkage of $B_5H_8^-$ ions has also been observed[9] to occur together with oxidative fusion (*vide infra*), generating 2,2'-$(B_5H_8)_2$ in contrast to the 1,2' isomer obtained with $PtBr_2$. In this case the metal salt, $FeCl_2$, functions as a stoichiometric rather than a catalytic agent.

2.2. Addition of Alkynes and Coupling of Alkenes to Boron Substrates

Wilczynski and Sneddon[10] have found that transition metal catalysts such as $Ir(CO)Cl(PPh_3)_2$ and $(RC{\equiv}CR')Co_2(CO)_6$ facilitate reactions of acet-

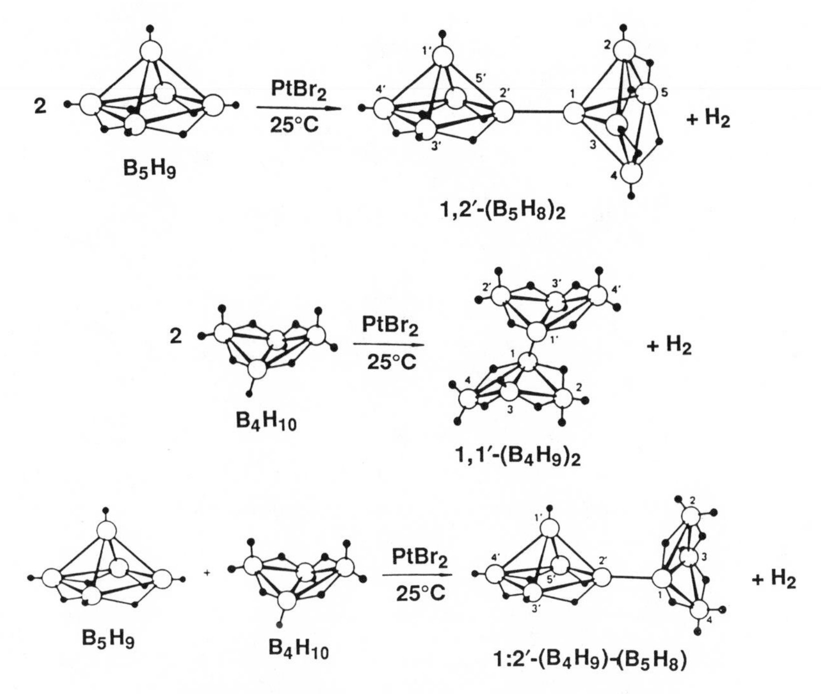

Figure 3. Dehydrocoupling reactions of boranes.[5b]

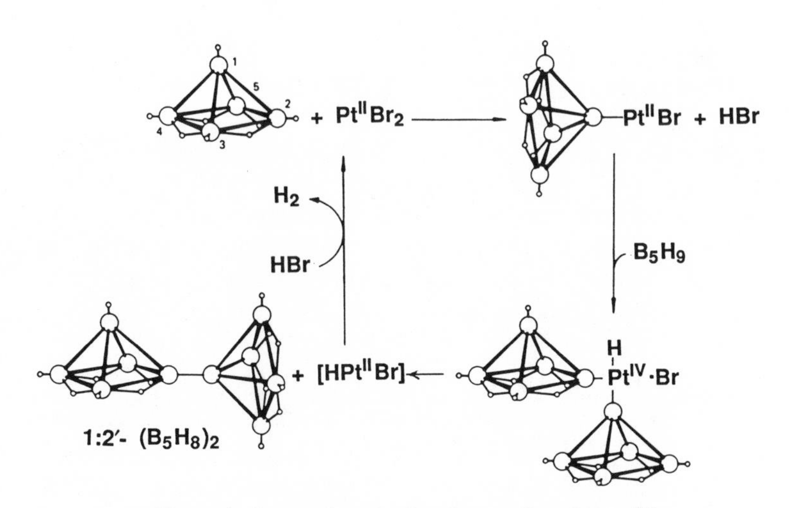

Figure 4. Proposed mechanism for coupling of B_5H_9.[5b]

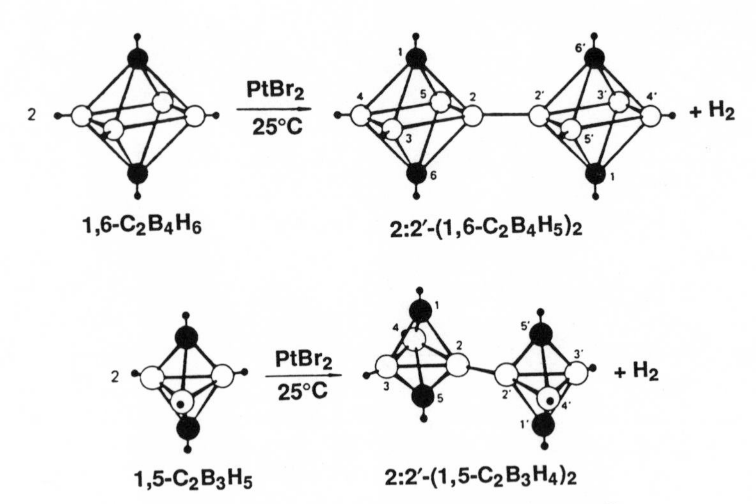

Figure 5. Dehydrocoupling reactions of carboranes.[8]

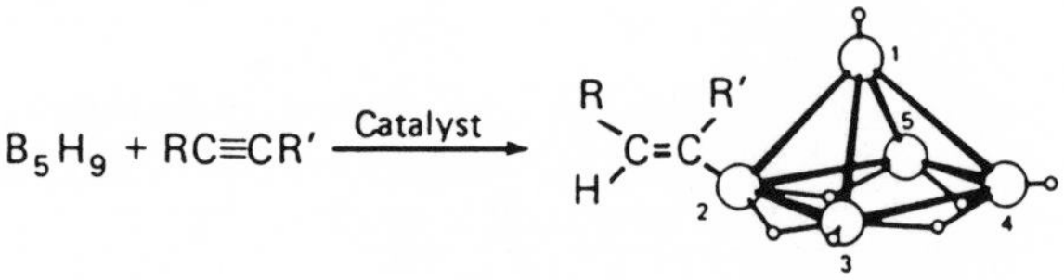

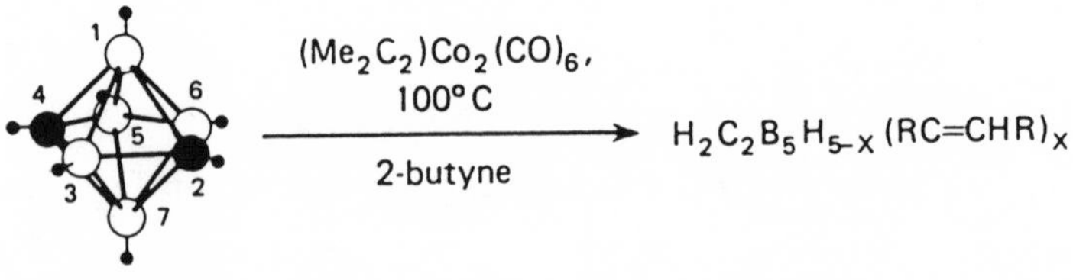

$B_5H_9 + RC{\equiv}CR' \xrightarrow{\text{Catalyst}}$ R(H)C=C(R′)–(B$_5$H$_8$)

2,4-$C_2B_5H_7$ $\xrightarrow[\text{2-butyne}]{(Me_2C_2)Co_2(CO)_6,\ 100°C}$ $H_2C_2B_5H_{5-x}(RC{=}CHR)_x$

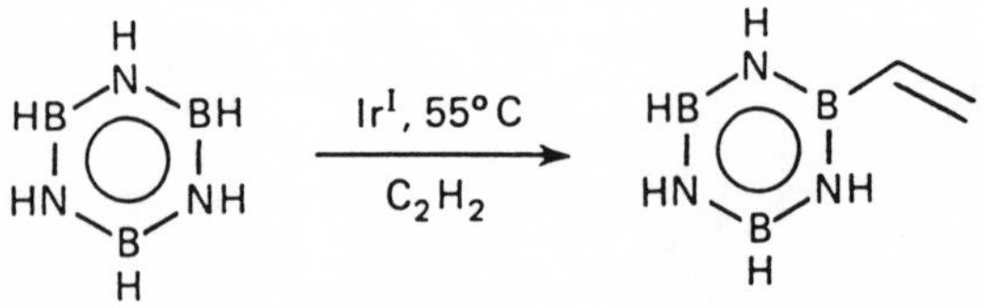

$B_3N_3H_6$ $\xrightarrow[C_2H_2]{Ir^{I},\ 55°C}$

Figure 6. Catalytic addition of alkynes to B—H bonds.[10]

ylenes with boron hydrides, carboranes, and borazines to give B-alkenyl products (Figure 6). This type of metal-catalyzed process strongly resembles the well-known hydrogenation reactions promoted by these catalysts, and a similar mechanistic scheme has been proposed (Figure 7).[5b] In the example shown, a triphenylphosphine group is replaced by an alkyne with subsequent oxidative addition at a basal B—H bond on pentaborane, insertion of the alkyne into the Ir—H and Ir—B bonds, and reductive elimination to give the alkenylboranes and regeneration of the catalyst. Only basal-substituted products are obtained in this case, an observation which is consistent with the reaction[11] of *trans*-Ir(CO)Cl$(PMe_3)_2$ and B_5H_9 to give only 2-[IrH(CO)Cl$(PMe_3)_2$]-B_5H_8, indicating that oxidative addition on pentaborane(9) takes place only at the basal borons.

A different type of transition metal catalyst, originally developed by Maitlis for use in organic synthesis,[12] contains Cp*Rh or Cp*Ir units (Cp* = C_5Me_5) and is useful in olefin hydrogenation and C—H bond activation.[13] Applied to boranes, these catalysts are reported[5b] to be the most active yet found, for example, promoting the addition of 1-butyne to B_5H_9 to generate 2-(*trans*-1-butenyl)-B_5H_8 efficiently. As compared to the iridium and rhodium carbonyl catalysts discussed above, the Cp*–metal compounds have the advantages that they do not release bases such as phosphines (which can react with the boron substrates and generate undesirable side products), and also are capable of effecting alkyne addition to large carboranes such as 1,2- and 1,7-$C_2B_{10}H_{12}$.[5b]

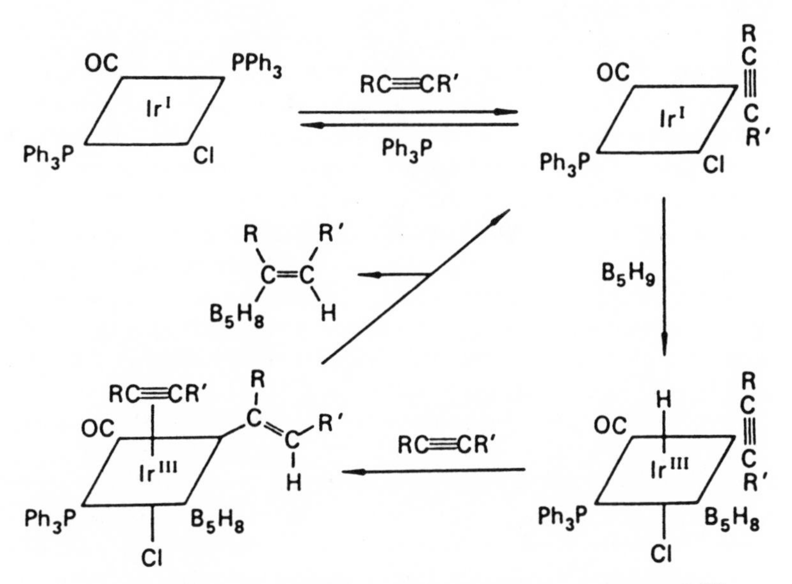

Figure 7. Proposed mechanism for alkyne addition to pentaborane(9).[5b]

Palladium salts have also been employed in the catalysis of olefin addition to boron hydrides. For example, propylene and B_5H_9 in the presence of $PdBr_2$ at 0 °C combine to form an isomeric mixture of propenylpentaboranes in high yield.[14] Similar olefin coupling to the small *nido*-carborane $Et_2C_2B_4H_6$ generates alkenyl-substituted derivatives (Figure 8).[5b] In contrast to arene–olefin couplings, these reactions do not require an additional oxidant. They are, however, nonregiospecific and produce isomer mixtures, which can present problems in separation and purification.

2.3. Acetylene Insertions into Carboranes

A particularly intriguing application of transition metal-promoted synthesis in boron chemistry is the incorporation of dicarbon alkyne units into two-carbon carborane cages to form four-carbon carboranes.[15,16] This type of reaction can be viewed as a cluster expansion and hence is related, at least conceptually, to the processes discussed in the following section; however, it represents a direct extension of the alkyne and alkene couplings just described and therefore will be outlined here. Several reagents including $NiCl_2$ and $Ru_3(CO)_{12}$ promote the interaction of $Et_2C_2B_4H_6$ with MeC≡CMe to form the four-carbon carborane $Et_2Me_2C_4B_4H_4$.[16] The nickel reaction is stoichiometric, but that employing triruthenium dodecacarbonyl is catalytic and can be viewed as analogous to reactions of trialkylsilanes with olefins to form alkenylsilanes, which are catalyzed by iron or ruthenium carbonyls.[16] A possible mechanism involves a proposed alkyne–nickel–carborane intermediate, as indicated in Figure 9.

The alkyne insertion can also be accomplished by a different route in which the metallovinylcarborane complex depicted in Figure 10 was prepared and subsequently thermolyzed to generate the C_4B_4 cage. The formation of the complex closely parallels organometallic reactions such as the activation of η^2-alkynes by ionic nucleophiles to give metallovinyl compounds, as studied, for example, by Reger *et al.*[17] The complex itself is of interest as the first such species in carborane chemistry, and has been shown by X-ray crystallography to have a structure in which one of the vinyl carbons bridges a B—B edge on the cage.[16] The conversion of this species to the $Et_2Me_2C_4B_4H_4$ product was proposed by Mirabelli and Sneddon[16] to take place via β-hydride elimination, a process which appears to be related to similar reactions of (η^1-2-butenyl)Ir(CO)L_2 studied by Schwartz *et al.*[18]

Somewhat surprisingly, no examples of metal-promoted alkyne insertion into binary boron hydride cages have been reported, despite the fact that thermal and base-promoted incorporation of C≡C units into boron hydrides are well known and, indeed, constitute the primary method for synthesis of carboranes.[19] (Alkenylboranes do undergo one-carbon thermal insertions to generate monocarbon carboranes, but no metals are involved.[10] There are,

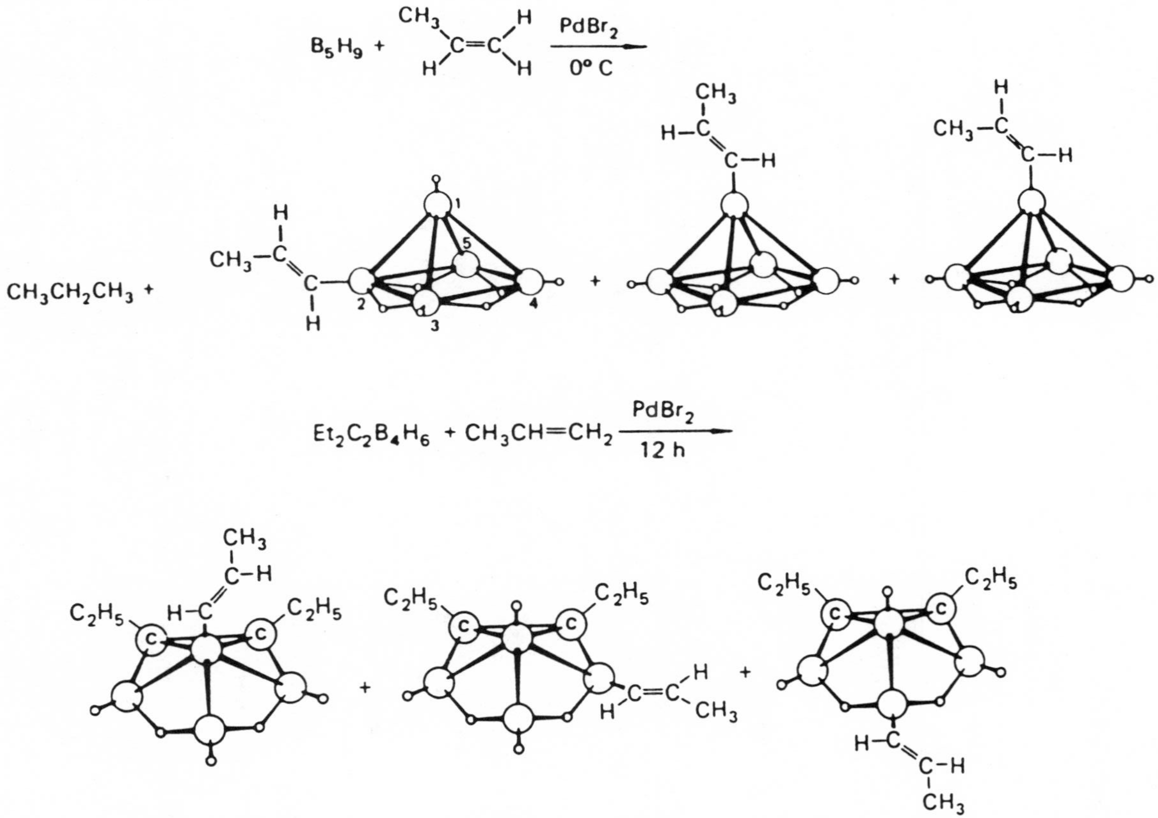

Figure 8. Palladium(II) bromide-catalyzed addition of propylene to B_5H_9 and to $Et_2C_2B_4H_6$.[14]

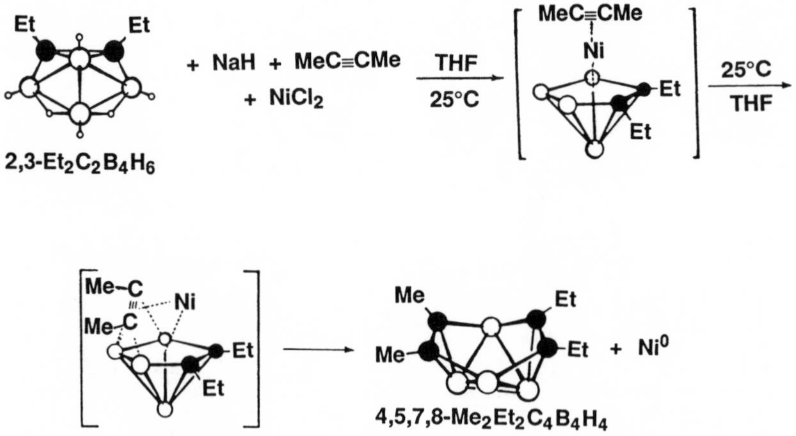

Figure 9. Proposed mechanism for the $NiCl_2$-promoted synthesis of $Me_2Et_2C_4B_4H_4$.[16]

however, examples of carbon incorporation into metallaborane frameworks to form metallacarboranes. Reactions of *nido*-2-$CpCoB_4H_8$ (an analog of B_5H_9) with alkynes generate pentagonal pyramidal *nido*-1,2,3-$CpCoRR'C_2B_3H_5$ complexes,[20] and the ferraborane *nido*-1-$(CO)_3FeB_4H_8$ (another B_5H_9 analog) interacts photolytically with Me—C≡C—Me to produce the ferracarborane $(CO)_3FeMe_4C_4B_4H_4$, which decomposes to give the carborane $Me_4C_4B_4H_4$.[21] Partial incorporation of a $C_5H_5^-$ ring into a cobalt–boron cage was observed in the reaction of $C_5H_5^-$, $CoCl_2$, and $B_5H_8^-$, from which the triple-decker complex μ-C_3H_4-$Cp_2Co_2C_2B_3H_3$ was isolated as a minor product.[22] In these cases the metal undoubtedly plays a role in the carbon insertion, although the mechanisms have not been established.

2.4. Cyclization Reactions

2.4.1. Boron–Nitrogen Rings

A class of reaction that can be viewed as analogous to organometallic cycloaddition is the conversion of iminoboranes to diazaboretidines. Processes of this type are catalyzed by transition metal reagents such as cymantrene ($CpMn(CO)_3$), but the most effective catalyst is reported to be *tert*-butylisonitrile.[23] An example of a metal-facilitated cyclodimerization[24] is shown in Figure 11a. Although the η^4-diazaboretidine ligand in the product is isoelectronic with square-planar R_4C_4, it is nonplanar owing to the different covalent radii of boron and nitrogen.[24] A related iminoborane–metal reaction[25] is the addition across the Co—Co bond in $Co_2(CO)_8$ to produce a Co_2BN cluster (Figure 11b).

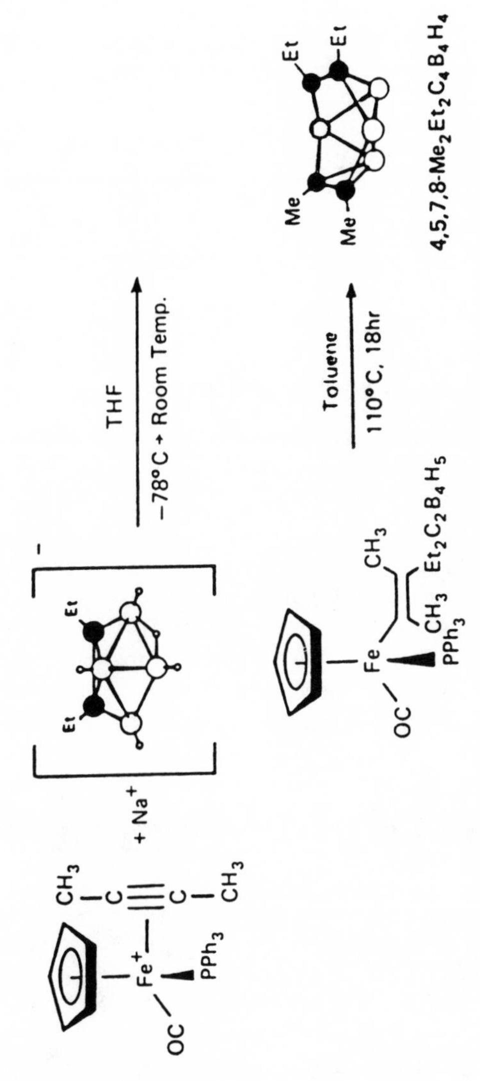

Figure 10. Synthesis of $Me_2Et_2C_4B_4H_4$ via an isolable metallovinylcarborane intermediate.[16]

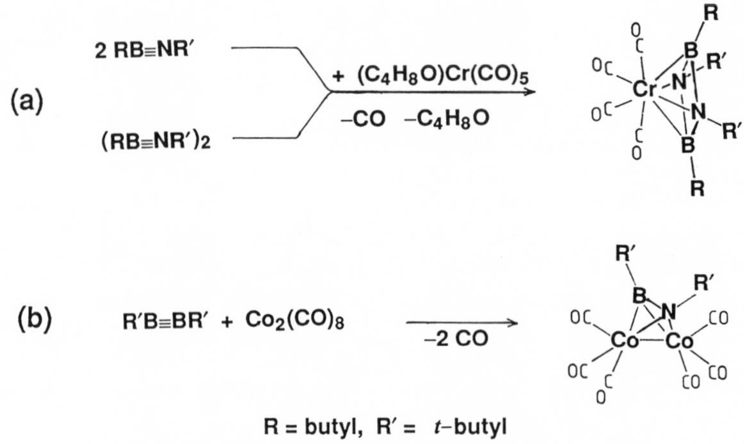

(a) 2 RB≡NR′ / $(RB{\equiv}NR')_2$ $\xrightarrow[-CO\ \ -C_4H_8O]{+\ (C_4H_8O)Cr(CO)_5}$ $(CO)_4Cr(RBNR')_2$

(b) R′B≡BR′ + $Co_2(CO)_8$ $\xrightarrow[-2\ CO]{}$ $(CO)_6Co_2(R'BNR')$

R = butyl, R′ = *t*-butyl

Figure 11. Metal-promoted cyclization of iminoboranes.[23a]

2.4.2. Polyphosphorus and Polyarsenic Rings

Among the more interesting examples of transition metal-stabilized main group cycles are the complexes of planar P_n or As_n rings[26] which are isoelectronic (in terms of valence electrons) with the corresponding cyclic planar C_nH_n molecules. While most such Group V(15) rings are not stable as free species (an exception[27] is P^-5), they are known as bridging ligands in triple-decker complexes, as for example $Cp^*Mo(P_6)MoCp^*$ (Figure 12a), which was prepared[28a] together with other products from $Cp^*(CO)_2Mo{\equiv}Mo(CO)_2Cp^*$ and white phosphorus at 140 °C. Analogous ditungsten and divanadium P_6 triple-deckers have also been synthesized.[28b] The hexaphosphabenzene ring in each of these complexes is a planar hexagon, isoelectronic with benzene.

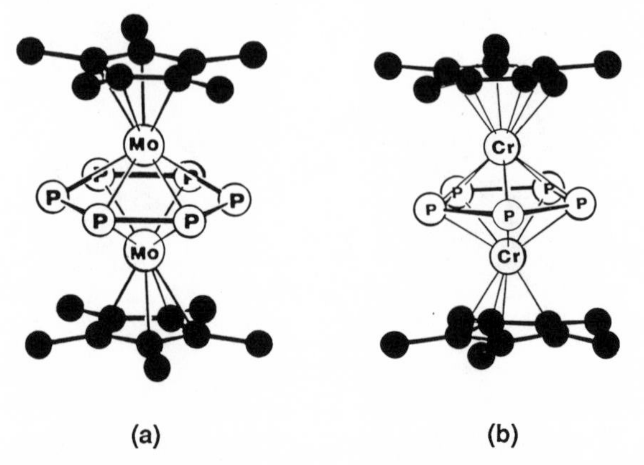

Figure 12. Structures of triple-decker complexes containing P_6 and P_5 bridging rings.[28a,28e]

The dimolybdenum species has been shown to form[(28a)] via net addition of a P_2 unit into $Cp^*(CO)MoP_4Mo(CO)Cp^*$, a process which is reminiscent of the metal-catalyzed synthesis of benzene from acetylene. The corresponding hexaarsabenzene sandwich $(C_5Me_4R)Mo(As_6)Mo(C_5Me_4R)$ (R = Me, Et) has been similarly prepared and structurally characterized.[(28c)] The P_6 and As_6 complexes are main-group counterparts of the benzene-bridged triple-decker sandwich[(28d)] $CpV(C_6H_6)VCp$, providing a further structural connection between inorganometallic and organometallic chemistry.

The formation of the P_5-bridged triple decker, $Cp^*Cr(P_5)CrCp^*$ (Figure 12b), via the interaction of $Cp^*(CO)_2Cr{\equiv}Cr(CO)_2Cp^*$ and P_4, provides another example of metal-promoted cyclization.[(28e)] Similar P_5-bridged sandwiches have been obtained from pentaphosphaferrocene, $Cp^*Fe(P_5)$, by reaction with $(C_5Me_4Et)Fe(P_5)$ under irradiation, giving the $[CpFe(P_5)Fe(C_5Me_4Et)]^+$ cation.[(28f)] Corresponding pentaarsacyclopentadienyl compounds are known; thus, $CpCr(As_5)CrCp$ was prepared via the interaction of $[Cp(Co)_3Cr]_2$ with yellow arsenic, As_4.[(28g)] In these complexes, the five-membered phosphorus or arsenic bridging rings are electronic surrogates for cyclopentadienyl (C_5H_5), and the complexes themselves can be viewed as analogous to cyclopentadienyl-bridged triple-decker sandwich compounds such as those recently reported by Kudinov *et al.*[(29)]

2.4.3. Nitrogen–Sulfur and Arsenic–Sulfur Rings

The extensive literature on sulfur–nitrogen clusters[(30)] includes a few cases in which transition metals are known to facilitate ring formation or ring expansion. One such instance is the preparation of $[S_4N_4Cl]FeCl_4$ from $(NSCl)_3$ and $FeCl_3$, which also yields $[S_4N_4]^+[FeCl_4]^-$ as a byproduct.[(30b)] A very recent example of metal-mediated ring expansion is the formation of the six- and eight-membered products *cyclo*-$(MeAsS)_n$ (n = 3,4) from sulfur and *cyclo*-$(MeAs)_5$ in the presence of $Mo(CO)_6$. Both the six- and the eight-membered ring products contain alternating As and S atoms and are puckered like their *cyclo*-C_nH_{2n} analogs.[(31)] A proposed mechanism for this reaction involves formation of the metal–$(MeAs)_n$ complex intermediates which are oxidized by sulfur, released from the metal, and subsequently undergo sulfur insertion into the ring. Indeed, two metal–arsathiane intermediate complexes have been isolated and structurally characterized, one of which is a $(CpMo)_2As_4S$ triple-decker sandwich.[(31)]

2.5. Cage Condensation and Cluster Building

As the above examples illustrate, metal-promoted cyclization can be applied to main-group elements in reactions which mimic, at least superficially, ring-forming reactions of hydrocarbons. However, in main-group chemistry one can carry this concept further, applying it in a *three-dimensional* sense to form polyhedral cages and clusters. Again, and not surprisingly, boron chemistry furnishes the most extensively developed examples. We first examine reactions involving homoatomic (X—X) bond formation.

2.5.1. Oxidative Fusion of Boron Clusters

A type of transition metal-promoted reaction, that has no precise counterpart in organometallic chemistry and yet has surprisingly broad versatility when applied to boron clusters, was discovered in 1974 by Maxwell, Miller, and Grimes.[32,33] It was found that metallacarborane complexes of the type $(R_2C_2B_4H_4)_2MH_x$, where M is a transition metal (usually Fe or Co), on exposure to air or other oxidants under mild conditions are essentially quantitatively converted to 12-vertex $R_4C_4B_8H_8$ tetracarbon carborane clusters (Figure 13).

The reactions occur easily in cold polar or nonpolar solvents and are noncatalytic, the metal ending up as a solvated complex, a hydroxide, or the free element depending on the conditions. The net process involves conversion of two *nido*-$R_2C_2B_4H_4^{2-}$ ligands to the neutral product, and hence is described as a four-electron oxidative fusion. As shown, the ligands are joined along B—B edges to form an open framework in which the C—C pairs are well separated; however, when the R groups are small (e.g., methyl, ethyl), the $R_4C_4B_8H_8$ cages undergo fluxional interconversion in solution between the "open" and "closed" forms as depicted above.[34] When the R groups are larger (e.g., benzyl, isopentyl, hexyl), the $R_4C_4B_8H_8$ cages are obtained only in the open form and are nonfluxional; even larger substituents, such as indenylmethyl or fluorenylmethyl, can sterically inhibit the fusion process and effectively prevent formation of the C_4B_8 cluster.[35]

Metal-promoted face-to-face oxidative fusion has been observed not only

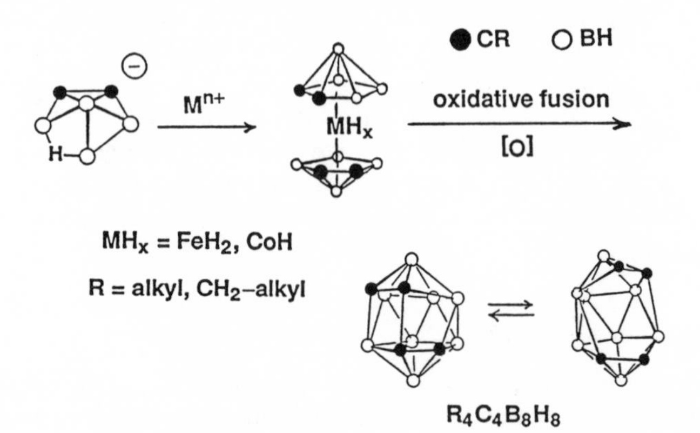

Figure 13. Synthesis of $R_4C_4B_8H_8$ carboranes via metal-promoted oxidative fusion of $R_2C_2B_4H_4^{2-}$ ligands.

with $R_2C_2B_4H_4^{2-}$ carborane ligands but also with boron hydride, metallaborane, and metallacarborane anionic ligands, as described below. However, it evidently does not occur with large substrates such as the dicarbollide ($C_2B_9H_{11}^{2-}$) ions.[33] The reactions involving $(R_2C_2B_4H_4)_2Fe^{II}H_2$ and its Co^{III} analog have some remarkable features. The bis(dimethylcarboranyl)iron dihydrogen complexes are tomato-red diamagnetic Fe(II) species whose two metal-bound hydrogens are acidic and apparently fluxional, migrating around the four equivalent Fe—B—B triangular faces. A number of facts have been established about this system, where R is methyl, ethyl, or propyl[36]:

(1) Fusion is observed only via metal complexation of the carborane ligand and does not occur in the absence of a transition metal.

(2) The process is intramolecular, as the oxidation of $(R_2C_2B_4H_4)_2$-$Fe^{II}H_2/(R'_2C_2B_4H_4)_2Fe^{II}H_2$ mixtures ($R \neq R'$) yields only homoligand ($R_4C_4B_8H_8$ and $R'_4C_4B_8H_8$) products.

(3) Fusion of the $(R_2C_2B_4H_4)_2MH_n$ complex occurs essentially instantaneously on exposure to oxygen, regardless of the solvent employed, but in the absence of air the results are strongly solvent-dependent. In deoxygenated nonpolar media (hexane, diethyl ether) no reaction is observed; in polar solvents (THF, dimethoxyethane), addition of a trace of Fe^{3+} catalyzes a conversion to a purple *dimetallic* species that is paramagnetic with four unpaired electrons. This complex has been characterized as $(L)Fe_2(Et_2C_2B_4H_4)_2$, where L is 2 THF or $C_2H_4(OMe)_2$; Figure 14 presents the structure of the dimethoxyethane complex.

(4) The diiron compound, on contact with air, $FeCl_3$, or other oxidants, undergoes fusion rapidly and quantitatively with formation of the C_4B_8 carborane. The extremely air-sensitive Fe_2 complex has been shown via ESR and Mössbauer spectroscopy to contain low-spin diamagnetic Fe(II) and high-spin paramagnetic (Fe(II), the latter occupying an exopolyhedral location in which it binds tetrahedrally to the centers of two carboranyl B—B edges.[36b] While the Fe_2 species is an isolable intermediate in this particular process, it is not known whether such complexes are common to all carborane or borane fusion reactions. However, there is evidence that interligand B—B bond

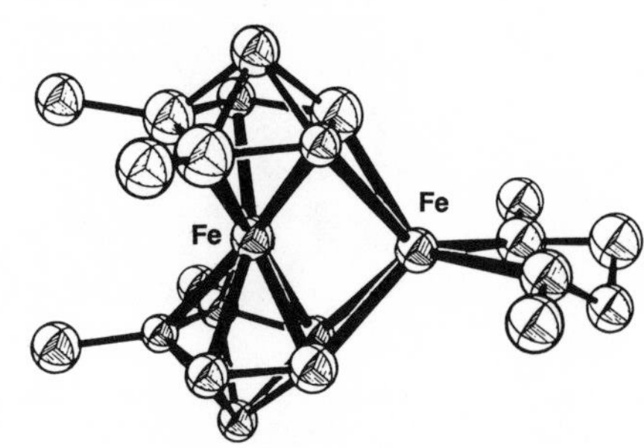

Figure 14. Structure of $[C_2H_4(OMe)_2]Fe_2$-$(Et_2C_2B_4H_4)_2$.[36b]

formation occurs at an early stage prior to expulsion of the metal, and that this is a general feature of fusion mechanisms. For example, bis(carboranyl) metal complexes, in which the ligands are linked directly or via bridging groups, have been isolated and structurally characterized;[37] in these cases it appears that fusion has been sterically blocked.

As mentioned earlier, oxidative fusion is also exhibited by other classes of boron clusters including boranes, metallaboranes, and metallacarboranes. It was noted above that large (e.g., icosahedral) clusters of the type $M(C_2B_9H_{11})_2$ have not been found to undergo this reaction; similarly, metallacarborane complexes of the class $M(R_4C_4B_8H_8)_2$ are stable toward oxidizers.[33,37b] In boron hydride chemistry, complexation of $B_5H_8^-$ ion with Fe^{2+} or Ru^{3+} followed by oxidation gave the fusion product, $B_{10}H_{14}$; the highest yields of decaborane were obtained with $RuCl_3$, which functions both as complexing agent and oxidant (Figure 15a).[9] When $FeCl_2$ was employed, both $B_{10}H_{14}$ and a coupling product, 2,2'-$(B_5H_8)_2$, were produced.

A mechanism for this reaction, suggested schematically in Figure 15b, involves base-to-base joining of two $B_5H_8^-$ substrates followed by further bond formation, so that the apex boron atoms (shown as solid circles) in the original B_5 units become the 2 and 4 vertices in the decaborane product. This scheme is supported by experiments involving apically substituted derivatives of $B_5H_8^-$, which produced the corresponding 2,4-substituted decaborane derivatives.[9] The same type of mechanism is evident in the metal-promoted fusion[9] of the cobaltaboranes 1- and 2-$CpCoB_4H_7^-$ ions (analogs of $B_5H_8^-$ in which CpCo replaces an apical or basal BH, respectively) as depicted in Figure 16.

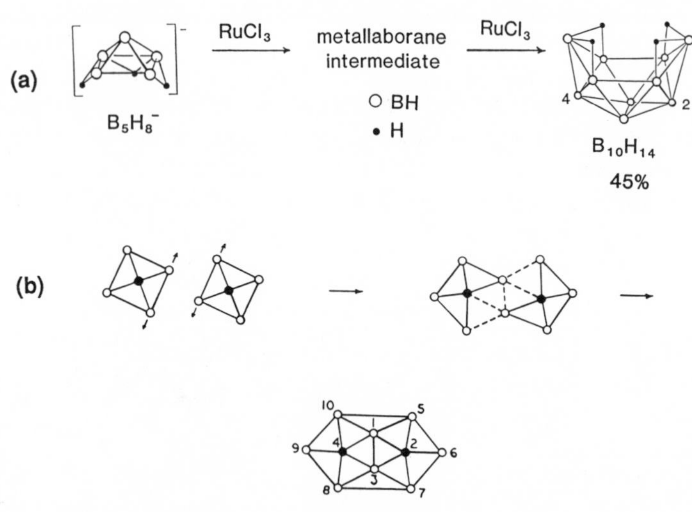

Figure 15. (a) Synthesis of $B_{10}H_{14}$ via $RuCl_3$-promoted fusion of $B_5H_8^-$. (b) Proposed pathway for fusion of square pyramidal B_5 or MB_4 units.[9]

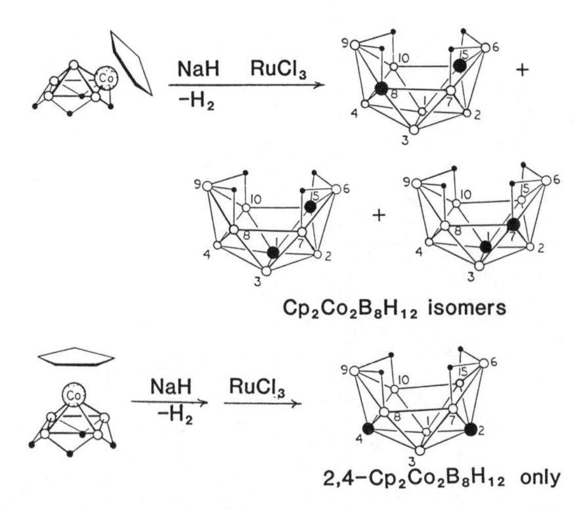

Figure 16. Synthesis of *nido*-$Cp_2Co_2B_8H_{12}$ isomers via $RuCl_3$-promoted fusion of 1- or 2-$CpCoB_4H_7^-$ ions.[9] (●) represent CpCo groups.

The metal-mediated conjoining of $B_6H_9^-$ ions to generate $B_{12}H_{16}$, a previously unknown borane, is an example of oxidative fusion as a tool in planned synthesis.[38] Although $B_6H_9^-$ is structurally and electronically analogous to the $R_2C_2B_4H_4^{2-}$ ligand, the geometry of the $B_{12}H_{16}$ product (Figure 17a) indicates that fusion of the B_6 units occurs edgewise rather than face-to-face as in the carborane reaction (Figure 13). The evident reason for this difference is that the bridging hydrogen atoms on the borane ions allow only edge-to-edge interaction.[38]

Further examples of transition metal-facilitated oxidative fusion of boron hydride substrates are given by the conversion of $B_9H_{12}^-$ to $B_{18}H_{22}$ in the presence of mercury(II) bromide,[39] and to its conjugate base anion $B_{18}H_{21}^-$, a process promoted by $[Os(CO)_3Cl_2]_2$.[40] The structure[41] of neutral n-$B_{18}H_{22}$ is shown in Figure 17b. Unlike some of the metal–carborane systems described above, in the fusion reactions involving boron hydrides and metallaboranes it has not been possible to isolate the metal complex intermediates owing to their instability. There is no doubt, however, that the transition metal ions play a role since fusion under mild conditions is not observed in their absence (the pyrolysis of small carboranes or boranes at high temperatures to give larger clusters is well known).

2.5.2. Condensation of Boron Clusters

A related type of metal-promoted reaction is the cage-dehydrocondensation of borane and carborane frameworks to produce larger clusters.[42] As

(a)

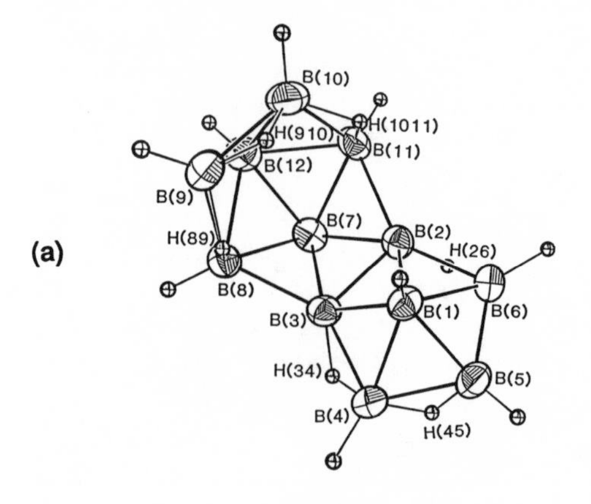

(b)

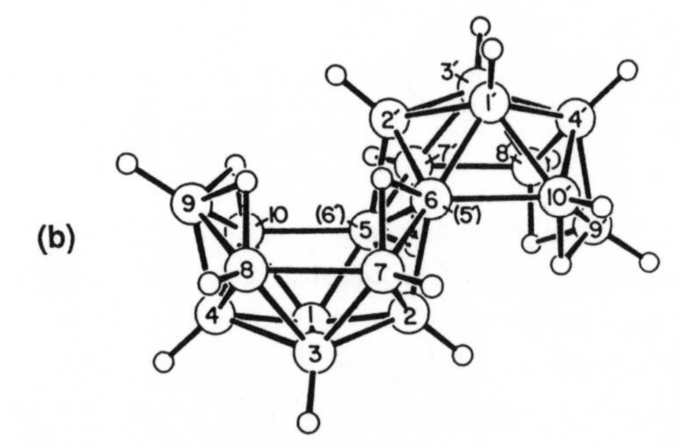

Figure 17. Structures of $B_{12}H_{16}$ (a)[38] and n-$B_{18}H_{22}$ (b).[41]

depicted in Figure 18, the small carborane 1,5-$C_2B_3H_5$ combines with diborane over $PtBr_2$ to generate an arachno-carborane, 5,6-$C_2B_6H_{12}$, nearly quantitatively. The proposed intermediate species depicted was not isolated, but support for its existence was obtained in the $PtBr_2$-promoted reactions of 1,6-$C_2B_4H_6$ and B_5H_9 with diborane which gave, respectively, the novel species 2:1′,2′-[1,6-$C_2B_4H_5$][B_2H_5] and 2:1′,2′-[B_5H_8][B_2H_5]. The proposed structures of these molecules, based on spectroscopic evidence,[42] are shown in Figure

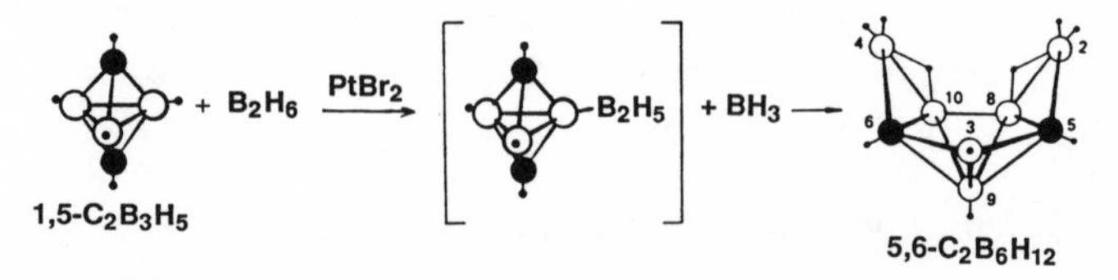

Figure 18. Synthesis of *arachno*-5,6-$C_2B_6H_{12}$ via $PtBr_2$-promoted dehydrocondensation of 1,5-$C_2B_3H_5$ and B_2H_6.[42]

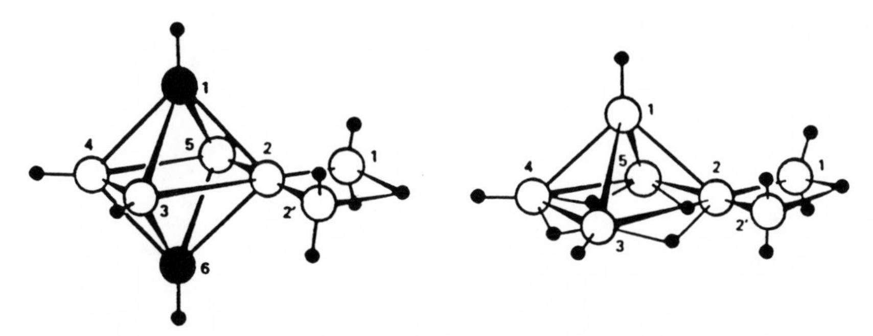

Figure 19. Proposed structures of intermediates in dehydrocondensation reactions.[42]

19. The scope of these heterogeneous "cage-growth" reactions has been little explored, but the method may prove to be complementary to the homogeneous oxidative ligand fusion processes described above, since the two approaches clearly follow different pathways and, not surprisingly, yield different kinds of products. At this writing, it is not clear whether the cage-condensation technique will prove to be broadly applicable, but it holds intriguing potential as a synthetic approach to cluster-building.

2.5.3. Stabilization of Main Group Rings and Clusters by Transition Metals

In the strict connotation of the term, transition metal-promoted reactions generate products in which the metal itself is not present. This is true by definition for metal-catalyzed processes, and it would normally be assumed for stoichiometric processes as well. From this point of view, the formation of main group–transition metal clusters would fall outside the scope of this chapter. However, certain of these clusters are structural and electronic analogs of organometallic complexes; as such, they fit nicely into the overall theme of this volume and hence will be selectively discussed. Further treatment of the structures and reactivity of many of these species will be found in other chapters of this book.

2.5.3.1. Metal–Boron Clusters. In an extensive series of studies, Fehlner and his students have explored the synthesis and properties of metal-rich cages containing one or two boron atoms,[43] which are direct analogs of corresponding metal–carbon cluster compounds. The triironborane $[HFe_3(CO)_9BH_3]^-$, prepared[44] by addition of $BH_3 \cdot THF$ to $[(CO)_4FeC(O)Me]^-$, has the structure shown in Figure 20a. This species in turn can be quantitatively converted to a tetraironborane, $[HFe_4(CO)_{12}BH]^-$ (Figure 20b), via treatment with $Fe_2(CO)_9$;[45] protonation of the latter cluster generates the neutral compound $HFe_4(CO)_{12}BH_2$ (Figure 20c). These complexes exhibit much intriguing chemistry, but from the standpoint of this chapter the focus of interest is

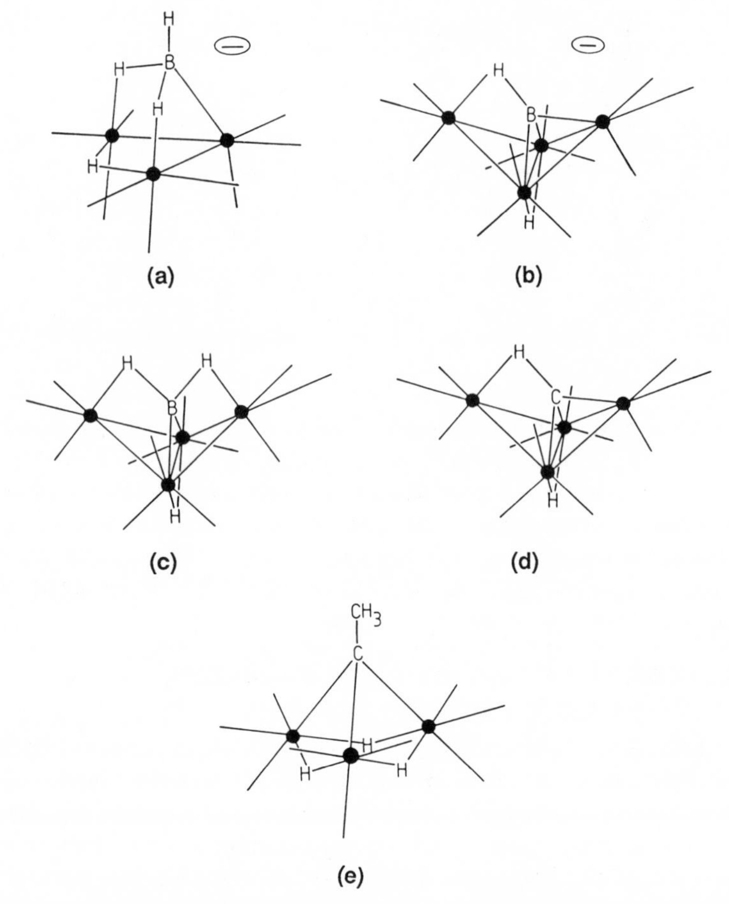

Figure 20. Structures of iron–boron and iron–carbon clusters (Fe atoms shown as solid circles:[43–45] (a) $[HFe_3(CO)_9BH_3]^-$; (b) $[HFe_4(CO)_{12}BH]^-$; (c) $HFe_4(CO)_{12}BH_2$; (d) $HFe_4(CO)_{12}CH$; (e) $H_3Fe_3(CO)_9CCH_3$.

in their analogy with the corresponding tri- and tetrairon–carbon clusters, which are isostructural with the boron species except for the presence of one less proton.[43]

The Fehlner group has pursued with notable success the idea that BH_n^- groups are useful models for isoelectronic CH_n units, from which it follows that detailed examination of the chemistry of clusters such as $[HFe_4(CO)_{12}BH]^-$ can yield new insight into the analogous carbon systems, such as the intensively studied[46] cluster $HFe_4(CO)_{12}CH$ (Figure 20d). For example, in $H_3Fe_3(CO)_9CCH_3$ (Figure 20e), the three cluster hydrogens bridge the Fe—Fe edges, while in the analogous neutral Fe_3B cluster they adopt

Fe—H—B bridge locations (protonated form of $[HFe_3(CO)_9BH_3]^-$, Figure 20a). This structural difference has been shown to reflect, in part, dissimilar charge distributions in the two cases.[47] This finding has guided studies of the behavior of these clusters, including their deprotonation and subsequent rearrangement. For example, when the Fe_3C species was deprotonated, the resulting anion was found to have an Fe—H—C bridge analogous to its Fe_3B counterpart.[48] Reprotonation of the anion at low temperature generates, on warming, *three* neutral $Fe_3(CO)_9CH_4$ clusters which exhibit temperature-dependent migration of protons from Fe—C to Fe—Fe edges. Consequently, the least stable (i.e., Fe—C protonated) form of the Fe_3C system corresponds to the most stable (Fe—B protonated) form of the Fe_3B cluster,[43,47,48] a finding which nicely illustrates the consequences of boron-vs-carbon replacement.

The iron clusters just discussed represent merely the tip of a very large iceberg, as this rapidly growing field has produced many other metal-rich metallaboron cluster families which are analogs of metal–carbon systems; a number of these have been reviewed by Fehlner[49] and by Housecroft.[50] In the context of this chapter, there are two relevant points about these clusters: one can view their synthesis as "metal-promoted" in the sense that the polymetal fragment provides a matrix for attachment of a borane moiety; furthermore, the reactivity at boron is profoundly influenced by the metal centers. Detailed discussions of the structural and bonding aspects of many of these species appear elsewhere in this volume.

2.5.3.2. Metal–Group V(15) Clusters. Another large class of transition element–main group element cage compounds which are geometric and electronic relatives of organometallic complexes consists of polyhedra incorporating P, As, Sb, or Bi atoms and a variety of metals from all three main transition series.[26] Among these are the so-called tetrahedranes, which are structural analogs of tetrahedral C_4H_4 (as well as the M_nC_{4-n}-type clusters) and are often prepared from di- or polymetallic transition metal carbonyls containing metal–metal bonds:[51]

$$P_4 + Co_2(CO)_8 \xrightarrow[\text{CO}]{\text{hexane, 10–50 °C}} P_n[Co(CO)_3]_{4-n} \qquad (n = 1\text{–}3)$$

$$AsCl_3 + Co_2(CO)_8 \rightarrow As_2[Co(CO)_3]_2$$

$$AsPh_6 + Cp(CO)_3Mo—Mo(CO)_3Cp \xrightarrow[\text{180 °C}]{\text{toluene}} As_2[Mo(CO)_2Cp]_2$$

The sandwich complexes of planar main-group ring ligands, described in Section 2.4, are also clusters and could as well have been discussed in this context. Many other types of inorganometallic ligands comprised of Group V(15) elements, including PN, E_2 (E = P, As, Sb, Bi), HP=PH, PhAs=AsPh, and RSb=SbR can be prepared and stabilized via transition metals,[26a] and this area of inorganometallic chemistry may well eventually rival boron cluster chemistry in its richness and variety.

3. PROMOTION OF X—X/X—Y BOND CLEAVAGE OR BOND ORDER LOWERING

3.1. Reactivity of Metal–N_2 Complexes

The dinitrogen molecule, N_2, is isoelectronic with acetylene and CO, and its reactivity toward transition metals is therefore a natural topic for discussion in the context of inorganometallic reactions. The strikingly low reactivity of N_2 compared to HC≡CH is usually attributed to the very high bond strength and low polarizability of the N≡N bond. From spectroscopic and structural evidence and theoretical calculations, it appears that N_2 *as an end-bonded ligand* is a weaker σ-donor and π-acceptor than CO, and that σ-donation plays a more important role in M—N bond formation than does π back-donation.[2b] Because of the relative inertness of N_2 it was not until 1965 that the first metal–N_2 complexes, of the type $[Ru(NH_3)_5(N_2)]^{2+}$, were prepared by Allen and Senoff.[52] Since then, over 200 dinitrogen complexes involving most of the transition elements have been prepared.[2b]

Metal–N_2 complexes exhibit a variety of reactions,[2a] including attack by electrophiles leading to N—H or N—C bond formation, nucleophilic attack resulting in addition to the N≡N bond, and addition of alkyl halides with loss of N_2. The first two types result in lowered N—N bond order, as for example in the interaction of $(Ph_2PCH_2CH_2PPh_2)_2Mo(N_2)_2$ with HCl, which proceeds through a $[(Ph_2PCH_2CH_2PPh_2)_2(H)Mo(N_2)_2]^+$ intermediate to give $[(Ph_2PCH_2CH_2PPh_2)_2(Cl)Mo(NNH_2)]^+$.

Interest in this area has been intensified by the search for metal complexes which can mimic the ability of certain biomolecules (e.g., nitrogenase) to catalyze the conversion of N_2 to ammonia. A number of dinitrogen metal complexes are readily hydrolyzed to NH_3, but most of these are not closely related to nitrogenase and do not contain sulfur. However, some sulfur-containing molybdenum–phosphine species such as *trans*-$Mo(N_2)_2(PMePh_2)_2$-$(PPh_2CH_2CH_2SMe)$ have been found to promote reduction of N_2 at room temperature under strongly reducing conditions.[53] The complex cited has two end-bonded N_2 ligands which occupy *trans* coordination sites on the metal; however, other metal–N_2 binding modes including side-bonding (η^2) and bridge-bonding are known.

Viewed as inorganometallic species, the metal–dinitrogen complexes are formally analogous to metal alkynyls, and in fact the two classes exhibit a similar range of metal–ligand coordination geometries. For example, both N_2 and RC≡CR′ ligands can adopt a variety of bridging modes with transition metals, including tetrahedral clusters having N_2M_2 or C_2M_2 cores.[54] In terms of reactivity there is less resemblance; for example, the facile conversion of metal-bound N_2 to NH_3 is not paralleled by production of methane from metal alkynyls. Similarly, the methathesis of alkynes, in which M≡CR and R′C≡CR″ are converted to M≡CR′ and RC≡CR″ via a square metallacyclobutadiene intermediate,[(55)] has no known counterpart in dinitrogen chemistry. On the other hand, nucleophilic attack on metal-coordinated acetylenes, which lowers the C—C bond order, can be compared to the reactions of metal dinitrogen complexes with alkylating agents such as methyllithium,[(2a)] which generate species such as L_nM—(Me)N=N—Me.

3.2. *Reactivity of Metal–NO Complexes*

Nitrogen monoxide (nitric oxide) can coordinate to transition metals in a variety of modes to form nitrosyl complexes, and an extensive chemistry has developed around these compounds. Here our focus is on reactions in which the metal facilitates conversion of NO to other nitrogen-containing products and is itself liberated in the process. Reactions with alkenes to generate amines provide an illustrative example:[(56)]

$$(CpCoNO)_2 + R_2C{=}CR_2 \rightarrow CpCo\left<\begin{matrix} \overset{O}{N}\text{—}CR^2 \\ | \\ \underset{O}{N}\text{—}CR_2 \end{matrix}\right. \xrightarrow{LiAlH_4} H_2N\text{—}CR_2\text{—}CR_2\text{—}NH_2$$

Intermolecular transfers involving NO ligands are well known, as in the process[(57)]

$$(Cp_2TiCl)_2 + 2NO \rightarrow (Cp_2TiCl)_2\mu\text{-}O + N_2O$$

Recent studies of reactions of electrophiles with the cluster anion $Ru_3(CO)_{10}(NO)^-$, in which an NO ligand bridges an Ru—Ru bond in the Ru_3 triangle, have revealed some interesting chemistry.[(58)] For example, treatment with CF_3SO_3Me results in O-methylation of the nitrosyl group, generating the neutral product $Ru_3(NOMe)(CO)_{10}$ in which the Ru_3 triangle is bridged by μ_3-CO and μ_3-NOMe ligands. An analogous complex containing NOH in place of NOMe is obtained via protonation of the original anionic cluster by CF_3SO_3H; however, treatment with weaker acids such as CF_3CO_2H proceeds differently, protonating an Ru—Ru bond to give $HRu_3(CO)_{10}NO$. Tautomerization of the proton from the bound NOH ligand to the Ru—Ru edge is induced by $(Ph_3P)^+$ salts. The finding that different sites on the cluster are protonated depending on the strength of the acid has been interpreted in terms of competing equilibria.[(58)]

This unusual behavior provides a nice illustration of transition metal-mediated main group chemistry which invites comparison with reactions of metal carbonyl cluster compounds, although broader studies of nitrosyl–metal cluster species (for example, with different metals) are clearly required.

3.3. Reactivity of Metal–O_2 Complexes

The interaction of dioxygen with transition metals has been extensively studied, and is undoubtedly the most important type of inorganometallic process from a biological viewpoint. Considerable effort has been directed to the synthesis of model compounds which can reversibly coordinate O_2 as in the Fe^{II} oxygen-carrying proteins hemoglobin and myoglobin. The difficulty lies in preventing irreversible oxidation of Fe(II) to Fe(III), and in fact much research in this area has utilized Co^{II} model complexes in order to take advantage of the versatile and rich cobalt–dioxygen coordination chemistry.[59] This exceedingly complex subject will not be explored here, but we will draw attention to some similarities between metal–O_2 and metal–alkene/alkyne bonding.

In general, reactions of metal complexes such as those of Co(II) with O_2 result in electron transfer to create a superoxide (O_2^-) species, which in turn can add a second metal group to generate a peroxide (O_2^{2-})-bridged bimetallic complex:

$$(L)Co^{II} + O_2 \rightarrow (L)Co^{III}O_2^-$$

$$(L)Co^{II} + (L)Co^{III}O_2^- \rightarrow (L)Co^{III}{-}(O{-}O)^{2-}{-}Co^{III}(L)$$

In some systems, the peroxide species can be reversibly oxidized to form a μ-superoxide complex:

$$(L)Co^{III}{-}(O{-}O)^{2-}{-}Co^{III}(L) \underset{+e}{\overset{-e}{\rightleftarrows}} (L)Co^{III}{-}(O{-}O)^{-}{-}Co^{III}(L)$$

Reactions of the latter type are of particular interest in a bioinorganic sense, since they are related to hemoglobin and myoglobin chemistry. Iron(II) complexes capable of reversible O_2 binding have also been prepared, notably the so-called "picket-fence" porphyrin complexes in which the irreversible formation of $Fe{-}O_2{-}Fe$ bridges is blocked. Changes in the $O{-}O$ bonding in these various species are reflected in X-ray-determined $O{-}O$ bond lengths, infrared stretching frequencies, ESR spectra, and other observations. Since O_2 donates electrons to metal centers via a σ orbital and accepts electrons from the metal into its π^* orbitals, weakening of the $O{-}O$ bond on metal complexation is expected, and this is in fact observed. However, there is still much debate concerning the mechanistic details of these interactions.[59]

Certain transition metal complexes react with O_2 with transfer of *two* electrons rather than one as shown above, and are therefore analogous to the oxidative-addition reactions of unsaturated hydrocarbons.[2a] A typical example involves a square-planar iridium complex:

$$trans - Ir^{I}Cl(CO)(PPh_3)_2 + O_2 \rightarrow Ir^{III}(O_2)Cl(CO)(PPh_3)_2$$

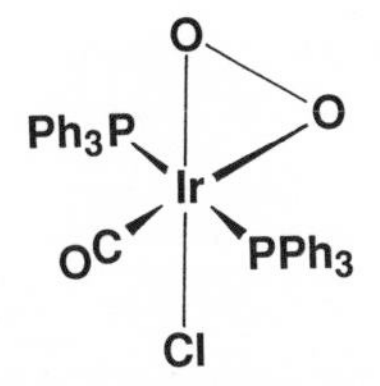

Figure 21. Structure of $Ir(O_2)Cl(CO)(PPh_3)_2$.

The product contains a peroxide unit which forms a three-membered ring with the metal, as depicted in Figure 21. A corresponding alkyne addition can be represented as shown:

$$(L)_nM + R{-}C{\equiv}C{-}R \rightarrow (L)_nM[R{-}C{=}C{-}R]$$

These processes produce a lowering of bond order in the X=X or X≡X unit; where the X—X (or X—Y) group is singly-bonded to begin with, oxidative addition results in cleavage:

$$trans - Ir^{I}Cl(CO)(PPh_3)_2 + HCl \rightarrow Ir^{III}(H)Cl_2(CO)(PPh_3)_2$$

3.4. Metal-Catalyzed Decomposition of H_2O_2

A well-known example of metal-promoted bond cleavage is the catalytic decomposition of hydrogen peroxide in aqueous solution in the presence of trace quantities of transition metal ions. In the case of ferric ion the net process is

$$2Fe^{3+} + H_2O_2 \xrightarrow{-2H^+} 2Fe^{2+} + O_2$$

Tracer studies of many such reactions show that the O_2 is generated solely from H_2O_2 and that oxygen exchange with H_2O is negligible; thus the O—O bond is not itself cleaved. However, there are exceptions such as Fenton's reagent (Fe^{2+} and H_2O_2), in which the reaction mechanism involves homolytic O—O bond-breaking and formation of OH· radicals.[2]

From the viewpoint of inorganometallic chemistry, one looks for parallel reactions in metal–hydrocarbon systems. Directly analogous processes are unlikely; as a consequence of the lower polarity of sp^3 C—H vs O—H bonds, and other factors, alkanes such as C_2H_6 (isoelectronic with H_2O_2) are far less reactive than hydrogen peroxide toward metal reagents. Nevertheless, complexes such as $Cp^*Re(PMe_3)_3$ have been shown to promote photolytic alkane C—H cleavage,[60] and, in the case of cyclopropane, even C—C cleavage.[61]

4. MAIN GROUP-PROMOTED REACTIONS OF TRANSITION METAL SPECIES: GENERAL OBSERVATIONS

Most reactions of transition metal complexes in solution involve p-block elements; indeed, classical coordination chemistry is dominated by metal bonding interactions with nitrogen, carbon, oxygen, phosphorus, or sulfur. Similarly, a plethora of solid-state materials (including some of the most interesting types, such as high T_c superconductors) incorporate transition and main group elements whose electronic interactions determine the bulk properties of the solid. Clearly, the overall subject of main group–transition metal chemistry is one of encyclopedic dimensions. The focus here is much narrower (as in other sections of this chapter) and our discussion will center on selected reactions of transition metal species which are facilitated by main group ligands or fragments *and which do not take place (or proceed differently) in their absence.*

5. REACTIONS INVOLVING M—X—M BRIDGES (X = MAIN GROUP ELEMENT)

An intensively studied type of electron transfer in transition metal chemistry is the ligand-bridged reaction, which occurs via a bimetallic M—X—M′ intermediate.[62] Many such processes are known, and detailed mechanistic information was developed in the pioneering work of H. Taube and his students. One of the best understood cases is the electron transfer from Cr^{II} to Co^{III}, as in the reaction

$$Cr(H_2O)_6^{2+} + Co(NH_3)_5Cl^{2+} \rightarrow (H_2O)_5Cr{-}Cl{-}Co(NH_3)_5^{4+} \rightarrow$$
$$Cr(H_2O)_5Cl^{2+} + Co(H_2O)_6^{2+} + 5NH_4^+$$

Here the bridging group is a chloride ion, but many others, including Cl^-, Br^-, I^-, PO_4^-, acetate, malonate, sulfate, or thiocyanate, can be used. The main point is that the bridging unit is essential, since the original Co^{III} complex is not labile, and formation of the Co—X—Cr bridge is required in order for transfer of the electron to occur. It is well established that the electron transfer occurs within the bridged species, but transfer of the X group itself need not always occur, although it most often does (in cases where X is not transferred, proof that the mechanism is via a bridged intermediate is less obvious). Usually the bridged complex is not isolable, but in certain cases it is; an example is the reaction of $Co(CN)_5^{3-}$ with $Fe(CN)_6^-$, in which the species $(CN)_5Co{-}CN{-}Fe(CN)_5^{6-}$ can be obtained as a heavy metal salt.[2a]

6. MAIN GROUP SPECIES AS ACTIVATING AGENTS IN METAL COMPLEXES

6.1. Lewis Acid/Base Ligands

In contrast to carbon, whose divalent species (e.g., CH_2) exist only transiently, those of the heavier Group IV(14) elements Ge, Sn, and Pb are stable as in neutral compounds such as $SnCl_2$. These species and their anionic derivatives (like $SnCl_3^-$) have proved to be highly versatile reagents, in part because of an electron lone pair on the metal which permits them to act as ligands toward transition metals; examples are complexes such as $Pt(SnCl_3)_2(PEt_3)_2H^-$ and $Ru(SnCl_3)_4Cl_2^{4-}$. The $SnCl_3^-$ ion in particular is a good leaving group, making it especially useful in homogeneous catalysis.[63]

In a different example of activation by p-block elements, Group III(13) halides promote migration of alkyl groups to CO on metal complexes, as in the overall reaction

$$(CH_3)Mn(CO)_5 + AlBr_3 \rightarrow \rightarrow \overset{CO}{\rightarrow} (CO)_5Mn{-}C(CH_3){=}OAlBr_3$$

Here the Lewis acid aluminum trihalide coordinates to the acyl oxygen, forming an intermediate containing both M—C=O—Al and M—Br—Al bridges, which adds CO to generate the product.[2a]

6.2. *Metallacarborane Catalyst Precursors*

A novel and imaginative application of main group chemistry to transition metal catalysis has been the development, by Hawthorne and co-workers, of metallacarboranes as effective agents in the hydrogenolysis of alkenyl acetates and the hydrogenation and isomerization of alkenes.[64] While several metals have been employed including ruthenium, iridium, and rhodium, the most extensively studied systems utilize rhodium complexes of the dicarbollide ion ($C_2B_9H_{11}^{2-}$) as catalyst precursors. The structure of the *closo*-3,1,2-$(Ph_3P)_2(H)RhC_2B_9H_{11}$ icosahedral species is depicted in Figure 22 together with one of the *exo–nido* tautomeric forms which are present in equilibrium with the *closo* form in solution. As shown, in the *exo–nido* tautomers, the metal is outside the cage and coordinated to it via two B—H—Rh 3-center, 2-electron bridges. The rhodium is in a formal +1 oxidation state in the *exo–nido* form and migrates on the boron framework, binding to different sets of adjacent BH vertices.

In the presence of catalytic quantities of this rhodacarborane complex (or its 2,1,7 or 2,1,12 isomers, which differ from the 3,1,2 compound in the

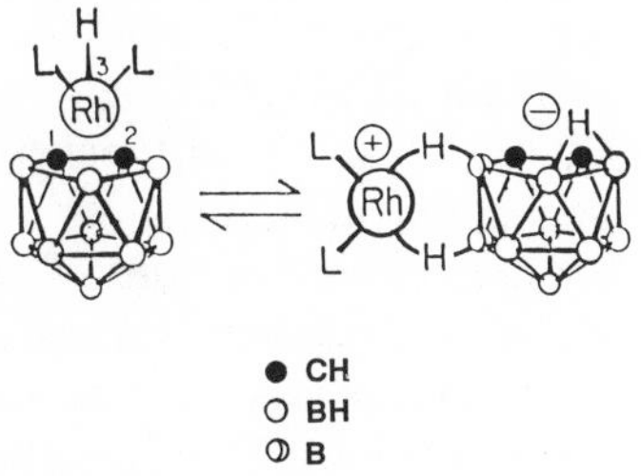

● CH
○ BH
◑ B

Figure 22. Equilibrium between *closo* and *exo–nido* rhodacarborane tautomers.[64]

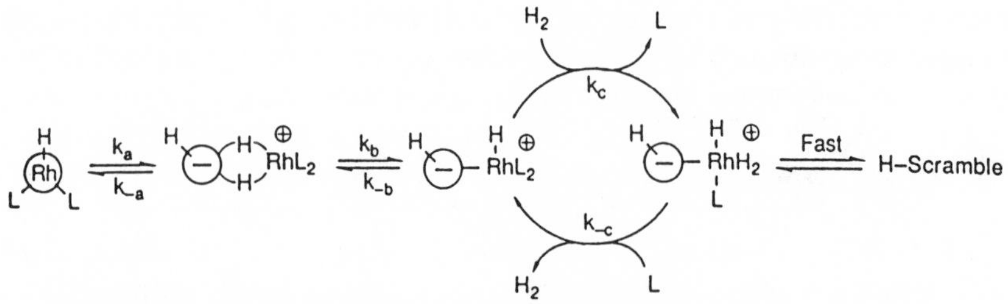

Figure 23. Simplified mechanistic scheme showing rhodacarborane redox equilibria.[64]

placement of cage heteroatoms), alkenyl acetates react with H_2 in THF solution to produce acetic acid and the corresponding alkene; when D_2 is used, isotopically pure CH_3COOD is obtained. Moreover, the hydride ligand which is coordinated to rhodium in the *closo* form is found to be inactive in catalysis. From detailed kinetic and deuterium-labeling studies, a mechanism has been proposed[64] which involves the slow formation (via regioselective oxidative addition of the Rh^I in the *exo-nido* tautomer to a cage B—H bond) of a rhodium monohydride species containing a $B—Rh^{III}—H$ array. This highly reactive metal center rapidly catalyzes the addition of hydrogen to the alkenyl substrate. A simplified scheme[64] outlining the proposed equilibria in the catalytic cycle is shown in Figure 23.

These recent findings have led to a revision[64] of earlier mechanistic proposals,[65] and support the conclusion that, in a wide range of reactions including hydrogenation and isomerization of alkenes as well as $B—H/D_2$ and B—H/C—D exchange, *the catalytically active species is derived from the exo-nido tautomer* which is in equilibrium with the *closo* form (Figure 22). It has been suggested[64] that a similar mechanism operates in the $(Ph_3P)_3RhCl$-catalyzed hydroboration of alkenes by 1,3,2-benzodioxaborole,[66] which is known to proceed via oxidative addition of a B—H bond to the rhodium complex to form a Rh—H bond into which the alkene is inserted.

It is clear that in these systems the boron hydride entity is directly involved in the catalytic process, and hence constitutes an unequivocal case of main group promotion of transition metal chemistry. The apparently broad scope of this type of transition metal–borane cluster interaction implies that much more is possible in this area, and that new families of useful catalysts may eventually emerge. A major advantage afforded by metallacarboranes as catalyst precursors is their inherent synthetic versatility. Using well-established preparative methods, a vast range of species is potentially accessible in which the nature and locations of metal and other heteroatoms in the cage can be varied, and organic functional groups can be attached. It should be possible to design catalysts which are "tuned" for specific activity via functionalization with electronically active groups that influence the charge distribution (hence the polarity of specific B—H bonds). This enormous versatility is a distinguishing attribute of boron chemistry, and may eventually have major impact on the methodology of homogeneous catalysis.

7. PROMOTION OF METAL CLUSTER FORMATION BY MAIN GROUP SPECIES

The scope and variety of known and structurally characterized molecular clusters is truly remarkable, and the pace of discovery in this area shows no sign of abating. Many clusters contain frameworks composed of both transition and p-block elements, including some types discussed in other sections of this review (e.g., metal–phosphorus, metal–arsenic, and metal–boron species). Many such compounds resist simple classification in terms of synthesis or structure, as they often can reasonably be viewed from more than one perspective, i.e., as main group-stabilized transition metal clusters or as main group clusters which are stabilized by transition elements. In this section our purpose is simply to draw attention to some of the more interesting examples of clusterification by transition metals in which p-block atoms or groups play a significant role.

The elements antimony and bismuth show a propensity to form stable polyhedral cages with iron or cobalt[(67)]:

$$Bi_2Fe_3(CO)_9 + Fe(CO)_4{}^{2-} \rightarrow Bi_2Fe_4(CO)_{13}{}^{2-}$$

$$Bi_2Fe_3(CO)_9 + Co(CO)_4{}^{-} \rightarrow Bi_2Fe_2Co(CO)_{10}{}^{-}$$

$$BiFe_3(CO)_{10}{}^{-} + CO \rightarrow Bi_4Fe_4(CO)_{13}{}^{2-}$$

The structure of the last cluster,[(67b)] which contains a Bi_4 tetrahedron capped on three faces by iron atoms, is shown in Figure 24a.

Sulfur also is highly effective in forming and stabilizing multimetallic clusters with a number of transition elements, as for example niobium and tantalum:[(68)]

$$M(OEt)_5 + (Me_3Si)_2S \rightarrow M_6S_{17}{}^{4-} \qquad (M = Nb, Ta)$$

In both anions (Figure 24b) the metal atoms adopt an open pentagonal pyramidal framework stitched together with bridging sulfurs to form a M_6S_{10} cluster to which additional sulfur ligands are attached; remarkably, one sulfur is located *inside* the cage. It is noteworthy, in the context of the theme of this chapter, that there appear to be no significant metal–metal bonding interactions, so that the sulfur not only participates in cluster-forming but plays a central role. A type of metal–sulfur cluster that is important in biochemistry is the M_4S_4 cubane-like unit, present in species such as $Cp_4Mo_4S_4$ and $(RS)_4Fe_4S_4{}^{n-}$, which are model compounds for ferredoxins.

(a)

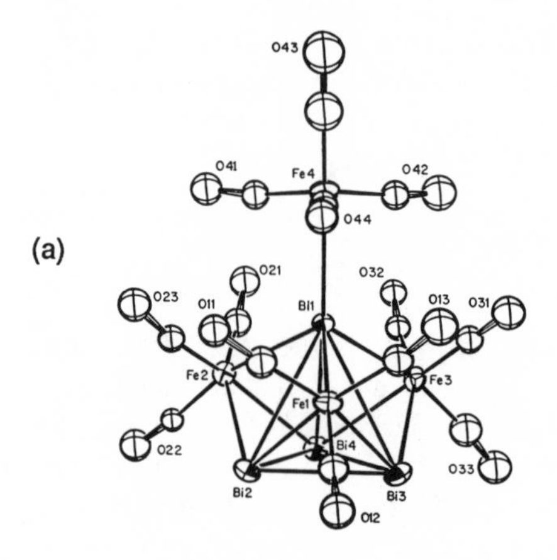

(b)

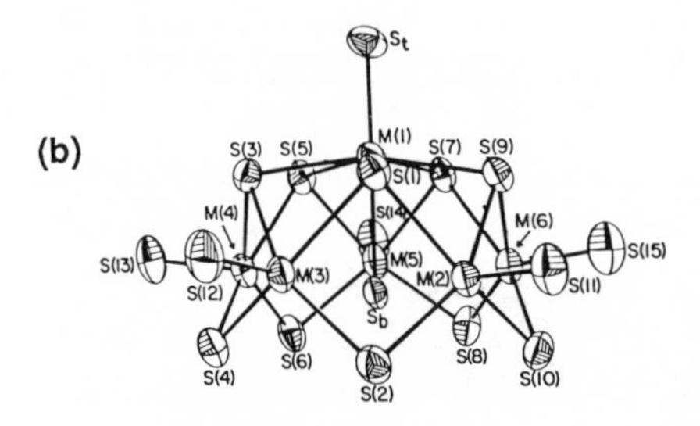

Figure 24. (a) Structure of the $Bi_4Fe_4(CO)_{13}^{2-}$-dianion.[67b] (b) Structure of the $M_6S_{17}^{4-}$ dianions (M = Nb, Ta).[68]

Niobium and tantalum are also prominently featured in another large class of metal–nonmetal clusters, the $M_6X_{12}^{n+}$ ions.[2] These species are formed in molten salt media, in reactions such as

$$14NbCl_5 + 16Nb + 20NaCl \xrightarrow{850\ °C} 5Na_4Nb_6Cl_{18}$$

Typically, these clusters consist of octahedra of metal atoms on which halide ligands are both face-capped and edge-capped as well as attached via terminal M—X bonds. Again, the point of interest here is that the nonmetal (halogen) units play an indispensable part in both the clusterification process itself and in the stabilization of the product species.

A different type of main group promoted incorporation of transition metals into clusters is seen in the preparation of boron-containing multidecker complexes in which the bridging rings are carborane (C_2B_3), borole (CB_4), diborole (C_3B_2), thiadiborole (C_2B_2S), or diborabenzene (C_4B_2).[69] Structurally related P_n- and As_n-bridged triple-decker sandwiches were discussed in Section 2.4, in the context of transition metal promoted cyclizations of main group

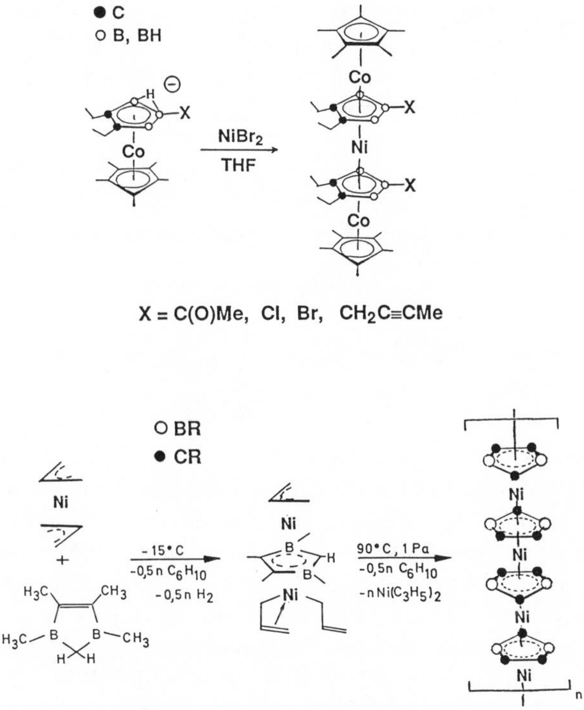

Figure 25. Synthesis of $Et_2C_2B_3H_3$-bridged tetradecker complexes (top)[72] and a nickel–$Me_2C_2B_2Me_2CH$-bridged polydecker sandwich (bottom).[73a]

elements. However, stacking reactions of the boron–carbon ligands are different in that they have been employed in the stepwise construction not only of triple-decker complexes[69,70] but also tetra- and higher-decker sandwiches,[69a,71,72] including polymers.[73] Two examples are depicted in Figure 25. In general, compounds of this class are air-stable, robust materials which can undergo oxidation and reduction without degradation. Moreover, electrochemical and ESR studies on many of them reveal extensive electron-delocalization between the metal centers.[74] Consequently, they provide an effective means of locking multiple transition metal atoms or ions into stable covalently bound stacks which may serve as precursors to materials having useful electronic and/or magnetic properties.

8. *MAIN GROUP REAGENTS IN METAL HYDRIDE SYNTHESIS*

As a final example of an important type of inorganometallic reaction, we briefly discuss the use of main-group compounds as hydride sources. Many kinds of reagents—inorganic, organic, and organometallic—have been employed for this purpose, among which species such as BH_4^- and AlH_4^- are particularly widely used. For example, while $NaBH_4$ is well known as a reducing agent in organic synthesis, it can also hydrogenate transition metal centers:

$$FeI_2(CO)_4 \xrightarrow{NaBH_4} FeH_2(CO)_4$$

Acids of p-block element reagents can protonate transition metal complexes[2a]:

$$Cp_2MoH_2 + H^+BF_4^- \rightarrow Cp_2MoH_3^+BF_4^-$$

$$Cp(CO)_3W—W(CO)_3Cp + H^+BF_4^- \rightleftharpoons Cp(CO)_3W \overset{H}{—} W(CO)_3Cp^+BF_4^-$$

$$trans\text{–}IrCl(CO)(PPh_3)_2 + HSiCl_3 \rightarrow IrCl(H)(SiCl_3)(CO)(PPh_3)_2$$

The coordination of transition metal atoms to hydrogen can occur via partial H transfer by forming M—H—X bridges. A novel occurrence of this, discovered in recent work by Reed and coworkers,[75] involves the icosahedral monocarborane anion $CB_{11}H_{12}^-$ and is an appropriate note on which to conclude this chapter, in view of the prominent role of boron clusters in the preceding discussions. In contrast to the usual type of silver salt metathesis, exemplified by the reaction of Vaska's compound ($IrCl(CO)(PPh_3)_2$) with silver perchlorate in toluene solution to form $Ir(OClO_3)(CO)(PPh_3)_2$ and AgCl, treatment of Vaska's complex with the Ag^+ salt of the *very weakly coordinating*

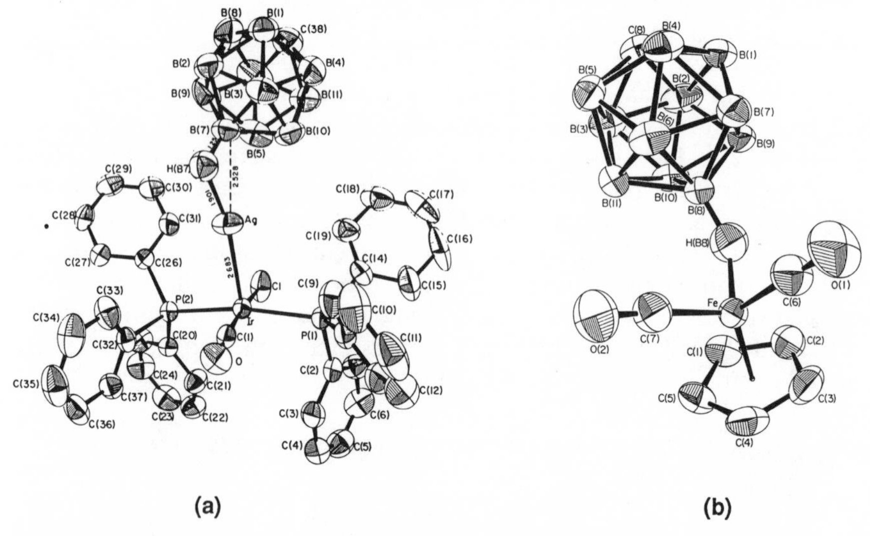

Figure 26. (a) Structure of $IrCl(CO)(PPh_3)_2 \cdot Ag(CB_{11}H_{12})$.[75] (b) Structure of $CpFe(CO)_2(CB_{11}H_{12})$.[76]

$CB_{11}H_{12}^-$ ion gave instead a 1:1 donor–acceptor complex, $IrCl(CO)(PPh_3)_2 \cdot Ag(CB_{11}H_{12})$, in which the iridium is strongly bound to the carborane via an Ir—Ag—H—B array (Figure 26a). No AgCl precipitates in this reaction, and the "unconsummated methathesis" has been taken as evidence of extremely low nucleophilicity on the part of the carborane ion arising from the charge delocalization and the absence of lone pairs on the cage framework.[75]

The reaction of $Ag^+CB_{11}H_{12}^-$ with $FeCp(CO)_2I$ gave an isolable iodide-bridged 1:1 adduct which, in turn, eventually did undergo metathesis to form $CpFe(CO)_2(CB_{11}H_{12})$ (Figure 26b). These observations reveal a previously unsuspected role of weakly coordinating counterions such as ClO_4^- and BF_4^- in silver methathesis reactions.[76]

REFERENCES

1. Davies, S. G. "Organotransition Metal Chemistry: Applications to Organic Synthesis"; Pergamon Press: Oxford, 1982; and references cited therein.
2. (a) Cotton, F. A.; Wilkinson, G. "Advanced Inorganic Chemistry", 5th ed.; Wiley–Interscience: New York, 1988. (b) Greenwood, N. N.; Earnshaw, A. "Chemistry of the Elements"; Pergamon Press: Oxford, 1984.
3. Coffy, T. J.; Medford, G.; Plotkin, J.; Long, G. J.; Huffman, J. C.; Shore, S. G. *Organometallics* 1989, **8,** 2404.
4. Ting, C.; Messerle, L. *J. Am. Chem. Soc.* 1989, **111,** 3449.
5. (a) Corcoran, E. W., Jr.; Sneddon, L. G. In "Advances in Boron and the Boranes [Mol. Struct. Energ. Vol. 5]"; Liebman, J. F.; Greenberg, A.; Williams R. E., Eds.; VCH Publishers, Inc.: New York, 1988; Chapter 4. (b) Sneddon, L. G. *Pure Appl. Chem.* 1987, **59,** 837.
6. (a) Grimes, R. N.; Lipscomb, W. N. *Proc. Natl. Acad. Sci. U.S.A.* 1961, **47,** 996. (b) Gaines, D. F.; Iorns, T. V.; Clevenger, E. N. *Inorg. Chem.* 1971, **10,** 1096.
7. Briguglio, J. J.; Carroll, P. J.; Corcoran, E. W., Jr.; Sneddon, L. G. *Inorg. Chem.* 1986, **25,** 4618.
8. Corcoran, E. W. Jr.; Sneddon, L. G. *J. Am. Chem. Soc.* 1984, **106,** 7793.
9. Brewer, C. T.; Grimes, R. N. *J. Am. Chem. Soc.* 1985, **107,** 3552.
10. (a) Wilczynski, R.; Sneddon, L. G. *Inorg. Chem.* 1981, **20,** 3955. (b) Wilczynski, R.; Sneddon, L. G. *Inorg. Chem.* 1982, **21,** 506.
11. Churchill, M. R.; Hackbarth, J. J.; Davison, A.; Traficante, D. D.; Wreford, S. S. *J. Am. Chem. Soc.* 1974, **96,** 4041.
12. (a) Maitlis P. M. *Acc. Chem. Res.* 1978, **11,** 301. (b) Gill, D. S.; White, C.; Maitlis, P. M. *J. Chem. Soc., Dalton Trans.* 1978, 617.
13. (a) Janowicz, A. H.; Bergman, R. G. *J. Am. Chem. Soc.* 1982, **104,** 352. (b) Hoyano, J. K.; Graham, W. A. G. *J. Am. Chem. Soc.* 1982, **104,** 3723.
14. Davan, T.; Corcoran, E. W., Jr.; Sneddon, L. G. *Organometallics* 1983, **2,** 1693.
15. Mirabelli, M. G. L.; Sneddon, L. G. *Organometallics* 1986, **5,** 1510.
16. Mirabelli, M. G. L.; Carroll, P. J.; Sneddon, L. G. *J. Am. Chem. Soc.* 1989, **111,** 592.
17. (a) Reger, D. L.; Coleman, C. J.; McElligott, P. J. *J. Organomet. Chem.* 1979, **171,** 73. (b) Reger, D. L., *Acc. Chem. Res.* 1988, **21,** 229; and references cited therein.
18. Schwartz, J.; Hart, D. W.; McGiffert, B. *J. Am. Chem. Soc.* 1974, **96,** 5613.
19. Grimes, R. N. "Carboranes"; Academic Press: New York, 1970.
20. Weiss, R.; Bowser, J. R.; Grimes, R. N. *Inorg. Chem.* 1978, **17,** 1522.
21. Fehlner, T. P. *J. Am. Chem. Soc.* 1980, **102,** 3424.

22. Pipal, J. R.; Grimes, R. N. *Inorg. Chem.* 1978, **17,** 10.
23. (a) Paetzold, P. *Adv. Inorg. Chem. Radiochem.* 1987, **31,** 123–170. (b) Paetzold, P. In "Boron Chemistry"; Hermanek, S., ed.; World Scientific: Singapore, 1987; pp. 446–475.
24. Delpy, K.; Schmitz, D.; Paetzold, P. *Chem. Ber.* 1983, **116,** 2994.
25. Paetzold, P.; Delpy, K. *Chem. Ber.* 1985, **118,** 2552.
26. (a) Scherer, O. J. *Angew. Chem., Int. Ed. Engl.* 1985, **24,** 924. (b) DiVaira, M.; Sacconi, L. *Angew. Chem., Int. Ed. Engl.* 1982, **21,** 330; and references cited therein.
27. Baudler, M.; Akpapoglou, S.; Ouzounis, D.; Wasgestian, F.; Meinigke, B.; Budzikiewicz, H.; Münster, H. *Angew. Chem., Int. Ed. Engl.* 1988, **27,** 280.
28. (a) Scherer, O. J.; Sitzmann, H.; Wolmershäuser, G. *Angew. Chem., Int. Ed. Engl.* 1985, **24,** 351. (b) Scherer, O. J.; Schwalb, J.; Swarowsky, H.; Wolmershäuser, G.; Kaim, W.; Gross, R. *Chem. Ber.* 1988, **121,** 443. (c) Scherer, O. J.; Sitzmann, H.; Wolmershäuser, G. *Angew. Chem.* 1989, **101,** 214. (d) Duff, A. W.; Jonas, K.; Goddard, R.; Kraus, H-J.; Krueger, C. *J. Am. Chem. Soc.* 1983, **105,** 5479. (e) Scherer, O. J.; Schwalb, J.; Wolmershäuser, G.; Kaim, W.; Gross, R. *Angew. Chem.* 1986, **98,** 293. (f) Scherer, O. J.; Brück, H.; Wolmershäuser, G. *Chem. Ber.* 1989, **122,** 2049. (g) Scherer, O. J.; Wiedemann, W.; Wolmershäuser, G. *J. Organomet. Chem.* 1989, **361,** C11.
29. Kudinov, A. R.; Rybinskaya, M. I.; Struchkov, Yu. T.; Yanovskii, A. I.; Petrovskii, P. V. *J. Organomet. Chem.* 1987, **336,** 187.
30. (a) Chivers, T. *Chem. Rev.* 1985, **85,** 341. (b) Woolins, J. D. "Non-Metal Rings, Cages, and Clusters", Wiley: New York, 1988.
31. DiMaio, A. J.; Rheingold, A. L. *Inorg. Chem.* 1990, **29,** 798.
32. (a) Maxwell, W. M.; Miller, V. R.; Grimes, R. N. *J. Am. Chem. Soc.* 1974, **96,** 7116. (b) Maxwell, W. M.; Miller, V. R.; Grimes, R. N. *Inorg. Chem.* 1976, **15,** 1343.
33. (a) Grimes, R. N. *Adv. Inorg. Chem. Radiochem.* 1983, **26,** 55; and references cited therein. (b) Grimes, R. N. *Acc. Chem. Res.* 1983, **16,** 22.
34. Venable, T. L.; Maynard, R. B.; Grimes, R. N. *J. Am. Chem. Soc.* 1984, **106,** 6187.
35. (a) Fessler, M. E.; Spencer, J. T.; Lomax, J. F.; Grimes, R. N. *Inorg. Chem.* 1988, **27,** 3069. (b) Whelan, T.; Spencer, J. T.; Pourian, M. R.; Grimes, R. N. *Inorg. Chem.* 1987, **26,** 3116. (c) Boyter, H. A., Jr.; Grimes, R. N. *Inorg. Chem.* 1988, **27,** 3075.
36. (a) Maynard, R. B.; Grimes, R. N. *J. Am. Chem. Soc.* 1982, **104,** 5983. (b) Grimes, R. N.; Maynard, R. B.; Sinn, E.; Brewer, G. A.; Long, G. J. *J. Am. Chem. Soc.* 1982, **104,** 5987.
37. (a) Finster, D. C.; Sinn, E.; Grimes, R. N. *J. Am. Chem. Soc.* 1981, **103,** 1399. (b) Wang, Z-T.; Sinn, E.; Grimes, R. N. *Inorg. Chem.* 1985, **24,** 826 and 834.
38. Brewer, C. T.; Grimes, R. N. *J. Am. Chem. Soc.* 1985, **107,** 3558.
39. Gaines, D. F.; Nelson, C. K.; Steehler, G. A. *J. Am. Chem. Soc.* 1984, **106,** 7266.
40. Bould, J.; Greenwood, N. N.; Kennedy, J. D. *Polyhedron* 1983, **2,** 1401.
41. Simpson, P. G.; Lipscomb, W. N. *J. Chem. Phys.* 1963, **39,** 26.
42. Corcoran, E. W. Jr.; Sneddon, L. G. *J. Am. Chem. Soc.* 1985, **107,** 7446.
43. Fehlner, T. P. In "Advances in Boron and the Boranes [Mol. Struct. Energ. Vol. 5]"; Liebman, J. F.; Greenberg, A.; Williams R. E., Eds.; VCH Publishers, Inc.: New York, 1988; Chapter 12, pp. 265–285.
44. Vites, J. C.; Housecroft, C. E.; Jacobsen, G. B.; Fehlner, T. P. *Organometallics* 1984, **3,** 1591.
45. Housecroft, C. E.; Fehlner, T. P. *Organometallics* 1986, **5,** 379.
46. Beno, M. A.; Williams, J. M.; Tachikawa, M.; Muetterties, E. L. *J. Am. Chem. Soc.* 1981, **103,** 1485.
47. Lynam, M. M.; Chipman, D. M.; Barreto, R. D.; Fehlner, T. P. *Organometallics* 1987, **6,** 2405.
48. Vites, J. C.; Jacobsen, G. B.; Dutta, T. K.; Fehlner, T. P. *J. Am. Chem. Soc.* 1985, **107,** 5563.
49. Fehlner, T. P. *New J. Chem.* 1988, **12,** 307.
50. Housecroft, C. E. *Polyhedron* 1987, **6,** 1935.

51. (a) Foust, A. S.; Foster, M. S.; Dahl, L. F. *J. Am. Chem. Soc.* 1969, **91,** 5633. (b) Foust, A. S.; Campana, C. F.; Sinclair, J. D.; Dahl, L. F. *Inorg. Chem.* 1979, **18,** 3047. Also see references in Ref. 23a.
52. Allen, A. D.; Senoff, C. V. *Chem. Commun.* 1965, 621.
53. Morris, R. H.; Ressner, J. M.; Sawyer, J. F.; Shiralian, M. *J. Am. Chem. Soc.* 1984, **106,** 3683.
54. (a) Dinitrogen complexes: Chatt, J.; Dilworth, J. R.; Richards, R. L. *Chem. Rev.* 1978, **78,** 589. Colquhoun, H. M. *Acc. Chem. Res.* 1984, **17,** 23. Leigh, G. J. *Transition Met. Chem.* 1986, **11,** 118; and references cited therein. (b) Alkyne complexes: Braunstein, P. *Chem. Rev.* 1983, **83,** 203. Palyi, G. *et al.* In "Stereochemistry in Inorganic and Organometallic Compounds", Vol. 1; I. Bernal, Ed.; Elsevier: Amsterdam, 1986.
55. Latham, I. A.; Sita, L. R.; Schrock, R. R. *Organometallics* 1986, **5,** 1508; and references cited therein.
56. (a) Bergman, R. G.; Becker, P. N. *J. Am. Chem. Soc.* 1983, **105,** 2985. (b) Connely, N. G.; Payne, J. D.; Geiger, W. E. *J. Chem. Soc., Dalton Trans.* 1983, 295.
57. Bottomley, F.; Lin, I. J. B. *J. Chem. Soc., Dalton Trans.* 1981, 271.
58. Stevens, R. E.; Guettler, R. D.; Gladfelter, W. L. *Inorg. Chem.* 1990, **29,** 451.
59. (a) Taube, H. *Prog. Inorg. Chem.* 1986, **34,** 607. (b) Klotz, I. M.; Kurtz, D. M., Jr. *Acc. Chem. Res.* 1984, **17,** 16. (c) Niederhoffer, E. C.; Timmons, J. H.; Martell, A. E. *Chem. Rev.* 1984, **84,** 137.
60. (a) Shilov, A. E. "Activation of Saturated Hydrocarbons by Transition Metal Complexes"; Reidel: Dordrecht; 1984. (b) Deem, M. L. *Coord. Chem. Rev.* 1986, **74,** 101.
61. (a) Crabtree, R. H. *J. Am. Chem. Soc.* 1986, **108,** 7222. (b) Bergman, R. G.; Periana, R. A. *J. Am. Chem. Soc.* 1986, **108,** 7346.
62. Haim, A. *Prog. Inorg. Chem.* 1983, **30,** 273.
63. Clark, H. C. *Organometallics* 1982, **1,** 64.
64. Belmont, J. A.; Soto, J.; King, R. E. III; Donaldson, A. J.; Hewes, J. D.; Hawthorne, M. F. *J. Am. Chem. Soc.* 1989, **111,** 7475; and references cited therein.
65. (a) Behnken, P. E.; Belmont, J. A.; Busby, D. C.; Delaney, M. S.; King, R. E. 3rd; Kreimendahl, C. W.; Marder, T. B.; Wilczynski, J. J.; Hawthorne, M. F. *J. Am. Chem. Soc.* 1984, **106,** 3011. (b) Behnken, P. E.; Busby, D. C.; Delaney, M. S.; King, R. E. 3rd; Kreimendahl, C. W.; Marder, T. B.; Wilczynski, J. J.; Hawthorne, M. F. *J. Am. Chem. Soc.* 1984, **106,** 7444.
66. (a) Männig, D.; Nöth, H. *Angew. Chem., Int. Ed. Engl.* 1985, **24,** 878. (b) Evans, G. A.; Fu, G. C.; Hovedya, A. H. *J. Am. Chem. Soc.* 1988, **110,** 6977.
67. (a) Whitmire, K. H.; Raghuveer, K. S.; Churchill, M. R.; Fettinger, J. C.; See, R. F. *J. Am. Chem. Soc.* 1986, **108,** 2778. (b) Whitmire, K. H.; Churchill, M. R.; Fettinger, J. C. *J. Am. Chem. Soc.* 1985, **107,** 1056. Churchill, M. R.; Fettinger, J. C. *J. Am. Chem. Soc.* 1985, **107,** 1056.
68. Sola, J.; Do, Y.; Berg, J. M.; Holm, R. H. *Inorg. Chem.* 1985, **24,** 1706.
69. (a) Siebert, W. *Angew. Chem., Int. Ed. Engl.* 1985, **24,** 943; *Pure Appl. Chem.* 1987, **59,** 947. (b) Grimes, R. N. *Pure Appl. Chem.* 1987, **59,** 847. (c) Herberich, G. E. In "Comprehensive Organometallic Chemistry"; Wilkinson, G., Stone, F. G. A.; Abel, E., Eds.; Pergamon Press: Oxford, 1982; Chapter 5.3. (d) Grimes, R. N. *Coord. Chem. Rev.* 1979, **28,** 47.
70. (a) Davis, J. H., Jr.; Sinn, E.; Grimes, R. N. *J. Am. Chem. Soc.* 1989, **111,** 4776. (b) Davis, J. H., Jr.; Sinn, E.; Grimes, R. N. *J. Am. Chem. Soc.* 1989, **111,** 4784; and references cited therein. (c) Fessenbecker, A.; Attwood, M. D.; Bryan, R. F.; Grimes, R. N.; Woode, M. K.; Stephan, M.; Zenneck, U.; Siebert, W. *Inorg. Chem.* 1990, **29,** 5157.
71. Fessenbecker, A.; Attwood, M. D.; Grimes, R. N.; Stephan, M.; Pritzkow, H.; Zenneck, U.; Siebert, W. *Inorg. Chem.* 1990, **29,** 5164.
72. Piepgrass, K. W.; Davis, J. H., Jr.; Sabat, M.; Grimes, R. N. *J. Am. Chem. Soc.* 1991, **113,** 681.

73. (a) Kuhlmann, T. H.; Roth, S.; Roziere, J.; Siebert, W. *Synth. Metals* 1987, **19,** 757. (b) Siebert, W. *Pure Appl. Chem.* 1988, **60,** 1345.
74. (a) Edwin, J.; Bochmann, M.; Boehm, M. C.; Brennan, D. E.; Geiger, W. E. Jr.; Kruger, C.; Pebler, J.; Pritzkow, H.; Siebert, W.; Swiridoff, W.; Wadepohl, H.; Weiss, J.; Zenneck, U. *J. Am. Chem. Soc.* 1983, **105,** 2582. (b) Merkert, J. M.; Geiger, W. E.; Davis, J. H., Jr.; Attwood, M. D.; Grimes, R. N. *Organometallics* 1989, **8,** 1580. (c) Merkert, J.; Davis, J. H., Jr.; Grimes, R. N.; Geiger, W. Abstracts of Papers, American Chemical Society National Meeting, Boston, MA, April 1990; Abstract INOR 92.
75. Liston, D. J.; Reed, C. A.; Eigenbrot, C. W.; Scheidt, W. R. *Inorg. Chem.* 1987, **26,** 2739.
76. Liston, D. J.; Lee, Y. J.; Scheidt, W. R.; Reed, C. A. *J. Am. Chem. Soc.* 1989, **111,** 6643.

7

The Metal–Nonmetal Bond in the Solid State

Timothy Hughbanks

1. TOOLS FOR UNDERSTANDING BONDING IN EXTENDED SYSTEMS

Undoubtedly, the most useful tools for understanding chemical bonds in molecules or solids are those that serve to *correlate* the wealth of structural and physical data known to chemists. This means that the Zintl concept and the simple ideas of the electron pair bond handed down by Lewis, Pauling and others will and should have a prominent role in the way chemists think about chemical bonds. Bearing this in mind our discussion below will begin with a review of the kinds of solid state compounds that can be best understood with the Zintl viewpoint.

Of course, molecular orbital theory has come into wide use as a powerful quantitative and qualitative tool for the understanding of bonding—one that has allowed chemists to generalize considerably beyond the often confining frame of mind that the elementary approach enforces. However, the discussion of bonding in extended crystalline systems presents difficulties if we wish to analyze such systems by making a detailed examination of the valence electronic structure. It is obvious that where extended systems are concerned, the

Timothy Hughbanks • Department of Chemistry, Texas A&M University, College Station, Texas 77843-3255.

Inorganometallic Chemistry, edited by Thomas P. Fehlner. Plenum Press, New York, 1992.

sheer number of "molecular" orbitals becomes so large that an orbital-by-orbital analysis becomes unwieldy.

Our motivation for developing new tools for understanding condensed systems arises from several sources. First is an intuition borne of our structural knowledge that crystalline solids are after all structurally *repetitive*. We therefore need to analyze the results of electronic structure calculations, or learn to make rough guesses as to what such calculations would yield, in such a way as to focus on the *local* information about the building blocks and push the *delocalized* molecular (band) orbitals into the background. Nevertheless, when thinking about the electronic structure of solids, it is a common error of chemists to push the mental localization of bonding too far into the localized extreme, and we must bear in mind that sometimes the correct and ultimately most useful description of electronic states involves a manifestly delocalized picture. (Solid state physicists have historically been at least as apt to make the opposite error—and until recent years have been reluctant to recognize that even delocalized states can be built up from a localized starting point.)

As is well known, band theory is the crystalline analog to molecular orbital theory. In discussing molecules, we have learned to think in terms of the crude product of molecular orbital calculations: the molecular orbital wavefunctions and corresponding one-electron energies (eigenvalues). The latter are conveniently represented in familiar MO energy diagrams. In band theory the distribution of one-electron energy levels is represented in two ways. First, one may plot the detailed *band structure* or *dispersion curves,* in which each electronic energy band's dispersion is plotted with respect to wavevector **k,** which carries information about the crystal orbitals' phase on translation from one unit cell to the next. Although plots of the band structure are useful for specific purposes, they can sometimes appear to be a complex jungle of information and for two- and three-dimensional extended systems are incomplete in that dispersion curves can only be plotted along lines in the two- or three-dimensional space over which **k** ranges. In a second method of presenting results of a band structure calculation, one may plot the *density of states* (DOS), which is simply the distribution of one-electron levels as a function of energy. One may think of full band structure plots and DOS plots as two ways of laying out molecular orbital diagrams for the giant *molecule* that is a perfect crystal.

In the DOS the "level diagram" is shorn of its symmetry labels (**k** may be thought of as just a translational symmetry label); one is looking at only the raw distribution of energy levels. In looking at DOS plots it is as if we are seeing a continuous analog of a molecular orbital diagram, without any symmetry labels to guide us as to the character of the orbitals whose energies are represented. The necessity for more detailed knowledge impels us to invent methods for more detailed analysis. By use of population analysis, we can extract much more chemical information from DOS plots. Contributions to the DOS made by specific atoms, atomic orbitals, or molecular fragment orbitals can be plotted in the same manner as the total DOS. Just as for molecular orbitals, we can qualitatively decide whether a particular band has predominantly metal d character, is built up from the π orbitals on a catenated nonmetal fragment, etc. Of course, chemists have an overriding interest in understanding *bonding.* A particularly useful tool in examining the bonding (or antibonding) nature of crystal orbitals has come with the development of *Crystal Orbital Overlap Population* (COOP) diagrams. COOP curves may be constructed for any given atom pair (bond) in the structure under examination. At any given energy, the DOS is weighted by the atom–atom overlap population contributed by crystal orbitals at that energy. For the pair of atoms in question, if the crystal orbitals at a given energy are *bonding* (i.e., have a *positive* overlap population between the two atoms) then the COOP takes on a positive value at that energy; if the crystal orbitals at that energy are *antibonding* (i.e., have a *negative* overlap population between the two atoms) then the COOP takes on a negative value at that energy. Examination of the variations in COOP curves as a function of energy gives us a simple graphical picture of the bonding nature of band orbitals.

In much of the remainder of this chapter we will be using COOP and DOS plots to illuminate bonding in condensed compounds. The use of these tools should become clearer when applied to real chemical systems and we will not go into any further detail at this juncture.

2. SOLID STATE AND MOLECULAR CHEMISTRY: ZINTL COMPOUNDS

2.1. Nets Built from Nonmetals[1,2]

The Zintl concept is a simple and powerful notion that allows a basic understanding of a large and important subclass of solid state compounds. Simply stated, Zintl showed that some compounds, say a binary with the generic formula A_xB_y, could be understood if we assumed that the electrons on the electropositive elements (A) were transferred to the electronegative elements (B) which used these excess electrons in forming (B—B) bonds or in the form of localized lone pairs. We expect that elements tend to do this in order to achieve closed-shell configurations and this means that Zintl's viewpoint often represents nothing more than the tendency for atoms within molecules or extended networks to conform to the "octet rule." Compounds with structures for which we can at least *think* that this occurs are referred to as "valence compounds."

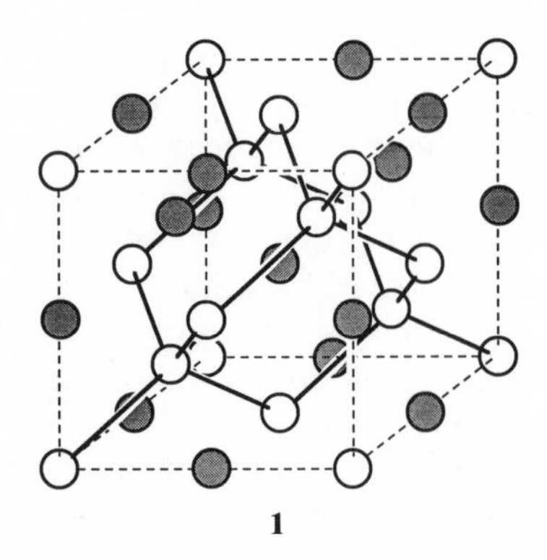

1

Let us consider some elementary examples. One may easily recognize the network of linked tetrahedra that constitute the familiar diamond structure embedded within the NaTl structure shown in **1.** Of course, this tetrahedral network structure is adopted by carbon (diamond), silicon, germanium, and gray Sn. The four-valence electrons are understood, within the valence bond picture, to occupy sp^3 hybrid orbitals that overlap to form $2e^-$ covalent bonds. From the isoelectronic NaIn, NaTl, LiAl, and LiGa compounds we draw the inference that the electropositive alkali metals have simply donated their lone

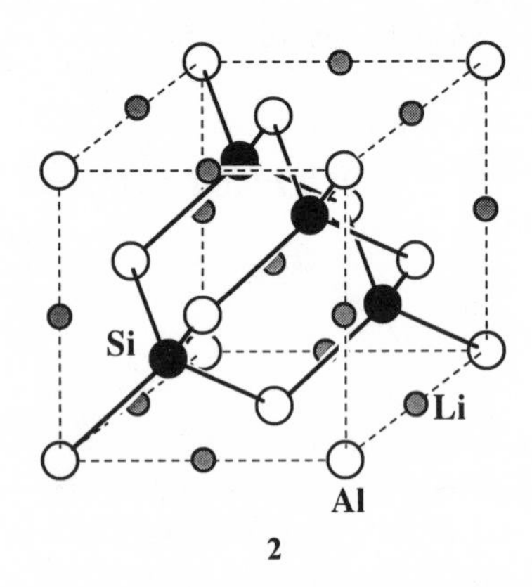

2

electron to the nonmetal–nonmetal bonds. The ternary LiAlSi (**2**) is just an intermediate case in which only half as many Li sites are filled, which corresponds to the electronic requirements of the sphalerite (GaAs-like) Al—Si net that underlies the structure. The diamond net need not be preserved to maintain a tetrahedral nonmetal network; the larger size of Sr^{2+} in $SrAl_2$ seems to force the Al network to open up without changing the fact that each Al center is involved in four Al—Al bonds (**3**).

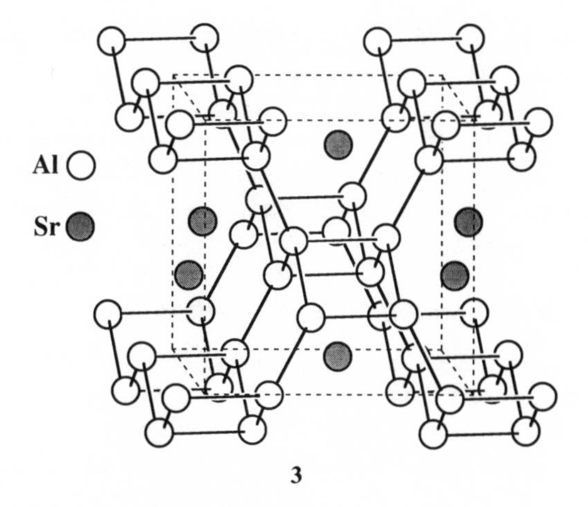

3

A three-dimensional network need not be preserved if there are more than four electrons available per electronegative atom. In this case we see the appearance of lone pairs, just as expected in main-group molecular structures. The elemental arsenic structure is easily understood as one in which the diamond-like structure is preserved in two dimensions but, because five electrons per As atom must be accommodated, the "fourth bond" is replaced by

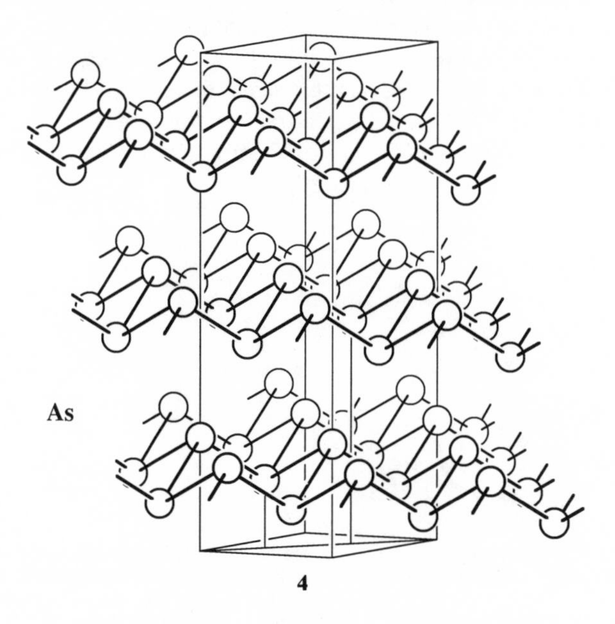

4

a lone pair (**4**). $CaSi_2$ is merely a stuffed structure in which puckered layers of silicon have the same form as the layers of the arsenic structure, between which Ca^{2+} ions are inserted. The dimensionality of the nonmetal network can be further reduced of course; one-dimensional chains are found in fibrous sulfur, α-selenium, α-tellurium as well as in LiAs, CaSi, $CaSb_2$, and many more materials. Tetrahedrane-like clusters are the structural basis for gas phase and white-phosphorus obtained by condensation of the vapor, but they are also found in the Zintl phases KGe, $BaSi_2$, and NaPb, making these compounds quasi-molecular.

2.2. Zintl-Like Compounds with Transition Metals

A Zintl-like approach can often be used in a broad analysis of transition metal chalcogenides and pnictides, as long as the metal-to-nonmetal ratio is not large. When we mentally disassemble transition metal polychalcides and polypnictides it is quite often simple to identify the transition metal as being in a common or "stable" oxidation state, and the nonmetals form polyanionic networks that are reasonable, "octet" structures. For example, the dichalcides $Pd(S,Se)_2$ form layer structures in which the puckered sheets shown in Figure 1 are found.[3] The chalcogens are present as dichalcide ions (X_2^{2-}; S—S, 2.13 Å; Se—Se, 2.36 Å) and the square-planar palladium is formulated, as usual, as Pd^{II}. Pd—X distances between layers are greater than 3.2 Å. The same sort of layers can be identified in PdPS and PdP_2, but in these compounds the layers are stitched together by P—P bonds. On paper, we can view PdPS as derived from PdS_2 by a one-electron oxidation of the X_2 unit (via replacement of sulfur by phosphorus), followed by homoatomic interlayer bond formation driven by the radical centers so generated (see Figure 2). In PdPS, the layers pair and $P_2S_2^{4-}$ units extend across the two layers of each double layer. PdP_2 is a fully three-dimensional structure in which polyphosphide chains serve to thread the layers together.

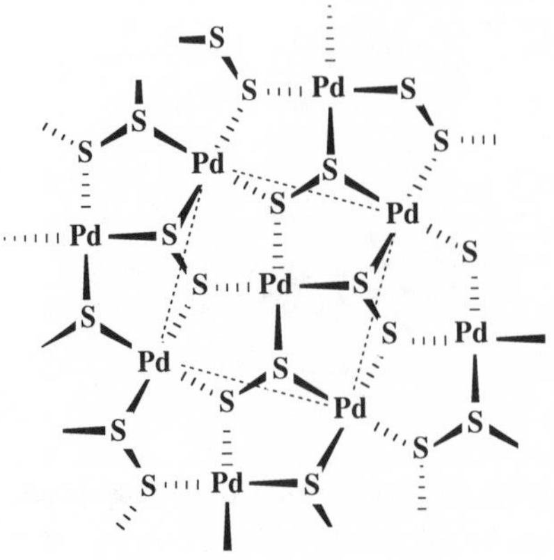

Figure 1. One puckered layer from PdS_2. The Pd^{II} centers have approximately square-planar coordination, as expected for a d^8 Pd center. The S_2^{2-} units are tilted with respect to the plane containing the Pd centers so that lone pairs on half the sulfurs project above the plane and half project below.

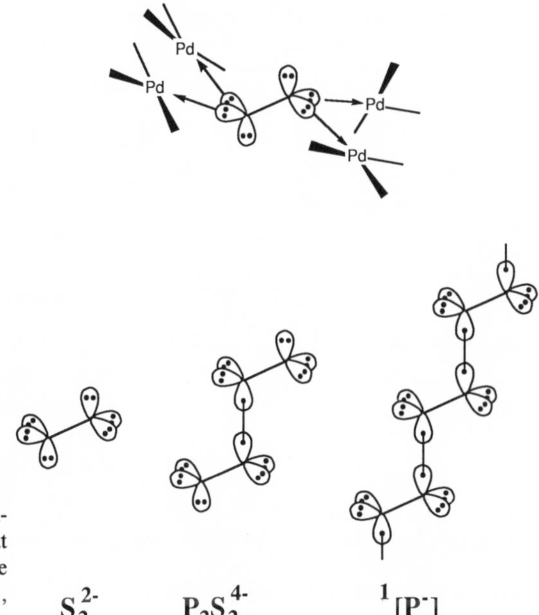

Figure 2. A view of the environment about a S_2^{2-} unit is given at left. A schematic view of the linkage between nonmetals found in PdS_2, PdPS, and PdP_2 is shown at right.

Polyphosphide networks can extend into two dimensions in more phosphorus-rich materials. In several compounds ten-membered phosphorus rings are fused into sheets in such a way that half of the phosphorus centers form bonds to three phosphorus neighbors and half form bonds to only two neighboring phosphorus centers. The ten-membered rings pucker in such a way as to form distorted octahedral pockets to accommodate divalent transition metals. The structure of MP_4 (M = Ru, Os), shown in Figure 3, gives an example of how the process described is accomplished in one instance.[4] FeP_4 and the long-known CdP_4 employ the same structural building blocks (puckered P_{10} rings centered by octahedral metals) but the ring fusion to form the final structures is different.[5,6] The verbal description above can be summarized by writing the chemical formulas for these compounds as $M_3P_{12} = (M^{II})_3(\mu_2\text{-}P^-)_6(\mu_3\text{-}P)_6$. In doing so we recognize that these are understandable materials in terms of the coordinate covalent metal–ligand bond. The nonmetal network collectively assumes the role of the "polydentate ligand" where each two-coordinate phosphido center has two lone pairs and each triply-bonded center donates one lone pair to the ensconced transition metals. The group 8 metals are low-spin, closed-shell d^6 (t_{2g}^6) ions and Cd is an even simpler d^{10} dication. King has discussed simple schemes for rationalizing structures and bonding of such transition metal polyphosphides.[7]

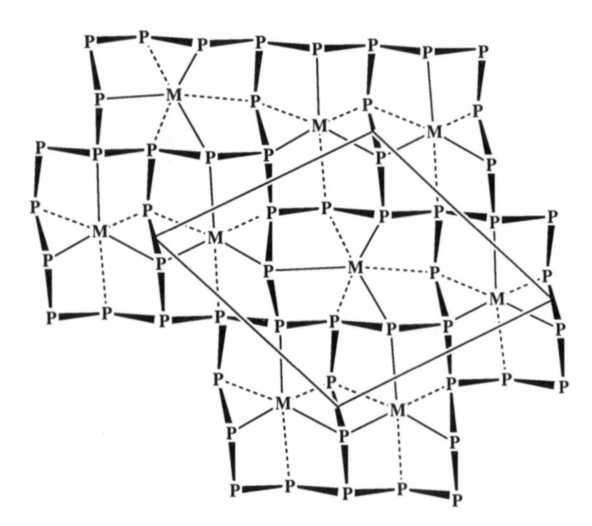

Figure 3. One puckered layer from structure of MP_4 (M = Ru,Os) compounds is depicted. Metals all sit in sites of approximate octahedral coordination, phosphorus centers bound to three adjacent phosphorus atoms form one bond to a metal, phosphorus centers bound to two adjacent phosphorus atoms form two bonds to metals.

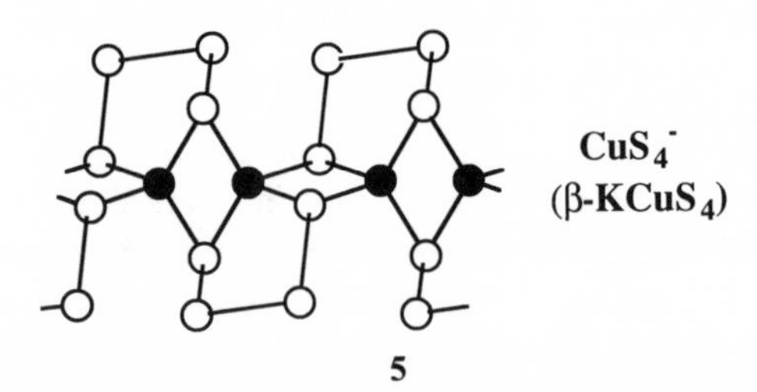

Much work has recently been reported on the synthesis of new chalcogen-rich materials by use of low-temperature polychalcogenide melts as reagents.[8–10] This synthetic approach has proven especially effective as a way of getting to materials with polychalcogenide ions as in $ACuS_4$ compounds (A = K, NH_4), illustrated in **5**.[11,12] In most such materials the bonding is no mystery; the polychalcogenide has two uninegative ends (such as $^-S—S—S—S^-$ in the ACu^IS_4 compounds. Many examples are given in Kanatzidas's recent review.[10]

2.3. The Skutterudites and Their Ternary Relatives

2.3.1. Electron-Precise Compounds

To convey some impression for the kind of information band-structure calculations can give in analyzing the bonding in the sort of materials we

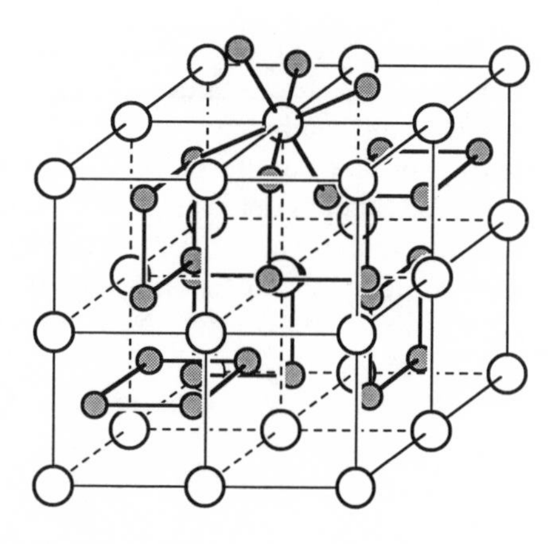

Figure 4. The structure of *skutterudite*, $CoAs_3$, is shown. Cobalt ions, shown as open circles, have octahedral coordination. The As atoms (shaded circles) are bound in As_4 squares and each bound to two adjacent metals. The structure can be seen as a simple cubic array of metals—As_4 squares (rectangles, actually) are inserted into three-quarters of the cubic boxes. The $Ln(M_4P_{12})$ compounds discussed in the text are obtained by inserting the lanthanides into the remaining unfilled boxes.

have discussed, let us consider some specific examples. We will take a more detailed look at the *skutterudites,* with the composition ME_3 (M = Co, Rh, Ir; E = P, As, Sb). The cubic structure adopted by these compounds, illustrated in Figure 4, consists of a network of octahedrally coordinated transition metals and square E_4 rings. If the E_4 rings are to be "saturated" species, we would naturally formulate the compounds as $(M^{III})_4(E_4^{4-})_3$ and we need only identify the group 9 transition metal cations as low-spin d^6 entities to fit these compounds into a Zintl-like scheme. These materials are diamagnetic semiconductors (e.g., CoP_3) and so this formalism is quite consistent with the materials' properties.[13–17] One can easily imagine that substitutional chemistry could be undertaken in these materials, and so it has. In the variants $M_2Ge_3E_3$ = $(M^{III})_4(Ge_2E_2)_3$ (M^{III} = Co, Rh, Ir; E = S, Se) germaniums and chalcogens

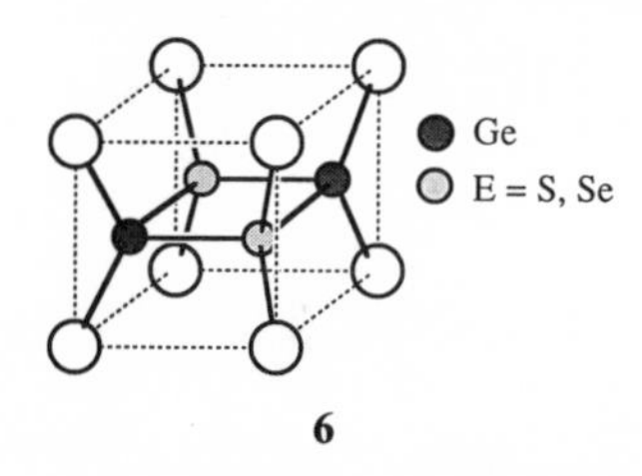

6

alternate within the squares (**6**), while in $FeNiSb_6$ the iron and nickel obviously substitute for cobalt.[18,19] All are isoelectronic and crystallize as skutterudite modifications.

2.3.2. Breaking the Counting Rules—More Interesting Properties

If the vacant "boxes" of the $CoAs_3$ structure are stuffed with lanthanide ions and the group 9 elements exchanged for group 8 atoms, one obtains the LnM_4E_{12} (Ln = lanthanide; M = Fe, Ru; E = P, As, Sb) series of compounds.[20–22] When the lanthanide is clearly trivalent, one has a $M_4E_{12}{}^{3-}$

framework that is one electron deficient compared to the parent skutterudites. Even more oxidized skutterudites, BaM_4Sb_{12}, were recently reported.[23,24] It is not immediately apparent whether the "holes" in these materials are primarily localized on the metal centers or on the E_4 rings—or whether they are best thought of as delocalized. This doubt surrounding the nature of the holes in skutterudites can be appreciated by considering the two alternative viewpoints implied by the E_4 ring diagrams in **7**. Is it more appropriate to consider these rings as conforming to the classical Zintl rules in which every E center

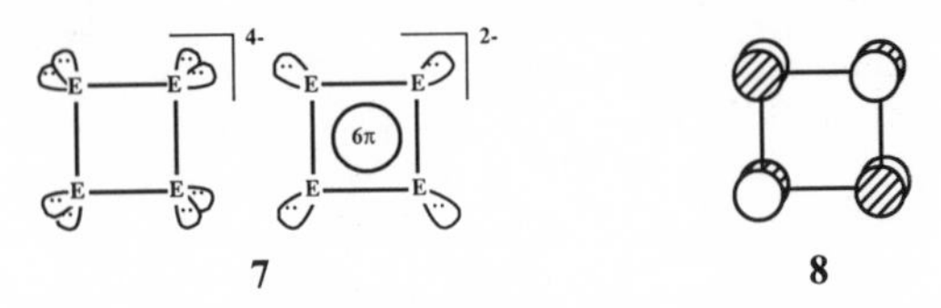

satisfies its valence by using sp^3 hybrids to form E—E bonds or to accommodate lone pairs? Or can the planar E_4 rings be oxidized by removal of electrons from the highest-lying π^* orbital (**8**) in order to become "aromatic"?

Band-structure calculations shed some light on these questions; consider the total density of states for CoP_3 obtained from an extended Hückel calculation plotted in Figure 5. In accordance with the semiconducting properties of the simple skutterudites,[14,15,17] a band gap is found at the point where one has 24 valence electrons per formula unit. This gap is considerably larger than is reasonable for such a material, but this seems to be an artifact of extended Hückel (EH) calculations in which higher-lying *d* orbitals are not included for main group elements. (The occupied bands are usually little

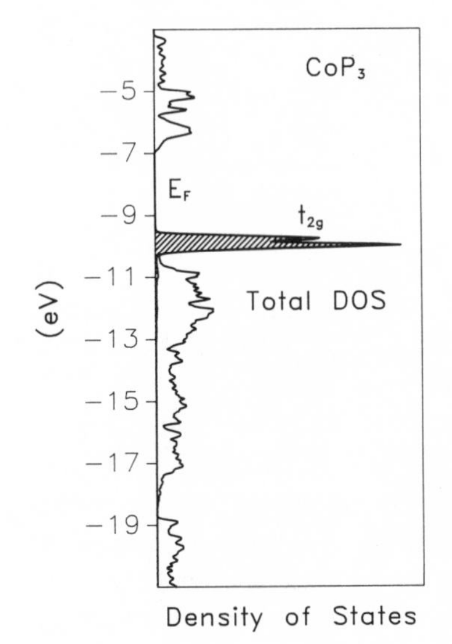

Figure 5. The Density of States (DOS) for CoP_3 is plotted. The sharply peaked shaded portion is the projection of the cobalt $d(t_{2g})$ orbitals. The Fermi level lies just above this peak in a gap between occupied and unoccupied levels.

affected by the inclusion of *d* orbitals on main group elements, but the *d* orbitals can mix more into the higher-lying conduction bands.) The projected DOS for the Co "t_{2g}" orbitals is indicated by the shaded region in the figure and clearly constitutes about 90% of the levels at the valence band edge. Thus, when holes are created in the valence band, these nonbonding *d* orbitals are implicated as the site of oxidation. The energy range near the Fermi level is shown in close-up in Figure 6, where it may be seen that oxidation should involve only metal localized levels. (Unfortunately this is a parameter-sensitive conclusion; the results of similar extended Hückel calculations have been recently published with another parametrization for phosphorus that give a different conclusion as to the nature of the band structure at the valence band edge.[25] The more general aspects of bonding discussed here are not altered.)

A search for some portion of the DOS that has primarily P_4 π^* character is futile. The projected DOS for the ring π^* orbitals is a distribution that spreads throughout the energy range shown in Figure 5. This simply means that the interactions between phosphorus and the metals are dominant over the π interaction between phosphorus centers within the rings.

The DOS can be interrogated to tell us about the character of the band orbitals with respect to Co—P or P—P bonding. In the two COOP plots shown in Figure 7, the orbital energy is increasing from left to right and the magnitude of each curve above (below) the baseline in each plot indicates the extent of the bonding (antibonding) character of the crystal orbitals. The COOP plot for the P—P bonds in the lower panel actually represents the average of two symmetry-inequivalent P—P bonds within the P_4 rings—this averaging has been performed for the sake of economy and because the COOP

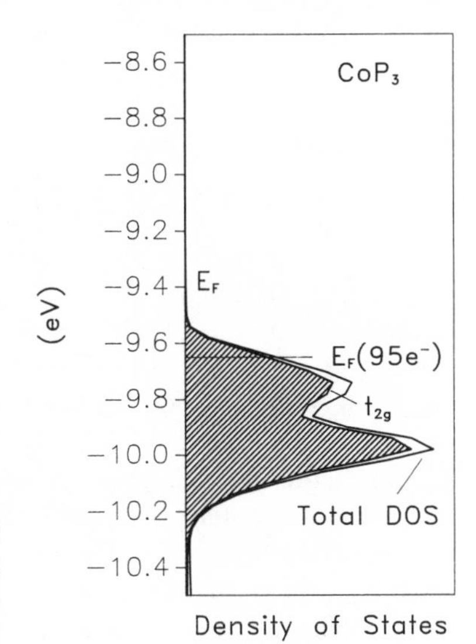

Figure 6. A "close-up" of the CoP_3 DOS at the valence band edge is shown. The Fermi level marked for a 95-electron case (as for $LaFe_4P_{12}$) shows that electrons in oxidized skutterudites are removed from levels with almost pure M–t_{2g} character.

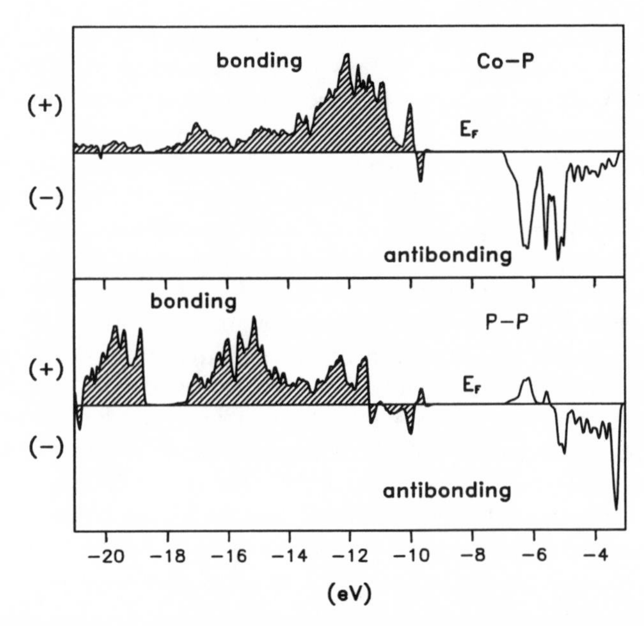

Figure 7. Crystal Orbital Overlap Population (COOP) curves are plotted for the Co—P (top) and P—P (bottom) in CoP_3. In these plots energy increases from left to right and the occupied levels are shown as shaded. Note that the t_{2g} peak that is so conspicuous in Figure 6 is much less prominent here. This is indicative of the nonbonding character of these levels.

plots for the two bonds are qualitatively similar anyway. We can see that the orbitals in the lower reaches of the displayed energy range (below about −14 eV) have predominantly P—P bonding character with some Co—P bonding character. The occupied band orbitals from −14 to about −10.5 eV are predominantly Co—P bonding, but with some P—P bonding admixture. What is most striking in comparing the DOS plot with the COOP plots is the near "disappearance" of the t_{2g} levels near the Fermi level in the latter. This illustrates graphically the expected nonbonding character of these levels.

The electron-deficient materials are more interesting precisely because they violate the simple electron counting rules that account for the properties of the simple binary skutterudites. If one thing is certain, it is that electron (or *hole*) localization does not occur; the materials LaM_4P_{12} (M = Fe, Ru, Os) are metallic and become superconducting at low temperatures (T_c = 4.1, 7.4, and 1.8 K, respectively).[26] But why are the holes not localized? To begin answering this question we must understand how electron–electron repulsion influences the system. The narrowness of the t_{2g} bands for the skutterudites is exemplified for CoP_3 in Figure 5, and understandable on the grounds that the transition metal t_{2g} orbitals are the nonbonding set for transition metals in an octahedral environment. Thus, localization might have been expected.

In such a situation the material is a true mixed-valence compound for which the best simple characterization would be $La^{III}[M_3{}^{II}M^{III}]P_{12}$. If there were *no* coupling between metal centers at all, this is certainly what we would expect since a localized state minimizes hole–hole repulsion. To put it another way, a localized state avoids contributions to the many-electron wavefunction that involve high-energy states in which two holes spend part of their time on the same ion. This means that a delocalized *band* description of a given pair of holes in the valence band implies some terms in the wavefunction that have $[M^{IV} + M^{II}]$ character—two holes on the same ion. The redox potentials of truly isolated ions would obviously disfavor such states. The quantum mechanical "truth" must be that the system manages to avoid much of the electron–electron repulsion implied by the simple band description but retains the essential delocalized nature that is characteristic of the band picture. Unfortunately, our ability to predict when localized states will prevail is still inadequate. For the LaM_4P_{12} materials, it would be interesting to know whether the question of such potential localization is involved in the materials' superconductivity.

3. METAL–NONMETAL MULTIPLE BONDING IN SOLIDS

3.1. Nitrides, Carbides

Outside of the oxides, unambiguous metal–nonmetal multiple bonds are a relatively rare occurrence in extended network solids. In the oxides, multiple bonding to metals with acceptor *d* orbitals is quite common, although the shortened bonds that occur in these cases have not always been understood as such by solid state chemists. Rather than discuss oxides here, we shall turn our attention to some newer examples of metal–nonmetal multiple bonding in solids that are being uncovered in the chemistry of the nitrides. As we look at some recent developments in this chemistry, the reader will note that the chemical transition from the molecular realm to the solid state is nearly seamless, and our understanding of the bonding in these materials carries directly over as well.

Consider, for example, the series of 16 valence electron molecules that includes $NO_2{}^+$, CO_2, N_2O, $N_3{}^-$, $CN_2{}^{2-}$. A solid state synthesis has produced the symmetrical $BN_2{}^{3-}$ species, isolated in alkali metal salts $M_3{}^IBN_2$ (M^I = Li, Na) has the expected short B—N bond length of 1.34 Å.[27–29] An even more remarkable addition to this molecular family is $ZnN_2{}^{4-}$, an ion found imbedded within the solid-state structure of Ca_2ZnN_2, prepared in DiSalvo's laboratories by a solid-state reaction of Ca_3Zn_2 and Zn powder under flowing N_2 at 680 °C.[30] The Ca_2ZnN_2 structure is shown in Figure 8; the Zn centers have a rigorous linear, symmetric coordination by nitrogen with short Zn—N distances of 1.84 Å. Each of the calcium and nitrogen centers have five neigh-

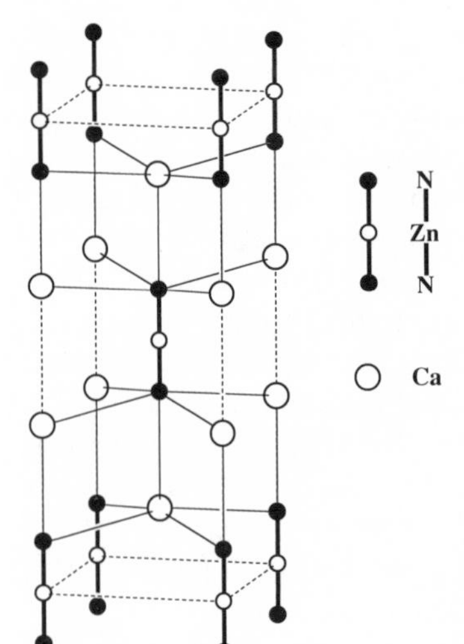

Figure 8. The structure of Ca_2ZnN_2 is shown. In the tightly bound ZnN_2^{4-} ions the Zn–N distances are a short 1.84 Å. The Ca–N distances average 2.51 Å.

bors [Ca—N = 2.54(×4), 2.41 Å]. How are we to think about this material? One approach to the bonding prompts us to write $(Ca^{2+})_2[N{=}Zn{=}N]^{4-}$ and the octet rule for zinc is obeyed. Of course, this structure only makes sense if the Zn $4p\pi$ orbitals are sufficiently low in energy to interact well with the nitrides' $2p\pi$ orbitals—recall that the Zn $3d$ orbitals are fully occupied and cannot participate in π bonding. A wholly ionic picture $[(Ca^{2+})_2Zn^{2+}(N^{3-})_2]$ seems unsatisfactory because the linear coordination of the zinc is then inexplicable. One may perform molecular orbital calculations on ZnN_2^{4-}, but the π donor strength of the nitrides would be exaggerated without accounting for the effect of the surrounding Ca^{2+} ions.

Ca_3CrN_3 is another quasi-molecular material in which Ca^{2+} ions serve to stabilize the unusual CrN_3^{6-} ion.[31a] The CrN_3^{6-} ions have C_{2v} symmetry, distorted from an ideal D_{3h} structure towards T-geometry (see the top of Figure 9). Magnetic data indicate that Cr^{III} has an unprecedented low-spin ($S = ½$) electronic configuration. The ion's shape is consistent with this and would be expected to distort in just the manner observed due to a Jahn–Teller instability for the D_{3h} geometry, as we discuss below. We should note that the CrN_3^{6-} ions would otherwise be expected to show *some* distortion since Ca_3CrN_3 crystallizes in an orthorhombic space group. However, in the observed structure not only is the angle distortion quite large, the Cr—N bond length asymmetry is that expected for a "T-distortion."

The above remarks are supported explicitly by calculation. Figure 9 shows an orbital diagram for the Cr d levels, derived from extended Hückel MO calculations for the D_{3h} and C_{2v} geometries. The lowest occupied d level is a s–d_{z^2} hybrid that is virtually Cr—N nonbonding. The s–d mixing in this MO serves to minimize the Cr-ligand overlap and the orbital's polarization in the

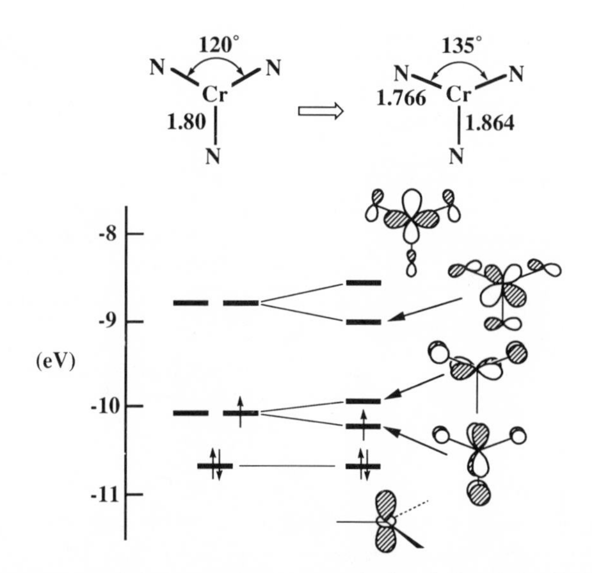

Figure 9. A molecular orbital diagram for the low-spin CrN_3^{6-} ions imbedded in Ca_3CrN_3 is given for D_{3h} and C_{2v} geometries. The lengths (in Å) of the Cr—N bonds used in the calculations are shown for the two geometries at top (those for the C_{2v} case are experimentally observed). Note that the observed Cr—N bond length asymmetry is that expected for the "T-shape" that is found.

region out of the molecular plane. Because of this and the strong π donation of the nitride ligands, this orbital lies below the Cr—N π^* set that constitutes a doubly degenerate pair for the D_{3h} geometry. Since we have a low-spin configuration, these levels hold only one electron between them and the ground state for the D_{3h} ion would have the Jahn–Teller unstable ^2E_g symmetry. (The FeN_3^{6-} ion should also be Jahn–Teller unstable, but has been reported to adopt a D_{3h} geometry.)[31b] As the symmetry is lowered from D_{3h} to C_{2v}, the orbital of the e_g set that is symmetric with respect to reflection in the plane perpendicular to the molecular plane becomes stabilized for the T-distortion. Note that the Cr—N bond length asymmetry is a result of the π bonding asymmetry in this singly occupied orbital. If CrN_3^{6-} had a *quartet* ground state, as is usual for Cr^{III}, then the ground configuration would not be Jahn–Teller unstable and we would expect the coordination environment to be more regular—with NCrN angles closer to 120° and less Cr—N bond length asymmetry. We shall comment on the potential role of the Ca^{2+} ions below.

Metal–nonmetal multiple bonding has been uncovered in truly extended compounds as well. Chern and DiSalvo discovered that CaNiN can be synthesized and is isotypic with the remarkable material YCoC uncovered by Gerss and Jeitschko.[32,33] CaNiN adopts the tetragonal structure illustrated

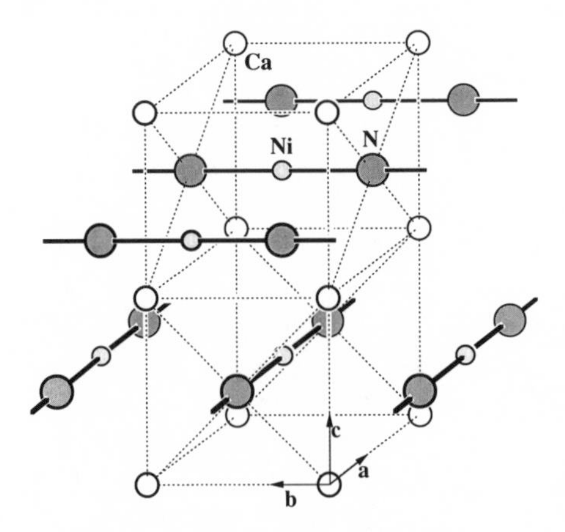

Figure 10. The tetragonal structure of CaNiN (YCoC-type) is depicted. The Ca^{2+} ions form an approximately cubic array (a = 3.581 Å, $c/2$ = 3.505 Å). Linear chains with alternating nickel and nitrogen atoms run through the center of the calcium "boxes" along the a and b directions. The Ni centers are just two-coordinate; the nitrogens have four additional Ca^{2+} neighbors at 2.505 Å.

in Figure 10. Isolated linear —Ni—N—Ni—N— chains run in the **a** and **b** directions with fairly short Ni—N distances within the chains (1.79 Å), but the closest contacts between chains are greater than 3.5 Å. Therefore, a simple one-dimensional model for the electronic structure of this compound should be quite reasonable if the Ca^{2+} orbital contributions can be safely neglected.

Hoffmann and coworkers presented a simple one-dimensional description of the electronic structure of YCoC that will be, if anything, even more appropriate for CaNiN after accounting for a change in electron count.[34] [Why *more* appropriate? Simply because Y is not as electropositive as Ca and the Y($4d$) and Y($5s$) orbital energies are such that they may make significant contributions to levels that we shall formally assume to be purely localized on the chains in the following discussion. We will have more to say on this point below.] The results of band structure calculations for a —M—E—M—E— (M = Ni or Co; E = N or C) are summarized in Figure 11, following Hoffmann's treatment. The figure shows the bands for all the chain's valence electrons except for the deeply occupied E localized $2s$ band. The band dispersions are easily understood by simply putting down the crystal orbital combinations for $k = 0$ and $k = \pi/a$ for each of the metal and nonmetal atomic orbitals and examining the metal–nonmetal interaction (see Figure 12). For example, the π orbitals for $k = 0$ show no mixing because, when the metal $d\pi$ orbitals are in phase as they are translated from site to site, they then form combinations that are antisymmetric with respect to planes perpendicular to the chain axis through the nonmetal atoms—and the converse applies to $k = 0$ combinations of the nonmetal $p\pi$ orbitals. At $k = \pi/a$, the metal–nonmetal interaction is maximized and the π and π^* bands (labeled xz,yz and N x,y in Figure 11) are respectively at their lowest and highest energies. The xy, $x^2 - y^2$ metal bands are virtually flat because the bridging nonmetals have no orbitals with which to make a δ overlap.

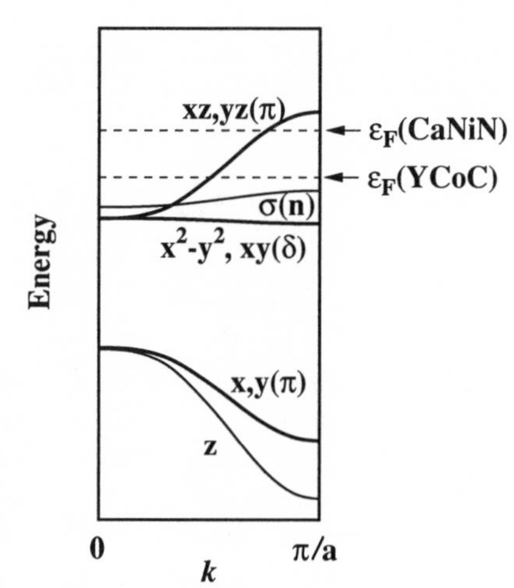

Figure 11. The band structure of $^{1}_{\infty}[CoC^{3-}]$ is shown as dispersion curves (redrawn from Ref. 34). The form of the band dispersions are understood by reference to Figure 12 and discussion in the text. Note that the Fermi level cuts at the three-quarters filled mark on the Ni—N π^* band for CaNiN and at the half-filled mark for YCoC.

There is a σ band that cuts across the metal d manifold that comes as some surprise if one thinks only in terms of the $3d$ levels on the transition metal. However, just as for the three-coordinate Cr center in the CrN_3^{6-} ion discussed above, the $4s$ orbitals play an important role. The low coordination number of the transition metal centers in these systems induces strong s–$d_z{}^2$

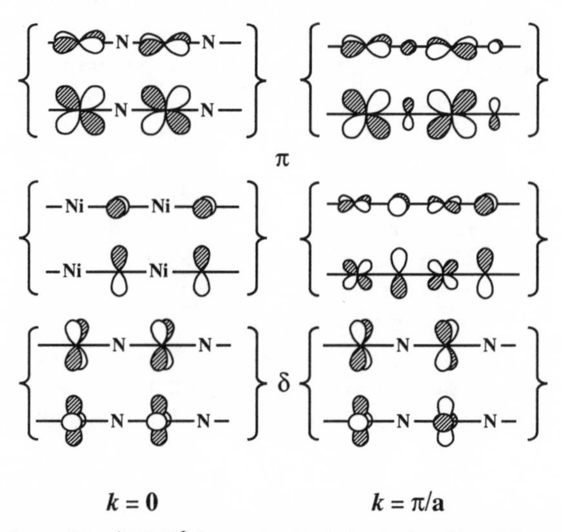

Figure 12. The form of the $^{1}_{\infty}[NiN^{2-}]$ crystal orbitals for the (π,π^*) and δ symmetry bands are shown for $k = 0$ and π/a. Note that the δ orbitals have no N character and are nonbonding for all k values. The Ni $d\pi$ and N $p\pi$ orbitals mix to an increasing extent as one goes from $k = 0$ to $k = \pi/a$.

mixing. For this case, it occurs throughout the Brillouin zone and the result may be easily understood by considering the hybridization scheme below:

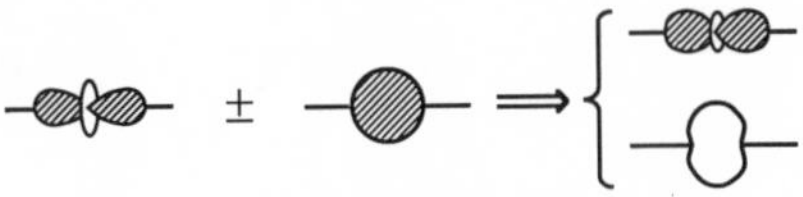

Scheme 1

The energy difference between the $3d$ level and the $4s$ level is small compared to the strength of interaction of both orbitals with the axial ligand orbitals surrounding them. It is then a reasonable approximation to freely combine the s and $d_z{}^2$ orbitals to produce one orbital that has maximum overlap with the ligand orbitals and one which has zero overlap. The former orbital mixes with the nonmetal s and p orbitals and this mixing causes the dispersion seen for the p_z band in Figure 11 (and in the s band not shown at lower energy). The latter hybrid gives rise to the flat, nearly nonbonding band that is labeled as σ(n) in Figure 11.

The net M—E bonding in the chains can be understood by reference to Figure 12 and the band structure diagram. Of course, all the bands with formally nonmetal character are in fact metal–nonmetal bonding in character. The δ symmetry bands are nonbonding and, because of s–d hybridization, the σ(n) band is mostly nonbonding as well. The xz,yz bands have π-antibonding character at the top of the bands (at $k = \pi/a$) but are nonbonding at the bottom (at $k = 0$). In other words, the fewer the number of electrons in the xz,yz bands, the stronger the π M—E bonding should be. It is surprising that the more electron-rich CaNiN was the next compound to be found with this structure type, since this bonding picture would predict weaker π bonding for the chains in this material.

Much remains to be understood with YCoC and CaNiN. If the electronic configuration of these compounds is such that all band orbitals are doubly occupied (as indicated by the Fermi level marks in Figure 11), then the one-dimensional chains would be structurally unstable with respect to a bond alternation (Peierls) distortions—though no evidence of such distortions has been uncovered. Hoffmann has interpreted the lack of a distortion as indicating a possible ferromagnetic ground state for the system, in which the xz,yz bands are fully occupied by spin-up electrons (analogous to a high-spin molecule):

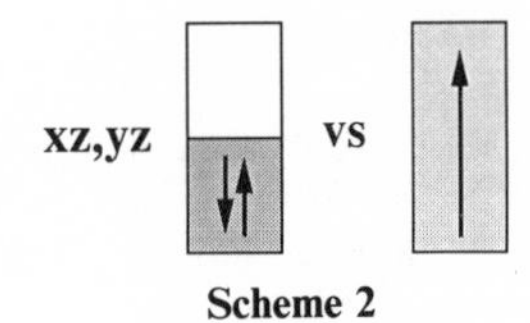

Scheme 2

With such a magnetic ground state the chains would be structurally stable, and of considerable interest since low-dimensional ferromagnets are quite rare. Measurements on CaNiN are not conclusive because, although there is some magnetic moment detected, it is small and impurity contributions cannot presently be ruled out.[33]

We have used a very simple approach for compounds like Ca_2ZnN_2, Ca_3CrN_3, CaNiN, and YCoC that neglects any covalency involving the most electropositive elements—Ca in the first three compounds, Y in the last. This is a conceptually uncomfortable position given the large formal charges that the more electronegative elements must bear in this view. For YCoC, this ionic approximation becomes even more suspect when we compare it with the bonding description we will develop below for reduced Y cluster compounds, where the assumption that yttrium is present as simple Y^{3+} is completely untenable. This may indeed complicate the simple one-dimensional picture presented for YCoC; yttrium may interact too strongly with the carbon centers in the ${}^{1}_{\infty}[CoC^{3-}]$ chains to be neglected. In Ca_3CrN_3, the unpaired electrons on the $CrN_3{}^{6-}$ ions are subject to rather strong antiferromagnetic coupling; the magnetic susceptibility declines from a maximum at 240 K. Such bulk antiferromagnetism is indicative of significant electronic communication between the $CrN_3{}^{6-}$ ions. The direct interionic contacts between $CrN_3{}^{6-}$ ions are long (Cr—Cr $\geq$ 4.06 Å; Cr—N $\geq$ 3.68 Å; N—N $\geq$ 3.58 Å) so that direct antiferromagnetic coupling would seem to be unlikely. The large negative formal charges on species such as $CrN_3{}^{6-}$, $ZnN_2{}^{4-}$, and ${}^{1}_{\infty}[CoC^{3-}]$ would also lead us to suspect that there must be some electron density donated back to the cations. Good estimates of the extent of "covalency" in the binding of the not so innocent cations in these materials are not at hand.

Finally, we should take note of fascinating materials containing a three-dimensional network of metal nitride multiple bonds uncovered in Jacobs's laboratory.[35] The structure of $M^{I}(Nb,Ta)N_2$ (M^{I} = K, Rb, Cs) consists of a network of tetrahedral transition metals linked together via two-coordinate bridging nitrogens. The structure is isotypic with $KGaO_2$, which may in turn

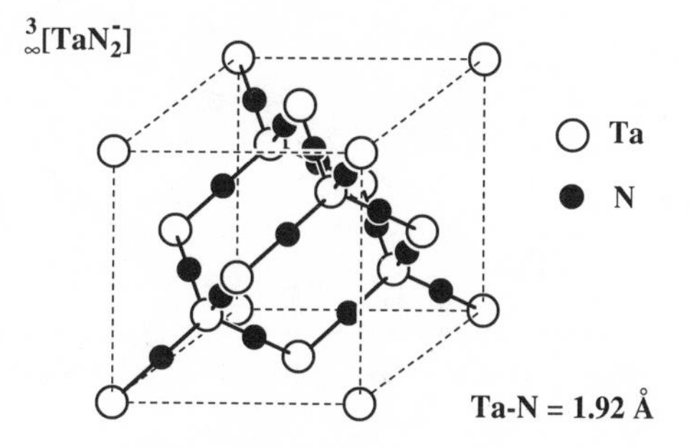

Figure 13. The ${}^{3}_{\infty}[TaN_2^-]$ net that ties the $CsTaN_2$ structure together is shown. The figure is somewhat idealized in that Ta—N—Ta angles are actually bent to about 164°. Some Ta—N multiple bonding would seem to be appropriate in describing this material.

be thought of as a cristobalite (quartz) structure in which cations are inserted. In Figure 13, we show an idealization of the ${}^{3}_{\infty}[TaN_2^-]$ network structure that weaves these materials together. The Cs ions are omitted for clarity and the Ta—N—Ta linkages are shown as if they are linear, although the Ta—N—Ta linkages are actually thought to be bent (∠Ta—N—Ta = 164°). Crystallographic disorder makes the nitrogen positioning somewhat uncertain, as is the case for the corresponding SiO_2 modification.

3.2. M—E Multiple Bonds Involving Heavier Nonmetals?

Solid-state compounds with unambiguous multiple bonding between transition metals and the heavier post-transition metals are nonexistent. Even in the molecular realm, this is still an emerging area of synthetic chemistry.[36] Three impressive examples from Huttner's laboratory are illustrated in Figure 14 along with diagrams that show the form of the molecular orbitals that describe the M—E π bonding.[37–39] The Mn—As—Mn angle in **a** is 176°, and this virtually linear molecule has a bonding pattern that has a direct topological equivalence with the 16 valence electron main group molecules discussed above. To put it another way, $As[MnCp(CO)_2]_2^+$ is isolobal with CO_2. Similarly, $Sb[Cr(CO)_5]_3^-$ is isolobal with CO_3^{2-}.

Solid-state chemists have not yet discovered heavier main group elements

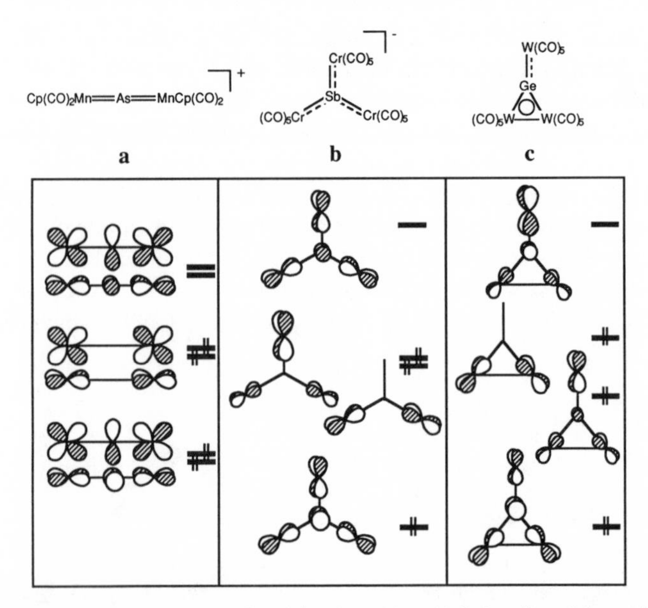

Figure 14. Some molecular examples involving M—E multiple bonding. Schematic diagrams describing the π bonding systems are given. These molecules have no close solid-state counterparts to date.

in coordination environments such as these. While it is tempting to dismiss such unsaturation as thermodynamically unstable (and thus likely to be inaccessible to solid-state synthetic chemists), the same might have been said about the nitrogen-rich nitrides discussed in the preceding paragraphs. Both the molecular examples illustrated in Figure 14 and the nitrides and carbides discussed above give hints as to the kind of materials synthetic solid-state chemists might see as attractive targets.

4. METAL–NONMETAL BONDING IN METAL-RICH COMPOUNDS

Understanding the bonding role that nonmetals play in metal-rich compounds is one of the most challenging areas for modern solid-state chemistry. The chemist's natural inclination is to first dismantle structures into digestible pieces and then to fit the pieces together using familiar bonding principles. In classical coordination chemistry our impulse is to divide metal–nonmetal bonded systems into "metals" + "ligands." Those who study molecules think of metals adorned with a rich array of potential *ligands:* CO, cyclopentadienyl, phosphines, amines, alkoxides, thiols, etc. In the inorganometallic solid-state chemistry discussed to this point we have implicitly performed the same categorization. However, this kind of partitioning of solid state compounds is not as generally appropriate as for molecular complexes. In a regime where extended metal–metal bonded networks coexist with metal–nonmetal bonding *and* nonmetal–nonmetal bonding, it is not yet clear how to implement this approach or even whether this approach will be generally useful. Nevertheless, such approaches are systematically useful when examining limited sets of materials and an examination of some examples provides some hints about how we might think about these complex systems. Of course, the discussion here suffers from the limitation that we may be unduly generalizing our understanding of rather unique systems to all the "slightly more complicated cases" simply because we have convinced ourselves that we do indeed *understand* them.

4.1. Interstitials in Clusters

Undoubtedly, one of the reasons that metal-rich compounds have presented such a challenge has been the lack of molecular paradigms that would allow localized or semilocalized bonding descriptions to be recognized as appropriate within the solids. For this reason, recent developments in the cluster chemistry of the early transition metals are particularly welcomed as a bridge between the molecular and solid state regimes. We present here some of the main points surrounding the bonding in these materials; the reader is referred elsewhere for more extensive treatments.[40–42]

4.1.1. The M_6X_{12} Cluster

A molecular model for the binding of main group interstitial atoms in metal-rich compounds can be found in recently synthesized $M_6X_{12}Z$ cluster compounds of zirconium and the rare-earths. The principal building block in these materials is shown in **9**. An octahedral M_6 cluster is bridged on all 12 edges by halides and centered by a wide variety of interstitials as described below (M = Zr, rare-earth element; X = halide; Z = interstitial).

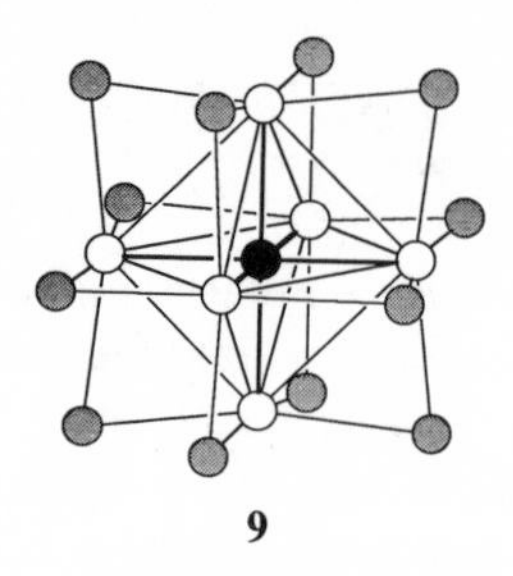

9

The parent Nb and Ta halide clusters of this now large family were discovered in the 1940s in Pauling's laboratories[43] but the breadth of M_6X_{12} cluster chemistry has become apparent only recently. The development of much of the lower oxidation state chemistry of zirconium and the rare-earth

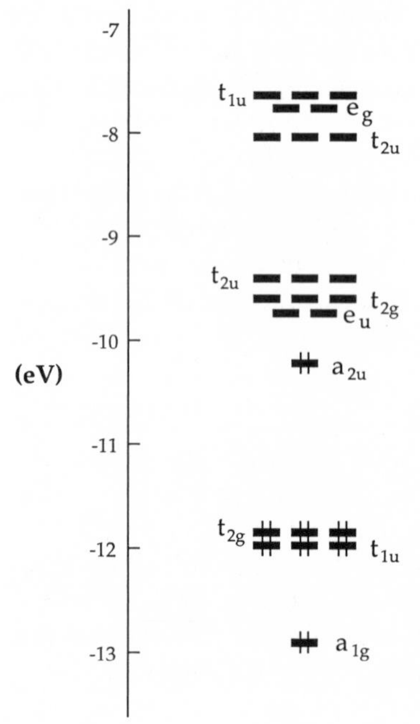

Figure 15. A molecular orbital level diagram for $Nb_6Cl_{18}^{4-}$ that shows the levels with primarily Nb d character; the occupancy indicated is appropriate for a 16 e^- cluster. The absolute position of the a_{2u} HOMO is somewhat dependent on computational parameters. Nevertheless, the gap below the a_{2u} orbital is significant in that 14 e^- clusters are quite commonly stable.

elements has been largely synonymous with the development of new cluster or condensed cluster compounds. The flowering of this cluster chemistry has come with an appreciation of the role played by "interstitials" (Z) which usually center clusters of these elements. Despite the unprecedented range of species that have been observed to act as interstitials, a fair degree of systematization of the chemistry emerges with an understanding of the clusters' electronic structure.

Bonding in the parent $(Nb,Ta)_6X_{12}$ clusters can be appreciated by the same sort of analysis we have discussed for M_6X_8 clusters above. In Figure 15 we show the manifold of metal-based levels obtained from an extended Hückel calculation on a $Nb_6Cl_{18}^{4-}$ cluster. A group of seven orbitals (a_{1g}, t_{2g}, t_{1u}) are clearly stabilized and have bonding character. Let us take each metal atom to reside in a local coordinate system in which the z axis runs through the center of the cluster and the x and y axes point at the bridging halides. Then the a_{1g} MO is an in-phase combination of z^2 orbitals with significant s and p admixture, the t_{2g} orbital has a predominantly (xz,yz) character, and the bonding t_{1u} orbital is a mixture of z^2 and (xz,yz). The x^2-y^2 orbitals are involved in bonding to the edge-bridging halides. The highest occupied MO for the $Nb_6Cl_{18}^{4-}$ cluster has a_{2u} symmetry and as such involves a pure com-

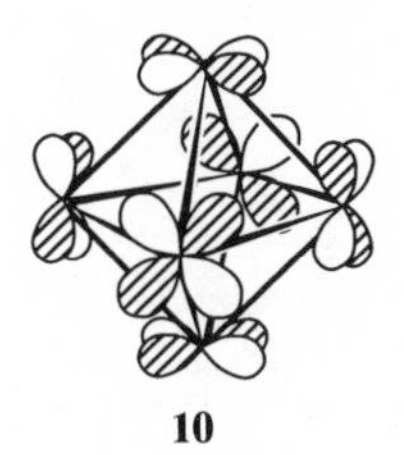

10

bination of xy orbitals (see **10**). The energy of this orbital is determined by a balance between its modest $(\pi + \delta)$ Nb—Nb bonding character and π^* Nb—Cl antibonding character. The positioning of the a_{2u} HOMO in the bonding–antibonding gap is therefore somewhat dependent on the method of calculation, but the general level ordering shown in Figure 15 is quite consistent.[44–46] EPR evidence quite unambiguously establishes that the a_{2u} level is half-occupied in $Nb_6Cl_{18}^{3-}$ (15 e^-) clusters.[47,48]

The general picture for M_6X_{12} clusters seems to be that one should expect to see clusters with at least 14 e^- available for metal–metal bonding and, depending on how accessible the a_{2u} orbital is for further reduction, 15 or 16 e^- species are possible. This picture is supported by the known Nb and Ta halide chemistry of these clusters. However, the breadth of M_6X_{12} cluster chemistry has been widened in recent years by the study of interstitially stabilized clusters, to which we now turn our attention.

4.1.2. Centered Clusters

In understanding the ordering of molecular orbitals, one naturally counts the number of bonding and antibonding interactions within each molecular orbital. However, in moving to a treatment of centered clusters, it is important to focus upon the symmetry and overlap characteristics of these orbitals with respect to the center of the cluster. The reader will recall that in octahedral environments the symmetries of atomic orbitals are as follows: $s(a_{1g}\colon \sigma)$, $p(t_{1u}\colon \sigma$ and $\pi)$, $d(e_g\colon \sigma$ and $t_{2g}\colon \pi)$, where we have classified the AOs according to both their irreducible representations and modes of bonding with surrounding atoms. We can divide interstitials into three obvious categories: hydrogen, with only a valence *s* orbital; main group atoms, with both *s* and *p* valence orbitals; and transition metals, with valence *d, s,* and *p* orbitals.

The addition of a hydrogen atom to its center is the simplest chemical modification we can make to a M_6X_{12} cluster and the change in the cluster's electronic structure is correspondingly simple. Since the hydrogen has only its valence 1*s* orbital available for bonding, the influence of centering by hydrogen on metal–metal bonding is quite straightforward: only the metal-based orbitals of a_{1g} symmetry are affected. Over the entire manifold of 24 metal bonding and antibonding levels only one, the lowest, has a_{1g} symmetry

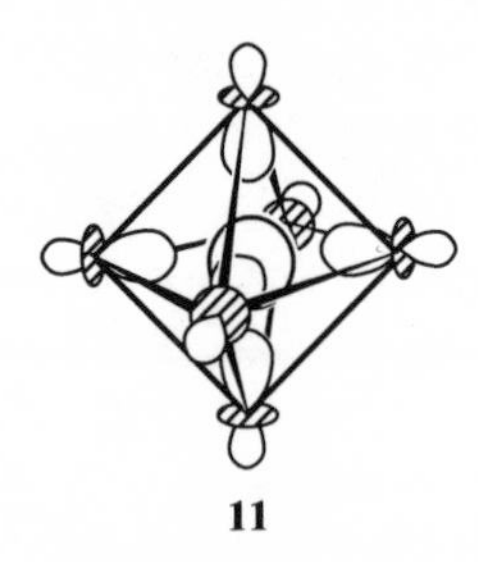

11

and we show the overlap of the hydrogen *s* orbital with this MO in **11.** The net effect on the hydrogen interstitial is the conversion of a pure metal–metal bonding orbital into one delocalized orbital with metal–metal and metal–hydrogen bonding character plus the addition of one unoccupied level that is just the antibonding combination of the hydrogen *s* orbital and the a_{1g} combination from the surrounding metals.

Note that the number of bonding orbitals does not change on the addition of the hydrogen interstitial. Reasoning by analogy with the empty Nb and Ta halide clusters, we anticipate that a cluster electron count of 14 is still indicated for hydrogen-centered clusters as long as the hydrogen's single electron is included in the total. We may reasonably expect that 14 electrons per cluster would be optimal for cluster bonding as long as inclusion of hydrogen has not destabilized the remaining orbitals by forcing the cluster to expand too much in accommodating it. The available evidence indicates that, if anything, the octahedral interstice offered by the Zr_6Cl_{12} cluster is too large for hydrogen. In $Zr_6Cl_{12}H$, solid-state NMR data suggest that the proton rapidly hops within its cage; the temperature dependence of the proton chemical shift for this 13 e^- cluster is consistent with the expected one unpaired spin per cluster.[49] Curiously, no clusters with the "optimal" electron count of 14 ($Zr_6X_{12}H^-$) have yet turned up.

M_6X_8 clusters (with X atoms capping the octahedral cluster faces) exhibit interstitial chemistry only for the inclusion of hydrogen in $Nb_6I_{11}H$ and $CsNb_6I_{11}H$. The Nb iodide clusters are unusual in that they may be prepared in the absence or presence of included hydrogen. A discussion of the bonding of hydrogen within Nb_6I_8 clusters must closely parallel that for encapsulation within M_6X_{12} clusters. In both cases the hydrogen 1s orbital can only interact with the a_{1g} cluster orbital to generate a delocalized totally symmetric bonding and antibonding pair of levels, the latter of which is occupied. The Nb_6I_{11} compounds are all quite electron deficient (from 19 e^- in Nb_6I_{11} to 21 e^- in $CsNb_6I_{11}H$) relative to the 24 e^- closed-shell configuration. This electron deficiency and its correlation with cluster distortions, temperature-dependent phase transitions, and unique spin-crossover transitions has been carefully studied.

4.1.3. *Transition and Post-Transition Elements in M_6X_{12} Clusters*[40]

The most important secret in the synthesis of octahedral zirconium and rare-earth metal clusters has been the incorporation of various species in the cluster centers. The list of species that have been included within these clusters now constitutes an impressive fraction of the periodic table: H, Be, B, C, C_2, N, Al, Si, P, K, Cr, Mn, Fe, Co, Ni, Ru, Rh, Os, Ir.[40,50] The modification of the cluster's electronic structure that such inclusion entails can be understood in a way that parallels our treatment of hydrogen-centered clusters as outlined above. In Figure 16, we show the effect of centering by both main group elements (with valence ns and np orbitals available for bonding) and transition elements (with valence nd, $(n+1)s$ and $(n+1)p$ orbitals—the last being relatively unimportant). In the center of Figure 16, we begin with the MO levels of the empty M_6X_{12} cluster. On the right-hand side of the diagram, the mixing of these orbitals with those of an enclosed main group atom is shown; at right is the interaction of an included transition metal with the orbitals of the empty cluster. The relative energetic disposition of the interstitials' atomic levels is as highly variable as the variety of species that can be encapsulated. It is the symmetries of the interacting levels on either interaction diagram that constitute the invariant core of our treatment.

Let us first take centering of the clusters by main group atoms. The dictates of symmetry demand that only the a_{1g} and t_{1u} cluster orbitals will mix with the central atom s and p orbitals. For pictorial clarity, we include

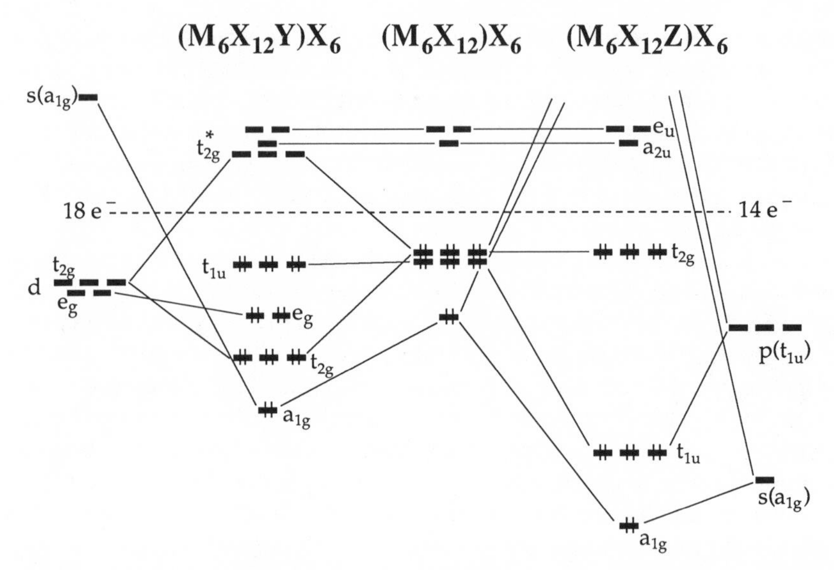

Figure 16. A double interaction diagram for the inclusion of s–p bonding main group elements, Z (right-hand side), and transition metals, Y (left-hand side), within a M_6X_{12} cluster. Note that the net number of occupied orbitals for the main-group-centered cluster remains at seven but is nine for the transition-metal-centered case. See text for details.

the centering atom s orbital on the same energy scale as the p orbitals and the manifold of metal bonding of the cluster. In actuality, the s–p energy gap will be greater than the ≈3–4 eV range that is a typical splitting of the metal levels shown. The p-orbital positioning will vary from well below the cluster bonding–antibonding gap (for electronegative interstitials like N) to a position that should be above the gap (for electropositive interstitials like K or Be). The relative positioning of these atomic levels does not have a net effect on the *number of levels with bonding character* that we obtain for the "complete" cluster in the interaction diagram. We continue to have bonding a_{1g} and t_{1u} orbitals that are occupied and below the "$14e^-$ line" in Figure 16. However, the picture does allow for a continuous shift of electron density from the metals to the interstitial as the latter becomes more electronegative. As the interstitial s and p levels drop in energy, the bonding a_{1g} and t_{1u} orbitals are increasingly localized on the centering atom. At the same time, orbitals which were strictly involved in metal–metal bonding in the empty cluster now have metal–interstitial bonding character in the centered clusters.

Now let us examine the encapsulation of transition metals by turning attention to the left-hand side of Figure 16. Once again, this diagram is a modestly schematic distillation of results from detailed calculations. The relative position of the valence nd and $(n+1)s$ orbitals at left on this diagram will vary downward as we move to more electronegative (late) transition metals. As for main-group-centered clusters, the shifting of the interstitial's orbital energies downward affects the mixed resultant bonding orbitals of the cluster by simply polarizing them more toward the interstitial atom. Aside from this, however, centering by transition metals constitutes an important qualitative change in the manifold of occupied levels by the addition of a degenerate pair of orbitals of e_g symmetry. Because there are no orbitals of e_g symmetry in the metal–metal bonding range for the *empty* cluster, these orbitals derive almost exclusively from the unchanged $d_z{}^2$ and $d_x{}^2 - y^2$ orbitals of the centering metal. To the extent that these levels are indeed purely localized on the interstitial, this is a remarkable result because it is precisely these d orbitals which one expects to be used for σ bonding with the surrounding cage metals. There are in fact cage orbitals of e_g symmetry, but they are well up into the antibonding block and the energy match with the relatively more electronegative interstitial metal is poor, thereby yielding only limited mixing into the e_g pair of orbitals. The $d\pi(t_{2g})$ and $s(a_{1g})$ orbitals of the centering metal mix more effectively with the cage orbitals of like symmetry to produce the bonding–antibonding splitting indicated in Figure 16.

It is surprising to see the range in both *size* and *nature* of interstitials that can be accommodated in zirconium and rare-earth clusters. The scheme given above can accommodate a range in electronegativity spanned by both main group and transition metal atoms; the polarization of the molecular orbitals resulting from interaction between the interstitial and the surrounding cage can vary with the changing electronegativity of the centering atom. The variability in interstitial size is more difficult to reconcile. As clusters become more distended in accommodating larger interstitials, geometry dictates that the direct interaction between the peripheral metals must significantly diminish. On moving from $CsZr_6I_{14}C$ to $CsZr_6I_{14}Mn$ and $Cs_{0.3}Zr_6I_{14}Si$ the average Zr — Zr separation goes from 3.30 Å to 3.55 and 3.58 Å. This opening of the cluster should make for a relative destabilization of the highest occupied molecular orbitals (HOMOs) in both main group (t_{2g}) and transition metal (t_{1u}) centered clusters. So far there seems to be little indication that the cluster expansion is particularly destabilizing. Perhaps the stability of rare-earth transition-metal-centered clusters with fewer than 18 cluster electrons is connected with the fact that the t_{1u} HOMO is virtually nonbonding for these expanded clusters.

4.1.4. Clusters as Models for Alloys

Some of the ideas we have used in discussing interstitially stabilized clusters are reminiscent of older thoughts about interstitial alloys. Goldschmidt considered interstitial alloys to be materials ". . .in which the metal–metal atom bond remains the dominant one, and the nonmetal atoms are sufficiently small to be accommodated within the metal lattice without, or with only a limited degree of, distortion of metal-type symmetry."[51] Included in his list of interstitials were ". . .the nonmetals H, C, N, O, B, Si; (conditionally) P and S, . . .metal atoms of their own or foreign kinds." Goldschmidt maintained that the ". . .effect of the interstitial atoms must be attributed basically to their acting as electron donors to the metal lattice . . . We may regard this as amounting, electronically, to an increase in the atomic number of the metal." This is clearly recollective of the parallels between the parent $(Nb, Ta)_6X_{12}$ clusters and the many interstially stabilized Zr and rare-earth cluster descendents. This approach is not really so different from that taken in accounting for included nonmetals in electron-counting schemes used by cluster chemists. Taking the focus away from the predominant weight in alloy chemistry that is traditionally placed on "radius ratios," Goldschmidt thought of ". . .the interstitial atom acting not so much through its mere volume as through its influence in filling the electron orbitals of the host metal atom."

Experience with cluster compounds indicates that it is a mistake to consider the metal–metal bond to be dominant in "interstitial compounds." The local perturbation of the electronic structure caused by the presence of an interstitial is quite important and there is every reason to believe that the same is true for interstitials in fully condensed phases. The study of interstitially stabilized clusters is important in this regard because we clearly see how it can *seem* that the effect of an interstitial is merely to increase the metal's atomic number while in fact strong bonds are formed between the interstitial and the surrounding metals. We can see how the qualitative picture for the electronic structure shows at least the plausibility of centering clusters by nonmetals or metals; the details of bonding can be quite different while the chemical environment surrounding the interstitial is identical. The initially surprising fact that intermetallic compounds can have very different chemical compositions and yet still have the same structures becomes a little more understandable.

4.2. Structure and Bonding in Metal-Rich Chalcogenides

The group 5 chalcogenides serve up timely examples of the kinds of problems faced by solid state chemists trying to understand bonding and its relation to structure in metal-rich compounds. Rather than discuss only successes in theory, we will use recent results in these systems to take a clear-eyed view of what *is* and is *not* understood of the bonding in metal-rich systems.

Sulfur forces a remarkable transformation in the tantalum's body-centered cubic structure. More than twenty years ago, Franzen and Smeggil discovered that Ta centered pentagonal antiprismatic chains, ${}^1_\infty[Ta_5Ta]$, occur in the Ta-rich compounds Ta_2S and Ta_6S (for which high-temperature monoclinic and low-temperature triclinic modifications have been uncovered).[52,53]

Recently, Ta_3S_2 ($=Ta_6S_4$) has been discovered and it has been determined that its structure incorporates $^1_\infty[Ta_5Ta]$ units linked in a way quite similar to Ta_2S.[54,55] New tantalum chalcogenides are turning up in a number of laboratories. We will see below the close similarity between the fivefold symmetric Ta centered chains in the binary Ta—S system and the fourfold symmetric $^1_\infty[Ta_4Z]$ chains found in the recently discovered Ta_4ZTe_4 (Z = Al, Si, Cr—Ni) series of compounds.[56,57] Tricapped trigonal prismatic Ta chains are centered with Fe, Co, or Ni (=M) atoms in $Ta_9S_6M_2$ and $Ta_{11}Se_8M_2$ ternaries.[58–60]

4.2.1. The Structures and Chemistry of Ta_3S_2 and Ta_2S[55,61]

Both of these compounds may be thought of as built from condensed Ta_6S_5 chains, to form a two-dimensional Ta—Ta bonded network in the first case and a three-dimensional network in the second. Fully intact Ta_6S_5 chains have not yet been observed as distinct structural units in any compound, but we introduce this system in Figure 17 as both a device for discussing these structures and in anticipation of our treatment of bonding below.

In the Ta_6S_5 chain, tantalum atoms are bound together in a chain of fused, centered pentagonal antiprisms. Alternatively, the chains are built up by stacking Ta_5 pentagons in a staggered fashion with the addition of tantalum atoms to the center of the pentagonal antiprisms so generated. The centering metal atoms are in a slightly compressed icosahedral environment. In the known binary tantalum sulfides the interlayer spacing between pentagons is

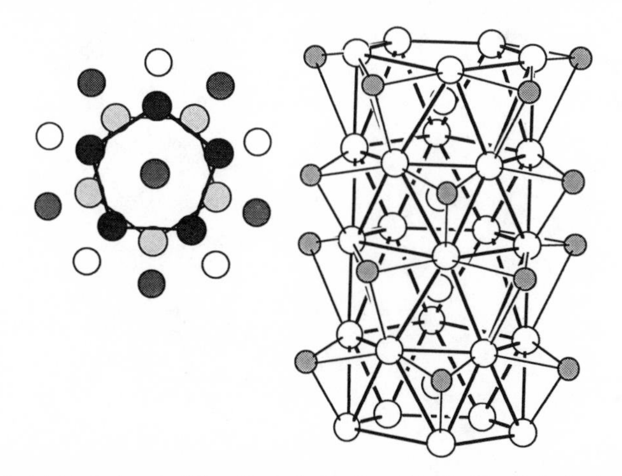

Figure 17. The Ta_6S_5 chain, a building block for Ta_2S and Ta_3S_2. The chain is viewed projected on the fivefold axis at upper left where the extent of shading is meant to indicate increasing heights of the atoms within one repeat-unit. In the projection, only Ta—Ta bonds involving the tantalums in the pentagonal antiprism are shown. In the vertical view at right, sulfurs are stippled and sulfur atoms at the back have been omitted for clarity.

in the range 2.62–2.80 Å (2.64 Å in monoclinic-Ta_6S,[52] 2.62 Å in triclinic-Ta_6S,[53] 2.79 Å in Ta_2S,[61] 2.80 Å in Ta_3S_2).[55] Sulfur surrounds the $^1_\infty[Ta_5Ta]$ chain so as to cap alternant exposed triangular faces, and construction of Ta_3S_2 entails the removal of a sulfur from each S_5 layer, each taken from the opposite sides of adjacent layers. In the Ta_5 rings, this leaves three tantalum centers with only two sulfur atom neighbors, while two tantalum atoms remain coordinated by sulfur as in the Ta_6S_5 chain. As depicted in Figure 18, these sulfur-deficient chains may then be linked by formation of interchain Ta—Ta bonds with the result that two-dimensional metal–metal bonded layers are formed. Figure 19 shows these layers in a vertical view, where one may see that interchain Ta—S bonds are formed as well. The net result is that sulfurs are four-coordinate, and sulfurs that cap triangular faces on one chain fill an open coordination site for tantalum atoms on adjacent chains. In Figure 20, these layers are seen to extend in the *ac* planes for Ta_3S_2 and in the *ab* planes for Ta_2S. These two-dimensional layers link together in Ta_3S_2 so that the remaining open coordination site at two tantalums in each Ta_5 ring is filled by sulfurs in the adjacent layers. Adjacent layers in Ta_3S_2 are held together solely by interlayer Ta—S bonds. Looking back to Figure 19, it is the three-coordinate sulfurs that cap triangular faces and do *not* participate in binding the chains together *within* the layers that are used

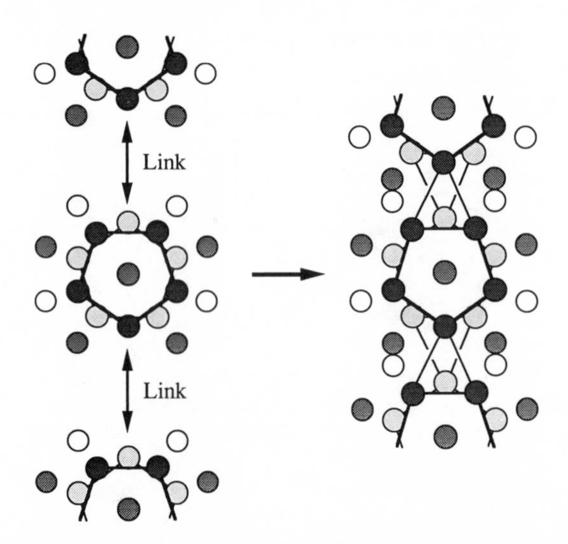

Figure 18. The linkage step for the construction of Ta_3S_2 layers is shown. A sulfur atom has been removed from each layer in the Ta_6S_5 chains and these chains are shown at left in the figure. All that remains is the coupling of these chains to form layers via the formation of interchain Ta—Ta bonds as shown. The "heights" of atoms are indicated by increasing shading of atoms, as in Figure 17.

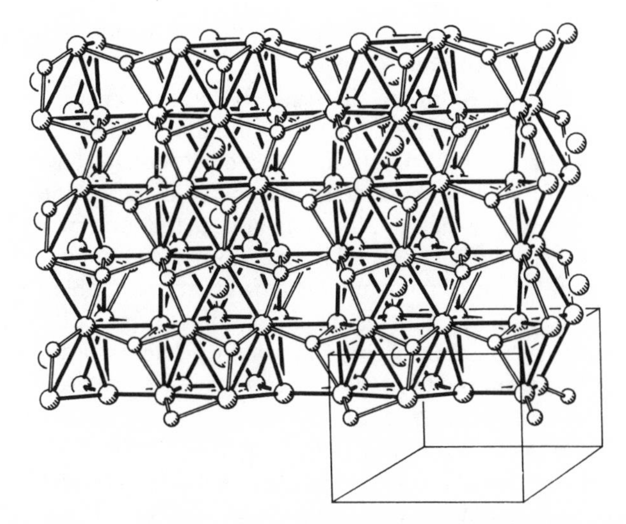

Figure 19. A vertical view of the "Ta_3S_2 layers" imbedded in Ta_3S_2 and Ta_2S. In addition to the formation of Ta—Ta bonds between the chains, Ta—S bonds are also present. For Ta_3S_2 this is a (010) projection, in Ta_2S it would be a (001) projection.

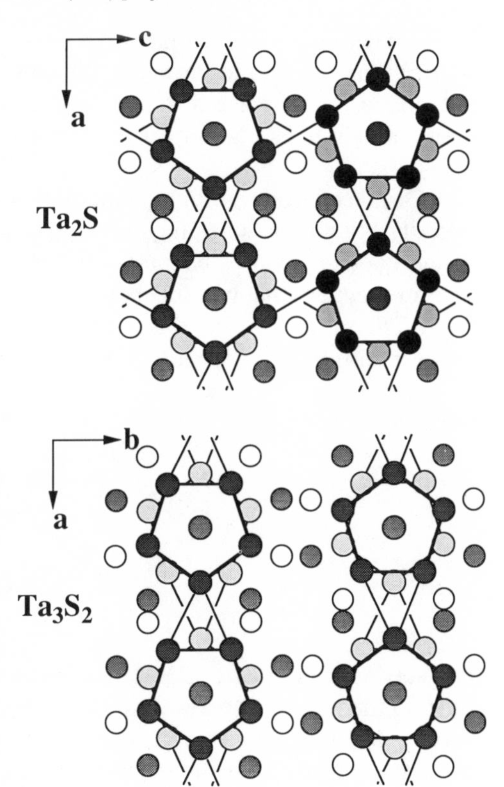

Figure 20. The Ta_2S and Ta_3S_2 structures, projected on planes normal to the pentagonal antiprismatic chains [(010) for Ta_2S, (001) for Ta_3S_2]. Shading of atoms is intended to indicate heights in the direction normal to the plane of projection (y values for Ta_2S, z values for Ta_3S_2; these values increase with the intensity of shading.) Only Ta—Ta bonds within and between the pentagonal antiprisms are indicated.

to form interlayer Ta—S bonds. Thus, in the full structure the sulfurs are all ultimately four-coordinate.

For Ta_2S the "Ta_3S_2 layers" are *fused* so that sulfurs are shared between chains in adjacent layers. This brings the tantalum atoms in adjacent layers closer together and Ta—Ta bonds now extend between the layers to complete the formation of a fully three-dimensional metal–metal bonded network. In the projection shown in Figure 20 it may appear that one may obtain Ta_3S_2 from Ta_2S by the insertion of an additional layer of sulfurs into the latter. In fact, more careful study of the structure shows that every other layer of linked pentagonal antiprismatic chains in Ta_2S has sulfurs capping the second set of alternant triangular tantalum faces in the chain. (Refer back to Figure 17 and move the sulfurs from the triangular faces where they are bound to the faces that are vacant—then construct the layer as indicated in Figures 18 and 19.)

Ta—Ta bonding dominates the structure of Ta_3S_2; among the shortest of these bonds are those involving the icosahedral tantalums that center the pentagonal antiprismatic chains. The average distance from the centering tantalums to their neighbors is 2.91 Å, which is just 0.02 Å shorter than what one calculates using Pauling's bond order relation and a Ta—Ta single bond distance of 2.71.[62] Relatively short Ta—Ta distances are observed for the contacts within the Ta_5 pentagons (average 3.02 Å); all of these distances are shorter than any found between pentagons. It is interesting to note that an attempt to link average distances to coordination numbers is useless since every tantalum is 12-coordinate if one counts both sulfur and tantalum neighbors.

Distances from sulfur to tantalum indicate that the sulfurs are more strongly bound to the triangular faces which they "cap" than they are to metals in other chains. The interlayer Ta—S distance of 2.68 Å is particularly long and suggests that bonding between layers is somewhat weaker than within them. This rather long interlayer Ta—S contact is dictated by steric crowding between sulfurs across the interlayer boundary, where very short 2.93 Å S—S

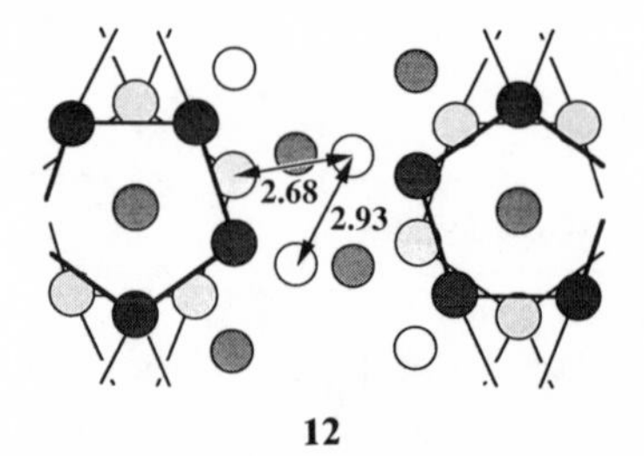

12

contacts are found (see **12**). This contact is considerably shorter than any other in the structure and is quite short when comparison is made with similar metal sulfides.[63] Band structure calculations (discussed further below) yield overlap populations for these close S—S contacts of −0.039, a value that is indicative of significant S—S repulsion. It is interesting to note that while there are close structural parallels between molybdenum cluster sulfides and selenides, this is not the case for the behavior of reduced tantalum chalcogenides. The well-known Chevrel phase sulfides and selenides have much in common, the Ta_2S and Ta_2Se structures are quite different. It appears that the close contacts between sulfides in the pentagonal antiprismatic chain structures (at least Ta_2S and Ta_3S_2) preclude these structure types for the selenides.

Ta_3S_2 was originally reported to be substoichiometric in sulfur due to the presence of vacancies.[54] The occurrence of sulfur *vacancies* would appear to be an unlikely prospect since the existence of vacancies implies the loss of four-coordinate sulfur and the creation of Ta centers that are coordinatively unsaturated (see above). It has since been established that the compound is in fact a stoichiometric line phase, but such nonstoichiometry does occur in more dense early metal sulfides wherein the creation of sulfide vacancies is compensated by a gain in metal–metal bonding surrounding the vacant site. Because of the close structural resemblance between the (010) planes of Ta_3S_2 and the (110) planes of Ta_2S, *intergrowth* of these two structures would appear to be a more structurally reasonable way to accommodate variable composition. Whether or not compounds intermediate in composition between Ta_3S_2 and Ta_2S can be prepared and will show such structural features remains to be seen.

4.2.2. Bonding in the Ta_6S_n Series

We now turn to a consideration of the bonding in these structurally remarkable materials. We will present results of band structure calculations for the Ta_2S, Ta_3S_2, and the hypothetical Ta_6S_5 systems and trace the evolution of these systems' electronic structures from bulk bcc Ta metal. We will stick to a presentation of the density of states (DOS) plots and the Crystal Orbital Overlap Populations (COOP) for the Ta—Ta bonds in these systems. In the complex structures we are considering here, we will present curves that are averaged over all the short Ta—Ta bonds in each structure, appropriately weighted to reflect the number of each bond type. This provides qualitative information about the metal–metal bonding in these materials without burdening us with too much detailed information.

Some comments on our Ta_6S_5 model chain are appropriate. The structures of Ta_3S_2 and Ta_2S were discussed using the hypothetical Ta_6S_5 chain as a starting point. This device was suggested by the structural trend that the former *known* tantalum-rich sulfides exhibited. Specifically, we wondered whether the pentagonal antiprismatic chains present in these compounds might be isolated as distinct one-dimensional entities. We speculate that the form that such one-dimensional chains will take can be surmised by careful inspection of the Ta_3S_2 structure and the ways that it differs from Ta_2S. This led us to focus on the way the additional sulfurs in Ta_3S_2 are accommodated and how the tantalum atoms to which they are bound are structurally affected. It was this process that led us to consider the Ta_6S_5 chain shown in Figure 17. Furthermore, we assume that in any real compound containing these chains that may be synthesized, the environment about the "outer" tantalums will be similar to the Ta centers in Ta_3S_2 that are involved in interlayer Ta—S bonding. This means that each of these metals will be bound to the three sulfurs that cap alternant triangular faces of the chain in which the metal in question resides *and* will also be bound to one "ausser" sulfur that fills the last vacant coordination site. Thus, we envision the chains formulated as $^1_\infty[Ta_5TaS_5{}^iS_5{}^a]$ systems where the ausser sulfurs may originate from neighboring chains or perhaps be replaceable by other donor ligands. The chain in the structural model used for calculations differs from that in Figure 17 only by the addition of "ausser" sulfides. In the DOS plots for this system, the contribution from these atoms has been subtracted to avoid the appearance of a spurious sulfur "lone-pair peak" at the bottom end of the plotted energy range.

Plots for the total DOS for Ta, Ta_2S, Ta_3S_2, and Ta_6S_5 are shown in Figure 21. In each case the range of energy is restricted so as to capture the Ta *d*-band region; the sulfur 3*s* and 3*p* bands lie lower in energy. For every panel, the shaded levels are occupied—the Fermi level chosen for the Ta_6S_5 chain is discussed below. The most important trend observed in these plots is the steady opening of a gap at the Fermi level as the sulfur content increases. There is a local minimum near the Fermi level for Ta_2S, a very deep minimum in Ta_3S_2, and a clear-cut band gap in Ta_6S_5. This sort of behavior seems to be a common characteristic of systems in which the dimensionality of metal–metal bonded networks are reduced. It is also a reasonable hallmark of stable systems that the bulk of the occupied levels are stabilized while unoccupied

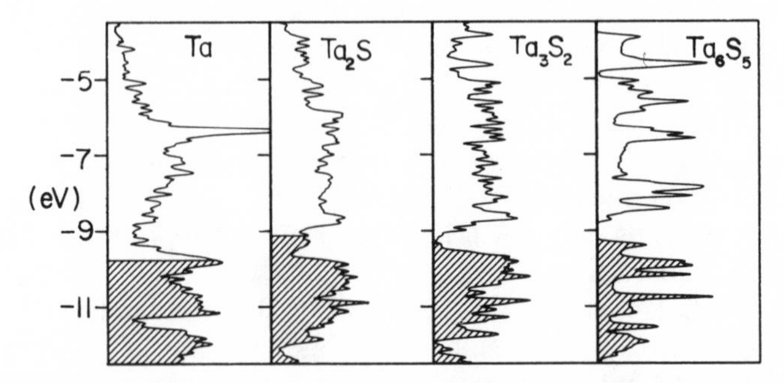

Figure 21. Density of states (DOS) plots are shown for elemental Ta, Ta_2S, Ta_3S_2, and our model $(Ta_6S_5)S_5$ chain. Shaded regions represent the occupied levels in each case. The energy range shows principally the Ta *d*-band region, the sulfur 3*p* bands overlap the Ta bands to some extent in lower parts of the range plotted.

levels are destabilized. From the molecular point of view this is the expected behavior as well; significant HOMO–LUMO gaps are a very useful qualitative guide to molecular stability.

COOP curves for these systems are no less enlightening. In Figure 22 we show averaged Ta—Ta COOP curves for the same four systems; the energy axes are horizontal and occupied levels are shaded. Each of the sulfides are striking in the extent to which their metal–metal bonding is optimized; for each case the Fermi level marks the dividing line between bonding and antibonding levels. It should be borne in mind that the number of electrons available for metal–metal bonding declines with increasing sulfur content. The COOP curves show that the number of metal–metal bonds lost as tantalum metal is effectively "broken up" by sulfur inclusion is exquisitely in tune with the declining number of available metal–metal bonding electrons.

The calculated electronic structure of these systems clearly suggests the need for further experimentation. A gap of 0.09 eV is calculated for Ta_3S_2, but the method of calculation is not of sufficient quantitative reliability for us to be confident of this result. It does suggest that Ta_3S_2 is likely to be a rather poor conductor or perhaps behave as a small-gap semiconductor. For structural reasons, anisotropy in the electrical conductivity is also to be expected, with conductivity along the [010] direction expected to be much lower. The significance of our computational results for the hypothetical Ta_6S_5

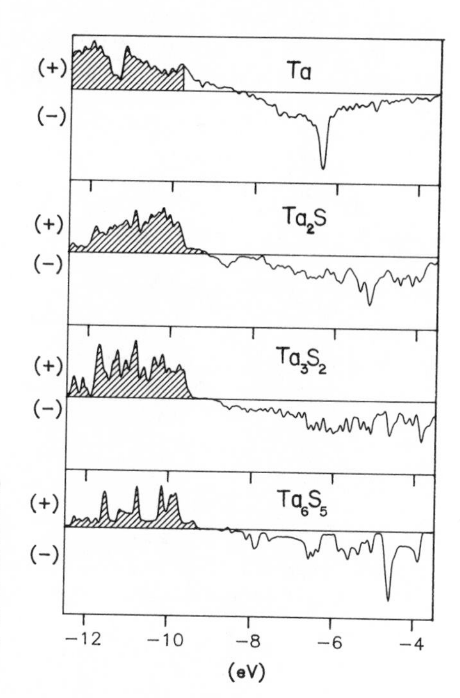

Figure 22. Averaged Crystal Orbital Overlap Population (COOP) curves are plotted from top to bottom for Ta, Ta_2S, Ta_3S_2, and our model $(Ta_6S_5)S_5$ chain, respectively. Energy increases from left to right and the occupied levels for each system are shown shaded. Note the extent to which the binary compounds optimize Ta—Ta bonding.

chain is amplified by considering the strong correlation between metal–metal "COOP optimization" (i.e., optimization of metal–metal bonding) and the discovery of so many new metal-rich compounds in the last 15 years.[40–42,64–68] This has been particularly true for compounds with low-dimensional metal–metal bonded arrays. If chains with fivefold rotational symmetry can somehow be crosslinked, our calculations indicate that $^1_\infty[Ta_5TaS_5{}^iX_5{}^a]$ systems should be stable when there are 20 metal-based electrons per Ta_6 unit. The calculated band gap for the model Ta_6S_5 chain system is considerably larger (0.43 eV) and, if efforts to synthesize compound(s) containing the neutral Ta_6S_5 chain are successful, *semiconducting* behavior can be expected. While this is a surprising conclusion since the chain contains an unbroken and undistorted metal–metal bonded thread and there are a nonintegral number of electrons (3⅓) per Ta, a similar situation obtains for the chain compound Gd_2Cl_3 where theory and experiment are in agreement on the semiconducting behavior of the compound.[69]

4.2.3. The Tantalum Sulfides Contrasted with Related Materials

Figure 22 and the accompanying discussion clearly implicate metal–metal bonding as a key factor in stabilizing these metal-rich phases. Does this mean that the bonding role of sulfur in these metal-rich sulfides is really so passive? We might be led to the conclusion that since optimizing the strength of metal–metal bonding in these phases is such an important constraint, the structures adopted should be fairly insensitive to the nature of the *nonmetal,* so long as the compounds considered are isoelectronic and otherwise not too dissimilar.

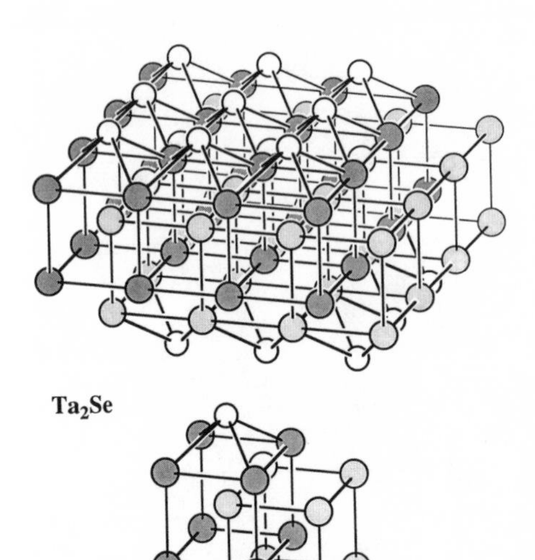

Figure 23. The thick sandwiches found in the structure of Ta_2Se are illustrated. Ta atoms are represented by large shaded circles, Se by small open circles. Every layer is a square net with a spacing of 3.375 Å; these are stacked upon each other in the order Se—Ta—Ta—Ta—Ta—Se in the same manner as layers are stacked in bcc—Ta. The layers are weakly bound together by Se—Se van der Waals interactions.

A comparison of the structures of Ta_2S, Ta_2Se, and Nb_2Se is sobering. The remarkable new compound Ta_2Se has a layer structure depicted in Figure 23.[70] The structure is quite simply related to the bcc-Ta; four square-net Ta layers ([100] planes) are inserted between two square-net Se layers. These thick sandwiches are then stacked with van der Waals gaps between them. Nb_2Se has a structure that has been described as built up from Nb_6Se_8 clusters, condensed into double chains as shown at right in Figure 24, that are in turn linked together in three dimensions as shown in two complementary views in Figure 24.[71]

The difference in the structures of Ta_2S and Ta_2Se is probably due to simple atomic size differences, as noted above. Indeed, high-temperature work on $Ta_2S_{1-x}Se_x$ solid solutions suggests that there is no significant incorporation of selenium into the Ta_2S structure, but a considerable amount of sulfur can be incorporated into the Ta_2Se structure.[72] $Ta_{2-z}Nb_zS$ phases can be prepared with the Ta_2Se structure and at least one Ta_2Se analog with a $Ta_{2.5-y}Nb_yS_2$ composition has been uncovered in which five (Nb,Ta) square net layers are sandwiched between sulfide layers.[73]

The difference in the Nb_2Se and Ta_2Se structures is presently inexplicable. In Figure 25 we show averaged M—M bond COOP plots for these two compounds. In each case nature has found structures that are virtually optimal in filling metal–metal bonding levels. Comparison of these COOP curves and that for Ta_2S in Figure 22 show that computations demonstrating "how good the bonding is" can still be inadequate for predicting structure. These refractory materials have large cohesive energies and both the theoretical analysis and experimental facts indicate how small the "structure determining" fraction of the cohesive energy may be.

The warning sounded above regarding the predictability of structure is even more serious when considering the issue of phase stability—where our ability to make confident predictions is practically nonexistent. In fact, care

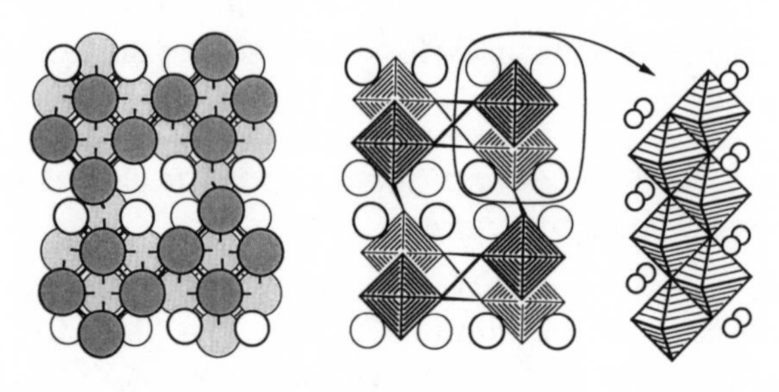

Nb_2Se

Figure 24. The complex structure of Nb_2Se is shown. As in the other metal-rich materials we have seen, the metals and nonmetals cluster together in different regions of the structure. More details are discussed in the text and elsewhere.[71]

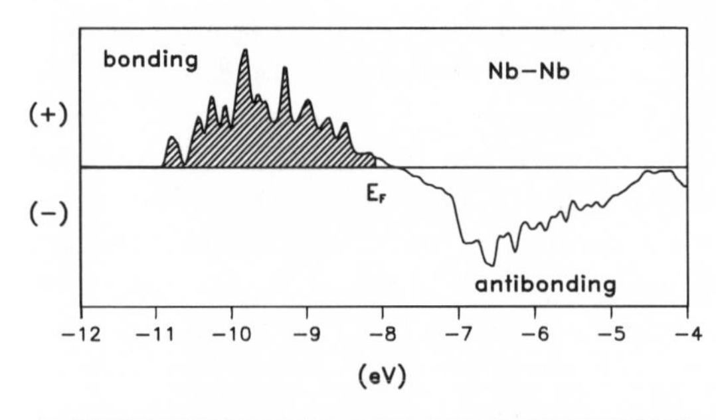

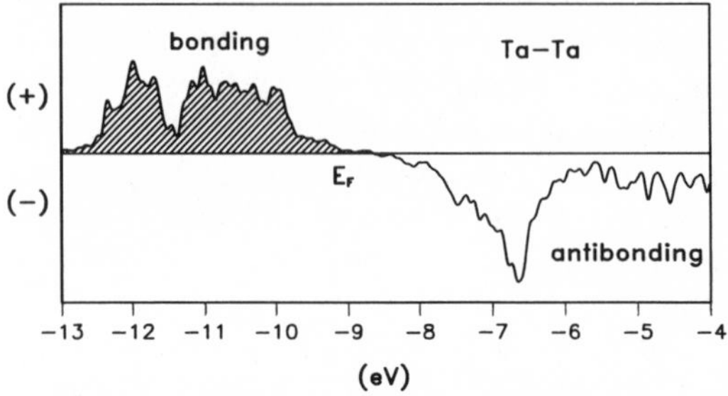

Figure 25. Crystal Orbital Overlap Population (COOP) curves are plotted for Ta_2Se (bottom) and Nb_2Se (top). Energy increases from left to right and the occupied levels for each system are shown shaded. Note the extent to which *both* compounds optimize metal–metal bonding, just as for the isoelectronic Ta_2S shown in Figure 22.

must be exercised in interpreting the existence of a material, even one with a well-defined structure such as Ta_2S, as evidence for that structure's ultimate "stability." Ta_2S, like *many* intermetallic compounds, is easily prepared at high temperatures from a melt of the appropriate composition ($TaS_2 + 3Ta$) and is cleanly obtained in virtually quantitative yield after quenching. Nevertheless, when this phase is annealed at 1000 °C (i.e., at *low* temperature), disproportionation occurs ($9Ta_2S \rightarrow 4Ta_3S_2 + Ta_6S$).[72] Thermodynamic measurements and calculations by Franzen and coworkers had anticipated the observed enthalpic instability of Ta_2S, though not with respect to the products ultimately found ($Ta_{1.35}S_2$ + Ta had been expected).[74] We must conclude that even if the practically impossible goal of calculating low-temperature stabilities (enthalpies) for systems such as these were within reach, many interesting materials would still be missed!

4.3. The Ta_4Te_4Z Series and Connections to "Mainstream" Intermetallic Compounds

DiSalvo and coworkers have uncovered a fascinating series of ternary tantalum tellurides with the formula Ta_4Te_4Z (Z = Al, Si, Cr— Ni) that are structural analogs to the fivefold symmetric chains we have discussed above.[57] Ta_4Te_4Z chains, depicted in Figure 26, have fourfold symmetry and consist of Z-atom centered square antiprisms that are fused on opposite square faces to form one-dimensional chains. The repeat distance along the chain dictates relatively short average Si–Si distances of 2.40 Å between adjacent Si centers.

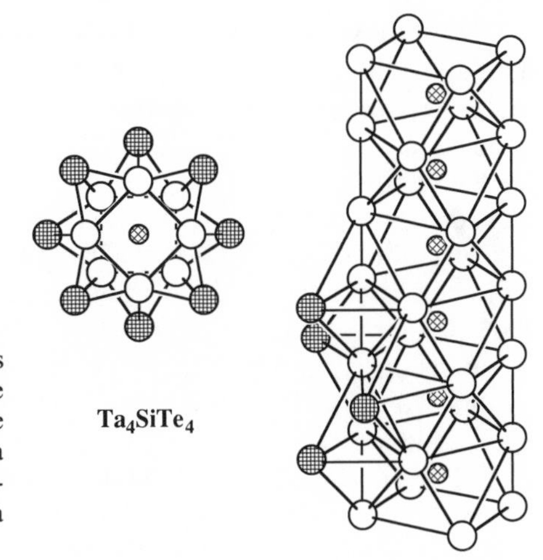

Figure 26. A projection down the chain axis of Ta_4SiTe_4 is shown at left and the entire chain is shown. At right we show only the $^1_\infty[Ta_4Si]$ chain with the axis vertical and a few of the outer Te ligands are shown to indicate the mode of Te binding and the Ta environment.

This is not much longer than the Si—Si single bonds in elemental silicon. These actual Si–Si distances are reported to alternate so that two inequivalent Si—Si bond lengths result, 2.35(2) Å and 2.45(2) Å. The rather large uncertainty in the refined Si position along the chain axis makes this alternation subject to some doubt, however.

A detailed analysis of the Ta—Ta and Ta—Si bonding in the $^1_\infty[Ta_4SiTe_4]$ chains has been presented by Li, Hoffmann, and coworkers using the same theoretical tools we have employed here.[56] The theoretical treatment suggests that despite the rather short Si–Si distance in the chain, Si—Si overlap populations are considerably smaller than for an unsupported Si—Si single bond. This result and overlap populations calculated for the Ta—Si bonds imply that the formation of strong Ta—Si bonds greatly stabilizes the structure. The iron-centered chain compound, Ta_4FeTe_4, was also studied and conclusions regarding Fe—Fe bonding were somewhat different. In this case, the Fe—Fe bonding was also diminished relative to a free standing $^1_\infty[Fe]$ chain, but this was predicted to be strongly dependent on the position of the Fermi level. The calculations indicated that if the chain could be oxidized, Fe—Fe bonding could be increased without loss of Ta—Fe bonding. Chain reduction was predicted to further weaken Fe—Fe bonding while modestly weakening the Ta—Fe bonding as well. The net effect of oxidation or reduction on Ta—Ta or Ta—Te bonding was predicted to be minor. Hopes that more electron-poor systems (such as Ta_4FeTe_4) can be realized through high-temperature synthesis should be tempered with the same caution we have used in discussing the binary compounds. Unfortunately, the very high air sensitivity of these materials make these materials less suitable as intercalation hosts.

Some links between the Ta_4SiTe_4 chain systems and established solid-state alloy chemistry can be seen. In the W_5Si_3 structure type (adopted by Mo_5Si_3 and a high-temperature modification of Nb_5Si_3, for example), the same square antiprismatic $^1_\infty[M_4Si]$ chains are recognizable, but condensed via the capping nonmetals (Si in this case) to form the three-dimensional array depicted in Figure 27.[75] The condensation is reminiscent of the cluster condensation we have discussed above (as in $Y_6I_{10}Ru$, for example) and even more closely related to the relationships we have pointed out for Ta_2S and Ta_3S_2. As for the Ta_4ZTe_4 system, the beginnings of a substitutional chemistry involving the centers of the square antiprisms have appeared in the literature; Nb_5Sn_2Si and Ta_5Sn_2Ga have structures in which this site is respectively filled by Si and Ga.[76,77]

How closely related are the electronic structures of Ta_4SiTe_4 and, say, Nb_5Si_3? In particular, is the binding of the Si centering atom as similar as the structures would suggest? Can the alloy phases be further derivatized to give ternaries in which the centering Si can be replaced by a range of interstitials like those reported for the Ta_4ZTe_4 system? An appreciation of the nature of the metal–nonmetal bond in these challenging systems will be crucial if we are to learn how to systematically control the structure and properties of metal-rich materials.

5. CONCLUDING REMARKS

The reader should now appreciate the unevenness of our understanding of bonding in solids. Unfortunately, when presenting theory that is so incomplete there is a great danger of selecting examples only because they are understood, not because they are representative. (Like the man who searches for his car keys under a lamppost, because the light is better, not because that

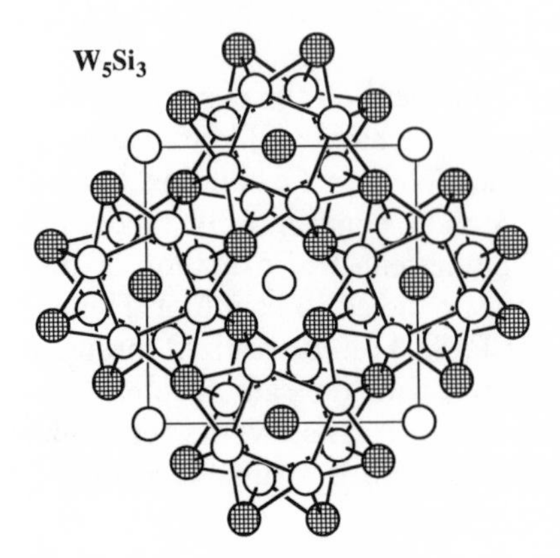

Figure 27. A projection down the c axis of the W_5Si_3 structure is shown. The $^1_\infty[W_4Si]$ chains imbedded in this structure are essentially the same as those in Ta_4SiTe_4.

is where he lost them.) In the present chapter we have succumbed to this temptation to an uncomfortable degree. Our discussion of multiple bonding in solids is highly selective in drawing a few unusual examples from nitride chemistry; the examples discussed are not representative of most transition metal nitrides.

While there is no need of a disclaimer for the examples discussed, it should be emphasized that most nitrides, carbides, silicides, phosphides, borides, etc of the transition metals are quite different.[78,79] In a compound with a simple stoichiometry like CrB, interatomic distances indicate the presence of B—B, Cr—B, and Cr—Cr bonding in a way that cannot be credibly disentangled. Solid-state chemists have found no way to use ideas from molecular chemistry to think about such systems, beyond the tenuous connections that are tentatively referred to in the foregoing discussion. On the other hand, tools for accomplishing at least the sorting of structure types using only a few atomic parameters (structure maps) have been developed and are quite powerful.[80,81] But while these methods demonstrate correlations between radii, valence electron concentration, and crystal structure, they are not detailed bonding pictures that serve to link these systems with molecules or to explain physical properties. Progress toward a clearer understanding of solid-state inorganometallic compounds is coming in fits and starts, but it is simply too soon to speak in more comprehensive terms.[82]

ACKNOWLEDGMENTS

Partial support of the research discussed in this chapter was derived from a grant (#010366-111) by the Texas Advanced Research Program. The support of the National Science Foundation through a Presidential Young Investigator Award (Grant DMR-8858151) and the Robert A. Welch Foundation for its support through grant A-1132 are gratefully acknowledged.

REFERENCES

1. Schäfer, H.; Eisenmann, B.; Müller, W. *Angew. Chem., Int. Ed. Engl.* 1973, **12,** 694.
2. Schäfer, H. *Ann. Rev. Mater. Sci.* 1985, **15,** 1.
3. Wells, A. F. "Structural Inorganic Chemistry," 5th; Oxford: New York, 1984.
4. Braun, P. J.; Jeitschko, W. *Z. Anorg. Allg. Chem.* 1978, **445,** 157.
5. Jeitschko, W.; Braun, P. J. *Acta Crystallogr.* 1978, **B34,** 3196.
6. Krebs, H.; Müller, K.-H.; Zurn, G. *Z. Anorg. Allg. Chem.* 1956, **285,** 15.
7. King, R. B. *Inorg. Chem.* 1989, **28,** 3048.
8. Sunshine, S. A.; Kang, D.; Ibers, J. A. *J. Am. Chem. Soc.* 1987, **109,** 6202.
9. Kang, D.; Ibers, J. A. *Inorg. Chem.* 1987, **27,** 549.
10. Kanatzidas, M. G. *Chem. Mater.* 1990, **2,** 353.
11. Burshka, C. *Z. Naturforsh.* 1980, **35B,** 1511.
12. Kanatzidas, M. G.; Park, Y. *J. Am. Chem. Soc.* 1990, **111,** 3767.
13. Ackermann, J.; Wold, A. *J. Phys. Chem. Solids* 1977, **38,** 1013.

14. Lutz, H. D.; Kliche, G. *Z. Anorg. Allg. Chem.* 1981, **480,** 105.
15. Kliche, G.; Lutz, H. D. *Infrared Phys.* 1984, **24,** 171.
16. Kliche, G.; Bauhofer, W. *Mater. Res. Bull.* 1987, **22,** 551.
17. Kliche, G.; Bauhofer, W. *J. Phys. Chem. Solids* 1988, **49,** 267.
18. Korenstein, R.; Soled, S.; Wold, A.; Collin, G. *Inorg. Chem.* 1977, **16,** 1977.
19. Kjekshus, A.; Rakke, T. *Acta Chem. Scand.* 1974, **A28,** 99.
20. Braun, D. J.; Jeitschko, W. *J. Less-Common Met.* 1980, **72,** 147.
21. Braun, D. J.; Jeitschko, W. *J. Solid State Chem.* 1980, **32,** 357.
22. Jeitschko, W.; Braun, D. *Acta Crystallogr.* 1977, **B33,** 3401.
23. Boonk, L.; Jeitschko, W.; Scholz, U. D.; Braun, D. J. *Z. Kristallogr.* 1987, **178,** 30.
24. Stetson, N. T.; Kauzlarich, S. M., *J. Solid State Chem.*, in press.
25. Jung, D.; Whangbo, M.-H.; Alvarez, S. *Inorg. Chem.* 1990, **29,** 2252.
26. DeLong, L. E.; Meisner, G. P. *Solid State Commun.* 1985, **53,** 119.
27. Yamane, H.; Kikkawa, S.; Koizumu, M. *J. Solid State Chem.* 1987, **71,** 1.
28. Yamane, H.; Kikkawa, S.; Horiuchi, H.; Koizumu, M. *J. Solid State Chem.* 1986, **65,** 6.
29. Evers, J.; Munsterkotter, M.; Oehlinger, G.; Polborn, K.; Sendlinger, B. *J. Less-Common Metals* 1990, **162,** L17.
30. Chern, M. Y.; DiSalvo, F. J. *J. Solid State Chem.* 1990, **88,** 528.
31a. Vennos, D. A.; Badding, M. E.; DiSalvo, F. J. *Inorg. Chem.* 1990, **29,** 4059.
31b. Cordier, G.; Höhn, P.; Kniep, R.; Rabenau, A. *Z. Anorg. Allg. Chem.* 1990, **591,** 58.
32. Gerss, M. H.; Jeitschko, W. *Z. Naturforsch.* 1986, **41b,** 946.
33. Chern, M. Y.; Disalvo, F. J. *J. Solid State Chem* 1990, **88,** 459.
34. Hoffmann, R.; Li, J.; Wheeler, R. *J. Am. Chem. Soc.* 1987, **109,** 6600.
35. Jacobs, H.; Pinkowski, E. V. *J. Less-Common Met.* 1989, **146,** 147.
36. Herrmann, W. A. *Angew. Chem., Int. Ed. Engl.* 1986, **25,** 56.
37. Huttner, G.; Weber, U.; Sigwarth, B.; Scheidsteger, O.; Lang, H.; Zsolnai, L. *J. Organomet. Chem.* 1985, **282,** 331.
38. Strube, A.; Huttner, G.; Zsolnai, L. *Angew. Chem., Int. Ed. Engl.* 1988, **27,** 1529.
39. Kostic, N. M.; Fenske, R. F. *J. Organomet. Chem.* 1982, **233,** 337.
40. The majority of the work discussed here has emerged from Corbett's laboratories. For a more complete review, see: Hughbanks, T. *Prog. Solid State Chem.* 1989, **19,** 329.
41. Simon, A. *J. Solid State Chem.* 1985, **57,** 2.
42. Simon, A. *Agnew. Chem., Int. Ed. Engl.* 1988, **27,** 160.
43. Vaughan, P. A.; Sturtivant, J. H.; Pauling, L. *J. Am. Chem. Soc.* 1950, **72,** 5477.
44. Cotton, F. A.; Haas, T. E. *Inorg. Chem.* 1964, **3,** 10.
45. Bursten, B. E.; Cotton, F. A.; Stanley, G. G. *Isr. J. Chem.* 1980, **19,** 132.
46. Wooley, R. G. *Inorg. Chem.* 1985, **24,** 3519.
47. Klendworth, D. D.; Walton, R. A. *Inorg. Chem.* 1981, **20,** 1151.
48. Mackay, R. A.; Schneider, R. F. *Inorg. Chem.* 1967, **6,** 549.
49. Chu, P. J.; Ziebarth, R. P.; Corbett, J. D.; Gerstein, B. C. *J. Am. Chem. Soc.* 1988, **110,** 5324.
50. Payne, M.; Dorhaut, P.; Corbett, J. D., personal communication.
51. Goldschmidt, H. J. "Interstitial Alloys"; Butterworths: London, 1967.
52. Franzen, H. F.; Smeggil, J. G. *Acta Crystallogr., Sect. B* 1970, **26,** 125.
53. Harbrecht, B. *J. Less Common Met.* 1988, **138,** 225.
54. Wada, H.; Onoda, M. *Mater. Res. Bull.* 1989, **24,** 191.
55. Kim, S.-J.; Nanjundaswamy, K. S.; Hughbanks, T. *Inorg. Chem.* 1991, **30,** 159.
56. Li, J.; Hoffmann, R.; Badding, M. E.; DiSalvo, F. J. *Inorg. Chem.* 1990, **29,** 3943.
57. Badding, M. E.; DiSalvo, F. J. *Inorg. Chem.* 1990, **29,** 3952.
58. Harbrecht, B.; Franzen, H. F. *J. Less-Common Met.* 1985, **113,** 349.
59. Harbrecht, B. *J. Less-Common Met.* 1986, **124,** 125.
60. Harbrecht, B. *J. Less-Common Met.* 1988, **141,** 59.

61. Franzen, H. F.; Smeggil, J. G. *Acta Crystallogr., Sect. B* 1969, **25,** 1736.
62. Pauling, L. "The Nature of the Chemical Bond;" 3, Cornell University Press: Ithaca, New York, 1960.
63. Corbett, J. D. *J. Solid State Chem.* 1981, **39,** 56.
64. Chevrel, R. In "Crystal Chemistry and Properties of Materials with Quasi-One-Dimensional Structures"; Rouxel, J., Ed.; Reidel: New York, 1986.
65. Adolphson, D. G.; Corbett, J. D. *Inorg. Chem.* 1976, **15,** 1820.
66. Torardi, C. C.; McCarley, R. E. *J. Am. Chem. Soc.* 1979, **101,** 1963.
67. Torardi, C. C.; McCarley, R. E. *Inorg. Chem.* 1985, **24,** 476.
68. McCarley, R. E. *Polyhedron* 1986, **5,** 51.
69. Lokken, D. A.; Corbett, J. D. *Inorg. Chem.* 1973, **12,** 556.
70. Harbrecht, B. V. *Angew. Chem., Int. Ed. Engl.* 1989, **28,** 1660.
71. Conard, B. R.; Norrby, L. J.; Franzen, H. F. *Acta Crystallogr.* 1969, **B25,** 1729.
72. Nanjundaswamy, K. S.; Hughbanks, T., unpublished research.
73. Franzen, H. F., personal communication.
74. Harbrecht, B. U.; Schmidt, S. R.; Franzen, H. F. *J. Solid State Chem.* 1984, **53,** 113.
75. Aronsson, B. *Acta Chem. Scand.* 1955, **9,** 1107.
76. Horyn, R.; Lukaszewicz, K. *Bull. Acad. Pol. Sci., Ser. Sci. Chim.* 1970, **18,** 59.
77. Ukei, K.; Shishido, T.; Fukuda, T. *Acta Crystallogr.* 1989, **C45,** 349.
78. A book that contains some of the best illustrations for intermetallic compounds is: Pearson, W. B. "The Crystal Chemistry and Physics of Metals and Alloys"; Wiley–Interscience: New York, 1972.
79. For a valuable compilation of data for intermetallics up to the mid-1980s, see: Villars, P.; Calvert, L. D. "Pearson's Handbook of Crystallographic Data for Intermetallic Phases"; American Society for Metals: Metals Park, Ohio, 1985.
80. For a derivation of pseudopotential radii, see: Zunger, A. In "Structure and Bonding in Crystals;" O'Keeffe, M., Navrotsky, A., Eds.; Academic Press: New York, 1981, and references cited therein.
81. A structure map example can be found in: Villars, P. *J. Less-Common Met.* 1983, **92,** 215.
82. For examples, see: Hoffmann, R. "Solids and Surfaces"; VCH Publishers Inc.: New York, 1988.

8

Molecular Precursors to Thin Films

M. L. Steigerwald

1. INTRODUCTION

1.1. Inorganic Thin Films

The chemistry of the inorganic metal–nonmetal bond is central to modern materials science. The modern electronics and computer hardware industries are based on inorganic materials such as compound semiconductors, which are embodiments of the metal–nonmetal bond. In view of the importance of such compounds, their synthesis offers important challenges and opportunities for reaction chemistry. This type of synthesis chemistry has a number of unique features, perhaps the most apparent of which is that the macroscopic form of the final product is as important as its microscopic chemical constitution. Depending on the ultimate use of the material it is required in forms as varied as large (dimensions of inches to feet) single crystals, powders (both polycrystalline and amorphous), monoliths, nanoscale materials, and thin films. Since so many different products are desired, a variety of synthesis techniques are required.

Inorganic thin films are used in a very large number of applications. Some examples follow: (1) Films of hard ceramic materials can be used as conformal protective coatings. (2) Semiconductor chip manufacture demonstrates thin-film technology in a variety of ways.[1] The actual semiconducting material (e.g., silicon) occurs in thin-film form, as do the metals which interconnect the electronic elements and the insulators which separate them. (3) The ultimate achievement in thin-film synthesis has been the quantum-

M. L. Steigerwald • AT&T Bell Laboratories, Murray Hill, New Jersey 07974.
Inorganometallic Chemistry, edited by Thomas P. Fehlner. Plenum Press, New York, 1992.

well superlattice.[2] A quantum well is a film of a semiconducting material which is so thin that some of the properties which are constants in the bulk solid (such as the bandgap, i.e., the HOMO–LUMO separation) cease to be constant and become instead functions of the film thickness. (The term quantum well is used to emphasize that this behavior is quantum mechanical in nature.) In a quantum-well superlattice a large number of quantum wells are stacked on top of one another in a regular way. The engineering of physical properties which accompanies the manipulation of physical and chemical structure in the nanometer-size regime has led to the fabrication of a large number of new electronic devices.

There are several very general guidelines which are applicable to most thin-film growth procedures. First, the films should show the appropriate stoichiometry. [For example, in the growth of indium phosphide (InP) films a deficiency of phosphorus is often a problem.] Second, the films should have the desired crystal structure. (For example, it is not sufficient to specify a carbon film since both graphite and diamond crystal structures have the same composition.) Third, the films should cover the substrate conformally and uniformly, and fourth, the film should adhere to the substrate. In addition to these general rules further constraints arise in certain applications. Single crystal films are necessary for many electronic materials. Crystallinity across an interface (epitaxy, see below) is often required. As alluded to above, it is also often necessary to have very precise control over film thickness. On top of these requirements are some obvious and other not so obvious practical considerations, such as restrictions on growth conditions—for example, it is not possible to use a very high temperature deposition process to coat a thermally delicate device with a material whose function is to protect the device from thermal damage.

Several general problems often plague thin-film growth. Crystallite nucleation in the vapor phase[3] and premature reactions between precursors (molecular or otherwise) can give both poor-quality films and contaminated growth apparatus. In conformal growth on three-dimensional "landscapes," shadowing of growth by pre-existing features gives poor device performance. Problems such as low throughput (particularly for MBE) and the danger associated with volatile source compounds (for CVD) are serious practical considerations.

One approach which has been used in the synthesis of inorganic solid-state materials is the molecular precursor method.[4] In a molecular precursor synthesis the elements which are to form the ultimate solid-state product are introduced in molecular reagents. These molecular reagents are induced to decompose, usually thermally, thereby liberating the masked central atoms such that they can add to the growing solid-state compound. At the same time the ancillary ligands, the complement of the precursor molecule, are removed and therefore not incorporated in the solid. It is this general molecular precursor method that is the subject of this and the following sections. In this section the primary focus will be the use of molecular reagents in the preparation of thin films of electronic materials.

The chemical reactions and physical processes that lead to inorganic thin films are extraordinarily complicated and for that reason are not particularly well understood. It is intriguing that the chemical reactions which lead to thin solid films nonetheless appear very simple when written, since the reagents are generally the simplest available and the reactions are presumed to be simple pyrolyses. One reason for the study of molecular precursor chemistry (particularly in the "inorganometallic" context) is the expectation that studying the reactions which take molecules to solids will lead to a clearer understanding of the growth process that will then lead to more efficient and effective use of the syntheses. A second reason for the study is the anticipation that, when particular problems arise in the fabrication of specific materials, specific precursor molecules can be designed which will aid in the solution. A third and most ambitious reason for the study is the hope that new precursors will lead to new solid-state materials which have not been prepared by known methods.

The attempt is made in this chapter to discuss the relationship between inorganic solid-state chemistry—particularly the fabrication of thin solid films—and the metal–nonmetal ("inorganometallic") bond. The subject is quite vast and no attempt is made at an exhaustive review of the literature. A number of recent texts and review articles have been useful in the preparation of this chapter and they are included in the list of references.[5]

1.2. A Molecular View of Solid State Chemistry

Many technologically important inorganic materials are solid state compounds. This term implies that the chemical forces holding solid state compounds together are stronger than the van der Waals forces which pack molecules to form crystals. In the limit, a solid state material such as a macroscopic piece of crystalline silicon is a single molecule in the sense that each silicon atom is covalently bound to the bulk lattice.

The molecular/solid-state dichotomy leads to some potentially confusing ideas. A molecular compound is said to be crystalline if a macroscopic number of molecules align and organize while precipitating. A solid state compound is said to be crystalline if the atoms internal to the macroscopic "single molecule" are regularly ordered. Given these definitions, it is possible to describe and prepare materials which are both crystalline and not crystalline at the same time. Materials which are intermediate in size between molecules and solids (nanoscale[6] materials) are crystalline in the solid state sense (each grain or cluster is internally regular) but not at all crystalline in the molecular sense (a macroscopic ensemble of clusters is not coherently arranged).

Chemical formula has a subtly different meaning in solid state chemistry than in molecular chemistry. A given molecular compound has a unique chemical formula; for a particular solid state compound this is not precisely true. A classic example of this is shown in the cobalt/tellurium phase diagram.[7] Cobalt telluride (CoTe) is a crystalline material having the so-called NiAs structure. This $CoTe_x$ structure is stable and crystalline for all values of x between 1 and 2. The structure of the NiAs lattice shows why this is so. Removal of every other Co atom in the CoTe lattice gives exactly the (high temperature) $CoTe_2$ lattice (CdI_2 structure). There is no particular stability inherent in the stoichiometric compounds. A related feature is relevant to the discussion of II–VI materials (12–16) (see below). Mercury telluride (HgTe) and cadmium telluride (CdTe) are isostructural and have essentially the same sized unit cell, and the lattice allows the random placement of Hg and Cd in the alloy (solid solution) mercury cadmium telluride [$Hg_xCd_{1-x}Te$, or equivalently, (Hg, Cd)Te].

A variety of synthesis techniques are used in inorganic materials chemistry, and the method chosen depends on the form of product desired.[8] The most familiar and widely used method of synthesis is the direct synthesis from the elements. In this process the pure elements are physically combined and heated to give interdiffusion and crystallization (in the solid state sense). This method generally gives polycrystalline powders. Direct synthesis from the elements is a tremendously powerful technique and is the workhorse of materials synthesis, but it does have limitations. The most obvious limitation is the high barrier to reaction intrinsic to the mixing of two solids. This can be the highest energetic barrier in the process, and is a reason that the products of a ceramic reaction are usually determined simply by thermodynamics rather than kinetics. The high barrier to interdiffusion is removed by dispersing the elements in solvent. Aqueous ionic precipitation of metal sulfides is a simple example of this. Very low temperature (ambient) precipitations generally give amorphous or small-grain polycrystalline powders. High temperature precipitation reactions, exemplified by hydrothermal growth and flux growth, allow the polycrystalline precipitates to dissolve and reform to give the growth of large single crystals.

When the product material is desired in the form of a film, a two-phase reaction system (liquid–solid or vapor–solid) is generally required. With the appropriate choice of reaction variables nucleation and growth of the solid state material will occur only at the interface (substrate surface) and not in the homogeneous phase. Since the entire interface is exposed, conformal coverage is favored.

2. CHEMICAL METHODS FOR THE SYNTHESIS OF THIN FILMS

2.1. Chemical Transport Reactions

The prototypical chemical transport reaction is shown in equation 1.

$$A_{(s)} + B \rightleftarrows C_{(v)} \tag{1}$$

In this reaction a solid state compound A reacts with a so-called transport agent, B, to form a volatile compound C. The only other requirement is that the reaction be truly reversible and that the equilibrium coefficient for the process be a function of temperature. With these specifications met, if the process is carried out in a closed tube which is in a temperature gradient, the solid material A will be transported through the gradient and redeposited. This reaction type has been developed and reviewed extensively by Schäfer.[4c]

Chemical transport is a very effective way to purify solids. One example is the extraction of elemental nickel from crude ore by the Mond process (equation 2).

$$Ni + CO \rightleftharpoons Ni(CO)_4 \quad (2)$$

Nickel in the crude ore reacts with CO to form $Ni(CO)_4$ at 50 °C. The volatile carbonyl moves (by convection in a closed system or with the stream in a flowing reaction) to the hot zone (180–200 °C) where the carbonyl decomposes by the reverse of 2 to give elemental nickel, generally as a mirror. A second example of the importance of chemical transport is the transport of tungsten by chlorine (equation 3).

$$W + 3\,Cl_2 \rightleftharpoons WCl_6 \quad (3)$$

Tungsten reacts with chlorine at fairly low temperature to give WCl_6, which moves to the hot zone of the reactor to give tungsten deposition by the thermolytic reverse of equation 3. This reaction is used elegantly in automobile "halogen" headlights. Since the halogen transports tungsten to the hot zone of the reactor, the elemental tungsten which is thermally evaporated from the filament is constantly returned, therefore the filament is continuously repaired and the lifetime of the bulb is extended.

Chemical transport reactions are generally conducted under conditions very close to equilibrium to insure that crystal growth is slow. This gives the sustained growth of only a few nuclei and therefore results in large single crystals. With the proper choice of growth conditions homogeneous films can be formed using identical chemical reactions. The reader is referred to Schäfer's monograph for a much more detailed discussion of chemical transport.

One feature of chemical transport is quite noteworthy in the present context. In the majority of cases the transport agent, B in equation 1, is very simple (halogen, O_2, H_2, H_2O, etc.) and the transport reactions rely on the volatility of the corresponding metal halides, oxides, and hydrides. Very little research has been done using traditional organometallic ligands as transport agents.

2.2. Chemical Vapor Deposition (CVD)

The chemical vapor deposition process is succinctly summarized by equation 1, reading right to left (equation 4).

$$C_{(v)} \rightarrow A_{(s)} + B \quad (4)$$

Table 1. Examples of CVD Reactions

Entry	Reaction		Reference
i	SiH_4	$\rightarrow Si + H_2$	5f
ii	$H_2 + SiCl_4$	$\rightarrow Si + HCl$	5f
iii	$Al(^{i}Bu)_3$	$\rightarrow Al$	*a*
iv	$Fe(CO)_5$	$\rightarrow Fe + CO$	*b*
v	TiI_4	$\rightarrow Ti + I_2$	*c*
vi	$Si(OEt)_4$	$\rightarrow SiO_2$	9b
vii	$M(BH_4)_{3,4}$	$\rightarrow MB_2$ M = Ti, Zr, Hf	12d
viii	$Fe_3(H)(CO)_9BH_4$	$\rightarrow Fe_{75}B_{25}$	12c
ix	Cp_2V	$\rightarrow VC$	10c
x	$GaMe_3 + AsH_3$	$\rightarrow GaAs$	18a
xi	$[(tBu)_2GaAs(tBu)_2]_2$	$\rightarrow GaAs$	32

[a] Green, M. L.; Levy, R. A.; Nuzzo, R. G.; Coleman, E. *Thin Solid Films* 1984, **114,** 367.
[b] Kaplan, R.; Bottka, N. *Appl. Phys. Lett.* 1982, **41,** 972.
[c] Campbell, L. E.; Jaffee, R. I.; Blocher, J. M.; Gurland, J.; Gosner, B. W. *J. Electrochem. Soc.* 1948, **93,** 271.

The equilibrium which makes chemical transport reactions function is not required since the volatile molecular precursor, C, is introduced directly as a previously isolated and purified reagent. Thus films of elemental nickel result from the thermolysis of $Ni(CO)_4$ in a flowing vapor stream.[4c] CVD reactions of the general form of equation 4 have been used to prepare films of a very large number of materials. Several examples are shown in Table 1.

The first entries in Table 1 establish that a wide variety of single-element films can be prepared by CVD. The phenomenological reactions that lead to the elemental depositions can be as apparently simple as the flow pyrolysis of silane (i). Slightly more complicated reactions such as the pyrolytic hydrogenolysis of $SiCl_4$ are common. [It is interesting to note that elemental silicon can be transported by HCl and therefore entry (ii) may be thought of as a combination of CVD and chemical transport reactions.] Organometallic reactions are clearly useful in CVD as shown by the deposition of aluminum with organoaluminum reagents.

The CVD method is clearly not restricted to elemental products. Materials such as metal oxides,[9] carbides,[10] nitrides,[11] borides,[12] and compound semiconductors[4d,5h] such as GaAs and CdTe are routinely prepared by CVD reactions. Processes can use unimolecular (single-source) reagents as in entry (iv) or multisource reagents as in entry (x). Reactions giving compound semiconductors will be discussed in greater detail below.

The essential features of a horizontal, atmospheric pressure CVD reactor are shown in Figure 1. A standard flow is established through the reactor, usually using a thermally conductive gas (H_2 or He). (In atmospheric pressure systems the flow is generally maintained by mass-flow controllers regulating the high-pressure carrier gas.) The substrate material is placed within the

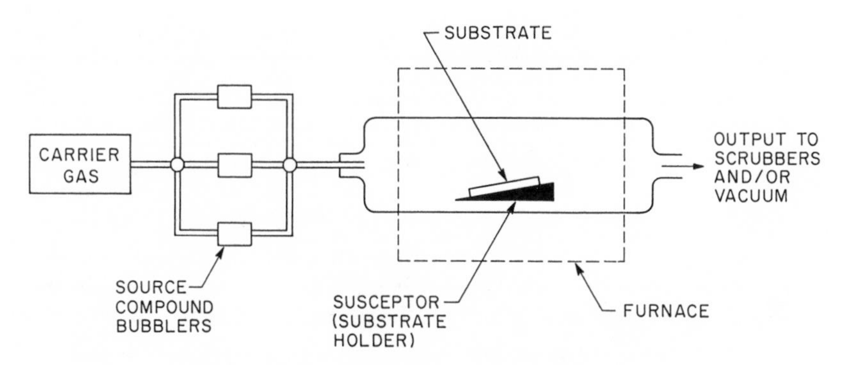

Figure 1. Schematic diagram of a typical CVD reactor.

main body of the reactor and, depending on the nature of the deposition, is supported on a wedge-shaped susceptor. The reactor tube is heated and the carrier gas is diverted through bubblers containing the source reagents. As the reagents mix and move into the hot zone of the reactor the chemical and physical reactions proceed and deposition occurs. The effluent gases are passed out of the system through the appropriate bubblers and scrubbers.

This very simple apparatus has been modified in a large number of ways. Reactor pressure can be varied from atmospheric (APCVD) to approx. 1 torr (LPCVD, low-pressure CVD)[4d,5f,13] to the very low pressure (10^{-4} torr) of gas-source molecular beam epitaxy.[14] Heating can be done by placing the body of the reactor in a furnace (hot-wall reactor) or by inductively heating the susceptor (generally graphite) by rf radiation. In the majority of cases the deposition reactions are thermally driven, but deposition has been enhanced using photochemistry[15] and plasma chemistry (photo- and plasma-enhanced CVD, respectively).

By comparison to the amount of effort devoted to the optimization of other reactor/reaction variables, little work has been done in optimizing the source compounds and/or their reactivities.

2.3. *Vapor Phase Epitaxy (VPE)*

Vapor Phase Epitaxy (VPE) is a special case of CVD that is very important to the electronics industries. The salient difference between CVD and VPE is shown diagramatically in Figure 2. In the general case of CVD crystallinity in the film is not a strict requirement. As implied in Figure 2a both the substrate and the film can be polycrystalline or amorphous. In VPE both the substrate and the film are, by definition, crystalline, with the crystallinity continuing across the interfacial plane as suggested by Figure 2b. This requirement of so-called epitaxial growth is an extreme one and with it comes a number of restrictions on the growth process and reactions.[4d]

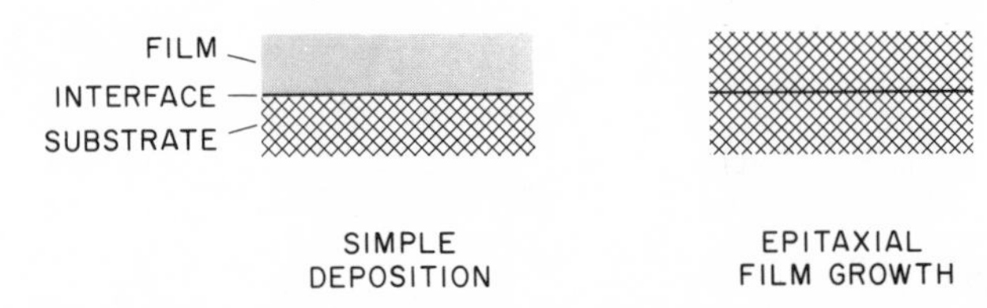

Figure 2. Simple deposition versus epitaxy. Cross-hatching denotes crystallinity.

The specification for epitaxial growth arises fundamentally from the need for reproducibility and high quality in the physical and electrical properties of thin films of electronic materials. The performance of electronic thin films is quite sensitive to structural and compositional irregularities so the only way to insure reproducible performance is to require crystalline perfection in the film. For the same reason the interface between the film and the substrate is necessarily as sharp and as close to perfectly crystalline as possible.

The first obvious restriction imposed by epitaxial growth is lattice matching. Crystal registration across the interface is clearly possible if the film and the substrate are the same material (homoepitaxy). If the two are different chemical compounds (heteroepitaxy) clean crystal growth across the interface is not possible if the separate lattices cannot fit together because of incommensurate size or shape. For example, HgTe and CdTe crystallize in the same structure (zinc blende) and have essentially the same-sized lattice (lattice parameter 6.453 Å and 6.481 Å, respectively)[16]; as a result heteroepitaxy of one on the other is not structurally difficult. In contrast GaAs and Si crystallize in the same cubic structure but have a much different sized lattice (5.653 Å vs 5.43 Å).[16] As a result epitaxial growth is much more difficult. The need for matching lattices insures that it is not possible to grow epitaxial films of all materials on all substrates.

The second requirement that epitaxial growth implicitly carries with it is the need for very high purity films. The random inclusion of impurities (due to the incomplete reaction of a precursor, for example) causes crystal defects which hamper subsequent growth. In addition, the presence of the impurity can drastically modify the properties of the film. The necessity for film purity has a considerable effect on the utility and choice of molecular precursors.

Processing variables must be controlled very carefully in epitaxial growth. Overall reactor pressure and the hydrodynamics[17] of the carrier gas flow are manipulated to guarantee that the film is both crystalline and homogeneous across the entire substrate. The temperature of the substrate during film growth is an important variable because of surface diffusion of the deposited atoms. In the generally held view of VPE the molecular precursor reacts with or at the substrate surface to deposit the growing film one atom at a time. The substrate must be hot enough to allow an adatom to migrate over the surface if the site at which it is initially deposited is incorrect. In other words a high substrate temperature allows the film to recrystallize during and subsequent to initial growth.

2.4. Organometallic Vapor Phase Epitaxy (OMVPE)

In the fabrication of epitaxial thin films of compound semiconductors (III–V compounds such as GaAs, InP, and AlGaAs; II–VI compounds such as HgTe and ZnSe) by VPE the most common molecular precursors are organometallic, hence the specialization of VPE to OMVPE. (The process is also known as metalorganic VPE, MOVPE.) The technique[18] was first implemented by Manasevit in 1968 using the process in equation 5.

$$Ga(CH_3)_3 + AsH_3 \rightarrow GaAs + 3CH_4 \quad (5)$$

There are two very attractive features of this scheme. The first is the very high volatility of trimethyl gallium (TMG). Elemental gallium has a very low vapor pressure. It can be chemically transported by halogen or HCl (see above) but gallium halides are not volatile enough for large-scale practical use. Trimethyl gallium, on the other hand, is among the most volatile gallium compounds known and was therefore a likely candidate for VPE regardless of reaction chemistry. However, the second attraction of equation 5 is the apparent simplicity of the reaction chemistry. The process as written is quite tidy in the sense that the two separate precursor molecules apparently just exchange ligands in order to generate the solid and stable molecular byproducts. As will be discussed below it is still not clear how simple equation 5 actually is.

2.5. Molecular Beam Epitaxy (MBE)

A very large amount of work has been done to study the microscopic mechanisms of thin-film growth. Many details are undetermined but the generally accepted picture emerging is that precursors decompose at or near the substrate surface to give atoms (or clusters of at most a few atoms) which add to the growing film. Molecular Beam Epitaxy (MBE)[19] essentially sidesteps the molecular precursor by using the atoms (or small clusters) directly as the chemical feedstock.

The essential features of an MBE reactor are sketched in Figure 3. The substrate is mounted within an ultrahigh vacuum chamber maintained at pressures low enough that the elements comprising the desired film can be evaporated to form a molecular beam which is directed at the substrate. Separate, shuttered source cells are used for each element and the constitution of the beam reaching the substrate is determined by which shutters are opened. The film grows by the surface diffusion of the beam-deposited adatoms to form the crystalline lattice.

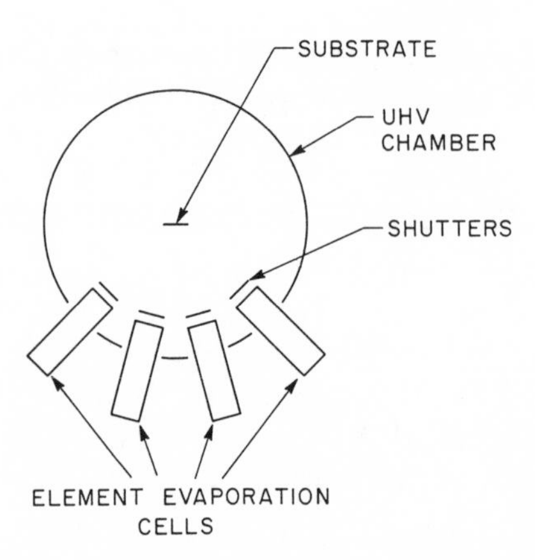

Figure 3. Schematic diagram of an MBE reactor.

Since chemical reactions seem to be irrelevant to the process, MBE is conceptually simpler than VPE. For the same reason control over all aspects of the deposition is more direct. Sharp interfaces between films of different compositions can be made by abruptly opening different source shutters. The substrate environment in MBE deposition is ultrahigh vacuum and is therefore very well defined. The most elaborate and precise thin-film structures to date have been prepared by MBE and it is considered the synthesis method of choice for thin-film research. Counterbalancing this technical excellence is the instrumental sophistication and associated expense of the method.

2.6. Gas-Source Molecular Beam Epitaxy (GSMBE)

This synthesis technique[14] [also known as Organometallic MBE (OMMBE), Metalorganic MBE (MOMBE), and Chemical Beam Epitaxy (CBE)[20]] is a hybrid of OMVPE and MBE. The basic design of a molecular beam apparatus is employed but the sources (precursors) are molecular, as in OMVPE, rather than elemental. In the case of GaAs film growth is conducted at $\sim 10^{-4}$ torr, the Ga is introduced as $Ga(C_2H_5)_3$, and the As is introduced as As_2 [from previous high-temperature cracking of $As(CH_3)_3$]. As in the other techniques substrate temperature is a critical experimental parameter.

This method has a number of potential advantages. Since the sources are gaseous the apparatus can in principle be operated uninterrupted for a longer time than an MBE system, which must have its solid source cells replenished periodically. Since the deposition in GSMBE is "line-of-sight" the hydrodynamics of the reactor are not as important as in OMVPE. The source compounds are clearly more volatile in GSMBE than in MBE. This allows more latitude in operating conditions but also raises questions of safety associated with the toxic and/or pyrophoric source compounds.

Gas Source MBE is one clear limit of low-pressure CVD and offers opportunities for more sophisticated (if less volatile) source compounds and the associated reaction chemistry (although in view of the currently believed mechanism[21] of film growth in GSMBE it is open to question whether such new sources would be important). The GSMBE growth mechanism has been most intensively studied for films of GaAs. Arsenic arrives at the surface of the film as As_2, which is incorporated directly. Gallium impinges on the surface as intact $Ga(C_2H_5)_3$. The organometallic compound then decomposes on the surface to give Ga, olefins, and alkyl radicals. The carbon-containing fragments do not disturb the film growth and are swept from the system *in vacuo.* Film growth must be conducted at substrate temperatures in excess of the decomposition temperatures for the Ga alkyls in order to realize a crystalline film, so it is not clear that the structure of the Ga alkyl is directly important although this has yet to be demonstrated.

3. EXAMPLES: MOLECULAR PRECURSOR ROUTES TO TETRAHEDRAL SEMICONDUCTORS

Tetrahedral semiconductors (silicon, III–V compounds and II–VI compounds) have been the most intensively studied electronic materials for the past several decades. There are quite a number of reasons for the interest in these materials but the most fundamental is that they are technologically useful. Molecular precursors have been employed extensively in the synthesis of the compounds and manufacture of the associated devices, therefore these materials offer some instructive case studies.

These semiconducting solids all have the same geometrical structure in which each of the atoms is coordinated tetrahedrally—hence the term tetrahedral semiconductors. Silicon crystallizes in the diamond structure.[16] This structure has been described in a number of remarkably different ways, but in the present molecules-to-solids context it is useful to view it as Si_6 rings in the "chair" conformation condensed into a regular three-dimensional polymer. The structure is cubic in the crystallographic sense ($a = b = c$, $\alpha = \beta = \gamma = 90°$). The so-called III–V compounds are binary solid-state materials having a one-to-one stoichiometry of a Group IIIB (Group 13) metal and a Group VB (Group 15) nonmetal. They crystallize in the sphalerite or zinc blende structure,[16] which is just the structure of silicon with alternate atoms being the metal and nonmetal. As in silicon the sphalerite structure has a cubic unit cell and all bonds are equivalent and all bond angles are equivalent. Moving the metal and nonmetal one column to the left and right respectively from the III–V compounds gives the II–VI compounds. These compound semiconductors have the formula MQ where M = Zn, Cd, Hg and Q = S, Se, Te. They also crystallize in the cubic, sphalerite[16] structure, having tetrahedrally coordinated atoms. The diamond (silicon) and sphalerite structures are very simple and this is one reason for the wide utility of these semiconducting materials.

Another reason for the popularity of these semiconductors is their chemical simplicity. In each III–V and II–VI system there is only one stable compound phase, that having the 1:1 stoichiometry. This should be contrasted with other binary systems[22] which show a large number of distinct and stable compounds. This insures that when the elements are combined, one and only one compound will result. (This is not strictly true for the lighter II–VI compounds. In addition to the sphalerite structure, the hexagonal, wurtzite structure[16,23] is stable in these materials. This is a potential complication in the growth of these materials.) The stoichiometric compounds are also thermodynamic sinks. This is important because it implies (as is found) that the compounds form readily and dissociate only slowly, if at all.

The structural simplicity and similarity of these tetrahedral semiconductors result in a chemical attribute which is quite important to their use in electronic devices. Since the binary compounds all have the same crystal structures, solid solutions (alloys) of two or more of them can be formed. An important example is mercury cadmium telluride, $Hg_xCd_{1-x}Te$. In this compound (alloy, solid solution) the tellurium atoms are arranged in a cubic close-packed lattice just as in the (sphalerite) crystals of HgTe or CdTe, but the cation sites are occupied randomly with mole fractions x Hg and $(1 - x)$ Cd. In this sense the otherwise separate HgTe and CdTe lattices are miscible, dissolving in one another to give the mercury cadmium telluride solid solution. The important result of this is that the physical properties of the solid solution vary with x. In the case of $Hg_xCd_{1-x}Te$ the band gap of the solution can be tuned from 1.4 eV (the band gap of CdTe) to 0 (HgTe is a semimetal) continuously by changing x. This allows the use of mercury cadmium telluride for infrared and longer-wavelength optical applications.

It was mentioned above that an essential feature of epitaxial growth of thin films is that the lattice parameters of the substrate and the film be matched. This is another reason that the sphalerite structures have been so thoroughly studied. For example, since GaAs and AlAs have the same structure type and the same lattice parameter (5.65 Å and 5.66 Å, respectively)[16] it is possible both to make solid solutions of the two ($Al_xGa_{1-x}As$) and to grow epitaxial films of varying compositions (and therefore varying properties) on one another. This allows the fabrication of devices such as GaAs/AlGaAs lasers.

3.1. Growth of III–V Thin Films by OMVPE

The technologically most significant III–V materials and the OMVPE reactions used to form them are listed in Table 2. The general theme is clear: the Group III element is introduced as the trialkyl and the Group V as the trihydride; the reagents are combined in a rapid flow of carrier gas and heated; and the solid compound deposits as a crystalline thin film. Despite the apparent simplicity the details of these processes are remarkably complicated.

Table 2. Reactions Used in III–V OMVPE

Reactants	Product
$GaMe_3 + AsH_3$	$\rightarrow$ GaAs
$GaEt_3 + AsH_3$	$\rightarrow$ GaAs
$GaMe_3 + H_2As^tBu$	$\rightarrow$ GaAs
$AlMe_3 + GaMe_3 + AsH_3$	$\rightarrow$ AlGaAs
$InMe_3 + PH_3$	$\rightarrow$ InP
$InEt_3 + PH_3$	$\rightarrow$ InP
$InEtMe_2 + PH_3$	$\rightarrow$ InP
$InMe_3 + H_2P^tBu$	$\rightarrow$ InP

The first obvious chemical complication is reagent stoichiometry. When the Group V hydrides are used, the input molar ratio of the Group V source to the Group III source (hereinafter the V/III ratio) is generally between 4 and 100.[4d,5b] The requisite V/III ratio is generally determined empirically based on the purity, morphology, and physical characteristics of the deposited film. At least two reasons for the high V/III ratio are given. The first is that since AsH_3 and PH_3 are more stable at the growth temperatures than are the Group III alkyls, the latter are more efficiently pyrolyzed than the former and therefore more of the former is required if a stoichiometric film is to result. The second suggestion is that the group V hydride (or its pyrolytic daughter radicals) help remove surface carbon radicals which are derived from the Group III alkyls. An excess of the Group V source is therefore required to remove the carbon which could be incorporated as a contaminant.

The identities of the Group III source compounds have impact on the OMVPE reactions and hint at some aspects of the mechanisms of the growth process. Owing to higher vapor pressure the trimethyl derivatives of the Group III elements are commonly used, but this is offset in part by the tendency of the higher alkyls to incorporate less carbon into the films.[24] This aftereffect has been interpreted as being due to the intrusion of a β-hydride elimination pathway into the decomposition of the Group III sources.[24] [It is interesting to note that $Ga(CH_3)_3$ and $Ga(C_2H_5)_3$ give essentially the same inorganic products on reaction with AsH_3 even though the mechanisms of pyrolysis are apparently quite different.]

While it is important that the source compounds decompose to the elements (or the equivalents) in the VPE reactor, it is equally important that they be stable enough to reach the hot zone. As discussed above the formation of a crystalline film demands some minimum substrate temperature and, if the source compound is not stable enough to reach the neighborhood of the substrate, epitaxy is clearly not possible. This adds another important ground rule to precursor design.

An important technical point is precursor volatility. For atmospheric pressure OMVPE source compound vapor pressures above 10 torr at room temperature are preferred.[24] For this reason a trimethyl compound may be more useful even though a triethyl compound would be chemically better suited. In addition, liquids are preferable to solids thus dimethylethylindium may prove more useful than trimethylindium.[25]

The carrier gas in OMVPE reactions is usually H_2 but He, N_2, and Ar are also used. Dihydrogen is prevalent because it can be very highly purified and has high thermal conductivity and low viscosity. High purity is a stringent requirement as discussed above and H_2 can be purified completely by diffusion through Pd. High thermal conductivity of the carrier gas is important—since the residence time in the OMVPE reactor is short, heating should be fast. Low viscosity is also desirable because it insures better flow,[17] giving a thinner boundary layer around the substrate.

The presence and nature of a substrate has an effect on the deposition process, which is in a sense related to the chemical transport process. A principal driving force for crystal growth by chemical transport is the lattice energy of the growing solid. On an atomic scale this is due to the formation of strong bonds on and to the crystal surface. Formation of the analogous bonds in the film-deposition process must have a corresponding kinetic effect, because it has been observed that the deposition of III–V compounds is accelerated by the presence of the III–V material in the reactor prior to deposition.[26] (This is related to the autocatalytic decomposition of many organometallic complexes. Viewed in solid-state terms these processes are not catalytic at all, but are self-promoted or product-promoted.) This is another indication that the mechanism of film deposition is quite complex.

At present there is no consensus on the fine details of the film growth process in III–V OMVPE. The most widely accepted view[4d] is that the separate precursor molecules arrive in the hot zone of the reactor intact and diffuse through the hydrodynamic boundary layer that covers the substrate. The precursors decompose independently at the surface to inorganic atoms and alkyl radicals. The alkyl radicals and hydrogen atoms recombine on the surface to give volatile hydrocarbons, which are removed in the carrier flow. The remaining adatoms migrate on the surface to form the crystalline lattice.

According to this proposed pathway there is no interaction between the different precursors prior to their decomposition, and a critical property of a molecular precursor is its decomposition as a function of temperature. Therefore the suitability of a precursor candidate is routinely assayed by independent pyrolysis studies of the following type.[27] A carrier gas flow is saturated with the compound of interest and passed through a heated tube. The composition of the effluent is determined (usually by mass spectroscopy) with the disappearance of input precursor being equated with decomposition. A good deal of information has been extracted from such pyrolysis experiments, and trends observed in pyrolysis studies have mapped onto actual film-growth studies. Chemical reaction mechanisms are also suggested by the appearance of pyrolysis products; e.g., the formation of ethylene in the pyrolytic decomposition of $Ga(C_2H_5)_3$ clearly implies a β-hydride process.[24]

While decomposition studies have been valuable in a survey sense they ignore the possible importance of intermolecular reaction chemistry. In fact copyrolysis [of, for example, $Ga(CH_3)_3$ and AsH_3] has shown that some sort of intermolecular process does occur under normal OMVPE conditions in as much as the disappearance of each precursor in the mixed flow occurs at substantially lower temperature than that required for the decomposition of each separately (all other variables being held constant).[28,29b,46]

The question of intermolecular chemistry was addressed directly by Stringfellow and co-workers.[29] They conducted the copyrolysis of $In(CH_3)_3$ and PH_3 in a carrier gas of D_2. They found that the decomposition temperature of each was lowered by the presence of the other, and that the methane by-product contained no deuterium. Since the CH bond in methane possesses the same strength as the bond in H_2, one would expect CH_3D if homolytic decomposition of $In(CH_3)_3$ were a dominant reaction in this process. This argues strongly for a 1,2 elimination reaction as in equation 6.

$$(CH_3)_3\,In + PH_3 \rightleftarrows (CH_3)_2\overset{\overset{\displaystyle CH_3}{|}}{In}-\overset{\overset{\displaystyle H}{|}}{P}H_2 \rightarrow \begin{array}{c} CH_4 \\ + \\ (CH_3)_2In-PH_2 \\ \mathbf{A} \end{array} \qquad (6)$$

From the perspective of molecular reaction chemistry, equation 6 makes sense. Eliminations of this type are very well documented,[30] and it would account for the products observed by Stringfellow *et al.* as well as for the lower activation barriers for the "decomposition" of both of the precursor compounds, since the elimination will have a lower barrier than will simple bond homolyses.

In spite of the ease of the intermolecular elimination reaction shown in equation 6, it is widely suggested[4d,5b] that the occurrence of such reactions, at least in the vapor phase, should be suppressed in the OMVPE process. This is because it is thought to be an example of a so-called parasitic reaction. Parasitic reactions are defined as undesired chemical reactions that prematurely remove feedstock precursors from the carrier flow. Parasitic reactions are not well characterized but usually arise from a gas-phase interaction between the different precursors upstream of the reactor hot zone and generate an involatile organometallic polymer either as a film on the reactor walls or as a fine dust.[3] These reactions are deleterious to film growth because they can disturb the reagent stoichiometry, they waste very expensive precursors, they cause tedious and troublesome reactor clean-up, and the gas-phase nucleation of dust can ruin the thin film. It is generally believed that parasitic reactions are condensation reactions such as in equation 6 that run out of control as in equation 7.

$$(CH_3)_2InPH_2 \rightarrow (CH_3)_2In\!-\!\underset{\underset{\displaystyle CH_3}{|}}{\overset{\overset{\displaystyle H}{|}}{P}}\!-\!In\!-\!PH_2 \rightarrow (CH_3)_2In(\underset{\underset{\displaystyle CH_3}{|}}{\overset{\overset{\displaystyle H}{|}}{P}}\!-\!In)_nPH_2 \qquad (7)$$

A **B**

It has not been proven that equation 6 is a parasitic reaction. It has been shown that[30a] when $In(CH_3)_3$ and PH_3 are combined in solution that an involatile polymer results, but this only means that the condensation reaction must be controlled in OMVPE, not necessarily eliminated. This raises a very important chemical question about the OMVPE growth process which has not been answered: namely, what fraction of the metal–nonmetal bonds can or must be formed in the vapor phase prior to irreversible attachment of the molecular fragments to the surface of the growing film? For large values of n the oligomer, **B**, in equation 7 would certainly not be important in film growth (first, because it would be heavy enough to "snow" out of the vapor phase, and second, because its large inorganic backbone might not easily conform to the film surface). At the other extreme in molecular size, Maury[31] and Cowley and Jones[32] have shown that so-called single-source precursors having metal–nonmetal covalent bonds (analogous to **A** in equations 6 and 7) can be used to make films of III–V compounds (see equation 8), therefore the complete suppression of equation 6 is certainly not required. Given this, it is interesting to speculate on which values of n make the oligomer **B** a possible, important or necessary reactive intermediate in the growth of thin films. (The relevance of inorganic clusters to the preparation of solid state compounds in general and thin films in particular is a question of increasing importance.)[33] A detailed understanding of growth mechanisms in OMVPE is not yet available.

$$[(CH_3)_2Ga\,As(C(CH_3)_3)_2]_2 \xrightarrow{\Delta} GaAs \qquad (8)$$

Several other significant practical problems arise in the chemical processes listed in Table 1. The first and most notorious is the extreme safety hazard associated with the Group V sources, PH_3 and AsH_3. These are two of the most toxic substances known, and to make the situation more dangerous they are both gaseous at standard temperature and pressure. They are used as gases and stored in high-pressure cylinders, therefore should there be an accident it would be difficult to contain. The safety issues have made the large-scale implementation of this technology difficult. One response has been to design very sophisticated computer-driven and safety-interlocked reactor systems[34] to contain an accident immediately. A second response has been to design alternative chemical source compounds.

The specifications for phosphine and arsine substitutes are simple and demanding. They must be less dangerous than the hydrides they are to replace, therefore they should be either intrinsically less toxic or they should not be gaseous. Just as for the Group III sources the compounds must nonetheless be volatile. (Less volatile sources can be used in low-pressure VPE or GSMBE, but most OMVPE is done at or near atmospheric pressure.) Since the exact regulation of input stoichiometry is desired, liquid sources are preferable to solids. The substitute compounds must be available in arbitrarily high purity and they must decompose and/or react with the Group III sources at temperatures no higher than PH_3 or AsH_3 to give the same III–V compounds without increased contamination.

As a rule alkyl substitution for H decreases the toxicity of arsine and phosphine. More importantly, most of the simple alkyl-substituted arsines and phosphines are liquids at room temperature. Trimethyl arsine, triethylarsine, dimethylarsine, diethylarsine, and *t*-butylarsine have all been examined as arsine substitutes. Trimethylarsine is not useful[35] as a source compound because it does not react as readily with $Ga(CH_3)_3$ as does AsH_3, and when films are grown using $As(CH_3)_3$ carbon contamination [from the $As(CH_3)_3$, not the $Ga(CH_3)_3$] is high. Triethylarsine reacts more efficiently with $Ga(CH_3)_3$ but C-incorporation is still a significant problem.[36] The dialkylarsines have not been studied in as much detail, but they appear more useful[37] than the corresponding trialkylarsines. The most work has been done with *t*-butylarsine, $H_2AsC(CH_3)_3$. It is quite volatile and reacts with $Ga(CH_3)_3$ at reasonable temperatures to give GaAs without carbon contamination.[38] It is a liquid that is less toxic as well and is currently a very good candidate to replace arsine.

The situation is similar with phosphine replacements, although the trialkylphosphines are comparatively less useful[39] because they are thermally more robust than the arsenic analogs. *t*-Butylphosphine is a less toxic volatile liquid that has been used[40] to make good-quality InP.

A second practical problem associated with OMVPE reactions to give III–V compounds is the pyrophoric nature of the Group III alkylcompounds. This poses a safety hazard, albeit not as severe as that posed by the Group V hydrides. An interesting solution to this problem has been the use of Lewis acid–base adducts.[41] The electron-deficient Group III alkyl can be treated with a base such as triethylphosphine or 1,2 bis (diphenylphosphine)ethane to form an air-stable solid. This solid can be handled more easily than the simple alkyl. With the appropriate base, heating breaks the complex to liberate the alkyl for the VPE process.

Lewis acid–base adducts have been used in an entirely different context in OMVPE growth of III–V compounds. Films of InP were grown using $InEt_3$ and PH_3 by Moss and co-workers,[42] but the reactions were plagued by parasitic reactions (see above). These workers believed the fundamental parasitic reaction to be that shown in equation 9a and reasoned (and found) that the parasitic reaction (elimination of ethane followed by condensation polymerization) could be thwarted by the addition of another base such as $P(CH_3)_3$. It was known that $P(CH_3)_3$ is stable under typical OMVPE conditions (it is a poor source of P itself) so it was the natural choice. It has subsequently been found that the parasitic reactions can be eliminated more easily by substituting $In(CH_3)_3$ for $In(C_2H_5)_3$, but in either case this shows the practical benefit of the simple reaction engineering that can be done in OMVPE.

$$(H_5C_2)_3In + PH_3 \longrightarrow (H_5C_2)_3In\text{—}PH_3 \begin{cases} \xrightarrow[a]{-C_2H_6} (H_5C_2)_2InPH_2 + \ldots \\ \xrightarrow[+P(CH_3)_3]{b} (H_5C_2)_3In\text{—}P(CH_3)_3 + PH_3 \end{cases} \quad (9)$$

To summarize this overview of III–V film growth by OMVPE: films of high quality can be made although the mechanisms of growth are not precisely understood. Several observations concerning the apparent reaction chemistry are particularly significant: (1) In the reaction of an organogallium reagent with an arsenic source compound, the incorporation of carbon into the growing film of GaAs arises from the arsenic source if a trialkylarsine is used, but from the gallium source if AsH_3 is used.[35a,43] In the latter case a large excess of AsH_3 is required to insure complete removal of the carbonaceous material. (2) Low V/III ratios (approximately unity) can give excellent films[38] of GaAs if (t-Bu)AsH_2 is used rather than AsH_3. In the same way high-quality films result from single-source precursors[31,32] (V/III ratio of exactly unity, equation 8). (3) H/D labelling studies[29] have shown that the $H_2(D_2)$ carrier gas is not chemically involved in the deposition process. Based on these points it seems to have been established that interprecursor reactions dominate OMVPE processes. It will be interesting to see to what extent these reactions occur in the vapor phase and to what extent condensation reactions (such as in equation 7) are important.

3.2. Growth of II–VI Thin Films by OMVPE

The II–VI material which has been the subject of the most intense study is mercury cadmium telluride, $Hg_xCd_{1-x}Te$(MCT). As mentioned above the band gap of this alloy is a function of x and spans the technologically important

1–12 μm wavelength range. Since the band gap is a sensitive function of x it is important to fabricate films of the alloy using a method which gives precise and reproducible control over metal stoichiometry. Superlattices composed of alternating layers of pure HgTe and pure CdTe have also been suggested[44] as important device materials. In the preparation of these materials it is not the (relative) metal stoichiometry that is critical (since each layer is a pure binary compound), but rather it is the ability to make sharp interfaces between the layers.

The II–IV materials, particularly those containing Hg, are not well suited for these two requirements (control of stoichiometry and sharpness of interfaces). Elemental Hg has a very high vapor pressure even at room temperature (in fact the element is the precursor of choice for Hg in atmospheric pressure OMVPE), and the Hg—Te bonds holding the HgTe lattice together are not strong.[45] As a result, when films of MCT are heated Hg tends to evaporate leaving a Cd surplus, therefore at sufficiently high temperatures[46] (above approx 250 °C) the Hg/Cd ratio is difficult to control. The II–VI materials also interdiffuse rapidly.[47] This implies that very sharp interfaces between HgTe and CdTe are hard to achieve and maintain, particularly at elevated temperatures.

The problems of evaporation and interdiffusion are much more pronounced for the II–VI compounds than for the III–V compounds, therefore the growth temperature is a much more important variable in the II–VI case. Reducing the reaction temperature has been a driving goal in II–VI OMVPE.

The original methods[46,48] for forming films of CdTe and HgTe are shown in equation 10.

$$\begin{aligned}&(a)\ Cd(CH_3)_2 + Te(C_2H_5)_2 \xrightarrow{400\,°C} CdTe\\&(b)\ Hg_{(v)} + Te(C_2H_5)_2 \xrightarrow{400\,°C} HgTe\end{aligned} \tag{10}$$

Although these reactions are conducted at much lower temperatures than are usual for III–V growth, these temperatures are nonetheless much too high.

The mechanisms of II–VI film growth are less certain than in the III–V case, but it has been established that the high reaction temperatures coincide with those required for the homolytic decomposition of diethyltellurium.[46] This leads to the hypothesis that weaker Te—C bonds will result in lower MCT growth temperatures. This has been observed. The temperature required for HgTe growth decreases[49] as $TeEt_2$ is replaced by Te^iPr_2, Te^tBu_2, and $Te(allyl)_2$. Similar reasoning[50] has led to the use of $Te(CH_3)(allyl)$ and dihydrotellurophene.[51] In view of the thermal instability of cadmium alkyls[52] the very simple mechanism for CdTe growth shown in equation 11 has been suggested. In this scenario the organometallic precursors independently decompose to give atoms which mitigate on the substrate surface to form the crystalline solid. Although this pathway is not elegant chemically, it does rationalize most of the experimental observations.

$$Cd(CH_3)_2 \xrightarrow[\Delta]{-2\ CH_3} \text{Cd ATOMS}$$

$$Te(C_2H_5) \xrightarrow[\Delta]{-2\ C_2H_5} \text{Te ATOMS}$$

$$\text{Cd ATOMS} + \text{Te ATOMS} \rightarrow \text{CdTe (ASSEMBLE ON SURFACE)} \qquad (11)$$

The mechanism of II–VI growth is at least slightly more complicated since the copyrolysis of $CdMe_2$ and $TeEt_2$ gives disappearance of the tellurium alkyl at a much lower temperature than pyrolysis of the alkyl separately.[46] The nature of the interaction between the Te compound and the Cd compound (or their pyrolysis products) has not been established.

A different approach to the Te-source problem is the use of ditellurides.[53] Assuming a pathway for II–VI growth that is similar to equation 11, the best "alkyl" source would be an alkyltelluryl radical, RTe, since it is well on the way to atomic Te. This reasoning suggests alkylditellurides. It has been found[53] that dimethylditelluride is suitable as a Te source compound and have shown that films of CdTe can be grown with this source at temperatures as low as 250 °C. In a companion study it was shown[54] that ditellurides react with elemental Hg to form $Hg(TeR)_2$, which disproportionates on mild heating to give HgTe and TeR_2 (equation 12).

$$RTe—TeR + Hg \rightarrow RTeHgTeR \rightarrow HgTe + TeR_2 \qquad (12)$$

It had been reported[55] that CdS could be prepared from $Cd(SR)_2$. We have since shown that ZnS, ZnSe, CdSe, and CdTe all can be formed from the corresponding organometallic precursors.[56]

It is important to note that the organic byproduct in equation 12 is a diorganotellurium compound, and that the analogous product was seen in the OMVPE reaction. The organotellurium byproduct is not suggested by equation 11 and implies that a different mechanism for film growth is operating in the ditelluride scheme. We have also followed the evolution of the solid state products in these reactions. That work will be described in the following section.

We developed another alternative tellurium source, which was based on the "atomic" nature of the deposition mechanism in equation 11. Phosphine tellurides were first described by Zingaro and co-workers,[57] who showed that the equilibrium in equation 13 is fast under very mild conditions.

$$Te + PR_3 \rightleftarrows TePR_3 \qquad (13)$$

A comparison with Section 2.1 above shows that this is exactly the type of equilibrium required for chemical transport reactions. Furthermore, if reaction 13 is as simple as written, then it should fit into equation 11 quite well by directly supplying Te atoms to the growth zone (especially in view of the known stability of trialkylphosphines under OMVPE conditions; see above). We demonstrated [58] that Me_3PTe could be used in vapor-phase processing by using PMe_3 to crystallize elemental tellurium (in analogy to equation 2). We further showed that HgTe could be prepared by allowing phosphine telluride to react with either Hg or diphenylmercury. In the attempt to further characterize the reactivity and utility of phosphine tellurides, we have prepared a number of simple solid-state tellurides and have intercepted a number of molecular intermediates between the starting molecular reagents and the ultimate solid-state products. [59]

Another proposed solution to the reduction in II–VI film growth temperature is photochemical. Molecular precursors for Cd, Hg, and Te decompose to liberate the elements when exposed to UV irradiation. This has been used by several workers [60] to successfully prepare MCT films at lower temperatures. This technique has recently been reviewed in depth. [61]

As in the case of the III–V materials, very high quality films of the II–VI compounds can be prepared by OMVPE. The introduction of a variety of new precursor compounds as well as the use of photochemical reactions have had a valuable impact on the fabrication technology. It is also the case that the microscopic details of the reactions by which the films are formed are largely unknown.

4. CLUSTERS AS INTERMEDIATES IN THE MOLECULAR PRECURSOR SYNTHESIS OF II–VI COMPOUNDS

In this chapter thus far there has been an absence in the discussion of the chemical intermediates which lie between the molecular starting materials and the solid state products. In this final section one family of such intermediates is discussed.

As written, reaction 12 is very simple; however, based on the desorption of solid state compounds given in Section 1.1 above it must be much more complicated. Bulk HgTe is a high molecular weight "polymer" of HgTe (or, more properly, of the HgTe unit cell) and the "polymerization" aspect of the HgTe synthesis is not contained in reaction 12. The questions are: what are the fundamental building blocks of the solid and how are they attached to the growing lattice? In order to address these points we have studied several reactions analogous to reaction 12.

Cadmium selenide, CdSe, is a tetrahedral semiconductor that crystallizes in both the zincblende and wurtzite structures and has a band gap of 1.7eV (near IR). We found that CdSe can be prepared in a bulk, polycrystalline form by the pyrolysis of $Cd(SePh)_2$ as a solid, [62] just as in equation 12. This reaction can be moderated by conducting it in solution. When $Cd(SePh)_2$ is dissolved in 4-ethylpyridine and the resulting solution heated to reflux, the initially colorless solution changes color as shown by the UV-visible absorption spectra in Figure 4. The reaction can be quenched at any point and examined by transmission electron microscopy (TEM). This analysis reveals that the intermediates in the pyrolysis of $Cd(SePh)_2$ are small (10–40 Å diameter) crystallites of CdSe. Nanometer-sized particles of several II–VI materials have been prepared in the same way.

The variation in color indicated by Figure 4 is due to the quantum size effect on semiconductor electronic structure and the progressive red-shift in the onset of absorption indicates that the particles are growing in size. The quantum size effect in semiconductor nanoclusters was first reported by Brus and co-workers,[63] who analyzed the electronic properties of colloidal suspensions of II–VI crystallites prepared by arrested precipitation. In this synthesis a solution of Cd^{2+} is treated with H_2Se in a polar solvent. Under dilute reaction conditions CdSe nucleates and forms small crystallites, but the crystallites do not fuse and precipitate. Using this technique Brus and co-workers were able to substantiate the quantum size effect, i.e., that the band gap (HOMO–LUMO excitation energy) of a semiconductor nanoscale particle decreases with increasing particle size. This effect has been verified and utilized in a number of laboratories, and has recently been reviewed extensively.[6]

We have been able to isolate "molecular" nanoclusters using reverse micelles as synthesis auxilliaries.[64] The reaction of Cd^{2+} with $Se(SiMe_3)_2$ in the water pools of a reverse micelle microemulsion gives a colloidal suspension in which each nanometer-sized particle is encapsulated by surfactant. Upon sequential addition of Cd^{2+} and $Se(SiMe_3)_2$ the particles grow (as suggested by absorption sepectroscopy and proven by TEM). We modified the

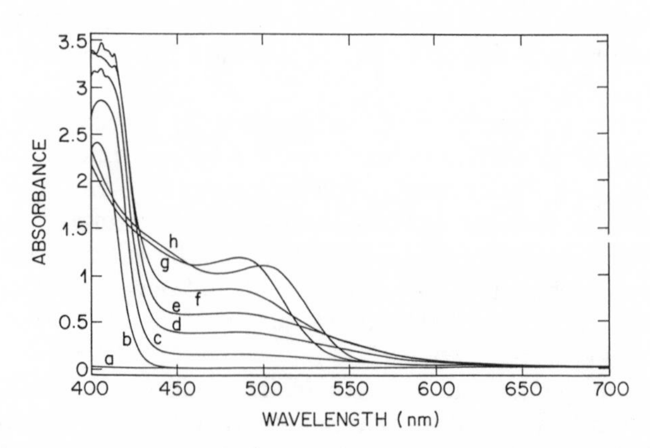

Figure 4. The time evolution of the visible absorption of a heated solution of $Cd(SePh)_2$ in 4-ethylpyridine.[62]

growth process by sequentially adding Cd^{2+} and $PhSeSiMe_3$ to preformed micelle-stabilized CdSe nanoparticles. This results in the termination of particle growth, passivation of the particle surface with phenyl groups, and the precipitation of the particles from the microemulsion. Particles formed by this method are dispersible in coordinating solvents and TEM verifies that the passivated particles have not fused. This process is summarized in Figure 5.

Analysis by X-ray diffraction[65] and TEM shows that the passivated nanoclusters are crystalline but have defects. These defects can be removed by heating the particles in dilute solution. By a process which is presumably related to surface diffusion in thin film growth, the nanoclusters recrystallize internally. It is interesting to note that the highest-temperature treatment causes the CdSe particles to crystallize internally in the wurtzite structure.

We have been able to establish that the nanoclusters formed by solution-phase pyrolysis of $Cd(SePh)_2$ are structurally identical to those prepared as in Figure 5. In this sense we have identified one type of "reactive intermediate" in the molecules-to-solids conversion, which is the topic of this chapter. As alluded to above it will be interesting to see what role clusters (as small as

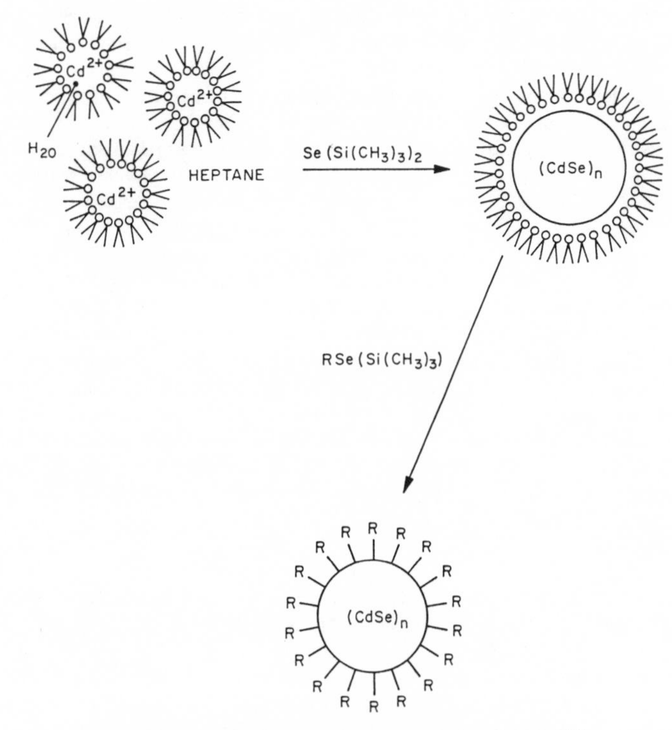

Figure 5. Synthesis of molecular particles of CdSe.[64]

just a few atoms or as large as several nanometers in diameter) can play in the synthesis of thin films and/or solid state materials in general.

5. CONCLUSION

In this chapter an attempt has been made to describe some of the general features of the inorganic chemistry of film growth. Particular examples have been the growth of compound semiconductors using organometallic vapor-phase epitaxy. Hopefully it has been made clear that film growth of this variety is a very interdisciplinary topic and that reaction chemistry can play a significant role in this field.

REFERENCES

1. Stroeve, P., Ed. "Integrated Circuits: Chemical and Physical Processing"; ACS Symp. Ser. 290, Am. Chem. Soc.: Washington, 1985.
2. (a) Seeger, K. "Semiconductor Physics: An Introduction"; Springer: Berlin, 1985; and, for example, (b) Kamimura, H.; Toyozawa, Y., Eds. "Recent Topics in Semiconductor Physics"; World Scientific Publishing: Singapore, 1983.
3. Murthy, T. U. M. S.; Miyamoto, N.; Shimbo, M.; Nishizawa, J. *J. Cryst. Growth* 1976, **33,** 1.
4. (a) West, A. R. "Solid State Chemistry and Its Applications"; Wiley and Sons: Chichester, 1986; (b) Rao, C. N. R.; Gopalakrishnan, J. "New Directions in Solid State Chemistry"; Cambridge University Press: Cambridge, 1986; (c) Schäfer, H. "Chemical Transport Reactions"; Academic Press: New York, 1964; (d) Stringfellow, G. B. "Organometallic Vapor Phase Epitaxy: Theory and Practice"; Academic Press: San Diego, 1989.
5. (a) Wells, A. F. "Structural Inorganic Chemistry"; Clarendon Press: Oxford, 1975; (b) Ludowise, M. J. *J. Appl. Phys.* 1985, **58,** R31; (c) Bryant, W. A. *J. Mater. Sci.* 1977, **12,** 1285; (d) Tiefjen, J. J. *Annu. Rev. Mater. Sci.* 1973; (e) Green, M. L.; Levy, R. A. *J. Met.* 1985, 63; (f) Jasinski, J. M.; Meyerson, B. S.; Scott, B. A. *Annu. Rev. Phys. Chem.* 1987, **38,** 109; (g) Klabunde, K. J., Ed. "Thin Films from Free Atoms and Particles"; Academic Press: Orlando, 1985; (h) Bonfils, J.-F.; Irvine, S. J. C.; Mullin, J. B., Eds. "Proceedings of the International Conference on Metalorganic Vapor Phase Epitaxy"; *J. Cryst. Growth* 1981, **55**; (i) Stringfellow, G. B., Ed. "Proceedings of the Third International Conference on Metalorganic Vapor Phase Epitaxy"; *J. Cryst. Growth* 1986, **77;** (j) Wanatabe, N.; Nakanisi, T.; Dapkus, P. D., Eds. "Proceedings of the Fourth International Conference on Metalorganic Vapor Phase Epitaxy"; *J. Cryst. Growth* 1988, **93;** (k) Cole-Hamilton, D. J.; Williams, J. O., Eds. "Mechanisms of Reactions of Organometallic Compounds with Surfaces"; Plenum Press: New York, 1989.
6. (a) Henglein, A. *Chem. Rev.* 1989, **89,** 1861; (b) Steigerwald, M. L.; Brus, L. E. *Annu. Rev. Mater. Sci.* 1989, **19,** 471.
7. Ref. 5a, p. 617.
8. Ref. 4a, Chapter 2.
9. (a) Eichorst, D. J.; Payne, D. A.; Wilson, S. R.; Howard, K. E. *Inorg. Chem.* 1990, **29,** 1458 and references cited therein; (b) Huppertz, H.; Engl, W. L. *IEEE Trans. Electron Devices* 1979, **ED-26,** 658 and references cited therein; (c) Jeffries, P. M.; Girolami, G. S. *Chem. Mater.* 1989, **1,** 8.
10. (a) Girolami, G. S.; Kaloyeros, A. E.; Allocca, C. M. *J. Am. Chem. Soc.* 1987, **109,** 1579; (b) Rutherford, N. M.; Larson, C. E.; Jackson, R. L. *Mater. Res. Soc. Symp. Proc.* 1987, **131,** 439; (c) Brown, G. M.; Maya, L. *Inorg. Chem.* 1989, **28,** 2007.

11. (a) Wu, H.-J.; Interrante, L. V. *Chem. Mater.* 1989, **1,** 564; (b) Boyd, D. C.; Haasch, R. T.; Mantell, D. R.; Schulze, R. K.; Evans, J. F.; Gladfelter, W. L. *Chem. Mater.* 1989, **1,** 119.
12. (a) Seyferth, D.; Rees, W. S., Jr.; Haggerty, J. S.; Lightfoot, A. *Chem. Mater.* 1989, 45; (b) Beck, J. S.; Albani, C. R.; McGhie, A. R.; Rothman, J. B.; Sneddon, L. G. *Chem. Mater.* 1989, **1,** 433; (c) Fehlner, T. P.; Amini, M. M.; Zeller, M. V.; Stickle, W. F.; Pringle, O. A.; Long, G. J.; Fehlner, F. P. *Mater. Res. Soc. Symp. Proc.* 1989, **131,** 413; (d) Jensen, J. A.; Gozum, J. E.; Pollina, D. M.; Girolami, G. S. *J. Am. Chem. Soc.* 1988, **110,** 1643.
13. Kern, W.; Schnable, G. L. *IEEE Trans. Electron Devices* 1979, **ED-26,** 647.
14. Panish, M. B.; Temkin, H. *Annu. Rev. Mater. Sci.* 1989, **19,** 209.
15. Irvine, S. J. C.; Mullin, J. B.; Giess, J.; Gough, J. S.; Royle, A. *J. Cryst. Growth* 1988, **93,** 732; (b) Irvine, S. J. C.; Mullin, J. B.; Hill, H.; Brown, G. T.; Barnett, S. J. *J. Cryst. Growth* 1988, **86,** 188.
16. Ref. 4a, Chapter 7.
17. Ref. 4d, Chapter 5.
18. (a) Manasevit, H. M. *Appl. Phys. Lett.* 1969, **116,** 1725; (b) Manasevit, H. M.; Simpson, W. I. *J. Electrochem. Soc.* 1968, **12,** 156.
19. Cho, A. Y. In Ref. 1, Chapter 8.
20. Tsang, W. T. *Appl. Phys. Lett.* 1984, **45,** 1234.
21. Robertson, A., Jr.; Chiu, T. H.; Tsang, W. T.; Cunningham, J. E. *J. Appl. Phys.* 1988, **64,** 877.
22. The Cu/S system is an interesting example. See Ref. 5a, p. 907.
23. Ref. 4b, Section 1.8.
24. Ref. 4b, Chapter 2.
25. (a) Fry, K. L.; Kuo, C. P.; Larsen, C. A.; Cohen, R. M.; Stringfellow, G. B.; Melas, A. *J. Electron. Mater.* 1986, **15,** 91; (b) Knauf, J.; Schmitz, D.; Jürgensen, H.; Heyen, M. *J. Cryst. Growth* 1988, **93,** 34.
26. (a) Schlyer, D. J.; Ring, M. A. *J. Organomet. Chem.* 1976, **114,** 9; (b) Larsen, C. A.; Stringfellow, G. B. *J. Cryst. Growth* 1986, **75,** 247.
27. For example: (a) Haigh, J.; O'Brien, S. *J. Cryst. Growth* 1984, **67,** 75; (b) Yoshida, M.; Watanabe, H.; Uesugi, F. *J. Electrochem. Soc.* 1985, **132,** 677; (c) Larsen, C. A.; Stringfellow, G. B. *J. Cryst. Growth* 1986, **75,** 247; (d) Buchan, N. I.; Larsen, C. A.; Stringfellow, G. B. *J. Cryst. Growth* 1988, **92,** 591.
28. (a) Nishizawa, J.; Kurabayashi, T. *J. Electrochem. Soc.* 1983, **130,** 413; (b) Nishizawa, J.; Kurabayashi, T. *J. Cryst. Growth* 1988, **99,** 525.
29. (a) Buchan, N. I.; Larsen, C. A.; Stringfellow, G. B. *Appl. Phys. Lett.* 1987, **51,** 1024; (b) Buchan, N. I.; Larsen, C. A.; Stringfellow, G. B. *J. Cryst. Growth* 1988, **92,** 605; (c) Larsen, C. A.; Buchan, N. I.; Stringfellow, G. B. *Appl. Phys. Lett.* 1988, **52,** 480.
30. (a) Didchenko, R.; Alix, J. E.; Toeniskoetter, R. H. *J. Inorg. Nucl. Chem.* 1960, **14,** 35; (b) Beachley, O. T., Jr.; Kopasz, J. P.; Zhang, H.; Hunter, W. E.; Atwood, J. L. *J. Organomet. Chem.* 1987, **325,** 69 and references cited therein.
31. Maury, F.; Combes, M.; Constant, G. In "Proceedings of European Conference on CVD-4, 1983"; Bloem, J.; Verspui, G.; Wolff, L. R., Eds. *Chem. Abstr.* **99:** 185120.
32. (a) Cowley, A. H.; Benac, B. L.; Ekerdt, J. G.; Jones, R. A.; Kidd, K. B.; Lee, J. Y.; Miller, J. E. *J. Am. Chem. Soc.* 1988, **110,** 6248; (b) Cowley, A. H.; Jones, R. A. *Angew. Chem., Int. Ed. Engl.* 1989, **28,** 1208.
33. See, for example, Andres, R. P.; Averback, R. S.; Brown, W. L.; Brus, L. E.; Goddard, W. A.; Kaldor, A.; Louie, S. G.; Moscovits, M.; Peercy, P. S.; Riley, S. J.; Siegel, R. W.; Spaepen, F.; Wang, Y. *J. Mater. Res.* 1989, **4,** 704.
34. See, for example, Hess, K. L.; Riccio, R. J. *J. Cryst. Growth* 1986, **77,** 95.
35. (a) Lum, R. M.; Klingert, J. K.; Kisker, D. W.; Tennant, D. M.; Morris, M. D.; Malm, D. M.; Kovalchick, J.; Heimbrook, L. A. *J. Electron. Mater.* 1988, **17,** 101; (b) Brauers, A.; Kayser, O.; Hall, R.; Heinecke, H.; Balk, P. *J. Cryst. Growth* 1988, **93,** 7.

36. Lum, R. M.; Klingert, J. K.; Wynn, A. S.; Lamont, M. G. *Appl. Phys. Lett.* 1988, **52,** 1475.
37. Bhat, R.; Koza, M. A.; Skromme, B. J. *Appl. Phys. Lett.* 1987, **50,** 1194.
38. (a) Lum, R. K.; Klingert, J. K.; Lamont, M. G. *Appl. Phys. Lett.* 1987, **50,** 284; (b) Chen, C. H.; Larsen, C. A.; Stringfellow, G. B. *Appl. Phys. Lett.* 1987, **50,** 218; (c) Kurtz, S. R.; Olson, J. M.; Kibbler, A. *J. Electron. Mater.* 1989, **18,** 15.
39. Moss, R. H.; Evans, J. S. *J. Cryst. Growth* 1981, **55,** 129.
40. Chen, C. H.; Larsen, C. A.; Stringfellow, G. B.; Brown, D. W.; Robertson, A. J. *J. Cryst. Growth* 1986, **77,** 11.
41. (a) Moore, A. H.; Scott, M. D.; Davies, J. J.; Bradley, D. C.; Faktor, M. M.; Chudzynska, H. *J. Cryst. Growth* 1986, **77,** 19; (b) Reier, F.-W.; Wolfram, P.; Schumann, H. *J. Cryst. Growth* 1986, **77,** 23; (c) Shenai-Khatkhate, D. V.; Orrell, E. D.; Mullin, J. B.; Cupertino, D. C.; Cole-Hamilton, D. J. *J. Cryst. Growth* 1986, **77,** 27.
42. (a) Moss, R. H.; Evans, J. S. *J. Cryst. Growth* 1981, **55,** 129; (b) Moss, R. H. *J. Cryst. Growth* 1984, **68,** 78.
43. Lum, R. M.; Klingert, J. K.; Kisker, D. W.; Abys, S. M.; Stevie, F. A. *J. Cryst. Growth* 1988, **93,** 120.
44. Smith, D. L.; McGill, T. C.; Schulman, J. N. *Appl. Phys. Lett.* 1983, **43,** 180.
45. Irvine, S. J. C.; Mullin, J. B. *J. Cryst. Growth* 1981, **55,** 107.
46. Mullin, J. B.; Irvine, S. J. C.; Ashen, D. J. *J. Cryst. Growth* 1981, **55,** 92.
47. Mullin, J. B.; Geiss, J.; Irvine, S. J. C.; Gough, J. S.; Royle, A. In "Materials for Infrared Detectors and Sources"; *Mater. Res. Soc. Symp. Proc. Vol.* 1987, **90,** 367 and references cited therein.
48. Manasevit, H. M.; Simpson, W. I. *J. Electrochem. Soc.* 1971, **118,** 644.
49. Hoke, W. E.; Lemonias, P. J.; Korenstein, R. *J. Mater. Res.* 1988, **3,** 329.
50. Korenstein, R.; Hoke, W. E.; Lemonias, P. J.; Kiga, K. T.; Harris, D. C. *J. Appl. Phys.* 1987, **62,** 4929.
51. Lichtman, L. S.; Parsons, J. D.; Cirlin, E.-H. *J. Cryst. Growth* 1988, **86,** 217.
52. Boersma, J. "Comprehensive Organometallic Chemistry," Wilkinson, G.; Stone, F. G. A.; Abel, E. W., Eds.; Pergamon Press: Oxford, 1982; Chapter 16.
53. Kisker, D. W.; Steigerwald, M. L.; Kometani, T. Y.; Jeffers, K. S. *Appl. Phys. Lett.* 1987, **50,** 1681.
54. Steigerwald, M. L.; Sprinkle, C. R. *J. Am. Chem. Soc.* 1987, **109,** 7200.
55. Osakado, K.; Yamamoto, T. *J. Chem. Soc., Chem. Commun.* 1987, 1117.
56. Brennan, J. G.; Siegrist, T.; Carroll, P. J.; Stuczynski, S. M.; Reynders, P.; Brus, L. E.; Steigerwald, M. L. *Chem. Mater.* 1990, **2,** 403.
57. Zingaro, R. A.; Stevens, B. H.; Irgolic, K. *J. Organomet. Chem.* 1965, **4,** 320.
58. Steigerwald, M. L.; Sprinkle, C. R. *Organometallics* 1988, **7,** 245.
59. (a) Steigerwald, M. L.; Rice, C. E. *J. Am. Chem. Soc.* 1988, **110,** 4228; (b) Steigerwald, M. L. *Chem. Mater.* 1989, **1,** 52; (c) Brennan, J. G.; Siegrist, T.; Stuczynski, S. M.; Steigerwald, M. L. *J. Am. Chem. Soc.* 1989, **111,** 9240.
60. (a) Kisker, D. W.; Feldman, R. D. *J. Cryst. Growth* 1985, **72,** 102; (b) Irvine, S. J. C.; Mullin, J. B.; Tunnidiffe, J. *J. Cryst. Growth* 1984, **68,** 188.
61. Irvine, S. J. C. *CRC Critical Rev. Solid State Mater. Sci.* 1987, **13,** 279.
62. Brennan, J. G.; Siegrist, T.; Carroll, P. J.; Stuczynski, S. M.; Brus, L. E.; Steigerwald, M. L. *J. Am. Chem. Soc.* 1989, **111,** 4141.
63. Rosetti, R.; Hull, R.; Gibson, J. M.; Brus, L. E. *J. Chem. Phys.* 1984, **82,** 552.
64. Steigerwald, M. L.; Alivisatos, A. P.; Gibson, J. M.; Harris, T. D.; Kortan, A. R.; Muller, A. J.; Thayer, A. M.; Duncan, T. M.; Douglass, D. C.; Brus, L. E. *J. Am. Chem. Soc.* 1988, **110,** 3046.
65. Bawendi, M. G.; Kortan, A. R.; Steigerwald, M. L.; Brus, L. E. *J. Chem. Phys.* 1989, **91,** 7782.

9

Ceramics

Robert T. Paine

1. INTRODUCTION

No, a thousand times no; there does not exist a category of science to which one can give the name applied science. There are science and applications of science, bound together as the fruit to the tree which bears it.

Louis Pasteur

In preceding chapters, synergistic aspects of the synthesis, structure, reactivity, and bonding of molecular inorganometallic compounds have been described. Fundamental parallels and distinctions between inorganometallics and the more highly systematized field of discrete organometallic compounds have been highlighted, and exciting frontiers for future study have been made apparent. In closing this book, some potential applications for fundamental inorganometallic chemistry are described, drawing particular attention to the field of ceramics—a field largely neglected until recently by molecular inorganic chemists. It is shown that designed, molecular inorganometallics offer an evolutionary, chemical bridge to commercially significant advanced ceramics, a functional fruit from the basic chemistry.

What are ceramics? Ceramics are traditionally defined as refractory, inorganic, solid-state materials. Compositionally, most are not new. Indeed, natural ceramics find their origins in the creation of the planet, and the first man-made materials may be traced to formation of pottery and porcelain by

Robert T. Paine • Department of Chemistry, University of New Mexico, Albuquerque, New Mexico 87131.

Inorganometallic Chemistry, edited by Thomas P. Fehlner. Plenum Press, New York, 1992.

prehistoric man. Most ancient materials, as well as many modern ceramics, are binary, and they are obtained in powder form from simple syntheses. Ceramic science is largely dominated by oxides in powder forms,[1] and a great deal is known about the chemistry, properties, and processing of these powders. In turn, oxides function as crucially important raw materials for major industrial products. Not to be ignored, beryllides, borides, carbides, nitrides, silicides, phosphides, and chalcogenides provide examples of useful nonoxide ceramics. However, less is known about these materials, and many fewer commercial applications have been realized.

Ceramics have attracted attention because of the special blend and diversity of properties they display. For example, many ceramics exhibit high thermal and chemical stability in extreme environments, and they are often mechanically strong, hard, and wear-resistant. They are less dense than metals, and they provide a broad range of thermal, electrical, magnetic, and optical properties. Unfortunately, they are also brittle. The powders are often difficult to fabricate into complex components, and properties are subject to variations brought about by largely undefined synthesis, purity, and processing factors. Nonetheless, by accentuating the positive, the conventional powder ceramics industry has steadily grown because of the efficacy of single function oxide components in simple devices.

Since ceramic components have been fabricated for years in adequate quality for many applications by unglamorous "heat and beat" approaches, many scientists perceive that ceramics science is a mature field with few exciting frontiers. Economy[2] has observed a similar notion, driven more by economic factors, among many industrial leaders. Several recent commentaries[2-11] provide excellent insights on this perception, especially for synthetic chemists whose background in materials science and engineering is typically weak. It may be immediately concluded from these articles that, although thoroughly developed in selected areas, e.g., classical oxide powders, the field of ceramics is far from mature. One indicator supporting this conclusion is that most matured technologies are well grounded in predictive theory. Such is not the state of affairs, particularly for advanced ceramics. Attempts have been made to conceptually separate relatively well understood "conventional" ceramics represented by crude oxide-based materials used in narrow high-temperature applications, from "advanced," "high-tech," or "high-performance" ceramics prepared from more complex chemical syntheses with increasingly sophisticated processing methods. Performing this artificial separation does not, however, address the source of the current intellectual attraction to ceramics, although it does emphasize how little is known at the molecular level about these materials.

The renewed interest in ceramics derives primarily from the recognition that industry is materials-limited in the development of many new technologies for the 21st century. Stated in another fashion, many futuristic technological problems have been identified, some conceptual solutions have been designed, but materials needed to implement solutions are not on the horizon. In this regard, Hondros and Bullock[4] accurately note that ". . . materials represent the rate limiting step in the emergence of future technologies and for the competitivity of many existing technologies." Ceramics, due to their unique properties and chemical diversity, represent the only materials that will provide solutions to many key materials problems.[8]

Unfortunately, if the grand design is to create sophisticated ceramic devices with properties tailored to specific applications, much of the fundamental data base for development is unavailable. Indeed, we find ourselves in a condition where critical applications of science are driving us to return to exploring basic science. Creative chemists and their conceptual molecular tool boxes are needed to provide new approaches and insights. The chemist's attention is particularly needed for the development of radically new ceramic materials, ceramic precursor synthons in the form of "active" powders, polymers, and composites, new chemical processing concepts, more thorough understandings of solid-state kinetic processes including sintering, and new advances in solid-state theory.

Due to the focus in preceding chapters on parallels between organometallics—compounds rich in metal–carbon bonds—and inorganometallics—compounds rich in metal–main group element bonds—it is appropriate to give primary attention in this chapter to selected binary, single-phase metal carbides, borides, nitrides, silicides, and phosphides. Oxide ceramics, though very important, will not be discussed. A brief overview of selected properties, applications, and traditional synthetic approaches will be given, followed by a general discussion of shortcomings inherent in materials produced by traditional methods. Several prospects for molecular and polymeric inorganometallic reagents as precursors for ceramics will be surveyed, along with needs for new processing approaches for these materials. For the most part, ternary and multiphasic materials are not discussed even though inorganometallic precursors hold exciting potential for future development of these compositions.

2. CLASSICAL ROUTES TO METAL CERAMICS

As is the case with most ceramics, binary metal carbides, borides, nitrides, silicides, and phosphides are most often prepared by thermodynamically driven, brute force, high-temperature reactions.[12–22] These materials are typically obtained as powders of adequate quality for many conventional ceramic applications. In addition, most of these ceramics have been produced as powders or thin films by vapor deposition processes. In the following sections, representative preparations for these metal ceramic families are summarized along with selected properties and applications.

2.1. Metal Carbides and Nitrides

Although nearly all of the transition metals, lanthanides, and actinides form carbides and nitrides, those containing Group 4, 5, and 6 transition metals and uranium have special industrial interest. These materials are difficult to obtain as very high purity powders or as dense articles due to their high melting points. They are often referred to as interstitial compounds since the parent metal, the carbide, and the nitride have many of the same properties, and structure determinations clearly indicate that the C and N atoms occupy octahedral interstitial sites in a metal close-packed or related lattice.[23] The common stoichiometry is 1:1; however, M_2C and M_2N compositions are known as well as a set of nonstoichiometric compositions, MC_x (x = 0.5–0.97) and MN_x (x = 0.37–1.3). The crystal chemistry for most compositions is well developed, and the C and N deficiencies often result from C and N atom vacancies in the fcc carbon/nitrogen sublattice in the NaCl structure.[12–21]

The metal carbides are readily prepared from direct combination of the elements at high temperature (>1200 °C) as illustrated in equation 1. This reaction is accomplished by poorly controlled arc heating methods or by resistive heating (carburization). Carburization reactions of metal oxides also produce metal carbides as shown in equation 2, although care must be taken to minimize oxide contamination.[12,13,15] Gas–solid phase carburizations with carbon monoxide (equation 3), acetylene, and hydrocarbons have also been reported,[12,15] and reductions employing metal halides and hydrides have been described.[13,15,20] Several carbides have been obtained by combination of the elements in a molten metal such as cobalt (menstruum method),[19] and this technique is used to obtain high stoichiometry single crystals.

$$2\,W + C \xrightarrow{\text{arc}} W_2C \tag{1}$$

$$ZrO_2 + 3\,C \xrightarrow{>1700\ ^\circ C} ZrC + 2\,CO \tag{2}$$

$$Ta + 2\,CO \xrightarrow{>1600\ ^\circ C} TaC + CO_2 \tag{3}$$

Vapor deposition approaches to metal carbides[22] most often employ a volatile metal halide (MX_6, MX_5, MX_4; X usually chloride) with a hydrocarbon and hydrogen reduction stream and substrate temperatures ranging from 600 °C to 2500 °C. Alternative metal sources include hot metal filaments and, more recently, metal carbonyl and organometallic compounds. Carbon sources include C-arc, CO, and saturated and unsaturated hydrocarbons.

$$2\,TiO_2 + 4\,C + N_2 \xrightarrow{1600\ ^\circ C} 2\,TiN + CO \tag{4}$$

$$2\,ZrCl_4 + 10\,NH_3 + H_2 \xrightarrow{1400\ ^\circ C} 2\,ZrN + 8\,NH_4Cl \tag{5}$$

$$VH_3 + NH_3 \xrightarrow{1000\ ^\circ C} VN + 3\,H_2 \tag{6}$$

The Group 4, 5, and 6 nitrides are often prepared by direct nitriding of the metal with N_2, NH_3, or NH_3/H_2 mixtures at 1100 °C to 1600 °C, or by nitriding a metal oxide in the presence of carbon as shown in equation 4. The latter technique is often troubled by the inclusion of oxide and carbide impurities. Nitrides may also be obtained by combination of NH_3 with metal amalgams, oxides, halides (equation 5), and hydrides (equation 6).[21] Vapor deposition approaches have generally relied on gas-phase reactions of volatile metal halides or organometallics and NH_3.[22] As with carbides, substrate temperatures and resultant product quality vary. Reports on the chemistry, properties, and commercial applications of later (Group 7 and higher) transition metal carbides and nitrides exist, but the information base is poorly developed. It has been concluded that these materials are relatively less stable in comparison to Group 4, 5, and 6 carbides and nitrides, and the Group 6 materials are relatively less stable than Group 4 and 5 analogs. Johansen[19] provides a brief discussion of observations involving the "chromium enigma." It should also be noted that a rich array of ternary compositions, MNX and MCX, exist,[12–21] but these are not considered here.

Since the composition, microstructure, and properties of carbides and nitrides obtained by metallurgical and vapor deposition techniques are dependent upon synthetic variables, care must be exercised in quantitative property comparisons.[12,19] Qualitatively, however, the Group 4, 5, and 6 carbides and nitrides are known for their very high melting points (TaC ~4000 °C, TiN and TiC ~3000 °C), hardness (TiN, TiC are hardest), chemical inertness at ordinary temperatures, low thermal conductivity, semimetallic electrical conductivity, and good mechanical strength.[12,17–20] The carbides and nitrides are susceptible to air oxidation at temperatures around 1000 °C and 800 °C, respectively. They are subject to brittle failure under stress and are not considered to have good thermal shock resistance. It has been reported that high-temperature strength and thermal shock resistance of carbides can be improved by inclusion of boron or graphite.[18,20] Extensive collections of phase, thermodynamic, and kinetic data for the materials exist.[12,15,17–20] Excellent descriptions of physical properties and bonding have been given recently by Williams[18] and Hoffmann *et al.*, respectively.[24]

A number of important applications for metal carbides have been realized,[12,13,20] but brittleness at room temperature and shape processing limitations presently hinder more extensive development.[18–20] Classical powder metallurgy techniques are typically employed to make shaped articles, but full density is rarely obtained. One approach involves addition of low melting metals, such as Co or Ni, that melt-bond the carbide grains. In this regard, the chief use of ZrC is as a component in cemented hard metal cutting tools that contain Co as a binder.[13,25] The primary commercial application for metal carbides is in machine tools where TiC, ZrC, VC, WC, and TaC find particular utility.[25] VC is also used in steel alloys, TaC as a coating to delay release of fission products, as a refractory for crucibles and incandescent filaments, and TaC and WC find applications as refractories and refractory coatings. There are also reports on the catalytic activity of metal carbides and "carbided" metals.[19]

Commercial applications for metal nitrides, relative to the corresponding carbides, are relatively few at the present time. TiN is used as a refractory hard metal in tool applications, often with a Mo or Ni binder phase, and there are reports that several nitrides are effective catalysts in ammonia synthesis.[26] Nitrides are also employed in hard cast alloys, rocket engine coatings, and refractory vessel materials. Several nitrides and carbonitrides display high T_c superconducting properties,[12,19–21] and efforts continue to improve their T_c and current-carrying capacity relative to oxide materials.

2.2. Metal Borides, Silicides, and Phosphides

Unlike carbides and nitrides, these materials are considered together—not because of atomic, structural, and property relationships, but because they are complex, and there is comparatively less systematic information available.[13,17,27–31] Boron forms an incredible array of metal-rich and boron-rich compositions. Greenwood and Earnshaw[13] list twenty-five stoichiometric compositions, and Thompson[27] gives seven general structural classifications for eleven of the composition types. The remaining compositions occur in structural subsets or in mixtures of the basic structures. No one metal displays all compositions or structures, and a useful periodic table summary of compositions is given by Aronsson *et al.*[28] As might be expected, the structures of the metal borides appear to change in a relatively regular fashion in response both to variation in metal atom size and concentration in the lattice.[23,27] The metal-rich compositions display lattices featuring isolated B atoms (no direct B—B bonding), and as metal size and concentration change, B—B units, single $(B—)_n$ zigzag chains, branched chains, fully crosslinked double chains, two-dimensional networks, and finally three-dimensional $(—B—)_n$ $(n = 4, 6, 12)$ frameworks appear.

Metal borides have been prepared in a number of high-temperature reaction schemes.[13,27,28,31] These include direct combination of the elements, reduction of an appropriate metal oxide and boric oxide with carbon, fused salt electrolysis of a metal oxide and boric oxide, reaction of a metal oxide with B_4C/C, and reaction of boron halide, hydrogen, and a metal oxide. The carbothermic reduction process and B_4C/C reaction with metals and metal oxides are used commercially as is a coreduction reaction employing metal oxide, boric oxide, and Al or Mg. All of the preparations must be accomplished in an inert or reducing (H_2) atmosphere because the borides are reactive at high temperature with oxygen, nitrogen, CO, and CO_2. Details of individual syntheses have been summarized by Thompson[27] and Matkovich.[31]

Several boride compositions have also been accessed by vapor deposition processes, and all but the most recent results have been summarized.[22] The primary limitation involves a general lack of appropriate volatile metal- and boron-containing reagents that do not leave residual substituent group or free metal impurities in the final composition. For the most part, the metallurgical and vapor deposition syntheses produce ceramic powders, although thin films may be obtained from gas-phase processes.

The primary properties that draw attention to metal borides are their high melting points (HfB_2, ~3250 °C is highest), hardness (TiB_2 is hardest), and, in many cases, electrical conductivity. They are typically metallic in appearance, and the Group 4, 5, and 6 borides are the most studied and utilized. Some thermodynamic data are available, although they are limited to the more studied compositions, e.g., MB_2 and MB_4. Thermal conductivity data are highly dependent on porosity and phase structure, and the materials are not especially resistant to thermal shock. In this regard, they resemble a metal oxide more than a metal; however, improvements can be realized by composite formation, e.g., with boron nitride.[27] As noted above, many of the metal borides, except boron-rich compositions, such as MB_{12}, are good electrical conductors at room temperature. Several boron-poor solids have been claimed to be superconductors.[27] As is the case with carbides and nitrides, brittleness accompanies the high hardness and high strength (high modulus of elasticity), and they are sintered with great difficulty.[32]

Due to the large range of compositions, it is difficult to make accurate generalizations on chemical and thermal reactivity of metal borides. Nonetheless, it appears that most of the borides are stable in air at 1000 °C, reactive with carbon only at high temperatures, and thermodynamically stable with respect to the nitride,[27] e.g., ZrN and BN at 1600 °C produce ZrB. As noted above, the Group 4, 5, and 6 borides are of commercial interest, and these materials find applications as high-temperature mechanical parts, e.g., turbine blades, rocket nozzles, engine liners, and ablation shields, as high-temperature reaction chambers and electrodes, neutron capture targets, and composite additives.

Metal silicides also reveal a wide diversity of compositions and structures.[13,17,28,33] Fourteen compositions spanning metal-rich M_6Si to silicon-rich MSi_6 are listed by Greenwood and Earnshaw,[13] and the structural complexity has general parallels with borides. In particular, metal-rich silicides contain isolated Si atoms in metal-like lattices, while compositions with greater quantities of silicon contain Si chains, layers, and tetrahedral and three-dimensional networks. The metal silicides are prepared by many of the same approaches given for borides, with direct combination of the elements representing the most utilized route for powder formation. Thin-film silicides have been commonly produced by hydrogen reduction of metal halide–silicon halide mixtures (600 °C) or by silane reduction of metal halides at 600 °C.[22] Such films are often contaminated with halide.

Transition metal silicides of Groups 4, 5, and 6 appear to be the most stable, and in these groups the silicides have comparable heats of formation to borides and carbides of the same stoichiometries.[28] It is interesting, however, that the melting points for the silicides are significantly lower and the hardness values are smaller. It also appears that the silicides may be slightly more reactive, although they still provide excellent corrosion-inhibiting layers. Applications for these materials are found in thermoelectric devices, thermocouples, wear resistant coatings, and in microelectronic device fabrication.[34]

Lastly, metal phosphides display a wide array of compositions[13,28] and structures.[23] The most notable distinction from carbides, nitrides, borides, and silicides is the larger number of phosphide compositions found with electron-rich Group 7–10 metals. The phosphides are most often prepared by direct combination of the elements or by electrolysis of fused salts. It should be noted that the volatility of elemental phosphorus at high temperatures requires that some reactions be performed in sealed reaction containers.

Since a broader range of metals form stable phosphides, a more complex set of properties appears. For example, the early metal phosphides form metal-rich compositions that are high melting, refractory, hard, metallic, and brittle. They have high electrical and thermal conductivity, and they are relatively inert. Hence, they share many of the same features with carbides, nitrides, borides, and silicides already discussed. All of the metals form a monophosphide, MP, and these materials show a "transition" in properties between the metal-rich and phosphorus-rich compounds. That is, their refractory nature decreases while they retain conduction properties. Finally, the phosphorus-rich compositions typically show much lower melting points, greatly reduced refractory properties, and they are relatively soft, semiconducting, and chemically more reactive.[28] Only a few commercial uses of early metal phosphides have been realized.

3. UNSOLVED PROBLEMS AND SHORTCOMINGS

Clearly, the classical powder synthetic chemistry and characterization data bases for binary early transition metal–nonoxide ceramics are well developed; however, given the important properties of the materials, their frequency of use in finished high-technology products is disappointing. In fact, there are many nonoxides with desirable properties that are underutilized due in part to an inability to process the materials into complex, integrated products in an economical fashion. It is fair then to ask the following questions: Can the application potential of known nonoxide compositions be better realized by developing molecular level syntheses and processing methods? Is there promise for new, more complex materials? Can chemists play a role?

The answer to these questions is an unqualified yes! Roy[11] and Westwood and Winzer[35] have outlined historical and current evidence in support of the conclusion that today the opportunities for atomic and molecular level manipulation of ceramics are "virtually limitless." Further, Rice[8] has recently reflected on these questions in a broader ceramics perspective. It is clear that chemistry is central to providing the answers to advanced ceramics issues, and some of Rice's points are considered here in the narrower context of inorganometallic chemistry as a molecular engineering bridge to new metal–nonoxide ceramics and processes. Working together, chemists and ceramists are in a position to respond to the challenge to produce the next generation of complex multicomponent structural, electronic, and optical ceramics.

As already pointed out, metal carbides, nitrides, borides, silicides, and phosphides are routinely produced in high-temperature reactions as crystalline refractory powders. Although there are needs on the horizon for less than fully dense articles with defined microstructures, the primary immediate needs for structural and electronic nonoxides are for fully dense articles with complex shapes. Currently it is difficult to manufacture such articles with existing powder technology, and entirely new chemical and processing approaches are needed.

Although powder processing has been the province of engineering and physics, inorganometallic chemistry has much to contribute to improved powder technology. For example, little is understood at the *molecular level* about processing reaction chemistry and sintering of nonoxide powders. Ceramists and chemists are attempting to shed light on these crucial events by, for example, modeling and performing sintering studies on very pure and monosized particles. In support of these studies, a goal to uncover low-temperature solution-based syntheses for pure, monosized ceramic powders has developed. This is a natural niche for synthetic chemists, but they should not focus all their efforts in this narrow area. Instead, as Rice points out,[8] it may be more useful to determine how to treat readily available, imperfect, high-temperature powders so they may be densified. The answers surely lie in the manipulation of the chemistry. Questions for the chemist include: Can powders be chemically "activated," or can powder interfacial chemistry be altered in order to reduce sintering barriers? Ceramists have empirically utilized sintering aids to accomplish this task with some success, but systematic chemistry that will support dramatic advances is missing.

Melt processing is an approach that has found applications in oxide systems[6] and the technique warrants attention for metal–nonoxide compositions. For example, polymer-based precursors may have some special utility in producing metal–nonoxide eutectics suitable for melt processing. This area, though difficult, is ripe for study by chemists.

Vapor deposition, already under active study by many chemists, requires additional efforts on less popular inorganometallic compositions. Vapor deposition has the particular advantages of being useful for producing thin films and near net shape articles for compositions that are difficult to process by standard methods. Attention should be directed to developing improved volatile precursors that permit production of consistent compositions with high purity, low stress features, fine and dense microstructures, and reproducible crystallinity. More attention to interfacial chemistry and mechanistic analysis of the gas-phase chemistry and solid-state growth processes is warranted.

Traditional metallurgical and vapor-phase powder forming of dense bodies will continue to occupy a central position in ceramic product development, but there are also increasing demands for metal–nonoxide ceramic forms that can not be obtained by classical powder processing: fibers, coatings, composites, controlled microstructures, e.g., honeycombs and other low-density forms. Here, solution inorganometallic sol-gel and polymer chemistry provide powerful alternatives. Indeed, polymer precursors for SiC, Si_3N_4, and BN have recently been developed,[36–39] yet there remains a lack of developed tractable polymers and polymer chemistry for most inorganometallic systems. This area requires devoted attention from synthetic chemists. Microengineered materials,[4,8,11] particularly new compositions, multiphasic materials, noncrystalline and nonequilibrium materials, composites, tailored microenvironment compositions, and artificially structured materials are increasingly considered in advanced designs, and chemistry will provide the appropriate precursors and processing advances to achieve the needed technology. Lastly, chemistry has much to offer in the development of more mature predictive theories and models for solid state processes.

In the following sections, attention will be directed to selected preparation and processing frontiers. Some molecular design objectives for the synthetic chemist are summarized along with illustrative examples of recently published attempts to generate metal–nonoxides via molecular and polymer reagents. No attempt has been made to be exhaustive in the literature review. New practitioners should carefully consult recent literature, including patents, since there is currently much activity in these areas.

4. NEW APPROACHES IN SYNTHESIS

As already described, metal–nonoxide ceramic powders are classically prepared from appropriate mixtures of simple, inexpensive solid reagents (often high melting themselves) by utilizing high-temperature reaction conditions. High temperatures are required to achieve interdiffusion of the elements between reagent grains, to gain access to the appropriate regions of phase diagrams, and to develop crystallinity. Although it is not certain that crystallinity is crucial in all applications, it is known that the magnitude of some physical properties are certainly degraded in amorphous solids or partially ordered solids compared to fully ordered materials. Clearly, if atomic scale mixing and composition control can be accomplished with molecular or polymeric inorganometallic reagents without the necessity of high-temperature processing, then considerable advantage would be realized for fabrication of advanced structure ceramic articles that cannot be obtained by classical approaches. It should be stressed that this must be accomplished without serious introduction of impurities, defects, or loss of targeted ceramic properties—structural, thermal, optical, or electronic. Herein lie the stimuli and hurdles for the current wave of interest in ceramic precursors.[36–45]

Steigerwald and co-workers[43–46] have summarized several chemical features of organometallic and inorganometallic compounds that may make them viable ceramic precursors. For example, if properly constructed, they have direct metal–nonmetal bonds of appropriate number and strength to provide the desired final composition. This design feature should offset some of the requirements for atomic diffusion required in classical syntheses to attain mixing. Further, these compounds can carry substituents or ligands that impart solubility and reactivity that insure facile conversion to solid state compositions. Unfortunately, few conversions of organometallics and inorganometallics to ceramics have been studied. Therefore, detailed steric and electronic substituent group features that may control the molecular to solid state transformations have not been elucidated. For example, it is not yet clear which substituents or ligands are most favorable as leaving groups or impart the least amount of residual impurities. It is also not yet possible to insure a final ceramic composition based on a given molecular composition, and it is not known how to systematically influence crystallization processes or the "selection" of a solid state phase. Nonetheless, some general design features are emerging that, for the most part, transfer as logical extensions of known trends in solution molecular or polymer chemistry.[38,44–48] A summary of these features is provided.

On paper, a "molecular engineered" ceramic synthesis may be written to proceed directly from monomeric reagents to a final desired solid-state composition. However, whether by design or natural tendency, most such schemes proceed by more tortuous routes through identified or undefined small-ring compounds, small or large cluster or cage compounds, chains, extended nets, two- or three-dimensional polymers, and finally to the solid-

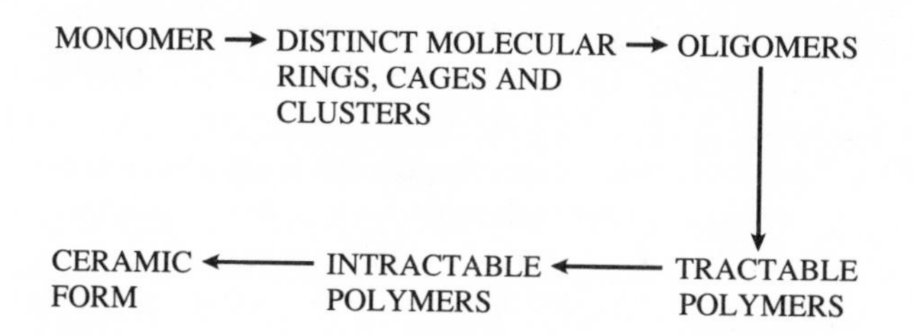

state ceramic form. In a few cases, a great deal is known about structures and mechanisms involved in these individual stages, while in most cases very little is known. It is particularly notable that the intractable polymer → crystalline ceramic step is especially mysterious. At modest pyrolysis temperatures, most polymers produce amorphous solids with very nearly the final ceramic composition. Unfortunately, high temperatures are usually required to achieve complete ordering (crystallization), and little is known on how to influence solid-state structure selection under mild conditions. New chemical schemes should include efforts to short-cut or reduce the energy of the ordering processes.

The considerable efforts spent on developing polymeric precursors for SiC and Si_3N_4 have taught us much about synthesis and process design, and it can be assumed, as a starting point, that features important in these systems will also be important in routes developed for other metal–nonoxide ceramics. For example, it is probably important to design early stage reagents, e.g., monomer, small-ring and cage compounds, that have strong bonds between the metal and nonoxide elements. This may include formation of reagents with formal metal–nonoxide multiple bonds and exclude utilization of reagents with weak Lewis acid–base coordinate bonds. Further, substituent–metal bonds involving "spectator" ligands required to stabilize early stage reagents should be as weak as possible, and the ligands should be subject to facile displacement. It is also desirable that facile ligand or substituent group displacement be accompanied by a process for carrying away the unwanted elements, e.g., formation of very stable and volatile byproducts. In the absence of this feature, substituent fragments may remain behind, trapped in an oligomer or polymer, to contribute atomic impurities in the final composition. Given the experience gained from vapor deposition of metal–nonoxide compositions, it is advisable to avoid substituents containing oxygen and halogen atoms. Finally, it is desirable that reagents be cheap, readily available, and have negligible vapor pressures at temperatures employed in precursor processing.

Once past these hurdles, a number of additional features should be addressed in order to keep the entire precursor conversion practical. Oligomer or polymer chemistry should be relatively straightforward and efficient, it should be adaptable to common process hardware, and it should have minimal environmental impact, such as byproduct toxicity and solvent recycle. Lessons learned from SiC and Si_3N_4 polymer processing suggest that nonlinear polymers are preferred over linear polymers, especially if linear polymers are likely to suffer from parasitic, thermally activated "back-biting" reactions that generate volatile, low molecular weight fragments. Instead, oligomers and polymers should be subject to thermal or photochemical crosslinking chemistry. It is important that the polymer be tractable at the latest possible stage of a processing scheme so that the precursor can be formed, as desired, into film, coating, fiber, and complex composite green bodies. After initial forming, the green polymer body should be susceptible to chemistry that creates an intractable (cured) form, then a final ceramic form, with a minimum overall weight loss (high ceramic yield), minimum shrinkage, and, in most cases, maximum density and minimum defects. It is also desirable that pyrolysis off-gases be nontoxic.

Clearly, the features described here present a tall chemical design order, and it may not be possible to achieve all of the goals in every precursor design. Nonetheless, it is imperative to keep these criteria and the practicalities of eventual applications in mind when working in this area. Otherwise, the likely outcome is frustration and papers describing the thermal bludgeoning of complex, expensive inorganometallic compounds. One final summary guide is offered: Keep it simple!

Recently, several groups have started to develop molecular reagent to ceramic conversion processes for several metal–nonoxides, and a few illustrative examples are provided.

4.1. Metal Chalconides

Although these compositions fall outside the primary field of this chapter, recent advances in solution chemical processing of metal chalconides dramatically illustrate several design criteria listed above. Further, the chemistry is definitely within the scope of inorganometallic chemistry; therefore, some highlights are summarized. Geoffroy and co-workers[49] have described the synthesis of crystalline ZnS powder and whiskers by reaction of the organozinc reagent $[EtZn(SBu^t)]_5$ and H_2S in CH_2Cl_2 at 25 °C followed by treatment of the solid product at 500 °C with H_2S. It was concluded that H_2S interacts first with weak Zn—Et bonds and then $Zn—SBu^t$ groups with formation of Zn—S—Zn units that evolve into stable, extended ring–chain ZnS arrays with morphologies related to the degree of thiolysis. The chemistry described here is potentially general for production of many sulfide solid-state materials.

$$CdMe_2 + Se(SiMe_3)_2 \rightarrow CdSe + 2\ Me_4Si \quad (7)$$

$$CdMe_2 + Te(SiMe_2Pr^i) \rightarrow CdTe + 2\ Me_3(Pr^i)Si \quad (8)$$

$$ZnEt_2 + Se(SiMe_3)_2 \rightarrow ZnSe + Me_3(Et)Si \quad (9)$$

Steigerwald and co-workers[50] have recently employed silane elimination reactions, well established in Group 13–15 chemistry, for the production of binary Group 12–16 combinations. In particular, they examined reactions described in equations 7–9. The reactions are performed in common organic solvents and elimination rates are influenced by solvent and steric effects of the silyl group. Weak CdR and Zn—R bonding and the high stability of R_4Si drive these reactions to completion. In a different scheme,[51] CdSe is formed by thermal decomposition of discrete molecules $Cd(SePh)_2$ and $[Cd(SePh)_2]_2 \cdot [Et_2PCH_2CH_2PEt_2]$ in which Ph_2Se and phosphine are lost at modest temperatures. As pointed out by these investigators, solution routes to *classical powders* have no real benefit over metallurgical routes; however, in this case the solution chemistry may be usefully transferred to a vapor-phase epitaxy process, and it is clearly useful for the generation of novel, soluble nanocluster particles.[52] The latter opens a new frontier that can be addressed only through molecular-solution chemistry.

$$Mn_2(CO)_{10} + 2\ Et_3PTe \xrightarrow[\Delta]{tol} [(Et_3P)_2(CO)_3MnTe]_2 \xrightarrow{300\ ^\circ C} MnTe + 6\ CO + 4\ Et_3P \quad (10)$$

$$Ni(COD)_2 + Et_3P + Et_3PTe \xrightarrow[\Delta]{tol} NiTe + 2\ Et_3P + 2\ COD \quad (11)$$

$$Pd(PPh_3)_4 + Et_3PTe \xrightarrow[\Delta]{tol} PdTe + Et_3P + 4\ PPh_3 \quad (12)$$

$$[CpFe(CO)_2]_2 + Et_3P + Et_3PTe \xrightarrow[\Delta]{tol} [Cp(CO)(PEt_3)Fe]_2Te \rightarrow FeTe \quad (13)$$

$$[CpFe(CO)_2]_2 + 2\ Et_3PTe \xrightarrow[\Delta]{tol} [Cp(CO)(PEt_3)FeTe]_2 \rightarrow FeTe_2 \quad (14)$$

In another series of studies, Steigerwald and co-workers[44,45,47] have produced a series of transition metal chalconides by utilizing unstable trialkylphosphine tellurides as soluble sources of reactive Te. Equations 10–14 describe formation of MnTe, NiTe, FeTe, $FeTe_2$, and PdTe. In each case, molecular metal–Te ring or cluster compounds have been isolated from the early stages of the elimination–condensation chemistry, and the development of the final metal–telluride solid-state structure may be visualized from molecular structure determinations for the intermediates. In the first three reactions, the byproducts are the expected volatile CO and phosphine ligands, and the chemistry is clearly driven by the inherent instability of the phosphine telluride and the loss of volatile byproducts. The latter two reactions feature a novel twist. Normally, a Cp ligand might be expected to be difficult to remove cleanly in the transformation of a precursor to its ceramic product. In the case of the iron reactions, however, Cp is scavenged by an incipient CpFe fragment and ferrocene is conveniently formed as a stable, easily removed byproduct along with CO and Et_3P.

Numerous extensions of this chemistry for other metal–group 16 element compositions are obvious, and additional concepts can be gleaned from the organometallic–main group element literature. A recent review by Whitmire[53] may be useful in this regard.

4.2. Metal Carbides

Surprisingly, molecular and polymer routes to metal carbides are not as well developed as those detailed by Steigerwald for metal chalconides. However, this deficiency opens a window of opportunity. Although little effort has been given to establishing appropriate molecular design criteria, evaluation of known vapor deposition processes and the established lore of organotransition metal chemistry provide a background for the formulation of guiding principles. For example, Laine and Hirschon[48] have suggested that metal–nonmetal multiple bonding or strong metal–nonmetal–metal bridge bonding in a precursor molecule may favor retention of these building blocks in a solid state product. For metal carbides, this suggests that metal alkylidenes and alkylidynes may be useful precursors. This prediction will prove true, however, only if extraneous substituents and ligands can be cleanly removed from the compounds. This may not be a trivial task since most of the synthetic achievements on these compounds have focused on development of stability, not instability. At this time, the author is not aware of successful alkylidene or alkylidyne molecular to polymer or ceramic transformations, but the well-developed chemistry of these reagents[54,55] suggests several potentially fruitful approaches.

Laine and Hirschon[48,56] have suggested that organo bimetallic and cluster compounds with strong metal–metal bonds may serve as useful carbide precursors, although removal of stabilizing ligands (e.g., CO, PR_3, and Cp) without resorting to harsh metallurgical conditions may again present some problems. Laine and Hirschon[48,56,57] have reported that pyrolysis of a bimetallic tungsten acetylene complex, $Cp_2W_2(CO)_4(DMAD)$ (DMAD = dimethyl acetylenedicarboxylate), in a quartz tube gives WO_2 and $W_2(C,O)$, while pyrolysis in a nickel boat results in W_2C, not WC, contaminated with WO_2 and graphite. It is worth recalling that a number of molecular metal carbido carbonyl cluster compounds have been prepared.[13,58] Unfortunately, most of these compounds are constructed with Group 7 or more electron-rich metals, and they contain only one carbide center in an array of four or more metal atoms. Still, the systematic chemistry of these clusters, as it develops, may point out useful precursor routes to metal carbides. In this regard, Laine[57] lists a series of known metal carbido carbonyls that may serve as potential precursors. The bias of this author is that carefully designed acetylide and polyacetylide clusters may offer a rational route for practical precursor development, but *great experimental care* must be exercised in exploring this field due to potential chemical hazards.

Some attempts have been made to pyrolyze metal carbonyl and substituted metal carbonyl compounds. It is known from vapor deposition studies that most metal carbonyls undergo facile thermal decarbonylation. In some cases, e.g., $Ni(CO)_4$, high-quality metal films are produced, while in other cases, e.g., $Mo(CO)_6$, pyrolysis leads to metal oxide and graphite, but not metal carbide. Although it appears that most carbonyl-rich molecules may not be ideal carbide precursors, Hurd and co-workers[59] have reported conditions for conversion of $W(CO)_6$ to W_2C, but not WC. Although additional studies are required to set boundaries, it appears that CO may be utilized as a dissociable π-acid stabilizing ligand in precursor design, but it is probably not an ideal source for retained carbide. Perhaps metal carbonyl fragments anchored on an appropriate organic polymer support can be manipulated to provide metal carbides, although high temperatures will probably be required to drive these transformations.

The oxide incorporation encountered with use of CO in a precursor serves to reemphasize the point made earlier that ligands or substituents containing oxygen or halogens should be avoided when possible. This is a consequence of the thermodynamic strengths of Group 4, 5, and 6 metal oxygen and halide bonds. Taken to the extreme, special care must also be taken when forming polymeric reagents in oxygenated or halogenated solvents since solvent retention in a polymer is hard to avoid, and it can have disastrous consequences during polymer processing.

Despite these shortcomings, there remain many classes of organometallics that have potential for polymer formation and conversion to metal carbides. For example, Girolami and co-workers(60) have observed in vapor deposition studies that volatile neopentyl titanium, $Ti[CH_2C(CH_3)_3]_4$, produces TiC films at $\sim$150 °C. It is not yet clear if other weakly bonded, early metal compounds [e.g., $M(CH_2Ph)_4$, $M(CH_2SiMe_3)_4$, $M(C_6H_6)_2$, Cp_2MR_2] can be employed in solution-based chemistry, but the potential exists. It does appear that alkyl substituents lacking β-hydride elimination pathways favor metal carbide formation over metal deposition. An intriguing variant on this scheme is suggested in a patent disclosure(61) in which a poly(zirconocarbosilane) is calcined and reported to produce a solid solution of ZrC/β-SiC. Presumably related materials, TiC/β-SiC, are obtained from a poly(titanocarbosilane).(62)

π-Allyl and cyclopentadienyl metal complexes have been examined in a number of vapor deposition processes in several laboratories, but details are relatively sparse, and transfer of this chemistry to solution polymer schemes apparently has not been reported. These ligands offer at least good potential as stabilizing/leaving groups, and if properly "engineered," they may be good carbide sources. Clearly, the imaginative organometallic chemist is left to develop this wide open area.

4.3. Metal Nitrides

Needs for tractable molecular and polymeric metal nitride precursors have intensified, and some efforts have been directed toward developing appropriate design chemistry, particularly for Group 4–6 metals.

Although carbon–nitrogen bonds are relatively strong, it would be expected that Group 4–6 metal amides might provide molecular or polymeric precursors for metal nitrides. Some contrasting reports over these reagents have appeared; however, a consistent picture has recently emerged. Sugiyama *et al.*[63] report vapor-phase thermal decomposition of $Ti(NR_2)_4$, $Zr(NR_2)_4$, and $Nb(NR_2)_5$ in Ar, H_2, and N_2 at temperatures as low as 300 °C. With H_2 and N_2 atmospheres, formation of TiN, ZrN, and NbN or Nb_4N_3 films is claimed; however, carbon and oxygen contaminates are apparent. Constant and co-workers[64] describe attempts to produce TiN from low-temperature (<600 °C), low-pressure CVD of $Ti(bipyr)_3$, but the resulting amorphous films contained not only Ti and N, but large amounts of carbon and hydrogen. Further, Girolami *et al.*[65] have noted that, in their laboratory, CVD of $Ti(NMe_2)_4$ produces TiC films as the primary product, and they point out that this is consistent with known thermal decomposition chemistry of metal amides.[66] The likely first step of the decomposition is shown in equation 15. Chlu and Huang[67] have also described formation of thin films with compositions TiNC and ZrNC from CVD of $Ti(NEt_2)_4$ and $Zr(NEt_2)_4$ with substrate temperatures above 600 °C.

$$M(NMe_2)_2 \longrightarrow M(CH_2NMe) + HNMe_2 \quad (15)$$

In related work, Laine and Hirschon[48] report that bulk pyrolysis of the bimetallic amide, $Mo_2(NMe_2)_6$, at 800 °C (atmosphere not defined) produces a partially crystalline mixture of products that contains MoC_2, but not MoN. This report was followed by a patent disclosure[68] that claims production of metal carbides by pyrolysis of a large number of transition metal organoamides including examples from Groups 4–6. Outside the claims, it appears that pyrolysis atmospheres containing N_2, NH_3, and Ar were used. Parkin and Chisholm,[69] on the other hand, have found that reaction of NH_3 with $W_2(NMe_2)_6$ in benzene solution produces an intermediate that, upon pyrolysis, gives WN in the temperature range 200–450 °C and W_2N in the temperature range 650–900 °C.

In another metal amide modification, Brown and Maya[70] observe that treatment of $Ti(NMe_2)_4$, $Zr(NEt_2)_4$, and $Nb(NEt_2)_5$ with liquid NH_3 give uncharacterized, insoluble polymeric ammonolysis products that, upon pyrolysis *in vacuo* or under He at 800 °C, provide ceramic products containing TiN and carbon (5.8–7.1%), ZrN, and carbon (2.5–3.1%) and an ill-defined niobium nitride and carbon (6.5–7.8%), respectively. The carbon impurities are reduced in the titanium and niobium samples by performing the pyrolyses under hydrogen. The ceramics were characterized by XRD and Raman and infrared spectroscopies, and attempts were made to follow the precursor transformation process by temperature-program pyrolysis mass-spectrometric analysis. Maya has also studied ammonolyses of $TiBr_4$ and $NbBr_5$ in the presence of borohydride and acetylide.[71] Characterization data are relatively few, but it appears that carbonitrides or nitrides are obtained.

Finally, Fix, Gordon, and Hoffman[72] have recently provided the most current and detailed account on the attempted conversion of titanium amides to TiN by vapor deposition methods. They conclude that TiN is formed in their approach with significant carbon and oxygen contamination. Furthermore, they show that the source of carbon is the amide organic groups. This is a particularly thorough study that provides insight to fruitful new avenues of approach for metal nitride syntheses.

Metal imido, metal hydrazido, and metal nitrido units are expected to have strong M—N bonds; therefore, it can be anticipated that these reagents may serve as useful metal nitride precursors if suitable substituent group and stabilizing ligand leaving chemistry can be designed. The chemistry of these types of reagents is not highly developed, but pertinent chemistry has been summarized by Nugent and Mayer.[55] Some further advances that generate enthusiasm are briefly described here.

Chisholm and co-workers[73] find that the compounds $(t\text{-BuO})_3M{\equiv}N$ (M = Mo, W) in the solid state form linear polymers with alternating long and short $M{\equiv}N{-}M$ bonds. These polymers are unlikely to serve as useful ceramic precursors due to the presence of strongly bonded alkoxide ligands. However, Denicke and Strähle[74] have been active in developing the chemistry of other neutral and ionic metal nitrido compounds, and that work appears to have stimulated extensions into successful preceramic polymer schemes. For example, Doherty and co-workers[75] have recently reported formation of a heterobimetallic μ_2-nitrido complex $(Me_3SiO)_3V{\equiv}N{-}Pt(Me)(PEt_3)_2$ via silyl fluoride elimination chemistry, as summarized in equation 16. They also observe that $Cl_3V{\equiv}NSiMe_3$, in the presence of base, undergoes rapid elimination of Me_3SiCl with formation of linear chain polymers linked by $V{\equiv}N{-}V$ units or molecular species, $Cl_2(L)_2V{\equiv}N$.[76] Lastly, they find that combinations of MCl_2L_4 (M = Mo, W; L = phosphines) with Me_3SiN_3 result in silylnitrido complexes, $M(NSiMe_3)Cl_2L_3$, and, under some conditions, a bis-silylnitrido complex, $W(NSiMe_3)_2Cl_2(PMePh_2)$.[77] Reports on conversion of these reagents to ceramic products have not appeared; however, it can be assumed that this is a likely goal in this work.

$$(Me_3SiO)_3V{\equiv}NSiMe_3 + FPt(PEt_3)_2Me \xrightarrow{-Me_3SiF} (Me_3SiO)_3V{\equiv}N{-}Pt(PEt_3)_2Me \quad (16)$$

Roesky and co-workers[78] have examined the reaction of Cp^*TaCl_4 and $N(SnMe_3)_3$. The stannyl amine apparently acts as a nitrido transfer agent with formation of a novel molecular triazatrimetalla benzene. This molecule undergoes ring opening in boiling xylene with formation of a soluble polymer whose conversion to TaN is apparently under study.[79]

In another approach involving ammonolysis chemistry of fragile metal alkyl bonds, Roesky and co-workers[80] report that reaction of Cp^*TiMe_3

with NH_3 produces a novel cluster $(CpTiNH)_3N$. It is likely that continuing studies of the thermal chemistry of this and related compounds will produce metal nitrides. This conclusion is validated by reports by Wolczanski and co-workers[81] on successful conversion of a molecular species, $Np_3Ta{=}CHBu^t$, under NH_3 to an oligomer that is further transformed at 400 °C to amorphous TaN. This solid is converted to crystalline, *cubic* TaN by heating at 820 °C. This phase has been obtained previously only under much more severe metallurgical conditions. The reaction of Cp^*TaMe_4 with NH_3 is described as a model, and the triazatrimetalla benzene, $[Cp^*(Me)TaN]_3$, is isolated. Single-crystal X-ray analysis reveals that this molecule is closely related to the chloro analog described by Roesky *et al.*[78] Further studies of the ammonolysis of $Np_3Ta{=}CHBu^t$ reveal the existence of a soluble, heat- and light-sensitive pentamer, $(NpTaN)_5$.[82] This compound displays a novel ladder structure that resembles the final structural motif of c-TiN more than h-TiN. Further extensions of these very encouraging results are in progress.

Although it falls at the edge of the general scope of the review, it is important to point out that Fehlner and co-workers[83] have examined the CVD process of $HFe_4(CO)_{12}N$, which appears to produce a mixture of α-Fe and γ'-Fe_4N at 160–180 °C. This is an intriguing result that suggests early metal carbonyl nitrides may be useful for forming metal nitrides by vapor deposition.

The reactions of few other Group 4–6 alkyls or aryls have been examined with attention to precursor applications; however, one further example is noteworthy. Berry and co-workers[84] find, in a vapor deposition process, that bis-toluene niobium and hydrazine produce multiphasic ceramic NbC_xN_{1-x} films that display superconducting transition temperatures in the range 2–11 K.

In a process related to a method developed for production of BN fibers,[85] Kamiya and co-workers[86] claim formation of TiN fibers by nitriding TiO_2 sol-gel prepared fibers at temperatures above 900 °C in NH_3. The nitridation process was followed by X-ray diffraction analysis. Although some details are lacking, it appears that the nitridation reaction is not complete even after 10 hours at 1100 °C. Further, the only other characterization data provided are color (golden) and a statement that samples are electrically conductive. Obviously, this approach suffers from solid-state diffusion limitations mentioned earlier.

Extending another precursor approach used to prepare BN,[87] Kuroda and co-workers[88] report nitridation, with N_2 at 1600 °C, of a glassy polymer obtained from combining $Ti(OPr^i)_4$ and triethanolamine. The ceramic product appears to contain TiN, Ti_3O_5, and a small amount of carbon. These initial studies indicate that the quality of a ceramic product depends greatly on precursor and processing conditions, and XRD analyses alone are not sufficient indicators of the successful transformation of a precursor to a pure ceramic material. Further compositional and microstructural analyses are required in order to judge the success and generality of these specific examples.

Two additional but innovative routes to nitrides have recently been reported. Bamberger and co-workers[89] report that heating TiO_2 or titanium phosphates with NaCN produces TiN powder. When sodium titanium bronze is employed, TiN whiskers are obtained. The mechanism for the intriguing growth of whiskers is not elucidated. Stacy and co-workers[90] observe that MoO_3 and several molybdates react with NH_3 at modest temperatures with formation of high surface area MoN or Mo_2N.

Lastly, Rüssel[91] has found that anodic dissolution of Ti in an organic electrolyte containing n-propylamine gives a presently uncharacterized liquid, which produces an amorphous solid upon heating. Calcination of this solid gives a TiN/TiC solid solution with varying amounts of carbon. This technique may have utility for the synthesis of a variety of metal nitride powders if practical details can be overcome.

4.4. Metal Borides, Silicides, and Phosphides

Although relatively well developed for electron-rich metals, the chemistry of metalloboranes, metallosilanes, and metallophosphines involving Group 4–6 metals is comparatively sparse. A few volatile molecular reagents have been examined as CVD precursors. For example, Girolami and co-workers[65,92] and Wayda and co-workers[93] report formation of TiB_2, ZrB_2, and HfB_2 films at low temperatures (100–270 °C) by vapor depositions from corresponding $M(BH_4)_4$ reagents. In addition, Rice and Woodin[94] have described ZrB_2 film formation from classical CVD and laser-assisted CVD, and they attempted to prepare ZrB_2 powder in a laser-driven process.

Several other groups are trying to develop the diversity of metalloboranes into useful precursor systems; however, published results are few. For example, Amini, Fehlner, and co-workers[95] have found that thermal decomposition of $HFe_3(CO)_9BH_4$ at 175–200 °C gives amorphous films of Fe_3B. Clues to additional fruitful avenues appear in earlier chapters in this text. It may prove possible to manipulate, for example, metalloborane cage expansion chemistry in a manner that desired MR_x fragments can be excised leaving stable neutral or anionic polyborane fragment byproducts that could be recycled. Metallocarboranes may also provide useful ceramic precursors, although the products will likely contain carbon or B_4C. Development of late metal–borane molecules into precursors holds much more promise, and the potential electronic ceramic applications for metal borides make exploratory research in this area worthwhile.

Aylett has provided guidelines for the development of metal silicide CVD precursors,[34,96] and some of these may be useful in polymer precursor design. For example, it is advisable to avoid alkyl and aryl–silicon bonds that may provide a source for SiC formation. On the other hand, compounds containing $M-SiH_3$ or $[M]_2Si$ units, though very reactive, are favored by elimination reactions of the Si—H bonds. Recent efforts to develop metal silylene chemistry and metal catalysts for polysilane synthesis have stimulated renewed attention to the area of metallosilanes. For example, Berry *et al.*[97] describe formation of $Cp_2W(Si_2Me_4)$, Harrod[98] reports isolation of $[CpTi((\mu\text{-}H)SiPhH)]_2$, and Henchen and Weiss[99] claim the existence of $Cp_2Ti_2(Si_2H_4)$. Additional attention to related early metal compounds, with the specific intent of generating preceramics, will certainly be successful.

The recent rapid evolution of "low coordinate" phosphorus ligand chemistry[100] provides signposts for the development of useful approaches to early transition metal phosphide precursors. A few illustrative examples are provided, and there are many others awaiting exploration. For example, Baker and co-workers[101] describe preparations of $Cp_2M(PR_2)_2$ (M = Hf, Zr; R = Ph, Cy), while Baker,[102] Jones[103] and their co-workers report formation of homoleptic metal phosphide compounds including $[Li(DME)][M(PCy_2)_5]$ (M = Zr, Hf), $[Li(DME)_n][M(PCy_2)_4]$ (M = Ti, V, Nb), $Mo(PCy_2)_4$, $[Li(DME)_n][M(PCy_2)_5]$ (M = Cr, W), and $Mo_2(PtBu_2)_4$. Terminal $M-PR_2$ distances in several of these examples are short, suggesting strong M—P bonding. In order to stabilize these compounds kinetically, bulky substituent groups with relatively strong P—C bonds have been chosen. Substituent group manipulation appears feasible and, for ceramic applications, it should be possible to derive monomers that shed organyl groups in facile displacement chemistry, thereby leaving processible oligomeric or cluster precursors. Availability of soluble precursors should permit ready formation of dense coatings and thin films of definite compositions. This has not been routinely possible with vapor deposition approaches.

Jeitschko,[104] von Schnering,[105] and their co-workers have provided a great deal of data on the preparation of metal phosphides by classical methods that employ elemental phosphorus as a reactant. Separately, Fritz and Härer,[106] Baudler and Glinka,[107] and von Schnering[105] have developed beautiful solution chemistry of poly-silylphosphines and poly-phosphide anions, and it now appears that these solution species may provide alternative routes to the metal phosphides. The chemistry will not be trivial, but it should be rewarding, and it is ripe for development for precursors. Some main group metal chemistry has been examined. For example, von Schnering reports reactions of $P_7(SiMe_3)_3$ with $ClPbMe_3$ and $ClSnMe_3$ that give $P_7(MMe_3)_3$ clusters and a reaction of Na_3P_7 with $ClGeMe_3$ that produces $P_7(GeMe_3)_3$. It is not known if these clusters can be polymerized or converted to solid-state phosphides.

Scherer and co-workers[100,108] have recently described reactions of P_4 with several early metal organometallic compounds, and these provide a host of novel compounds containing P_6, P_4, and P_2 fragments. Products include $[Cp^*Mo]_2P_6$, $[Cp^*Ti]_2P_6$, $(Cp^*Nb)_2P_6$, $[Cp^*Nb(CO)_2]P_4$, and $[Cp^*Mo(CO)(P_2)]_2$. None of these compounds has been converted to ceramics; however, it is likely that judiciously chosen reaction chemistry will lead to desired precursors. Phosphinidene complexes have also attracted attention,[109,110] and compounds of the general type $L_nM-P(R)-ML_n$ should provide useful precursors. In all of these compounds, a key will be designing facile substituent group-leaving chemistry that operates at temperatures sufficiently low that elemental phosphorus will not form as a stable phase and be removed by sublimation or vapor transport.

5. NEW APPROACHES IN CHEMICAL PROCESSING

There is a high probability that a number of processible molecular and polymeric precursors for advanced nonoxide ceramic forms will appear in the next few years and their origin will probably be traceable to fundamental inorganometallic chemistry. In order to realize the potential of these precursors and their ceramic products, chemists should also devote attention to new processing concepts specifically suited to these precursors. This is an exciting frontier, and it should not be overlooked.

Modern polymer science will certainly play an important role in processing metal nonoxide ceramic polymers, and the area is attracting attention from that community. Sol-gel techniques have found tremendous applications for preparation of fine, advanced oxide ceramic compositions. In these cases, hydroxylic solvents, typically alcohols, are employed as the solvent system. For the most part, these solvents will not be applicable to the synthesis and manipulation of nonoxide preceramics. On the other hand, some nonhydroxylic solvent systems may permit sol-gel processing of nonoxide precursors. For example, we have found several organic solvents, $NH_3(l)$ and $CO_2(l)$ appropriate for processing boron nitride preceramic polymers, and similar successes can be expected for metal–nonoxide compositions.

A benefit realized from sol-gel processing of oxide precursors is the attendant flexibility for producing various final ceramic forms. For example, with tetraethyl orthosilicate, it is possible to obtain silica as fibers, dense coatings (xerogels), and low-density aerogels. It is also possible to make tailored "composites." This flexibility should carry forward to the processing of metal–nonoxide precursors. In our own work, we have found this to be the case for nonoxide ceramics, including BN and its composites. It should also be possible to form aerosols from the preceramic polymer solutions that, in turn, may be converted to nano-crystallites. These powders will display unique processing features.

In Section 3, polymer melt processing was briefly mentioned. This technique should prove especially powerful for precursors that remain liquid at modest temperatures after most of the substituent elimination/gas forming chemistry is complete or for cases where precursor/substrate reactions result in a liquid state at elevated temperatures. That this is possible is demonstrated by our own results with BN precursors that "flux" with selected oxide substrates and deposit tightly bonded h-BN coatings upon cooling.[111] With careful attention to interfacial chemistry, it may prove feasible to accomplish related chemistry with metal–nonoxide coating precursors, and this will provide a powerful method for preparing complex multilayer architectures. There should also be some attention given to the development of colloids from metal–nonoxide precursors.

The chemistry outlined in Section 4 was narrowly confined to binary compositions. More complex precursor chemistry will evolve to generate currently unknown ceramic compositions, e.g., ternary and composite materials. The possibilities are immense, but there are needs on the immediate horizon for microengineered ceramic–ceramic and metal–ceramic composites and organic–inorganic composites with controlled properties and microstructures. There are needs for more chemistry and chemical understanding in the interfacial region in composites, and this will be achieved only by close work between synthetic chemists, surface chemists, and microanalytical ceramists. There are also needs for the development of theories that *predict* composition–property relationships including compositional variation (doping) effects on microstructure, ceramic compatibility, mechanical and thermal responses, corrosion mechanisms, cracking phenomena, and composite reinforcement mechanisms. The field of study appears limitless at this stage, and it becomes increasingly challenging as we begin to develop "high knowledge content" materials for the next century.

6. CONCLUSIONS

Turning back to questions posed earlier, we can anticipate that the chances for realizing the vast potential of early metal–nonoxide ceramics will be improved by the development of processible molecular and polymer precursors and improved chemically controlled processing of these reagents. In addition, extensions of newly developed chemistry should provide new compositions with unique multifunctional properties. It is also likely that the efforts in these areas will lead to formulation of predictive theories for tailored property modification by composition variation.

As we approach the 21st century and confront its technological problems, there will be increasing needs for high-performance structural and electronic ceramics obtained from complex, tailored chemical processes. These precursors will evolve from well-planned fundamental synthesis programs that keep an open eye for applications. Inorganometallic chemistry will play crucial roles in this evolutionary process.

REFERENCES

1. Kingery, W. D.; Bowen, H. K.; Ullman, D. R. "Introduction to Ceramics," 2nd ed.; Wiley: New York, 1976.
2. Economy, J. *Angew. Chem., Int. Ed. Engl.* 1989, **28,** 229.
3. "Materials Science and Engineering for the 1990s: Maintaining Competitiveness in the Age of Materials"; National Academy of Sciences: Washington, D.C., 1990.
4. Hondros, E. D.; Bullock, E. *Angew. Chem., Int. Ed. Engl.* 1989, **28,** 1088.
5. Robinson, A. L. *Science* 1986, **233,** 25.
6. Aldinger, F.; Kalz, H.-J. *Angew. Chem., Int. Ed. Engl.* 1987, **26,** 371.
7. Ulrich, D. R. *Chem. Eng. News.* 1990, **68,** 28, Uhlmann, D. R.; Zelinski, B. J. J.; Wnek, G. E. *Mater. Res. Soc. Symp. Proc.* 1984, **32,** 59.
8. Rice, R. W. In "Design of New Materials," Cocke, D. L.; Clearfield, A. Eds.; Plenum: New York, 1987; pp. 169–194.
9. Schneider, S. J. Ed., "A National Prospectus on the Future of the U.S. Advanced Ceramics Industry" NBS Report IR 85-3240, National Bureau of Standards, Washington, D.C., 1985.
10. Bowen, H. K. *Mater. Res. Soc. Symp. Proc.* 1984, **24,** 1.
11. Roy, R. *J. Am. Ceram. Soc.* 1977, **60,** 350.
12. Toth, L. E. "Transition Metal Carbides and Nitrides"; Plenum: New York, 1971.
13. Greenwood, N. N.; Earnshaw, A. "Chemistry of the Elements"; Pergamon: London, 1984.
14. Frad, W. E. *Adv. Inorg. Chem. Radiochem.* 1968, **11,** 153.
15. Kosolapova, T.Ya., "Carbides Properties, Production and Applications"; Plenum: New York, 1971.
16. Storms, E. K. "The Refractory Carbides"; Academic: New York, 1967.
17. Hausner, H. H.; Bowman, M. G. Eds. "Fundamentals of Refractory Compounds"; Plenum: New York, 1968.
18. Williams, W. S. *Prog. Solid State Chem.* 1971, **6,** 57; Williams, W. S. *Mater. Sci. Eng.* 1988, **A105/106,** 1.
19. Johansen, H. A. *Survey Prog. Chem.* 1977, **8,** 57.
20. Kendall, E. G. In "Ceramics for Advanced Technologies"; Hove, J. E.; Riley, W. C. Eds.; Wiley: New York, 1965; Chapter 5.
21. Juza, R. *Adv. Inorg. Chem. Radiochem.* 1966, **9,** 81.
22. Powell, C. F.; Oxley, J. H.; Blocker, J. M. "Vapor Deposition"; Wiley: New York, 1966.
23. Wells, A. F. "Structural Inorganic Chemistry," 5th ed.; Oxford: London, 1984.
24. Wijeyesekera, S. D.; Hoffmann, R. *Organometallics* 1984, **3,** 949; Li, J.; Hoffmann, R. *Chem. Mater.* 1989, **1,** 83; Wheeler, R. A.; Hoffmann, R.; Strähle, J. *J. Am. Chem. Soc.* 1986, **108,** 5381.
25. Almond, E. A. In "Transformation of Organometallics into Common and Exotic Materials: Design and Activation"; Laine, R. M., Ed.; M. Nijhoff: Dordrecht, 1988; pp. 32–48.
26. Mittasch, A. *Adv. Catal.* 1980, **2,** 81.
27. Thompson, R. In "Progress in Boron Chemistry"; Brotherton, R. J.; Steinberg, H., Eds.; Pergamon: London, 1970; Vol. 2, pp. 173–230.
28. Aronsson, B.; Lundström, T.; Rundquist, S. "Borides, Silicides and Phosphides"; Wiley: New York, 1965.
29. Greenwood, N. N. "Boron"; Pergamon: London, 1975.
30. Hoard, J. L.; Hughes, R. E. In "The Chemistry of Boron and Its Compounds"; Muetterties, E. L., Ed. Wiley: New York, 1967; pp. 99–140.
31. Matkovich, V. I. "Boron and Refractory Solids"; Springer-Verlag: Berlin, 1977.
32. Telle, R.; Petzow, G. *Mater. Sci. Eng.* 1988, **A105/106,** 97.
33. Kieffer, R.; Benesovsky, F. "Hartstoffe"; Springer-Verlag: Wien, 1963.
34. Aylett, B. J. In "Transformation of Organometallics into Common and Exotic Materials: Design and Activation"; Laine, R. M., Ed.; M. Nijhoff: Dordrecht, 1988; pp. 165–177.

35. Westwood, A. R. C.; Winzer, S. R. In "Advancing Materials Research"; Psaras, P. A.; Langford, H. D. Eds.; National Academy Press: Washington, D.C., 1987; pp. 225–244.
36. Wynne, K. J. *Annu. Rev. Mater. Sci.* 1984, **14,** 297.
37. Rice, R. W. *Am. Ceram. Soc. Bull.* 1983, **62,** 889.
38. Seyferth, D. In "Transformation of Organometallics into Common and Exotic Materials: Design and Activation"; Laine, R. M., Ed.; M. Nijhoff: Dordrecht, 1988; pp. 133–154.
39. Paine, R. T.; Narula, C. K. *Chem. Rev.* 1990, **90,** 73.
40. Foise, J.; Kim, K.; Covino, J.; Dwight, K.; Wold, A.; Chianelli, R.; Passaretti, J. *Inorg. Chem.* 1983, **22,** 61; Pasquariello, D. M.; Kershaw, R.; Passaretti, J. D.; Dwight, K.; Wold, A. *Inorg. Chem.* 1984, **23,** 872.
41. West, A. R. "Solid State Chemistry and Its Applications"; Wiley: New York, 1984.
42. Rao, C. N. R.; Gopalakrishnan, J. "New Directions in Solid State Chemistry"; Cambridge University Press: Cambridge, UK, 1986; Nanjundaswamy, K. S.; Vasanthacharya, N. Y.; Gopalakrishnan, J.; Rao, C. N. R. *Inorg. Chem.* 1987, **26,** 4286.
43. Steigerwald, M. I.; Kometani, T. Y.; Jeffers, K. S. *Appl. Phys. Lett.* 1987, **50,** 1681; Steigerwald, M. L.; Sprinkle, C. R. *J. Am. Chem. Soc.* 1987, **109,** 7200.
44. Steigerwald, M. L. *Chem. Mater.* 1989, **1,** 52.
45. Brennan, J. G.; Siegrist, T.; Stuczynski, S. M.; Steigerwald, M. L. *J. Am. Chem. Soc.* 1989, **111,** 9240.
46. Steigerwald, M. I.; Rice, C. F. *J. Am. Chem. Soc.* 1988, **110,** 4228.
47. Brennan, J. G.; Siegrist, T.; Stuczynski, S. M.; Steigerwald, M. L., submitted for publication.
48. Laine, R. M.; Hirschon, A. S. In "Transformation of Organometallics into Common and Exotic Materials: Design and Activation"; Laine, R. M., Ed.; M. Nijhoff: Dordrecht, 1988; pp. 21–31.
49. Guiton, T. A.; Czekaj, C. I.; Rau, M. S.; Geoffroy, G. L.; Pantano, C. G. *Mater. Res. Soc. Symp. Proc.* 1988, **121,** 503.
50. Stuczynski, S. M.; Brennan, J. G.; Steigerwald, M. L. *Inorg. Chem.* 1989, **28,** 4431.
51. Brennan, J. G.; Siegrist, T.; Carroll, P. G.; Stuczynski, S. M.; Brus, L. E.; Steigerwald, M. L. *J. Am. Chem. Soc.* 1989, **111,** 4141.
52. Steigerwald, M. L.; Alivisatos, A. P.; Gibson, J. M.; Harris, T. D.; Kortan, R.; Muller, A. J.; Thayer, A. M.; Duncan, T. M.; Douglass, D. C.; Brus, L. F. *J. Am. Chem. Soc.* 1988, **110,** 3046; Steigerwald, M. L.; Brus, L. F. *Annu. Rev. Mater. Sci.* 1989, **19,** 471; Kortan, A. R.; Hull, R.; Opila, R. I.; Bawendi, M. G.; Steigerwald, M. I.; Carroll, P. J.; Brus, L. E. *J. Am. Chem. Soc.* 1990, **112,** 1327.
53. Whitmire, K. H. *J. Coord. Chem.* 1988, **17,** 95.
54. Collman, J. P.; Hegedus, I. S.; Norton, J. R.; Finke, R. G. "Principles and Applications of Organotransition Metal Chemistry"; University Science Books: Mill Valley, CA, 1987.
55. Nugent, W. A.; Mayer, J. M. "Metal–Ligand Multiple Bonds"; Wiley: New York, 1988.
56. Laine, R. M.; Hirschon, A. S. *Mater. Res. Soc. Symp. Proc.* 1986, **73,** 373.
57. Laine, R. M. U.S. Patent 4,826,666, May 2, 1989.
58. Tochikawa, M.; Muetterties, E. L. *Prog. Inorg. Chem.* 1981, **28,** 203.
59. Hurd, D. T.; McEntee, H. R.; Brisbin, P. H. *Ind. Eng. Chem.* 1952, **44,** 2432.
60. Girolami, G. S.; Jensen, J. A.; Pollina, D. M.; Williams, W. S.; Kaloyeros, A. E.; Alloca, C. M. *J. Am. Chem. Soc.* 1987, **109,** 1579.
61. National Institute for Research in Inorganic Materials, Ube Industries, Ltd. *Jpn Kokai Tokkyo Koho* JP 58,132,026 [83,132026] Aug. 6, 1983.
62. Research Foundation for Special Inorganic Materials, Ube Industries, Ltd. *Jpn Kokai Tokkyo Koho* JP 58,132,025 [83,132,025] Aug. 6, 1983.
63. Sugiyama, K.; Pac, S.; Takahashi, Y.; Motojima, S. *J. Electrochem. Soc.* 1975, **122,** 1545.
64. Morancho, R.; Petit, J. A.; Dabosi, F.; Constant, G. *J. Electrochem. Soc.* 1982, **129,** 854; Morancho, R.; Constant, G.; Ehrhardt, J. J. *Thin Solid Films* 1981, **77,** 155.

65. Girolami, G. S.; Jensen, J. A.; Gozumn, J. E.; Pollina, D. M. *Mater. Res. Soc. Symp. Proc.* 1988, **121**, 429.
66. Takahashi, Y.; Onoyama, N.; Ishikawa, Y.; Motojima, S.; Sugiyama, K. *Chem. Lett.* 1978, 525.
67. Chlu, H. T.; Huang, C. C. Abstr. Inorg. 253 American Chemical Society National Meeting, Washington, D.C., Aug. 1990.
68. Laine, R. M. U.S. Patent 4,789,534, Dec. 6, 1988.
69. Chisholm, M. H.; Parkin, I. P. Abstr. Inorg. 188, American Chemical Society National Meeting, Washington, D.C., Aug. 1990.
70. Brown, G. L.; Maya, L. *J. Am. Ceram. Soc.* 1988, **71,** 78.
71. Maya, L. In "Transformation of Organometallics into Common and Exotic Materials: Design and Activation"; Laine, R. M., Ed.; M. Nijhoff: Dordrecht, 1988; pp. 49–55.
72. Fix, R. M.; Gordon, R. G.; Hoffman, D. M. *Chem. Mater.* 1990, **2,** 235.
73. Chisholm, M. H.; Hoffman, D. M.; Huffman, J. C. *Inorg. Chem.* 1983, **22,** 2903; Chan, D. M. T.; Chisholm, M. H.; Folting, K.; Huffman, J. C. *Inorg. Chem.* 1986, **25,** 4170.
74. Denicke, K.; Strähle, J. *Angew. Chem. Int. Ed. Engl.* 1981, **20,** 413.
75. Doherty, N. M.; Critchlow, S. C. *J. Am. Chem. Soc.* 1987, **109,** 7906.
76. Critchlow, S. C.; Lerchen, M. E.; Smith, R. C.; Doherty, N. M. *J. Am. Chem. Soc.* 1988, **110,** 8071.
77. Lichtenhan, J. D.; Critchlow, S. C.; Doherty, N. M. *Inorg. Chem.* 1990, **29,** 439.
78. Plenio, H.; Roesky, H. W.; Noltenmeyer, M.; Sheldrick, G. M. *Angew. Chem., Int. Ed. Engl.* 1988, **27,** 1331.
79. Roesky, H. W.; Lücke, M. *J. Chem. Soc., Chem. Commun.* 1989, 748.
80. Roesky, H. W.; Bai, Y.; Noltenmeyer, M. *Angew. Chem., Int. Ed. Engl.* 1989, **28,** 788.
81. Banaszak Holl, M. M.; Kersting, M.; Pendley, B. D.; Wolczanski, P. T. *Inorg. Chem.,* in press.
82. Banaszak Holl, M. M.; Wolczanski, P. T.; Van Duyne, G. D., submitted for publication.
83. Fehlner, T. P.; Amini, M. M.; Stickle, W. F.; Pringle, D. A.; Long, G. L.; Fehlner, F. P. *Chem. Mater.* 1990, **2,** 263.
84. Cukauskas, E. J.; Holm, R. T.; Berry, A. D.; Kaplan, R. *IEEE Trans. Magn.* 1987, **23,** 999.
85. Economy, J.; Anderson, R. *Inorg. Chem.* 1966, **5,** 989.
86. Kamiya, K.; Yoko, T.; Bessho, M. *J. Mater. Sci.* 1987, **22,** 937.
87. Wada, H.; Nojima, K.; Kuroda, K.; Kato, C.; Yogyo-Kyokai-Shi *J. Ceram. Soc. Jpn.* 1987, **95,** 130.
88. Kuroda, K.; Tanaka, Y.; Sugahara, Y.; Kato, C. *Proc. Mater. Res. Soc.* 1988, **121,** 575.
89. Bamberger, C. E.; Angelini, P.; Nolan, T. A. *J. Am. Ceram. Soc.* 1989, **72,** 587.
90. Jaggers, C. H.; Michaels, J. N.; Stacy, A. M. *Chem. Mater.* 1990, **2,** 150.
91. Rüssel, C. *Chem. Mater.* 1990, **2,** 241.
92. Jensen, J. A.; Gozum, J. E.; Pollina, D. M.; Girolami, G. S. *J. Am. Chem. Soc.* 1988, **110,** 1643.
93. Wayda, A. L.; Schneemeyer, L. F.; Opila, R. L. *App. Phys. Lett.* 1988, **53,** 361.
94. Rice, G. W.; Woodin, R. L. *J. Am. Ceram. Soc.* 1988, **71,** C181.
95. Amini, M. M.; Fehlner, T. P.; Long, G. L.; Politowski, M. *Chem. Mater.* 1990, **2,** 432.
96. Aylett, B. J. In "Silicon Chemistry"; Corey, F. R.; Corey, J. Y.; Gaspar, P. P. Eds.; Wiley: New York, 1988; Chapter 33.
97. Berry, D. H.; Chey, J. H.; Zipin, H. S.; Carroll, P. J. *J. Am. Chem. Soc.* 1990, **112,** 452.
98. Harrod, J. F. In "Transformation of Organometallics into Common and Exotic Materials: Design and Activation"; Laine, R. M., Ed.; M. Nijhoff: Dordrecht, 1988; pp. 103–115; Aitken, C. T.; Harrod, J. F.; Samuel, E. J. *J. Am. Chem. Soc.* 1986, **108,** 4059.
99. Henchen, G.; Weiss, E. *Chem. Ber.* 1973, **106,** 1747.
100. Scherer, O. J. *Angew. Chem., Int. Ed. Engl.* 1985, **24,** 924.

101. Baker, R. T.; Whitney, J. F.; Wreford, S. S. *Organometallics* 1983, **2,** 1049.
102. Baker, R. T.; Krusic, P. J.; Tulip, T. H.; Calabrese, J. C.; Wreford, S. S. *J. Am. Chem. Soc.* 1983, **105,** 6763.
103. Jones, R. A.; Lasch, J. G.; Norman, N. C.; Whittlesey, B. R.; Wright, T. C. *J. Am. Chem. Soc.* 1983, **105,** 6184.
104. Jeitschko, W.; Möller, M. H. *Phosphorus Sulfur* 1987, **30,** 413; Jeitschko, W.; Flörke, U.; Scholz, U. D. *J. Solid St. Chem.* 1984, **52,** 320.
105. von Schnering, H. G. *Angew. Chem., Int. Ed. Engl.* 1981, **20,** 33; Mujica, C.; Weber, D.; von Schnering, H. G. *Z. Naturforsch.* 1986, **41b,** 991.
106. Fritz, G.; Härer, J. In "The Chemistry of Inorganic Homo- and Heterocycles"; Haiduc, I.; Sawerby, D. B. Eds.; Academic Press: New York, 1987; Vol. 1, pp. 277–286; Fritz, G. *Adv. Inorg. Radiochem.* 1987, **31,** 171.
107. Baudler, M.; Glinka, K. In "The Chemistry of Inorganic Homo- and Heterocycles"; Haiduc, I.; Sowerby, D. B. Eds.; Academic Press: New York, 1987; Vol. 2, pp. 423–466; Baudler, M. *Angew. Chem., Int. Ed. Engl.* 1987, **26,** 419.
108. Scherer, O. J.; Sitzmann, H.; Wolmershäuser, G. *Angew. Chem., Int. Ed. Engl.* 1985, **24,** 351; Scherer, O. J.; Swarowsky, H.; Wolmerschäuser, G.; Kaim, W.; Kohlmann, S. *Angew. Chem., Int. Ed. Engl.* 1987, **26,** 1153; Scherer, O. J.; Vondung, J.; Wolmerschäuser, G. *Angew. Chem., Int. Ed. Engl.* 1989, **28,** 1355.
109. Schmidt, U. *Angew. Chem., Int. Ed. Engl.* 1975, **14,** 523.
110. Huttner, G.; Evertz, K. *Acc. Chem. Res.* 1986, **19,** 406.
111. Paine, R. T.; Narula, C. K.; Schaeffer, R.; Datye, A. K. *Chem. Mater.* 1989, **1,** 486.

Index

Compounds are listed alphabetically according to (a) principal metal, (b) principal main group atom, and (c) metal nuclearity. Many of the solid-state compounds will be found under the generic description M *where* M *refers to the transition elements.*

BGSU Libraries
A11320295543